"Laminated composite with anisotropy is a wonderful world of mechanics to visit. This book is a result of Prof. Hwu's long-term dedication. It can be used as a textbook for senior undergraduate and graduate students, and a great reference book for engineers in composite modeling and design."
Professor Ernian Pan, *National Yang Ming Chiao Tung University*

"This new textbook presents the *Mechanics of Laminated Composite Structures* in a very nice integrative manner, connecting the underlying theory of elasticity, the basic structural elements, the practical engineering applications, and the available numerical methods. An ideal, well-balanced compilation for university students as well as practising engineers!"
Professor Wilfried Becker, *Technical University of Darmstadt*

"The author is a world-renowned expert in the theoretical analysis and computational modelling of anisotropic solids and composite structural elements. This book is a unique combination of the multi-field continuum mechanics of anisotropic solids and the structural analysis of composite beams, plates, shells, sandwiches, wings, and smart composites."
Professor Vladislav Mantic Lescisin, *University of Seville*

Mechanics of Laminated Composite Structures

In this textbook for students of laminated composite materials, composite structures, and anisotropic elasticity, Chyanbin Hwu draws on more than three decades of research and applications experience to provide a leading resource on many unique topics related to laminated composite structures.

This book introduces the mechanical behavior of laminated composite materials and provides related theories and solutions. All basic structural elements such as beams, plates, and shells are described in detail. Further contents include composite sandwich construction and composite wing structures. To connect with practical engineering applications and analyze more complicated real structures, numerical methods and their theoretical basis in anisotropic elasticity are also included. Advanced topics addressed include solutions for magneto-electro-elastic laminated plates; Green's functions for thick laminated plates and beams; typical thick laminated beams; theory for general laminated composite shells; sandwich beams, plates, and cylindrical shells as well as delaminated composite sandwich beams; modeling and analysis of composite wing structures; complex variable theories of anisotropic elasticity and the related Green's functions; and numerical methods such as finite element method, boundary element method and meshless method. Through this book, readers will learn not only the mechanics of laminated composite structures but also anisotropic elasticity and some popular numerical methods.

This textbook is vital for advanced undergraduate and graduate students interested in the mechanics of composite materials, composite structures, and anisotropic elasticity, such as aerospace, mechanical, civil, and naval engineering; applied mechanics; and engineering science. It is also useful for engineers working in these fields and applied mathematicians and material scientists.

Chyanbin Hwu is Chair Professor in the Department of Aeronautics and Astronautics at National Cheng Kung University, Tainan, Taiwan. He earned a BS at National Taiwan University, ROC, in 1981; an MS at National Tsing-Hua University, ROC, in 1985; and a PhD in engineering mechanics at the University of Illinois, Chicago, United States, in 1988. He was elected the president of the Society of Theoretical and Applied Mechanics (ROC) in 2008. He is a fellow of the Aeronautical and Astronautical Society (ROC) and the Society of Theoretical and Applied Mechanics (ROC).

Mechanics of Laminated Composite Structures

Chyanbin Hwu

CRC Press
Taylor & Francis Group
Boca Raton London New York

CRC Press is an imprint of the
Taylor & Francis Group, an **informa** business

Designed cover image: Concept of self-healing material. 3D illustration isolated on a white background.

First edition published 2024
by CRC Press
2385 NW Executive Center Drive, Suite 320, Boca Raton FL 33431

and by CRC Press
4 Park Square, Milton Park, Abingdon, Oxon, OX14 4RN

CRC Press is an imprint of Taylor & Francis Group, LLC

ISBN: 978-1-032-74694-4 (hbk)
ISBN: 978-1-032-74695-1 (pbk)
ISBN: 978-1-003-47046-5 (ebk)

DOI: 10.1201/9781003470465

Typeset in Minion
by SPi Technologies India Pvt Ltd (Straive)

Contents

Preface, xiv

Chapter 1 ▪ Introduction to Composites 1

1.1 CLASSIFICATION AND CHARACTERISTICS OF COMPOSITE MATERIALS 3

1.2 FABRICATION OF FIBER-REINFORCED COMPOSITE MATERIALS 4

1.2.1 Fibers 4

1.2.2 Matrix 5

1.2.3 Fabrication 5

1.3 MECHANICAL BEHAVIOR OF COMPOSITE MATERIALS 6

1.4 COMPOSITE STRUCTURES 7

NOTES 8

REFERENCES 9

Chapter 2 ▪ Mechanical Behavior of Laminated Composite Materials 10

2.1 LAMINA ELASTIC BEHAVIOR 11

2.1.1 Specially Orthotropic Lamina 12

2.1.2 Generally Orthotropic Lamina 12

2.2 MICROMECHANICS PREDICTION OF LAMINA STIFFNESS 15

2.2.1 Mechanics of Materials Approach 17

2.2.2 Elasticity Approach 19

2.3 LAMINA STRENGTH 21

2.3.1 Longitudinal Tensile Strength 22

2.3.2 Longitudinal Compressive Strength 23

2.3.3 Transverse Strength 24

2.3.4 Shear Strength 25

2.3.5 Lamina Failure Theories 25

2.4 MECHANICAL BEHAVIOR OF A LAMINATE 28

2.4.1 Classical Lamination Theory 29

2.4.2 Hygrothermal Effects 33

2.5 LAMINATE STIFFNESS 35

2.5.1 Symmetric Laminates 36

2.5.2 Anti-Symmetric Laminates 37

2.5.3 Cross-Ply Laminates 37

2.5.4 Angle-Ply Laminates 37

2.5.5 Balanced Laminates 38

2.5.6 Specially Orthotropic Laminates 38

2.5.7 Quasi-Isotropic Laminates 38

2.6 LAMINATE STRENGTH ANALYSIS 38

REFERENCES 41

CHAPTER 3 ■ Laminated Composite Plates 42

3.1 GENERAL FORMULATION 43

3.1.1 Thin Plates 43

3.1.2 Effects of Transverse Shear Deformation 49

3.1.3 Higher-Order Plate Theory 50

3.2 SYMMETRIC LAMINATED PLATES 50

3.2.1 Static Analysis 52

3.2.2 Buckling and Free Vibration 57

3.3 MAGNETO-ELECTRO-ELASTIC LAMINATED PLATES 59

3.3.1 Governing Equations 60

3.3.2 Navier's Solution 67

3.3.3 Levy's Solution 73
3.3.4 Numerical Illustration 80
3.4 THICK LAMINATED PLATES 85
3.4.1 General Formulation 85
3.4.2 Green's Functions 93
REFERENCES 106

CHAPTER 4 ■ Laminated Composite Beams 109

4.1 GENERAL FORMULATION OF THIN LAMINATED BEAMS 110
4.1.1 Narrow Beams 112
4.1.2 Wide Beams 114
4.2 SYMMETRIC LAMINATED BEAMS 116
4.2.1 Static Analysis 116
4.2.2 Buckling and Free Vibration 118
4.3 THICK LAMINATED BEAMS 120
4.3.1 General Formulation 120
4.3.2 Green's Functions 127
4.4 TYPICAL THICK BEAM PROBLEMS 130
4.4.1 Explicit Solutions 130
4.4.2 Numerical Results 140
REFERENCES 141

CHAPTER 5 ■ Laminated Composite Shells 142

5.1 COORDINATES 143
5.1.1 Curvilinear Coordinates 143
5.1.2 Shell Coordinates 146
5.2 GENERAL FORMULATION 150
5.2.1 Displacement Fields 150
5.2.2 Strain–Displacement Relations 150
5.2.3 Stress Resultants and Stress Couples 151
5.2.4 Constitutive Laws 153

5.2.5 Equilibrium Equations 154
5.2.6 Governing Equations 155
5.2.7 Boundary Conditions 156
5.3 SHELLS OF REVOLUTION 157
5.3.1 Conical Shells 159
5.3.2 Circular Cylindrical Shells 161
5.3.3 Spherical Shells 162
5.3.4 Shallow Spherical Shells 163
5.3.5 Solution Procedure and Examples 166
5.4 MEMBRANE SHELLS 169
5.4.1 Membrane Shells of Revolution 171
5.4.2 Axisymmetrical Load 173
5.4.3 Conical Membrane Shells 174
5.4.4 Cylindrical Membrane Shells 175
5.5 VIBRATION OF SHELLS 176
REFERENCES 179

CHAPTER 6 ▪ Composite Sandwich Construction 180

6.1 COMPOSITE SANDWICH PLATES 181
6.1.1 General Formulation 181
6.1.2 Buckling Analysis 188
6.1.3 Free Vibration 193
6.2 COMPOSITE SANDWICH BEAMS 203
6.2.1 General Formulation 203
6.2.2 Buckling Analysis 206
6.2.3 Free Vibration 211
6.2.4 Forced Vibration 218
6.2.5 Vibration Suppression 220
6.3 COMPOSITE SANDWICH CYLINDRICAL SHELLS 225
6.3.1 General Formulation 226
6.3.2 Free Vibration 230

6.4 DELAMINATED COMPOSITE SANDWICH BEAMS 232
6.4.1 Buckling Analysis 233
6.4.2 Postbuckling Analysis 239
6.4.3 Free Vibration 242
REFERENCES 249

CHAPTER 7 ▪ Composite Wing Structures 251
7.1 COMPREHENSIVE BEAM MODEL 252
7.1.1 Static Analysis 253
7.1.2 Dynamic Analysis 258
7.1.3 Matrix Form 260
7.1.4 Some Reductions 262
7.2 WINGS WITH UNIFORM CROSS-SECTION 265
7.2.1 Free Vibration 265
7.2.2 Forced Vibration 270
7.2.3 Aeroelastic Divergence 272
7.3 TAPERED WINGS 276
7.3.1 Comprehensive Finite Element Model 280
7.3.2 Static Analysis 283
7.3.3 Free Vibration 284
7.3.4 Aeroelastic Divergence 285
7.4 VARIABLE THICKNESS PLATE MODEL 286
7.4.1 General Formulation 286
7.4.2 Finite Element Method 289
REFERENCES 292

CHAPTER 8 ▪ Anisotropic Elasticity 294
8.1 TWO-DIMENSIONAL ANALYSIS 295
8.1.1 Stroh Formalism for Anisotropic Elastic Materials 295
8.1.2 Stroh Formalism in Laplace Domain for Viscoelastic Materials 297

8.1.3 Expanded Stroh Formalism for Piezoelectric Materials 298
8.1.4 Expanded Stroh Formalism for MEE Materials 300
8.1.5 Extended Stroh Formalism for Thermoelastic Problems 302
8.2 COUPLED STRETCHING–BENDING ANALYSIS 303
8.2.1 Stroh-Like Formalism for General Laminated Plates 304
8.2.2 Specialization to Plate Bending Analysis 307
8.2.3 Expanded Stroh-Like Formalism for Electro-Elastic Laminates 308
8.2.4 Expanded Stroh-Like Formalism for MEE Laminated Plates 311
8.2.5 Extended Stroh-Like Formalism for Thermal Stresses in Laminates 314
8.3 THREE-DIMENSIONAL ANALYSIS 317
8.3.1 Radon–Stroh Formalism for Anisotropic Elastic Materials 317
8.3.2 Radon–Stroh Formalism for Anisotropic Piezoelectric and MEE Materials 324
8.4 GREEN'S FUNCTIONS FOR TWO-DIMENSIONAL PROBLEMS 324
8.4.1 An Infinite Anisotropic Elastic Plane 325
8.4.2 An Anisotropic Elastic Half-Plane 325
8.4.3 An Anisotropic Elastic Bi-Material 326
8.4.4 A Plane with an Elliptical Hole 327
8.4.5 A Plane with a Straight Crack 328
8.4.6 A Plane with an Elliptical Inclusion 329
8.5 GREEN'S FUNCTIONS FOR COUPLED STRETCHING–BENDING DEFORMATION 331
8.5.1 An Infinite Laminated Plate 331
8.5.2 An Infinite Laminated Plate with an Elliptical Hole 333

8.5.3 An Infinite Laminated Plate with a Straight Crack 336
8.5.4 An Infinite Laminated Plate with an Elliptical Inclusion 337
8.6 GREEN'S FUNCTIONS FOR THREE-DIMENSIONAL PROBLEMS 341
8.6.1 Derivatives of the Green's Functions 343
8.6.2 Computation of Green's Functions and Their Derivatives 344
REFERENCES 347

CHAPTER 9 ■ Numerical Methods 349
9.1 FINITE ELEMENT METHOD 349
9.2 BOUNDARY ELEMENT METHOD 353
9.2.1 Beam Analysis 353
9.2.2 Two-Dimensional Analysis 361
9.2.3 Coupled Stretching–Bending Analysis – Thin Plate 365
9.2.4 Coupled Stretching–Bending Analysis – Thick Plate 368
9.2.5 Three-Dimensional Analysis 369
9.2.6 Boundary-Based Finite Element Method 370
9.3 MESHLESS METHOD 372
9.3.1 Element-Free Galerkin (EFG) Method 372
9.3.2 Meshless Local Petrov–Galerkin (MLPG) Method 376
9.4 NUMERICAL EXAMPLES 379
9.4.1 Laminated Composite Beams 380
9.4.2 Laminated Composite Plates 385
REFERENCES 390

AUTHOR INDEX, 392
SUBJECT INDEX, 395

Preface

WHILE CURRENTLY THERE ARE several textbooks dealing with the subject of composite materials, due to their diversity and versatility, none of them can include everything about the composites. Some texts focus on the introduction of material properties, some others focus on their mechanical behavior. As to the latter, most of the texts introduce the mechanical behavior of composite materials, while some others deal with the behavior of structures composed of composite materials. Due to the nature of anisotropy, composite materials are usually modeled as anisotropic elastic solids. Therefore, some texts will also briefly introduce anisotropic elasticity. For practical engineering, how to use the numerical methods to solve the problems related to composites is also important. With the above considerations, I try to build a bridge to connect composite materials, structures, and anisotropic elasticity theoretically and numerically.

This book introduces the mechanical behavior of laminated composite materials and provides the theories and solutions of laminated composite structures. To cover most of the structures, all the basic structural elements such as beams, plates, and shells are stated in detail. Further contents include composite sandwich construction and composite wing structures. To connect with practical engineering applications and analyze more complicated real structures, numerical methods (including finite element method, boundary element method, and meshless method) and their theoretical basis anisotropic elasticity are also included. Through this book, the readers will learn not only the mechanics of composite structures but also anisotropic elasticity and some popular numerical methods. Moreover, the readers will also learn about how to model practical composite structures such as composite wings.

In this book, the basic contents are included in Chapters 1 and 2 for the description of the mechanical behavior of laminated composite materials (including micromechanics, lamina and laminate stiffness, lamination theory, failure and strength, etc.), and in the frontal parts of Chapters 3 and 4 for the basic analysis of plates and beams. Most of the other contents of this book are different from the ones presented by the other textbooks. For example, Section 3.3 presents the solutions for magneto-electro-elastic (MEE) laminated plates; Sections 3.4 and 4.3 present the solutions of Green's functions for thick laminated plates and beams; Section 4.4 presents explicit solutions for some typical thick laminated beams; Chapter 5 presents the theory for the general laminated composite shells instead of a specific cylindrical shell; Chapter 6 presents some analytical solutions for sandwich beams, plates, and cylindrical shells as well as delaminated composite sandwich beams; Chapter 7 presents the modelling and analysis of composite wing structures; Chapter 8 presents the complex variable theories of anisotropic elasticity and the related Green's functions; Chapter 9 briefly introduces not only the finite element method but also the other important numerical methods such as boundary element method and meshless method, as well as the numerical illustration done by the boundary element method for laminated composite beams and plates.

This book is appropriate to be a university textbook for the courses related to mechanics of composite materials, composite structures, and anisotropic elasticity, which are generally offered to the senior undergraduate and graduate students majoring in aerospace, mechanical, civil, and naval engineering, applied mechanics, and engineering science. It will also be of interest to aeronautical engineers, mechanical engineers, civil engineers, ocean engineers, applied mathematicians, and material scientists.

I wish to express my gratitude to my PhD thesis adviser, Prof. T.C.T. Ting, and my mentors, Prof. W.H. Chen of National Tsing-Hua University and the late Prof. C.S. Yeh of National Taiwan University for their guidance during my studies. I also like to thank my friends, Professors K. Kishimoto (Tokyo Institute of Technology), M. Omiya (Keio University), N. Miyazaki (Kyoto University), T. Ikeda (Kagoshima University), Y.W. Mai (Sydney University), T. Aoki (Tokyo University), T. Yokozeki (Tokyo University), C. Zhang (Siegen University), W. Becker (Technical University of Darmstadt), and V. Mantic (University of Seville) who have helped me during my visiting in their departments. Special thanks also to my former assistant H. Shen and my present students C.W. Hsu, M.L. Shieh, W.T. Chang, Y.S. Wei, Y.J. Shi, S.Y. Wu, J.S. Yu, J.S. Hsu, J.S. Jen, J.H. Tao, who helped me drawing part of figures presented in this book, and my former students J.S. Hu, J.S. Moh, M.C. Yu, Z.S. Tsai, H.S. Gai, W.C. Chang,

H. W. Hsu, Y.H. Lin, W.Y. Huang, etc. who put their efforts on the related works of this book. I acknowledge the National Science and Technology Council of Taiwan for the support of my research in the area of composite materials and structures.

Finally, I would like to dedicate this book to my wife, Wenling, and my daughters, Frannie and Vevey, with thanks for their constant support and encouragement in everything.

Chyanbin Hwu
Tainan, Taiwan
November 12, 2023

CHAPTER 1

Introduction to Composites

Composite materials are composed of two or more distinct constituents. Most of the natural objects such as the human body, plants, and animals are composites. Modern engineering use of composites such as fiber-reinforced composite materials has created a revolution in high-performance structures in recent years. Relative to conventional metallic materials, advanced composite materials offer significant advantages in the ratios of strength and stiffness to weight. Moreover, the interesting aspect of composite materials is the freedom to select the precise form of the materials to suit the application. For example, in using fiber-reinforced composites, one has the ability to specify the type of fibers, their amount and orientation, and the matrix materials to get an optimum performance for the composites. Modern composites have therefore been described as being revolutionary in the sense that the material can be designed as well as the structure. Along with the freedom to design both the material and the structure comes additional responsibility for the designer as well. Hence, certain knowledge on the part of the designer to understand how to use the composite materials in a rational manner is necessary.

Due to the nature of anisotropy, composite materials are usually modeled as anisotropic elastic solids. Extensive research papers and books with names related to *mechanics of composite materials*, *mechanics of composite structures*, or *anisotropic elasticity* appear in the literature such as Jones (1974), Vinson and Chou (1975), Christensen (1979), Tsai (1980),

DOI: 10.1201/9781003470465-1

Hoskin and Baker (1986), Vinson and Sierakowski (1986), Agarwal and Broutman (1990), Gibson (1994), Powell (1994), Hull and Clyne (1996), and Herakovich (1998) for mechanics of composite materials; Kollar and Springer (2003), Reddy (2003), Vasiliev and Morozov (2013), Kassapoglou (2013), and Altenbach et al. (2018) for mechanics of composite structures; and Lekhnitskii (1963, 1968), Ting (1996), and Hwu (2010, 2021) for anisotropic elasticity. Perhaps because of different training or favorites, it is difficult to find one book discussing these three topics simultaneously. This book attempts to bring together these three seemingly connected and different topics.

Like most of the theories in structural mechanics, mechanics of composite structures usually assumes that the structures composed of composite materials are in the form of beams, plates, shells, or their combinations. When describing the mechanical behavior of composite structures, sectional responses like the mid-surface deflection, stress resultants, and bending moments are usually used. The associated boundary conditions for each boundary value problem of composite structures are therefore described using the sectional responses, while for anisotropic elasticity, every point inside the solids is considered in mechanical analysis. Therefore, for a simple structure like a cantilever beam, although it is easy to find an analytical solution satisfying all the sectional properties and boundary conditions, it is usually difficult to find an analytical solution satisfying all the point properties and boundary conditions. Because the failure of a structure usually initiates from a point, not a section, to understand the failure mechanism the approach of anisotropic elasticity is more suitable than the mechanics of composite structures. Furthermore, to see more clearly it is better to focus our attention on the locations which we are concerned, like the cracks, holes, or inclusions, and disregard all the other boundary effects by considering infinite domain. In this way, it is possible for us to get an analytical solution which satisfies point properties and boundary conditions. Moreover, this solution is usually useful for engineers to understand the failure of complex structures and is also a suitable guidance for engineering design.

From the above statements, we see that both of mechanics of composite structures and anisotropic elasticity are important for understanding the mechanical behavior of composite materials and structures. Moreover, they are related through integration or differentiation since one is concerned with the sectional responses and the other is concerned with the point responses. We hope that the attempt to bring them together in this book will be useful for the engineers and researchers who are involved in design and analysis with composite materials and structures.

1.1 CLASSIFICATION AND CHARACTERISTICS OF COMPOSITE MATERIALS

A composite material is a material system composed of a mixture or combination of two or more macro-constituents differing in form and/or material composition and that are essentially insoluble in each other. Most composite materials developed thus far have been fabricated to improve mechanical properties such as strength, stiffness, toughness, and high-temperature performance. The strengthening mechanism strongly depends on the geometry of the reinforcement. Therefore, it is quite convenient to classify composite materials on the basis of the geometry of a representative unit of reinforcement. There are three commonly accepted types of composite materials: *fibrous composites*, *laminated composites*, and *particulate composites*. Fibrous composites consist of fibers in a matrix; laminated composites consist of layers of various materials; particulate composites are composed of particles in a matrix. Numerous multiphase composites exhibit more than one characteristic of the classes. For example, reinforced concrete is both particulate and fibrous; and laminated fiber-reinforced composites are obviously both laminated and fibrous composites.

From the application point of view, one of the important composite material types is the fibrous composite. Such material system has the desirable properties of having high strength and/or stiffness as well as being light in weight. Fibrous reinforcement is so effective because many materials are much stronger and stiffer in fiber form than they are in bulk form, and the binder material usually called a matrix can take the responsibility of supporting, protecting, and transferring stresses. Naturally, fibers are of little use unless they are bound together to take the form of a structural element which can take loads. Typically, the matrix is of considerably lower density, stiffness, and strength than the fibers.

The need for fiber placement in different directions according to the particular application has led to various types of composites (Gibson, 1994). Among all the different types of fibrous composites, the *continuous fiber reinforcement* is used extensively. The major problem for this kind of material is the potential for delamination or the separation of the laminae. *Woven fiber composites* do not have distinct laminae and are not susceptible to delamination, but strength and stiffness are sacrificed due to the fact that the fibers are not so straight as in the continuous fiber laminates. *Chopped-fiber* (sometimes referred to as *whisker* or *short-fiber*) *composites* are used extensively in high-volume applications due to low manufacturing cost, but their mechanical properties are considerably poorer than those of continuous fiber-reinforced composites. *Hybrid composites* may consist of mixed chopped and continuous fibers, or mixed fiber types. Finally, the *composite sandwiches* consist of high-strength composite facing sheets and a lightweight foam or honeycomb core. The composite sandwiches have extremely high flexural stiffness-to-weight ratios and are widely used in aerospace structures.

The most common application of composites in aircraft structures is for the skin of wings, tails, and control surfaces. Consider the skin of a main wing box for a straight-wing aircraft. This is primarily required to carry direct stresses in the spanwise direction due to the wing bending more or less as a cantilever beam and shear stresses caused by the wing twisting. The simplest design approach is to use a laminate pattern consisting of 0° plies to carry the spanwise direct stresses, ±45° plies to carry the shear stresses, and 90° plies to carry the chordwise direct stresses. As an example, the basic pattern for the F/A-18 wing skin comprises 46% of 0° plies, 50% of ±45 plies, and 4% of 90° plies (Hoskin and Baker, 1986).

1.2 FABRICATION OF FIBER-REINFORCED COMPOSITE MATERIALS

A broad classification of the composites and its sub-category, fibrous composites, is presented above. Following is a brief introduction of the fibers, matrix materials, and the fabrication of the fibrous composites.

1.2.1 Fibers

The most common reinforcing fibers for polymer matrix composites are glass fibers. The principal advantages of glass fibers are their low cost and high strength. Their disadvantages are low modulus, poor abrasion resistance, and poor adhesion to polymer matrix resins. In addition to glass fibers, the other commonly used fibers are graphite, carbon, aramid, boron, ceramic, alumina, silicon carbide, etc. Among them, graphite fibers are the predominant high-strength, high-modulus reinforcement used in the fabrication of high-performance resin-matrix composites. The term "graphite fiber" is used to describe fibers that have a carbon content in excess of 99% whereas the term "carbon fiber" describes fibers that have a carbon content of 80–95%.

1.2.2 Matrix

The matrix has a strong influence on several mechanical properties of the composites such as transverse modulus and strength, shear properties, and properties in compression. The commonly used matrix materials consist of polymers (commonly called *plastics*), metals, ceramics, etc. Polymers are the most widely used matrix material for fibrous composites. Their chief advantages are low cost, easy processibility, good chemical resistance, and low specific gravity. On the other hand, low strength, low modulus, and low operating temperatures limit their use. According to their structure and behavior, polymers can be classified as thermoplastics (e.g., PI, PS, PEEK, PPS) or thermosets (e.g., epoxy, polyester, phenolic). The polymers that soften or melt on heating and can be reshaped by the application of heat and pressure are called thermoplastic polymers. Those that do not soften but decompose on heating are called thermosetting plastics. As a result, thermosets cannot be reshaped once they are solidified by a curing process. As to the metals, their chief advantages include high strength, high modulus, high toughness and impact resistance, and relative insensitivity to temperature changes. Thus, in applications that require exposure to high temperatures and other severe environmental conditions, metal matrices possess the greatest advantage over polymer matrices. The factors that limit their usage include their high density, high processing temperatures, reactivity with fibers, and attack by corrosion. The most commonly used metal matrices are based on aluminum and titanium.

1.2.3 Fabrication

The formation of composites involves the combination of matrix and fiber such that the matrix impregnates, surrounds, and wets the fibers. The choice of a fabrication process is strongly influenced by the chemical nature of the matrix and the temperature required to form, melt, or cure[1] the matrix. For *thermosetting resin matrix composites*, the fabrication processes can be broadly classified as wet forming processes and processes using premixes or prepreg[2]. In the wet forming processes, the final product is formed while the resin is quite fluid and the curing process is usually completed in one step. The wet forming processes include hand lay-up, filament winding, pultrusion, and bag molding. In the processes using premixes, compounding is separated from lay-up or molding. Compounding is done to make such premixes as bulk molding compounds, sheet molding compounds and prepregs. For *thermoplastic resin matrix composites*, the principle method used for the production of parts with short-fiber-reinforced thermoplastics is injection molding. As to the fabrication of *metal matrix composites*, significantly higher process temperatures are needed. Since fibers may react with the metal matrix at such higher temperatures, sufficient care must be exercised to limit the interaction. They are most often fabricated by liquid infiltration or hot pressing of solid matrix on fibers.

The special nature of composite materials makes it essential for designers to be aware of certain types of material defects that occur during manufacture or fabrication. Such defects are intimately related to the achievement of design strengths and stiffnesses of the finished product. Some of the common defects that must be controlled and subsequently related to the performance of the structural element are: voids, delaminations, disbands, foreign body inclusions, resin-starved areas, resin-rich areas, incomplete resin cure, incorrect fiber orientation, wavy fibers, incorrect ply sequence, fiber gaps, wrinkled layers, poor surface condition, etc.

1.3 MECHANICAL BEHAVIOR OF COMPOSITE MATERIALS

Most types of the composite materials described above are both inhomogeneous and nonisotropic. Because of the inherent heterogeneous nature of composite materials, they are conveniently studied from two points of view: micromechanics and macromechanics. *Micromechanics* is the study of composite material behavior wherein the interaction of the constituent materials is examined on a microscopic scale. Specifically speaking, micromechanics is aimed at providing an understanding of the behavior of composites in terms of the properties and interactions of the fibers and matrix. Approximate models are used to simulate the microstructure of the composite and hence predict its *average* properties such as strength and stiffness in terms of the properties and behaviors of the constituents. *Macromechanics* is the study of composite material behavior wherein the material is *presumed homogeneous* and the effect of the constituent materials are detected only as *average* apparent properties of the composite.

The inherent anisotropy of composite materials leads to mechanical behavior characteristics that are quite different from those of conventional isotropic materials. The primary advantage of composites is the ability to control anisotropy by design and fabrication. For isotropic materials, normal stress causes extension in the direction of the applied stress and contraction in the perpendicular direction. Also, shear stress causes only shearing deformation. While for anisotropic materials, application of a normal stress leads not only to extension in the direction of the stress and contraction perpendicular to it but to shearing deformation. Conversely, shearing stress causes extension and contraction in addition to the distortion of shearing deformation. To describe this kind of mechanical behavior, the mathematical models for general anisotropic elastic solids and the complex variable formulations for anisotropic elasticity will be introduced in Chapter 8. To focus on the laminated composite materials, the mechanical behavior of a lamina and the prediction of its stiffness, strength and fracture toughness as well as the behavior of a laminate will be introduced in Chapter 2. For the other kinds of composite materials, one may refer to the relevant textbooks mentioned at the beginning of this chapter.

1.4 COMPOSITE STRUCTURES

The basic structural form in which composite materials are most generally employed consists of beams, rods, columns, plates, panels, shells, sandwich structures, tubes, and some other complex forms. A beam, column, or rod is a long thin structural component. The term *beam* is utilized when the structure is subjected to a lateral load such that bending occurs. The term *rod* is used when the structure is loaded in the axial direction by tensile forces which try to stretch the structure. The term *column* is used when the structure is subjected to compressive forces in the axial direction, which results in compressive stresses and/or elastic instability (buckling). Combination of these loads may occur, such as when the lateral and compressive loads occur simultaneously, resulting in the structure being referred to as a *beam column*. *Plates* are initially straight and flat surface structures whose thickness is much smaller compared to their other dimensions. The static or dynamic loads carried by plates are predominantly perpendicular to the plate surface. A *shell* is a thin-walled structure whose middle surface is curved in at least one direction. *Sandwich structures* usually consist of two thin sheets of high-strength material between which a thick layer of low average strength and density is sandwiched. The two thin sheets are called faces, and the intermediate layer is the core of the sandwich.

To analyze the structural responses, the constitutive laws of composite materials, the equations of motion, the kinematic relations, and the boundary and initial conditions of composite structures should all be described by a suitable mathematical modeling. Detailed description of the mathematical modeling for composite structures will be given in Chapters 3–7.

The design procedure to establish the static strength of a composite structure generally involves, first, a detailed theoretical structural analysis (now almost always done by using a finite element structural model) and second, a large amount of structural testing on specimens of varying degrees of complexity. The latter is outlined as follows:

1. Coupon tests: These are tests on small, plain specimens used to establish the basic material properties (e.g., the ultimate tensile strength).
2. Structural detail tests: Structural details are also small specimens but they contain typical elementary structural features. Examples might be tension specimens with open or filled bolt holes, simple lap joint specimens, etc.
3. Subcomponent test: A subcomponent is a full-scale representation of a moderate-size region of a structure.
4. Component test: A component is a full-scale representation of a major region of a structure.
5. Full-scale article test: This is a test on a virtually complete structure. Almost certainly, only one test specimen will be available.

Coupon tests are done to establish some design allowables, such as "knock-down" factors (i.e., reduction factors). A comparison of the "RT/dry" (RT stands for room temperature) value and "hot/wet" value of an allowable provides an environmental knockdown factor. Furthermore, because the stresses in a multidirectional laminate can vary markedly from ply to ply, allowable values are likely to be cited in terms of strain rather than stress.

NOTES

1 Curing is the drying or polymerization of the resinous matrix material to form a permanent bond between fibers and between laminae.

2 Prepreg is a form of preimpregnated fibers of which the fibers are saturated with resinous material such as polyester or epoxy resin that holds the fibers in position and serves as the matrix material.

REFERENCES

Agarwal, B.D. and Broutman, L.J., 1990, *Analysis and Performance of Fiber Composites*, 2nd Ed., John Wiley & Sons, Inc., New York.

Altenbach, H., Altenbach, J., and Kissing, W., 2018, *Mechanics of Composite Structural Elements*, 2nd ed., Springer, Singapore.

Christensen, R.M., 1979, *Mechanics of Composite Materials*, Wiley-Interscience, New York.

Gibson, R.F., 1994, *Principles of Composite Material Mechanics*, McGraw-Hill Int., New York.

Herakovich, C. T., 1998, *Mechanics of Fibrous Composites*, John Wiley and Sons, Inc., New York.

Hoskin, B.C. and Baker, A.A., 1986, *Composite Materials for Aircraft Structures*, AIAA Education Series, AIAA, New York.

Hull, D. and Clyne, T.W., 1996, *An Introduction to Composite Materials*, 2nd Ed., Cambridge University Press, Cambridge.

Hwu, C., 2010, *Anisotropic Elastic Plates*, Springer, New York.

Hwu, C., 2021, *Anisotropic Elasticity with Matlab*, Springer, Cham.

Jones, R.M., 1974, *Mechanics of Composite Materials*, Scripta, Washington, D.C.

Kassapoglou, C., 2013, *Design and Analysis of Composite Structures: With Applications to Aerospace Structures*, 2nd ed., John Wiley and Sons Inc., West Sussex.

Kollar, L. P. and Springer, G. S., 2003, *Mechanics of Composite Structures*, Cambridge University Press, Cambridge.

Lekhnitskii, S.G., 1963, *Theory of Elasticity of an Anisotropic Body*, MIR, Moscow.

Lekhnitskii, S.G., 1968, *Anisotropic Plates*, Gordon and Breach Science Publishers, New York.

Powell, P.C., 1994, *Engineering with Fibre-Polymer Laminates*, Chapman and Hall, London.

Reddy, J. N., 2003, *Mechanics of Laminated Composite Plates and Shells: Theory and Analysis*, 2nd. ed., CRC Press, Boca Raton.

Ting, T.C.T., 1996, *Anisotropic Elasticity: Theory and Applications*, Oxford Science Publications, N.Y.

Tsai, S.W., 1980, *Introduction to Composite Materials*, Technomic Pub., Westport.

Vasiliev, V.V. and Morozov, E.V., 2013, *Advanced Mechanics of Composite Materials and Structural Elements*, 3rd ed., Elsevier, Oxford.

Vinson, J.R. and Chou, T.W., 1975, *Composite Materials and Their Use in Structures*, Applied Science Pub., Essex.

Vinson, J.R. and Sierakowski, R.L., 1986, *The Behavior of Structure Composed of Composite Materials*, Martinus Nijhoff Pub., Dordrecht.

CHAPTER 2

Mechanical Behavior of Laminated Composite Materials

Laminated composites consist of layers of various materials. Most laminates made of fibrous composites consist of several distinct layers of unidirectional fiber-reinforced composites. Each layer is usually made of the same constituent materials. In this chapter, we consider *laminated composites* as a material consisting of layers of various unidirectional fiber-reinforced composites made of the same constituents, as shown in Figure 2.1. A single layer of a laminated composite material is generally referred to as a *ply* or *lamina*. A single lamina is generally too thin to be directly used in most of the engineering applications. Several laminae are bonded together to form a structure termed a *laminate*. Properties of a lamina may be predicted by knowing the properties of its constituents, that is, fibers, matrices, and their volume fractions. Properties and orientation of the laminae in a laminate are chosen to meet the laminate design requirements. Properties of a laminate may then be predicted by knowing the properties of its constituent laminae. Behavior of the laminate is governed by the behavior of individual laminae. Thus, analysis or design of a laminate requires a complete knowledge of the behavior of the laminae. In this chapter, we will therefore first discuss the behavior of a lamina and

DOI: 10.1201/9781003470465-2

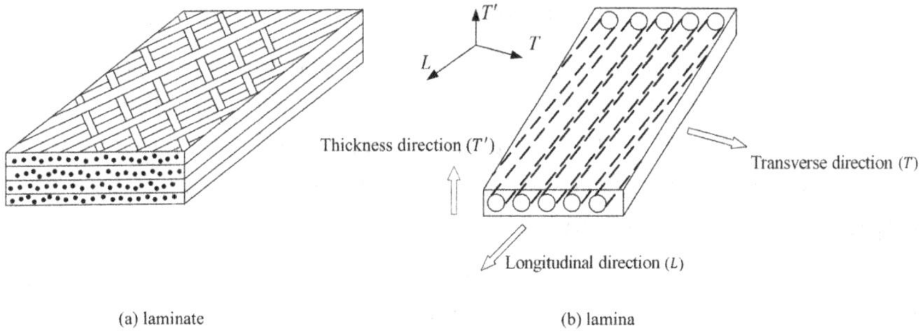

FIGURE 2.1 Laminated composites: (a) laminate; (b) lamina.

the prediction of its stiffness, strength, and fracture toughness, then discuss the behavior of a laminate and the prediction of its strength.

2.1 LAMINA ELASTIC BEHAVIOR

As shown in Figure 2.1, each laminae of the laminated composites is considered to be a *unidirectional fiber-reinforced composite* which consists of parallel fibers embedded in a matrix. The direction parallel to fibers is generally called the *longitudinal direction* and is referred to as L. The direction perpendicular to the fibers is called the *transverse direction* and is referred to as T. These axes are also referred to as the *material axes* of the ply. The ply depicted schematically in Figure 2.1(b) shows only one fiber through the ply thickness. In practice, this may be true only for large-diameter fibers such as boron. Plies formed by other fibers may have several fibers through the actual ply thickness. The fibers are usually randomly distributed throughout the cross-section and may be in contact with each other in some locations. With this particular structure, a unidirectional fiber-reinforced composite shows different properties in the longitudinal and transverse directions, and nearly identical properties in all directions of the cross-section due to the random fiber distribution. Along the longitudinal and transverse directions, two orthogonal planes of material symmetry can be constructed for the unidirectional fiber-reinforced composite. Such materials are said to be *orthotropic*. Moreover, rotating about the axis along the fiber direction, there is one plane in which the mechanical properties are equal in all directions. In this sense, the unidirectional fiber-reinforced composite also belongs to the category of *transverse isotropic materials.*

2.1.1 Specially Orthotropic Lamina

If the reference coordinate axes coincide with the principal material directions, the lamina is called *specially orthotropic lamina.* By using the subscripts L and T to denote the components related to the principal material directions, the two-dimensional stress–strain relation for specially orthotropic lamina can be written as

$$\boldsymbol{\sigma}^* = \mathbf{Q}\boldsymbol{\varepsilon}^*, \tag{2.1a}$$

where

$$\boldsymbol{\sigma}^* = \begin{Bmatrix} \sigma_L \\ \sigma_T \\ \tau_{LT} \end{Bmatrix}, \quad \mathbf{Q} = \begin{bmatrix} Q_{11} & Q_{12} & 0 \\ Q_{12} & Q_{22} & 0 \\ 0 & 0 & Q_{66} \end{bmatrix}, \quad \boldsymbol{\varepsilon}^* = \begin{Bmatrix} \varepsilon_L \\ \varepsilon_T \\ \gamma_{LT} \end{Bmatrix}. \tag{2.1b}$$

$\mathbf{Q}$ is generally called *stiffness matrix* and its expressions in terms of engineering constants can be written as

$$\mathbf{Q} = \begin{bmatrix} \dfrac{E_L}{1-\nu_{LT}\nu_{TL}} & \dfrac{\nu_{LT}E_T}{1-\nu_{LT}\nu_{TL}} & 0 \\ & \dfrac{E_T}{1-\nu_{LT}\nu_{TL}} & 0 \\ \text{symm.} & & G_{LT} \end{bmatrix}, \text{ generalized plane stress} \tag{2.2}$$

Here, E, G and ν denote, respectively, Young's modulus, shear modulus, and Poisson's ratio, and the subscripts L and T stand for the longitudinal and transverse directions. Due to the symmetry of compliances, $\nu_{LT}/E_L = \nu_{TL}/E_T$.

2.1.2 Generally Orthotropic Lamina

If the principal material directions do not coincide with the reference coordinate directions that are geometrically natural to the solution of the problem, an orthotropic lamina appears to be anisotropic in reference coordinate direction, which is called a *generally orthotropic lamina.* Since a laminated composite is usually constructed by stacking several unidirectional laminae in a specified sequence of orientation, the principal directions of each lamina make a different angle with a common set of reference coordinate axes. Although each lamina is orthotropic and obeys the previously described stress–strain relation (2.1) referred to its

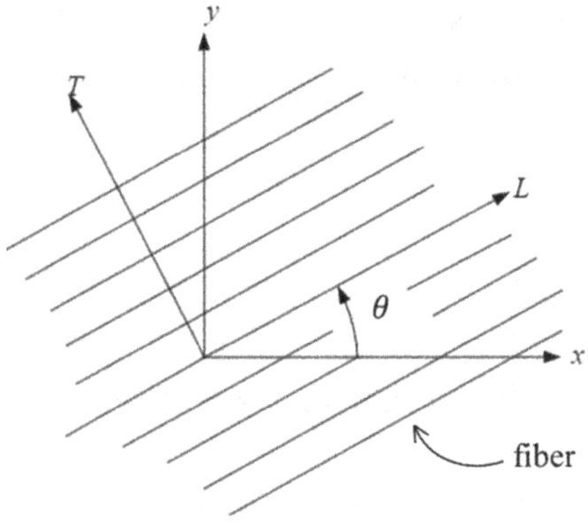

FIGURE 2.2 Transformation between x–y and L–T coordinate systems.

principal material axes, for the purpose of analysis and synthesis of laminated structures it is necessary to refer the stress–strain relation to a common reference coordinate axes. To know this relation, we now derive the stress–strain relation of a lamina with arbitrary orientation as follows.

It's known that the stresses and strains are tensors of rank 2. If the angle θ between these two coordinate systems is taken positive when the angle of the L–T axes measured from x–y axes is in the counterclockwise direction (see Figure 2.2), the transformation relations of the stresses and strains can be written as

$$\sigma^* = \boldsymbol{T}\sigma, \quad \varepsilon^* = \boldsymbol{RTR}^{-1}\varepsilon, \tag{2.3a}$$

where

$$\boldsymbol{T} = \begin{bmatrix} \cos^2\theta & \sin^2\theta & 2\sin\theta\cos\theta \\ \sin^2\theta & \cos^2\theta & -2\sin\theta\cos\theta \\ -\sin\theta\cos\theta & \sin\theta\cos\theta & \cos^2\theta - \sin^2\theta \end{bmatrix},$$

$$\boldsymbol{R} = \begin{bmatrix} 1 & 0 & 0 \\ 0 & 1 & 0 \\ 0 & 0 & 2 \end{bmatrix}, \quad \sigma = \begin{Bmatrix} \sigma_x \\ \sigma_y \\ \tau_{xy} \end{Bmatrix}, \quad \varepsilon = \begin{Bmatrix} \varepsilon_x \\ \varepsilon_y \\ \gamma_{xy} \end{Bmatrix}. \tag{2.3b}$$

Note that the appearance of the matrix $\boldsymbol{R}$ is because the transformation rule of second-order tensor is applied for the elastic strains while the constitutive relation (2.1) is written for the engineering strains, and the engineering shear strain is twice of the elastic shear strain, that is, $\gamma_{LT} = 2\varepsilon_{LT}$ and $\gamma_{xy} = 2\varepsilon_{xy}$. Substituting (2.3a) into (2.1a), we obtain

$$\sigma = \boldsymbol{Q}^{*}\varepsilon, \tag{2.4a}$$

where

$$\boldsymbol{Q}^{*} = \boldsymbol{T}^{-1}\boldsymbol{QRTR}^{-1} = \boldsymbol{T}^{-1}\boldsymbol{QT}^{-T}. \tag{2.4b}$$

Usually, it will take time to calculate the inverse of a matrix. However, since the matrix $\boldsymbol{T}$ stands for the transformation matrix by rotating angle θ, its inverse $\boldsymbol{T}^{-1}$ can be obtained directly from $\boldsymbol{T}$ by replacing the angle θ with $-\theta$. By simple matrix multiplication shown in (2.4b), the relation between the transformed stiffness **Q*** and the stiffness **Q** can be written as

$$\begin{aligned}
Q_{11}^{*} &= Q_{11}\cos^4\theta + 2\left(Q_{12} + 2Q_{66}\right)\sin^2\theta\cos^2\theta + Q_{22}\sin^4\theta,\\
Q_{12}^{*} &= \left(Q_{11} + Q_{22} - 4Q_{66}\right)\sin^2\theta\cos^2\theta + Q_{12}\left(\sin^4\theta + \cos^4\theta\right),\\
Q_{22}^{*} &= Q_{11}\sin^4\theta + 2\left(Q_{12} + 2Q_{66}\right)\sin^2\theta\cos^2\theta + Q_{22}\cos^4\theta,\\
Q_{16}^{*} &= \left(Q_{11} - Q_{12} - 2Q_{66}\right)\sin\theta\cos^3\theta + \left(Q_{12} - Q_{22} + 2Q_{66}\right)\sin^3\theta\cos\theta,\\
Q_{26}^{*} &= \left(Q_{11} - Q_{12} - 2Q_{66}\right)\sin^3\theta\cos\theta + \left(Q_{12} - Q_{22} + 2Q_{66}\right)\sin\theta\cos^3\theta,\\
Q_{66}^{*} &= \left(Q_{11} + Q_{22} - 2Q_{12} - 2Q_{66}\right)\sin^2\theta\cos^2\theta + Q_{66}\left(\sin^4\theta + \cos^4\theta\right).
\end{aligned} \tag{2.5}$$

The matrix **Q*** is now fully populated and similar in appearance to the ***Q*** matrix for an anisotropic material. There is coupling between shear strain and normal stresses as well as between shear stress and normal strains. However, because the lamina does have orthotropic characteristics in principal material directions, its mechanical behavior is still governed by only *four* independent material constants, not six elastic constants fully populated in **Q***.

Similar to the transformation of the stiffness matrix **Q**, the transformed compliance matrix **S*** of the *x*-*y* coordinate axes can be found to be related to the compliance matrix **S** of the *L*-*T* coordinate axes by

$$\boldsymbol{S}^{*} = \boldsymbol{T}^{T}\boldsymbol{ST}. \tag{2.6}$$

By comparing the transformed compliances with the anisotropic compliances in terms of engineering constants (Hwu, 2010), the apparent engineering constants for an orthotropic lamina in x–y coordinates can be obtained as

$$
\begin{aligned}
\frac{1}{E_x} &= \frac{1}{E_L}\cos^4\theta + \left(\frac{1}{G_{LT}} - \frac{2\nu_{LT}}{E_L}\right)\sin^2\theta\cos^2\theta + \frac{1}{E_T}\sin^4\theta,\\
\nu_{xy} &= E_x\left[\frac{\nu_{LT}}{E_L}\left(\sin^4\theta + \cos^4\theta\right) - \left(\frac{1}{E_L} + \frac{1}{E_T} - \frac{1}{G_{LT}}\right)\sin^2\theta\cos^2\theta,\right]\\
\frac{1}{E_y} &= \frac{1}{E_T}\cos^4\theta + \left(\frac{1}{G_{LT}} - \frac{2\nu_{LT}}{E_L}\right)\sin^2\theta\cos^2\theta + \frac{1}{E_L}\sin^4\theta,\\
\frac{1}{G_{xy}} &= 2\left(\frac{2}{E_L} + \frac{2}{E_T} + \frac{4\nu_{LT}}{E_L} - \frac{1}{G_{LT}}\right)\sin^2\theta\cos^2\theta + \frac{1}{G_{LT}}\left(\sin^4\theta + \cos^4\theta\right),\\
\eta_{xy,x} &= E_x\left[\left(\frac{2}{E_L} + \frac{2\nu_{LT}}{E_L} - \frac{1}{G_{LT}}\right)\sin\theta\cos^3\theta - \left(\frac{2}{E_T} + \frac{2\nu_{LT}}{E_L} - \frac{1}{G_{LT}}\right)\sin^3\theta\cos\theta\right],\\
\eta_{xy,y} &= E_y\left[\left(\frac{2}{E_L} + \frac{2\nu_{LT}}{E_L} - \frac{1}{G_{LT}}\right)\sin^3\theta\cos\theta - \left(\frac{2}{E_T} + \frac{2\nu_{LT}}{E_L} - \frac{1}{G_{LT}}\right)\sin\theta\cos^3\theta\right].
\end{aligned}
\tag{2.7}
$$

2.2 MICROMECHANICS PREDICTION OF LAMINA STIFFNESS

In this section, we will *predict* lamina properties by the procedures of *micromechanics* in which the constituents of the composites, fibers and matrix, are separately modeled. The predicted properties are average apparent properties of the lamina and can be verified by measuring lamina properties with physical means. Moreover, they can be used in the macromechanical analysis of composite materials wherein the material is presumed homogeneous.

In most simple micromechanical models, the fibers are assumed to be homogeneous, linearly elastic, isotropic, regularly spaced, perfectly aligned, and of uniform length. The matrix is assumed to be homogeneous, linearly elastic, and isotropic. The fiber/matrix interface is assumed to be perfectly bonded with no voids or disbonds. More complex models representing more realistic situations may include voids, disbonds, flawed fibers, wavy fibers, nonuniform fiber dispersions, fiber length variations, and residual stresses.

There are two basic approaches to the micromechanics of composite materials. One is the *mechanics of materials approach*, the other is the *elasticity approach*. The former employs vastly simplifying assumptions by considering the mechanical behavior from the sectional viewpoint. While in the latter approach, most of the mechanical restraints should be satisfied point by point, not only the entire section. Typical models and representative elements considered by these two approaches are shown in Figure 2.3. Thus, in the mechanics of material approach, only the properties of fiber and matrix and their volume fractions are considered. The predicted properties will then be nothing to do with the arrangements of fibers neither with the sectional geometry of the fibers, etc. In the elasticity approach, the effects of fiber arrangements and fiber cross-sectional geometry as well as the degree of contiguity of fibers may be considered in the specific models.

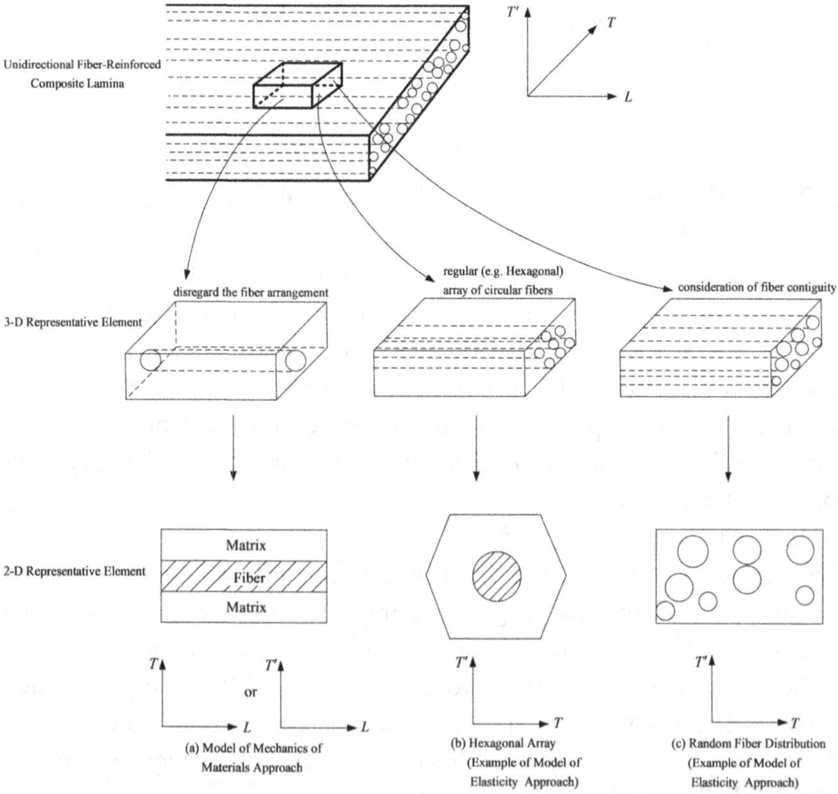

FIGURE 2.3 Model of a unidirectional fiber-reinforced composite lamina.

2.2.1 Mechanics of Materials Approach

The key feature of the mechanics of materials approach is that certain simplifying assumptions are made regarding the mechanical behavior of a composite material. As indicated in Figures 2.4(a) and (c), the longitudinal strain in the fiber is assumed to be the same as that in the matrix when the element is stretched by a longitudinal stress. When the representative element is subjected to a transverse stress or pure shear, the transverse stress or pure shear stress is assumed to be applied to both fiber and matrix as shown in Figures 2.4(b) and (d). By these assumptions, the elastic stiffness

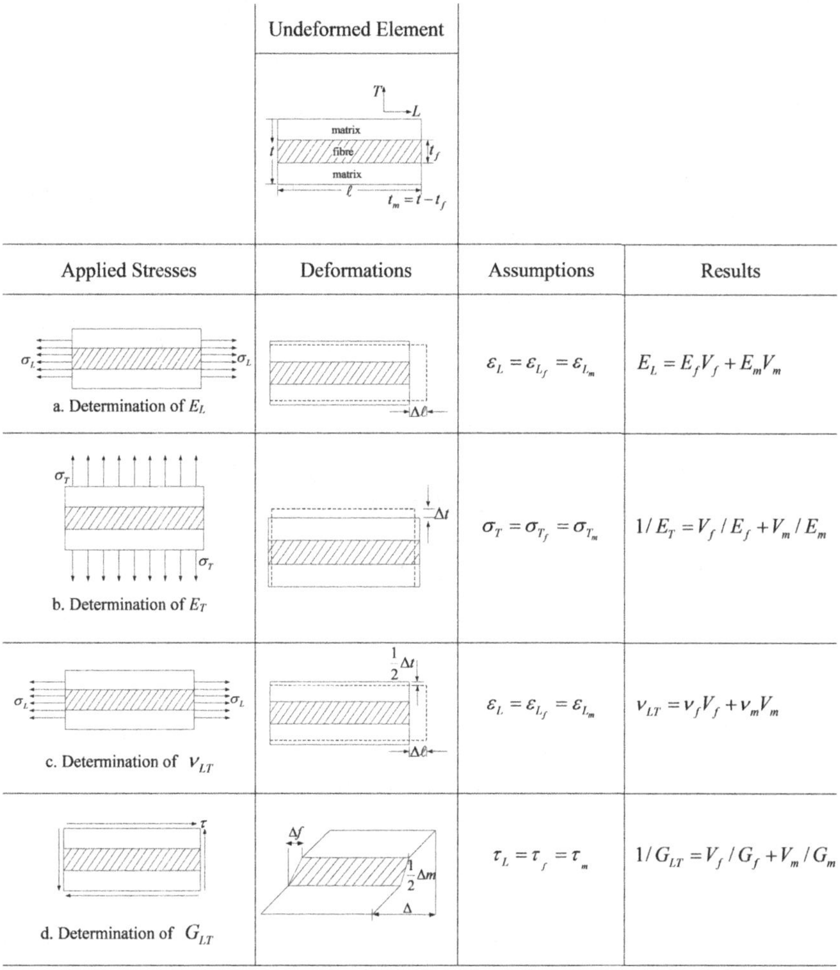

FIGURE 2.4 Models for determination of elastic constants by "mechanics of materials approach."

of a unidirectional fiber-reinforced composite material can be derived and the results are shown in Figure 2.4. Since most of the textbooks of mechanics of composite materials have detailed descriptions about the derivation of elastic stiffness by mechanics of materials approach, such as Jones (1974), to save space only brief derivation is given below:

i. E_L, *Young's modulus in the longitudinal direction*
Consider the assumption given in Figure 2.4(a). The resultant longitudinal force on the element leads to

$$\sigma_L t = \sigma_f t_f + \sigma_m t_m \Rightarrow E_L \varepsilon_L = E_f \varepsilon_{Lf} \frac{t_f}{t} + E_m \varepsilon_{Lm} \frac{t_m}{t}$$
$$\Rightarrow E_L = E_f V_f + E_m V_m. \tag{2.8}$$

ii. E_T, *Young's modulus in the transverse direction*
Consider the assumption given in Figure 2.4(b). The elongation in the transverse direction leads to

$$\varepsilon_T t = \varepsilon_f t_f + \varepsilon_m t_m \Rightarrow \frac{\sigma_T}{E_T} = \frac{\sigma_{Tf}}{E_f}\frac{t_f}{t} + \frac{\sigma_{Tm}}{E_m}\frac{t_m}{t} \Rightarrow \frac{1}{E_T} = \frac{V_f}{E_f} + \frac{V_m}{E_m}. \tag{2.9}$$

iii. ν_{LT}, *Major Poisson's ratio*
Consider the assumption given in Figure 2.4(c). The contraction in the transverse direction leads to

$$\Delta t = \Delta t_f + \Delta t_m \Rightarrow -(\nu_{LT}\varepsilon_L)t$$
$$= -(\nu_f \varepsilon_{Lf})t_f - (\nu_m \varepsilon_{Lm})t_m \Rightarrow \nu_{LT} = \nu_f V_f + \nu_m V_m. \tag{2.10}$$

iv. G_{LT}, *Shear modulus*
Consider the assumption given in Figure 2.4(d). The shear deformation in the longitudinal direction leads to

$$\Delta = \Delta_f + \Delta_m \Rightarrow \gamma t = \gamma_f t_f + \gamma_m t_m \Rightarrow \frac{\tau}{G_{LT}}$$
$$= \frac{\tau_f}{G_f}\frac{t_f}{t} + \frac{\tau_m}{G_m}\frac{t_m}{t} \Rightarrow \frac{1}{G_{LT}} = \frac{V_f}{G_f} + \frac{V_m}{G_m}. \tag{2.11}$$

In the above the subscripts f and m denote, respectively, the properties pertaining to fiber and matrix, and V_f and V_m are the volume fractions of fiber and matrix.

2.2.2 Elasticity Approach

The elasticity approach depends to a great extent on the specific geometry of composite material as well as on the characteristics of fibers and matrix. The fibers can be hollow or solid, circular cross-section or rectangular cross-section, isotropic or transversely isotropic, or can be regularly spaced in the matrix or randomly spaced even to touch each other. Depending upon the complexity of the modeling, different approaches may be applied, which can be divided into two classes: bounding techniques and direct methods.

i. *Bounding Techniques*

Consider an *admissible stress field* which satisfies the stress equations of equilibrium and the specified boundary conditions. The *principle of minimum complementary energy* states that the actual strain energy due to the specified loads should be the minimum and will not exceed the strain energy due to any other admissible stress field. By this way a *lower bound* on the apparent Young's modulus can be determined.

Similarly, the *upper bound* on the apparent Young's modulus can be determined by the application of the *principle of minimum potential energy* which can be stated as: the actual strain energy due to the specified displacements should be the minimum and will not exceed the strain energy due to any other admissible strain field. The *admissible strain field* is any compatible state of strain that satisfies the specified displacement boundary conditions.

The usefulness of the results obtained by the bounding techniques depends upon the closeness of the bounds. The bounding solutions are not very useful if the bounds are too far apart.

ii. *Direct Methods*

Various representative models by treating the fibers as the elastic inclusions embedded in the matrix have been employed to find the stiffness of the unidirectional fiber-reinforced composites. The problems of finding the analytical solutions to various cases of elastic inclusions in an elastic matrix can be found in Hwu (2010). In addition, numerical techniques such as finite element method and boundary element method can also be used to study the stiffness properties of the composites. In many cases the analytical solutions can be found only for the simplest cases such as one inclusion embedded in an infinite matrix. This solution is useful for the understanding of the effects of fiber reinforcement. However, to know the effects such as the fiber arrangements and the degree of contiguity of fibers, some special models and semi-empirical approximations such as the self-consistent model and Halpin–Tsai equations, etc. are presented in the literature. Detailed discussion of these methods can be found in Jones (1974). Since the Halpin–Tsai equations (Halpin and Tsai, 1969) have the beauty of simple and generality, in the following we like to show its formula for the readers' reference.

$$\begin{aligned} E_L &\cong E_f V_f + E_m V_m, \\ \nu_{LT} &= \nu_f V_f + \nu_m V_m, \\ \frac{M}{M_m} &= \frac{1+\xi\eta V_f}{1-\eta V_f}, \end{aligned} \tag{2.12a}$$

where

$$\eta = \frac{(M_f / M_m) - 1}{(M_f / M_m) + \xi} \tag{2.12b}$$

In the above, M is the composite modulus E_T, G_{LT}, or $\nu_{TT'}$; M_f is the corresponding fiber modulus E_f, G_f, or ν_f; M_m is the corresponding matrix modulus E_m, G_m, or ν_m; and ξ is a measure of fiber reinforcement of composite that depends on the fiber geometry, packing geometry, and loading conditions. The values of ξ are obtained by comparing (2.12b) with exact elasticity solutions and assessing a value of ξ by curve fitting techniques.

Note that the expressions for E_L and ν_{LT} are the generally accepted rule of mixtures obtained from mechanics of materials approach discussed in (2.8) and (2.10). The Halpin–Tsai equations are equally applicable to fiber, ribbon, or particulate composites. The only difficulty in using the Halpin–Tsai equations seems to be the determination of a suitable value for ξ.

Some physical insight into the Halpin–Tsai equations can be gained by examining their behavior for the range of values of ξ. When $\xi = 0$,

$$\frac{1}{M} = \frac{V_f}{M_f} + \frac{V_m}{M_m}, \tag{2.13}$$

which is the series-connected modal generally associated with a lower bound of a composite modulus. When $\xi = \infty$,

$$M = M_f V_f + M_m V_m, \tag{2.14}$$

which is the parallel-connected model, known as the rule of mixtures, generally associated with an upper bound of a composite modulus. Thus, ξ is a measure of the reinforcement of composite by the fibers. For small values of ξ, the fibers are not very effective whereas for large values of ξ, the fibers are extremely effective in increasing the composite stiffness above the matrix stiffness.

From the vast discussion presented in the literature, it can be generally stated that the constituent material properties have the following effect on the properties of the composite: (a) E_f makes a main contribution to E_L; (b) E_m makes a main contribution to E_T and G_{LT}; (c) ν_f and ν_m have little effect on E_T and G_{LT} and have no effect on E_L.

2.3 LAMINA STRENGTH

To predict the strength of a lamina, it is better to know what kind of failure may occur when the applied load exceeds the lamina strength. The micro-level failure mechanisms of the fibrous composite materials include fiber fracture, fiber buckling, fiber splitting, fiber pullout, fiber/matrix debonding, matrix cracking, etc. Due to the heterogeneous characteristics of the fibrous composites, the failure of a lamina may exhibit many local failure mechanisms prior to rupture. Thus, the first failure does not necessarily correspond to the final failure. The local failures are usually referred to as *damage*, and the development of additional local failures with increasing load or time is called *damage accumulation*. The lamina strength is better predicted by the consideration of final failure. The simplest model for the strength prediction may therefore be the one based upon the assumption that all fibers have the same strength. A more realistic model may consider the statistical strength distribution. There are many interesting physical models discussed in the literature; however, only the well-recognized models presented in most of the textbooks will be illustrated in this section.

2.3.1 Longitudinal Tensile Strength

When a composite lamina is subjected to longitudinal tensile loads, whether fracture occurring as a fiber failure or a matrix failure depends on the relative ductility of the fibers and matrix. The usual failure modes are: (a) brittle failure, (b) brittle failure with fiber pullout, and (c) brittle failure with debonding and/or matrix failure (Agarwal and Broutman, 1990). In most cases of the fibrous composites the fiber is relatively brittle in comparison to the matrix and failure initiates when the fibers are strained to their fracture strain. If we assume that all fibers fail at the same strain and the matrix will not be able to support the entire load when all the fibers break, composite failure will then take place instantly. Under these conditions, the ultimate longitudinal tensile strength of the composite can be predicted by the rule of matrix as

$$\sigma_{Lu} = \sigma_{fu} V_f + \left(\sigma_m\right)_{\varepsilon_{fu}} \left(1 - V_f\right), \tag{2.15}$$

where σ_{Lu} is the longitudinal tensile strength of the composite, σ_{fu} is the ultimate strength of the fibers, and $(\sigma_m)_{\varepsilon_{fu}}$ is the matrix stress at the fiber fracture strain ε_{fu}.

When the fiber volume fraction is too small, most of the loads are undertaken by the matrix. Therefore, when all the fibers break, the composite will still sustain until the matrix reaches its ultimate strength. Under this condition, the ultimate strength of a composite is given by

$$\sigma_{Lu} = \sigma_{mu} \left(1 - V_f\right), \tag{2.16}$$

where σ_{mu} is the ultimate strength of the matrix. Thus, the minimum fiber volume fraction $V_{\min}$ that ensures fiber-controlled composite failure can be obtained by equating right-hand sides of (2.15) and (2.16), which is

$$V_{\min} = \frac{\sigma_{mu} - \left(\sigma_m\right)_{\varepsilon_{fu}}}{\sigma_{fu} + \sigma_{mu} - \left(\sigma_m\right)_{\varepsilon_{fu}}}. \tag{2.17}$$

The longitudinal tensile strength versus fiber volume fraction as predicted by (2.15) and (2.16) is shown in Figure 2.5. It is observed that the

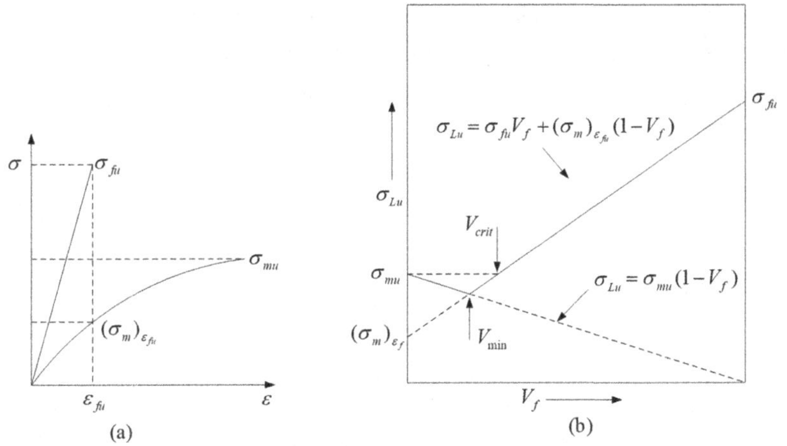

FIGURE 2.5 Longitudinal tensile strength versus fiber volume fraction.

longitudinal tensile strength predicted by (2.16) is always less than the strength of unreinforced matrix. For strengthening the matrix by fibers, a critical fiber volume fraction V_{crit} must be exceeded, which can be calculated by letting the predicted value of (2.15) be greater than σ_{mu}. Thus,

$$V_{\text{crit}} = \frac{\sigma_{mu} - (\sigma_m)_{\varepsilon_{fu}}}{\sigma_{fu} - (\sigma_m)_{\varepsilon_{fu}}}. \tag{2.18}$$

The foregoing analysis of tensile strength assumes simultaneous fracture of equal-strength fibers in one plane. In reality, the situation is much more complex, because of the factors such as misorientation of fibers, fibers of nonuniform strength, discontinuous fibers, fiber/matrix interfacial condition and interaction, and residual stresses, etc. Consideration of all these possible influencing factors, the models based on experimental observation and statistical analysis have been developed, such as Rosen's model of cumulative damage (Rosen, 1964).

2.3.2 Longitudinal Compressive Strength

When a composite lamina is subjected to longitudinal compressive loads, several different failure modes such as (a) transverse tensile splitting, (b) fiber microbuckling in extensional mode, (c) fiber microbuckling in shear mode, and (d) shear failure may occur (Agarwal and Broutman, 1990). The observation of failure modes in compression has been used to formulate theoretical expressions for the prediction of longitudinal compressive strength. For example, if fiber microbuckling occurs, the longitudinal compressive strength has been predicted by the energy method as (Jones, 1974)

$$\sigma'_{Lu} = 2\left[V_f + \left(1 - V_f\right)\frac{E_m}{E_f}\right]\sqrt{\frac{V_f E_m E_f}{3\left(1 - V_f\right)}}, \text{ buckling in extensional mode,}$$

$$\sigma'_{Lu} = \frac{G_m}{1 - V_f}, \text{ buckling in shear mode,} \tag{2.19}$$

where σ'_{Lu} is the longitudinal compressive strength of the composite. If the transverse splitting occurs, a simple theoretical expression for the longitudinal compressive strength has also been obtained as (Agarwal and Broutman, 1990)

$$\sigma'_{Lu} = \frac{\left(E_f V_f + E_m V_m\right)\left(1 - V_f^{1/3}\right)\varepsilon_{mu}}{\nu_f V_f + \nu_m V_m}, \text{ transverse splitting,} \tag{2.20}$$

where ε_{mu} is the matrix ultimate strain.

2.3.3 Transverse Strength

When a unidirectional composite lamina is subjected to transverse loads, the fibers are unable to take as large a proportion of the load as they do in the case of longitudinal loads. The constraints placed on the matrix by the fibers cause stress concentrations in the matrix adjacent to the fibers and thus result in composite failure at a much lower load than the load applied on the unreinforced matrix material. Therefore, opposite to the improvement of the longitudinal strength, the transverse strength is reduced because of the presence of fibers. The strength-reduction factor *S*, which depends on the relative properties of the fibers and the matrix and their volume fractions, is usually assumed to be the stress concentration factor. Thus, the transverse strength can be written as

$$\sigma_{Tu} = \sigma_{mu} / S. \tag{2.21}$$

One may refer to Agarwal and Broutman (1990) for the other approaches to predict transverse strength.

2.3.4 Shear Strength

When a unidirectional composite lamina is subjected to shear loads, the failure could take place by matrix shear failure, constituent debonding, or a combination of these two. Since little work has been done on the prediction of shear strength, no representative formula is given in this section.

2.3.5 Lamina Failure Theories

In the previous discussion, the lamina strengths are predicted in the longitudinal and transverse directions. The verification of the prediction can be done by appropriate experiments. Usually, the strengths of a material are experimentally measured by subjecting suitable specimens to loads that produce simple stress fields in the test specimen and by determining the load at which failure occurs (Carlsson and Pipes, 1987). For example, longitudinal tensile strength of the unidirectional composite lamina is obtained through test that produces longitudinal uniform tensile stress in the test section of the specimen.

Since the strengths in the longitudinal and transverse directions of the unidirectional composite lamina are usually different, it is reasonable to say that the lamina strengths are direction dependent and there may be infinite number of strength values corresponding to infinite number of directions. Unlike the elastic constants whose values in different directions can be calculated according to the transformation law of fourth rank tensor, it is difficult to determine the strengths in different directions by using the transformation law because they are not low-order tensors. In order to predict the strength of arbitrary directions by using the strengths in the principal material directions (i.e., longitudinal tensile strength, longitudinal compressive strength, transverse tensile strength, transverse compressive strength, and shear strength), many failure theories for orthotropic materials have been developed from the failure theories of isotropic materials. In the following, three widely used failure theories: maximum stress, maximum strain, and maximum work theories, are discussed.

Maximum Stress Theory

This theory states that failure will occur if any of the stresses in the principal material axes exceeds their corresponding strengths. Note that in this theory we consider the stresses in the principal material axes, which are not necessary to be the principal stresses. In other words, not the highest stress is considered in this theory because the strengths in the axes of the principal stress are also not necessary to be the highest strength of the orthotropic lamina. Thus, the off-axis strength can be predicted from the following inequalities:

$$\sigma_L < \sigma_{Lu} \text{ or } \sigma'_{Lu}, \quad \sigma_T < \sigma_{Tu} \text{ or } \sigma'_{Tu}, \quad |\tau_{LT}| < \tau_{LTu}. \qquad (2.22)$$

Note that the tensile and compressive strengths of materials are generally recognized to be different, while the shear strength is independent of the sign of shear stress. According to this theory, when one of the inequalities of (2.22) is not satisfied, the material is considered to have failed by the failure mechanism associated with the corresponding strength. There is no interaction between modes of failure in this criterion. There are actually three subcriteria.

To illustrate the application of this theory to the prediction of off-axis strength, consider a unidirectional reinforced composite lamina subjected to uniaxial tensile load σ_x at angle θ to the fibers as shown in Figure 2.6(a). The stresses in the principal material direction are obtained by transformation law $(2.3)_1$ as

$$\sigma_L = \sigma_x \cos^2\theta, \quad \sigma_T = \sigma_x \sin^2\theta, \quad \tau_{LT} = -\sigma_x \sin\theta\cos\theta. \qquad (2.23)$$

Substituting (2.23) into (2.22), we obtain

$$\sigma_x < \frac{\sigma_{Lu}}{\cos^2\theta}, \quad \sigma_x < \frac{\sigma_{Tu}}{\sin^2\theta}, \quad \sigma_x < \frac{\tau_{LTu}}{\sin\theta\cos\theta}. \qquad (2.24)$$

The smallest of (2.24) will then give us the maximum uniaxial tensile load σ_x allowed in this lamina, that is, the off-axis tensile strength.

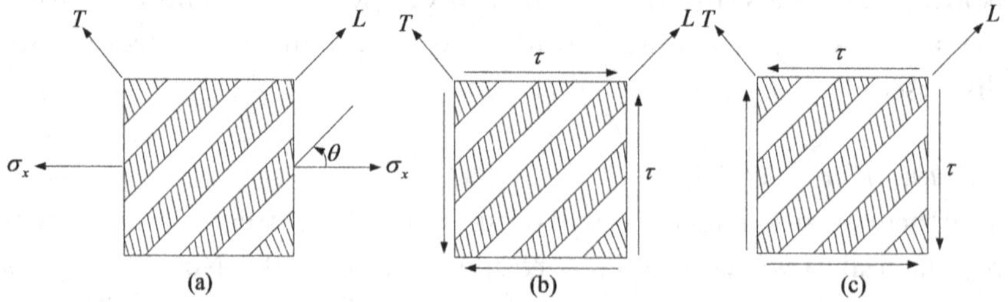

FIGURE 2.6 A unidirectional composite lamina subjected to off-axis loads: (a) uniaxial tensile stress; (b) positive shear stress; (c) negative shear stress.

Similarly, if we want to predict the off-axis shear strength by the maximum stress theory, a positive or negative shear stress may be applied on the unidirectional lamina as shown in Figures 2.6(b) and (c). The stresses in the principal material direction may then be obtained as

$$\sigma_L = \tau \sin 2\theta, \quad \sigma_T = -\tau \sin 2\theta, \quad \tau_{LT} = \tau \cos 2\theta. \tag{2.25}$$

Substituting (2.25) into (2.22), the off-axis shear strength can be predicted as follows:

If τ is positive and $\theta < 45^\circ$, or τ is negative and $\theta > 45^\circ$:

$$\tau < \frac{\sigma_{Lu}}{\sin 2\theta}, \quad \tau < \frac{\sigma'_{Tu}}{\sin 2\theta}, \quad \tau < \frac{\tau_{LTu}}{\cos 2\theta}. \tag{2.26a}$$

If τ is positive and $\theta > 45^\circ$, or τ is negative and $\theta < 45^\circ$:

$$\tau < \frac{\sigma'_{Lu}}{\sin 2\theta}, \quad \tau < \frac{\sigma_{Tu}}{\sin 2\theta}, \quad \tau < \frac{\tau_{LTu}}{\cos 2\theta} \tag{2.26b}$$

Again, the smallest of (2.26a) or (2.26b) will give us the maximum allowable shear stress τ applied on the lamina, that is, the off-axis shear strength. Due to the difference of the tensile and compressive strengths, the results of (2.26) show that the off-axis shear strength of a lamina depends not only on the fiber orientation but also on the sign of applied shear stress.

Maximum Strain Theory

The maximum strain theory is basically the same as the maximum stress theory except strains are limited rather than stresses. Thus, the off-axis strength can be predicted from the following inequalities:

$$\varepsilon_L < \varepsilon_{Lu} \text{ or } \varepsilon'_{Lu}, \quad \varepsilon_T < \varepsilon_{Tu} \text{ or } \varepsilon'_{Tu}, \quad |\gamma_{LT}| < \gamma_{LTu}, \tag{2.27}$$

where ε_{Lu}, ε'_{Lu}, ε_{Tu}, ε'_{Tu}, and γ_{LTu} denote, respectively, the maximum longitudinal tensile strain, maximum longitudinal compressive strain, maximum transverse tensile strain, maximum transverse compressive strain, and maximum shear strain. Since this theory is quite similar to the maximum stress theory, no illustration will be given here. One may refer to the textbooks such as Jones (1974) and Agarwal and Broutman (1990) for examples.

Maximum Work Theory

Instead of considering the maximum stress or strain, this theory considers the maximum strain energy. Hill (1950) proposed a yield criterion for anisotropic elastic materials, which is an extension of von Mises' isotropic yield criterion (Mendelson, 1970). Based upon Hill's criterion, Tsai (1968) derived the following failure criterion for orthotropic lamina in plane stress state:

$$\frac{\sigma_L^2}{\sigma_{Lu}^2} - \frac{\sigma_L \sigma_T}{\sigma_{Lu}^2} + \frac{\sigma_T^2}{\sigma_{Tu}^2} + \frac{\tau_{LT}^2}{\tau_{LTu}^2} < 1, \tag{2.28}$$

which was called *Tsai–Hill theory* and was later improved by *Tsai–Wu tensor theory* (1971). Unlike the maximum stress and strain theories both of which consist of three subcriteria and do not consider the interaction between modes of failure, the maximum energy theory provides a single function to predict strength and does take into consideration the interaction between longitudinal, transverse, and shear strengths. The maximum energy theory has found wider acceptability compared to the other two theories primarily because of the smooth variation of strength according to a single function. Experimental support for this theory has been reported by many investigators (Agarwal and Broutman, 1990).

Application of this theory to the prediction of off-axis strength can be shown by considering the stress field (2.23) induced by the uniaxial tensile load σ_x at angle θ to the fibers. Substituting (2.23) into (2.28), we obtain

$$\frac{\cos^2\theta\cos 2\theta}{\sigma_{Lu}^2} + \frac{\sin^4\theta}{\sigma_{Tu}^2} + \frac{\sin^2\theta\cos^2\theta}{\tau_{LTu}^2} < \frac{1}{\sigma_x^2}, \tag{2.29}$$

which is one criterion, not three as in the previous criteria. Results of (2.29) will then provide the prediction of the off-axis tensile strength.

2.4 MECHANICAL BEHAVIOR OF A LAMINATE

In the previous sections, the mechanical behavior as well as the stiffness and strength of a unidirectional fiber-reinforced composite lamina was discussed. In reality, most of the composite structures are comprised of numerous laminae that are bonded together to act as an integral structural element which is called a laminate. The overall properties of the laminates can be designed by changing the fiber orientation and the stacking sequence of laminae.

2.4.1 Classical Lamination Theory

To describe the overall properties and macromechanical behavior of a laminate, the most popular way is the *classical lamination theory* (Jones, 1974). According to the observation of actual mechanical behavior of laminates, the following assumptions are made in this theory. (a) The laminate consists of perfectly bonded laminae and the bonds are infinitesimally thin as well as non-shear-deformable. Thus, the displacements are continuous across lamina boundaries so that no lamina can slip relative to another. (b) A line originally straight and perpendicular to the middle surface of the laminate remains straight and perpendicular to the middle surface of the laminate when the laminate is deformed. In other words, the transverse shear strains are ignored, that is, $\gamma_{xz} = \gamma_{yz} = 0$. (c) The normals have constant length so that the strain perpendicular to the middle surface is ignored, that is, $\varepsilon_z = 0$.

Based upon the above assumptions, the laminate displacements u, v, and w in the x, y, and z directions can be expressed as

$$
\begin{aligned}
u(x,y,z) &= u_0(x,y) - z\frac{\partial w(x,y)}{\partial x}, \\
v(x,y,z) &= v_0(x,y) - z\frac{\partial w(x,y)}{\partial y}, \\
w(x,y,z) &= w_0(x,y), \qquad (2.30)
\end{aligned}
$$

where u_0, v_0 and w_0 are the middle surface displacements. If small deformations are considered, the laminate strains can be written in terms of the middle surface displacements as

$$
\begin{aligned}
\varepsilon_x &= \frac{\partial u}{\partial x} = \frac{\partial u_0}{\partial x} - z\frac{\partial^2 w}{\partial x^2}, \\
\varepsilon_y &= \frac{\partial v}{\partial y} = \frac{\partial v_0}{\partial y} - z\frac{\partial^2 w}{\partial y^2}, \\
\gamma_{xy} &= \frac{\partial u}{\partial y} + \frac{\partial v}{\partial x} = \frac{\partial u_0}{\partial y} + \frac{\partial v_0}{\partial x} - 2z\frac{\partial^2 w}{\partial x \partial y}. \qquad (2.31)
\end{aligned}
$$

Or, in matrix notation,

$$\boldsymbol{\varepsilon} = \boldsymbol{\varepsilon}_0 + z\boldsymbol{\kappa}, \tag{2.32a}$$

where $\boldsymbol{\varepsilon}$, $\boldsymbol{\varepsilon}_0$ and $\boldsymbol{\kappa}$ denotes, respectively, strain vector, midsurface strain vector, and plate curvature vector, which are defined as

$$\boldsymbol{\varepsilon} = \begin{Bmatrix} \varepsilon_x \\ \varepsilon_y \\ \gamma_{xy} \end{Bmatrix}, \quad \boldsymbol{\varepsilon}_0 = \begin{Bmatrix} \varepsilon_x^0 \\ \varepsilon_y^0 \\ \gamma_{xy}^0 \end{Bmatrix} = \begin{Bmatrix} \dfrac{\partial u_0}{\partial x} \\ \dfrac{\partial v_0}{\partial y} \\ \dfrac{\partial u_0}{\partial y} + \dfrac{\partial v_0}{\partial x} \end{Bmatrix}, \quad \boldsymbol{\kappa} = \begin{Bmatrix} \kappa_x \\ \kappa_y \\ \kappa_{xy} \end{Bmatrix} = -\begin{Bmatrix} \dfrac{\partial^2 w}{\partial x^2} \\ \dfrac{\partial^2 w}{\partial y^2} \\ 2\dfrac{\partial^2 w}{\partial x \partial y} \end{Bmatrix}. \tag{2.32b}$$

Substituting (2.32a) into the stress–strain relation (2.4a) for each lamina, the stresses in the kth lamina can be written in terms of the laminate middle surface strains $\boldsymbol{\varepsilon}_0$ and curvatures $\boldsymbol{\kappa}$ as

$$\begin{Bmatrix} \sigma_x \\ \sigma_y \\ \tau_{xy} \end{Bmatrix}_k = \boldsymbol{\sigma}_k = \mathbf{Q}_k^* \left(\boldsymbol{\varepsilon}_0 + z\boldsymbol{\kappa} \right), \tag{2.33}$$

where $\mathbf{Q}_k^*$, as shown in (2.5) for a lamina with its principal material axes oriented at an angle θ with the reference coordinate axes, is the transformed stiffness matrix of the kth lamina. Since $\mathbf{Q}_k^*$ may be different for each lamina, the stress variation through the laminate thickness is not necessarily linear even though the strain variation is linear. In other words, the stresses calculated from (2.33) may be discontinuous at the interface of two laminae due to the discontinuity of the material properties of laminae.

Like classical plate theory, the thickness of laminate is considered to be small compared to its other dimensions. Therefore, instead of dealing the stress distribution across the laminate thickness, an integral equivalent system of forces and moments acting on the laminate cross-section is used in the classical lamination theory. By integration of the stresses in each lamina through the laminate thickness, the resultant forces $\boldsymbol{N}$ and moments $\boldsymbol{M}$ acting on a laminate cross-section are defined as follows:

$$\mathbf{N}=\begin{Bmatrix}N_x\\N_y\\N_{xy}\end{Bmatrix}=\int_{-h/2}^{h/2}\begin{Bmatrix}\sigma_x\\\sigma_y\\\tau_{xy}\end{Bmatrix}dz=\sum_{k=1}^{n}\int_{h_{k-1}}^{h_k}\begin{Bmatrix}\sigma_x\\\sigma_y\\\tau_{xy}\end{Bmatrix}_k dz=\sum_{k=1}^{n}\int_{h_{k-1}}^{h_k}\sigma_k dz,$$

$$\mathbf{M}=\begin{Bmatrix}M_x\\M_y\\M_{xy}\end{Bmatrix}=\int_{-h/2}^{h/2}\begin{Bmatrix}\sigma_x\\\sigma_y\\\tau_{xy}\end{Bmatrix}zdz=\sum_{k=1}^{n}\int_{h_{k-1}}^{h_k}\begin{Bmatrix}\sigma_x\\\sigma_y\\\tau_{xy}\end{Bmatrix}_k zdz=\sum_{k=1}^{n}\int_{h_{k-1}}^{h_k}\sigma_k zdz, \tag{2.34}$$

where h_k and h_{k-1} are defined in Figure 2.7. Substituting (2.33) into (2.34), the resultant forces $\boldsymbol{N}$ and moments $\boldsymbol{M}$ can be written in terms of the laminate middle surface strains $\boldsymbol{\varepsilon}_0$ and curvatures $\boldsymbol{\kappa}$ as

$$\begin{Bmatrix}\boldsymbol{N}\\\boldsymbol{M}\end{Bmatrix}=\begin{bmatrix}\boldsymbol{A}&\boldsymbol{B}\\\boldsymbol{B}&\boldsymbol{D}\end{bmatrix}\begin{Bmatrix}\boldsymbol{\varepsilon}_0\\\boldsymbol{\kappa}\end{Bmatrix}, \tag{2.35}$$

where $\boldsymbol{A}$, $\boldsymbol{B}$, and $\boldsymbol{D}$ are called the *extensional*, *coupling*, and *bending stiffness matrices*, respectively, and are determined by

$$\mathbf{A}=\sum_{k=1}^{n}\int_{h_{k-1}}^{h_k}\mathbf{Q}_k^* dz=\sum_{k=1}^{n}\mathbf{Q}_k^*\left(h_k-h_{k-1}\right),$$

$$\mathbf{B}=\sum_{k=1}^{n}\int_{h_{k-1}}^{h_k}\mathbf{Q}_k^* zdz=\frac{1}{2}\sum_{k=1}^{n}\mathbf{Q}_k^*\left(h_k^2-h_{k-1}^2\right),$$

$$\mathbf{D}=\sum_{k=1}^{n}\int_{h_{k-1}}^{h_k}\mathbf{Q}_k^* z^2dz=\frac{1}{3}\sum_{k=1}^{n}\mathbf{Q}_k^*\left(h_k^3-h_{k-1}^3\right). \tag{2.36}$$

From (2.35), we see that if the coupling matrix $\boldsymbol{B}$ is a zero matrix, the resultant forces $\boldsymbol{N}$ will induce only the midsurface strains while the resultant moments $\boldsymbol{M}$ will induce only the plate curvatures. The presence of the matrix $\boldsymbol{B}$ implies coupling between bending and extension of a laminate. Thus, when a laminate is subjected to an extensional force or a bending moment, it may suffer extensional as well as bending and/or twisting deformations at the same time. By $(2.36)_2$, we also know that the presence of a nonzero coupling matrix is not attributable to the orthotropy or anisotropy of the layers but rather to the nonsymmetric stacking of laminae.

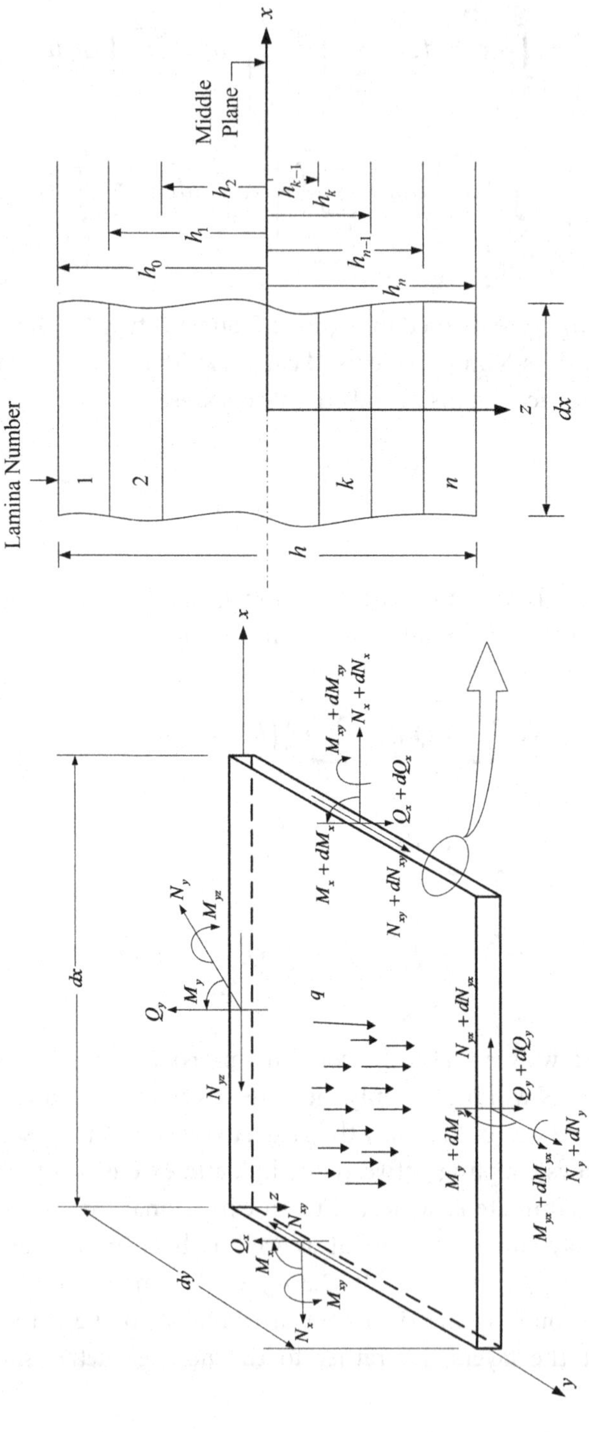

FIGURE 2.7 Loading and geometry of a laminate.

The aim of analysis of laminated composites is to determine the stresses and strains in each of laminae forming the laminate. These stresses and strains can be used to predict the load at which failure initiates. If the resultant forces $\boldsymbol{N}$ and moments $\boldsymbol{M}$ are known at a particular cross-section through the structural analysis, the midsurface strains and curvatures at this cross-section may then be determined by the inversion of (2.35), that is,

$$\begin{Bmatrix} \varepsilon_0 \\ \kappa \end{Bmatrix} = \begin{bmatrix} \boldsymbol{A} & \boldsymbol{B} \\ \boldsymbol{B} & \boldsymbol{D} \end{bmatrix}^{-1} \begin{Bmatrix} \boldsymbol{N} \\ \boldsymbol{M} \end{Bmatrix} = \begin{bmatrix} \widehat{\boldsymbol{A}} & \widehat{\boldsymbol{B}} \\ \widehat{\boldsymbol{B}}^T & \widehat{\boldsymbol{D}} \end{bmatrix} \begin{Bmatrix} \boldsymbol{N} \\ \boldsymbol{M} \end{Bmatrix}, \tag{2.37a}$$

where

$$\begin{aligned} \widehat{\boldsymbol{A}} &= \boldsymbol{A}^{-1} + \boldsymbol{A}^{-1}\boldsymbol{B}\widehat{\boldsymbol{D}}\boldsymbol{B}\boldsymbol{A}^{-1}, \\ \widehat{\boldsymbol{B}} &= -\boldsymbol{A}^{-1}\boldsymbol{B}\widehat{\boldsymbol{D}}, \\ \widehat{\boldsymbol{D}} &= \left(\boldsymbol{D} - \boldsymbol{B}\boldsymbol{A}^{-1}\boldsymbol{B}\right)^{-1}. \end{aligned} \tag{2.37b}$$

The stresses and strains in each lamina can therefore be determined from (2.33) and (2.32).

To summarize the laminate stress analysis described in this section, a flowchart illustrating the analysis procedure is now shown in Figure 2.8.

2.4.2 Hygrothermal Effects

The lamination theory discussed previously does not account for the effects of temperature or moisture change. It is known that a change in temperature or moisture content of a body will cause a change in its dimensions proportional to the change in temperature or moisture content and its initial dimensions. The hygrothermoelastic stress–strain relations for each lamina are

$$\sigma_k = \boldsymbol{Q}_k^* \left(\varepsilon_0 + z\kappa - \alpha_k \Delta T - \beta_k \Delta m\right), \tag{2.38}$$

where ΔT and Δm are the changes in temperature and moisture content; $\boldsymbol{\alpha}_k$ and $\boldsymbol{\beta}_k$ are the vectors containing the transformed coefficients of thermal and moisture expansions of the kth lamina, which are related to the coefficients of thermal and moisture expansion in the longitudinal and transverse directions, α_L, α_T, β_L, and β_T, by

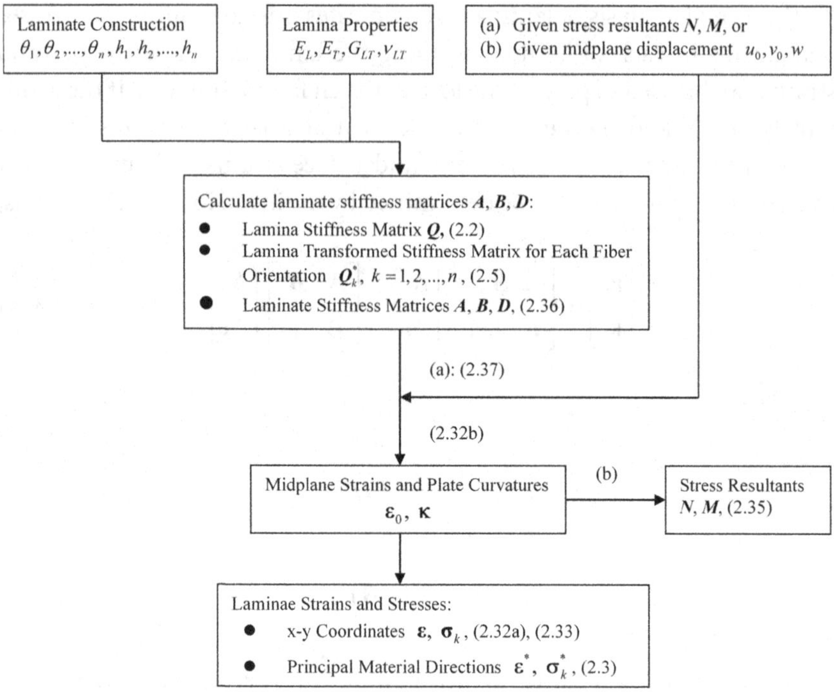

FIGURE 2.8 Flowchart for laminate stress analysis.

$$
\boldsymbol{\alpha}_k = \begin{Bmatrix} \alpha_x \\ \alpha_y \\ \alpha_{xy} \end{Bmatrix}_k = \boldsymbol{R}\boldsymbol{T}^{-1} \begin{Bmatrix} \alpha_L \\ \alpha_T \\ 0 \end{Bmatrix} = \boldsymbol{R}\boldsymbol{T}^{-1}\boldsymbol{\alpha},
$$

$$
\boldsymbol{\beta}_k = \begin{Bmatrix} \beta_x \\ \beta_y \\ \beta_{xy} \end{Bmatrix}_k = \boldsymbol{R}\boldsymbol{T}^{-1} \begin{Bmatrix} \beta_L \\ \beta_T \\ 0 \end{Bmatrix} = \boldsymbol{R}\boldsymbol{T}^{-1}\boldsymbol{\beta}, \tag{2.39}
$$

in which the matrices $\boldsymbol{R}$ and $\boldsymbol{T}$ are given in (2.3b) for the purpose of transformation. Substituting (2.38) into the definitions of resultant forces and moments given in (2.34), we obtain

$$
\begin{Bmatrix} \boldsymbol{N} \\ \boldsymbol{M} \end{Bmatrix} = \begin{bmatrix} \boldsymbol{A} & \boldsymbol{B} \\ \boldsymbol{B} & \boldsymbol{D} \end{bmatrix} \begin{Bmatrix} \boldsymbol{\varepsilon}_0 \\ \boldsymbol{\kappa} \end{Bmatrix} - \begin{Bmatrix} \boldsymbol{N}_T \\ \boldsymbol{M}_T \end{Bmatrix} - \begin{Bmatrix} \boldsymbol{N}_m \\ \boldsymbol{M}_m \end{Bmatrix}, \tag{2.40a}
$$

where

$$\boldsymbol{N}_T = \sum_{k=1}^{n}\left\{\mathbf{Q}_k^*\boldsymbol{\alpha}_k \int_{h_k}^{h_{k-1}} \Delta T dz\right\}, \quad \boldsymbol{N}_m = \sum_{k=1}^{n}\left\{\mathbf{Q}_k^*\boldsymbol{\beta}_k \int_{h_k}^{h_{k-1}} \Delta m dz\right\},$$

$$\boldsymbol{M}_T = \sum_{k=1}^{n}\left\{\mathbf{Q}_k^*\boldsymbol{\alpha}_k \int_{h_k}^{h_{k-1}} z\Delta T dz\right\}, \quad \boldsymbol{M}_m = \sum_{k=1}^{n}\left\{\mathbf{Q}_k^*\boldsymbol{\beta}_k \int_{h_k}^{h_{k-1}} z\Delta m dz\right\}. \quad (2.40b)$$

When temperature and moisture change are independent of z in the laminate,

$$\boldsymbol{N}_T = \Delta T\sum_{k=1}^{n}\mathbf{Q}_k^*\boldsymbol{\alpha}_k\left(h_k - h_{k-1}\right), \quad \boldsymbol{N}_m = \Delta m\sum_{k=1}^{n}\mathbf{Q}_k^*\boldsymbol{\beta}_k\left(h_k - h_{k-1}\right),$$

$$\boldsymbol{M}_T = \frac{\Delta T}{2}\sum_{k=1}^{n}\mathbf{Q}_k^*\boldsymbol{\alpha}_k\left(h_k^2 - h_{k-1}^2\right), \quad \boldsymbol{M}_m = \frac{\Delta m}{2}\sum_{k=1}^{n}\mathbf{Q}_k^*\boldsymbol{\beta}_k\left(h_k^2 - h_{k-1}^2\right). \quad (2.41)$$

If only temperature and moisture change in the laminate and no external forces and moments are applied, $\boldsymbol{N} = \boldsymbol{M} = \mathbf{0}$. Thus, (2.40a) can be rearranged as

$$\begin{Bmatrix} \boldsymbol{N}_T \\ \boldsymbol{M}_T \end{Bmatrix} + \begin{Bmatrix} \boldsymbol{N}_m \\ \boldsymbol{M}_m \end{Bmatrix} = \begin{bmatrix} \boldsymbol{A} & \boldsymbol{B} \\ \boldsymbol{B} & \boldsymbol{D} \end{bmatrix}\begin{Bmatrix} \boldsymbol{\varepsilon}_0 \\ \boldsymbol{\kappa} \end{Bmatrix}, \quad (2.42)$$

where $\boldsymbol{N}_T$, $\boldsymbol{N}_m$ and $\boldsymbol{M}_T$, $\boldsymbol{M}_m$ are apparent forces that produce the midsurface strains $\boldsymbol{\varepsilon}_0$ and plate curvatures $\boldsymbol{\kappa}$, and are therefore called *hygrothermal forces and moments.*

2.5 LAMINATE STIFFNESS

In the classical lamination theory, the overall laminate stiffness is represented by the extensional, coupling, and bending stiffness. By the definitions given in (2.36), we know these stiffness matrices are related to the stiffnesses and thickness of each lamina. In addition, the stacking sequence of the laminae will also influence these stiffness matrices. In general, the laminate stiffness matrices will be fully populated. That is, all the terms of the stiffness matrices will generally be nonzero.

In engineering application, it is sometimes desired to make certain terms zero to avoid undesirable coupling. On the opposite, there could be some structural applications where the coupling effects can be used to benefit the design. For example, the forward swept wing design of an aircraft uses coupling between bending and torsion to achieve aeroelastic stability, and the design probably could not function using metallic wings. The couplings are features of composite structures that have no counterpart in isotropic materials and can be taken advantage to achieve unique characteristics or can be undesirable due to unwanted deformations and stresses. In general, the nonzero $\boldsymbol{B}$ matrix will induce coupling between bending and stretching responses; the nonzero A_{16} and A_{26} will induce coupling between normal forces and shear strain as well as between shear force and normal strains; and the nonzero D_{16} and D_{26} will induce coupling between normal moments and twist curvature as well as between twist moments and normal curvatures.

Because of the need for adequate description of many possible combinations of ply orientation and stacking sequences in laminates, a standard laminate code has evolved and is defined as follows (Agarwal and Broutman, 1990): (a) Each lamina is denoted by a number representing the angle in degrees between its fiber direction and the x-axis. (b) Individual adjacent laminae are separated in the code by a slash if their angles are different. (c) The laminae are listed in sequence from the top surface to the bottom surface, with brackets indicating the beginning and end of the code. (d) Adjacent laminae of the same orientation are denoted by a numerical subscript. (e) The laminates possessing symmetry of laminae orientations about the geometric midplane require specifying only half of the stacking sequence. The subscript s to the bracket indicates that only one-half of the laminate is shown, with the other half symmetric with respect to the midplane. (f) Repeating sequences of laminae are called sets and are enclosed in parentheses with the number of sets as subscript.

In the following, we discuss some important classes of laminates which are designed and constructed to possess special decoupling characteristics.

2.5.1 Symmetric Laminates ($\boldsymbol{B} = \mathbf{0}$)

The laminates, which are constructed by placing the laminae symmetrically with respect to the midplane, are called *symmetric laminates*. From $(2.36)_2$, we see that the coupling stiffness $\boldsymbol{B}$ will be identically zero for symmetric laminates since for each ply above the midplane there is an identical ply placed an equal distance below the midplane.

2.5.2 Anti-symmetric Laminates ($A_{16} = A_{26} = D_{16} = D_{26} = 0$)

If each ply of the laminates above the midplane is matched by a ply with orientation of opposite sign placed an equal distance below the midplane, the laminates are called *anti-symmetric laminates.* To cover the cases of cross-ply discussed next, the matched opposite orientation of 0° is considered to be 90°. From $(2.36)_{1,3}$, it can be found that $A_{16} = A_{26} = D_{16} = D_{26} = 0$ for anti-symmetric laminates.

2.5.3 Cross-Ply Laminates ($A_{16} = A_{26} = D_{16} = D_{26} = 0$)

The laminates consisting of layers with fiber orientations of 0° and 90° only are called *cross-ply laminates.* From (2.5), we know that Q^*_{16} and Q^*_{26} are zero for fiber orientations of 0° or 90°. By (2.36), we have $A_{16} = A_{26} = D_{16} = D_{26} = 0$. Thus, the cross-ply laminate is one kind of *specially orthotropic laminates.* Examples of cross-ply laminates include (but not limited to): $[0_n/90_n]$ (*balanced anti-symmetric*), $[0/90_2/0_2/90]$(*balanced anti-symmetric*), [0/90/90/0](i.e., $[0/90]_s$, *balanced symmetric*), $[(0/90)_n]_s$ (*balanced symmetric*), [0/90/0] (*unbalanced symmetric*), $[0_5/90_2]$ (*unbalanced non-symmetric*). Note that the definition of balanced and unbalanced will be given in the later subsection.

2.5.4 Angle-Ply Laminates (with equal number of equal-thickness laminae at $\pm\theta$ angles: $A_{16} = A_{26} = 0$)

The laminates consisting of layers at $+\theta$ and $-\theta$ fiber orientations are called *angle-ply laminates.* From (2.5), we know that layers of opposite sign will have Q^*_{16} and Q^*_{26} values of opposite sign. Hence, the sum of $+\theta$ and $-\theta$ layers of the same thickness will always lead to the results that $A_{16} = A_{26} = 0$. Thus, if an angle-ply laminate has an equal number of equal-thickness at $+\theta$ and $-\theta$ fiber orientations, the laminate will be specially orthotropic, that is, $A_{16} = A_{26} = 0$. Examples of this kind of angle-ply laminates include (but not limited to) $[\pm\theta]$(*balanced anti-symmetric*), $[+\theta/-\theta/+\theta/-\theta]$(i.e., $[(\pm\theta)_2]$, *balanced anti-symmetric*), $[+\theta/-\theta/-\theta/+\theta]$ (i.e., $[\pm\theta/\mp\theta]$ or $[\pm\theta]_s$, *balanced symmetric*, $[(\theta/-\theta)_n]_s$, *balanced symmetric*). According to (2.36), the values of D_{16} and D_{26} of the *anti-symmetric angle-ply laminates* will also be zero. Other examples of the angle-ply laminates are $[\theta/-\theta/\theta]$ (*unbalanced symmetric*), $[\theta_2/-\theta]$ (*unbalanced non-symmetric*), etc.

2.5.5 Balanced Laminates ($A_{16} = A_{26} = 0$)

The laminates in which all off-axis $+\theta_i$ and $-\theta_i$ layers are present in equal thickness t_i are called *balanced laminates.* This definition includes the cases of cross-plies, that is, if a cross-ply laminate has equal number of 0° and 90° layers it is called balanced cross-ply laminates. On the opposite, the unbalanced cross-ply laminates means a cross-ply laminate with different A_{11} and A_{22} as well as different D_{11} and D_{22}, but their A_{16} and A_{26} are still zero. Examples of balanced laminates include (but not limited to) $[+\alpha/-\alpha/-\beta/+\beta]$, $[\pm\alpha_5/\pm\beta_8]_s$, $[(0/90)_n]$.

2.5.6 Specially Orthotropic Laminates ($A_{16} = A_{26} = 0$)

Laminates which do not exhibit coupling between in-plane extensional and shear responses are called *specially orthotropic laminates.* By this definition, $A_{16} = A_{26} = 0$ but D_{16} and D_{26} are not necessarily zero. Thus, cross-ply laminates, angle-ply laminates, and any combination of them as well as unidirectional laminates and balanced laminates are all specially orthotropic. Examples are: $[\theta/0/-\theta]_s$, $[0/\pm\theta/90]_s$, $[\theta/90/-\theta]$, $[0_n]$, and $[\pm\alpha_n/0/90/\pm\beta_m]_s$, etc.

2.5.7 Quasi-Isotropic Laminates (A: isotropic)

A laminate whose in-plane elastic response is isotropic is called *quasi-isotropic laminate.* In other words, the extensional stiffness matrix $\boldsymbol{A}$ is isotropic which is independent of orientation in the plane. In general, the coupling and bending stiffnesses ($\boldsymbol{B}$ and $\boldsymbol{D}$) are not necessarily isotropic for quasi-isotropic laminates. A quasi-isotropic laminate can be constructed by meeting the following conditions: (a) The total number of layers must be at least three. (b) The individual layers must have identical stiffness and thickness. (c) Each layer in a set of layers must be oriented at an angle $\theta_k = \pi(k-1)/n$, $k = 1, 2, \ldots, n$, where n is the number of the set of layers. Examples are: $[0/\pm 60]$, $[0/\pm 45/90]$, etc.

2.6 LAMINATE STRENGTH ANALYSIS

As discussed in Section 2.3, the strength of a lamina is a function of fiber orientation; it is therefore expected that all laminae in the laminates will not fail at the same load. Thus, for a laminated composite, failure of one ply does not necessarily imply failure of the entire laminate. If a ply fails, its associated elastic properties may be set equal to zero to represent its malfunction. However, different failure modes interact in many cases, so that it is difficult to accurately ascertain which properties of the failed plies should be set to zero. As a conservative approach, it is usually suggested that all the elastic properties of failed plies are set equal to zero. Then, the analysis after plies failure is calculated based upon the modified stiffnesses of the laminates in which all the stiffnesses of the failed plies are set equal to zero. With this concept, the laminate strength analysis procedure is described in Figure 2.9. For example, if we want to predict the tensile strength in x-axis of the laminates, a unit load $N_x = 1$ is assumed to be

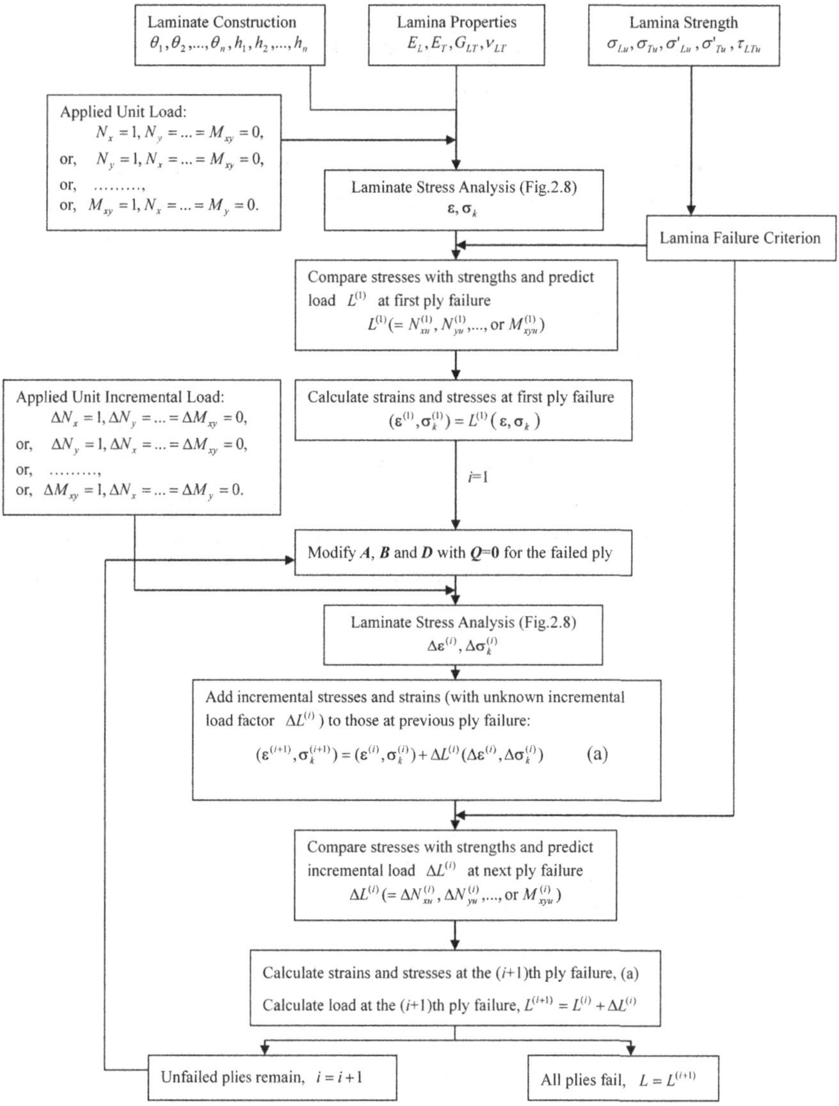

FIGURE 2.9 Flowchart for laminate strength analysis.

applied on the laminated specimen. For a macroscopically homogeneous material, the stress resultants will then be uniformly distributed within the specimen, that is, $N_x = 1$, $N_y = N_{xy} = M_x = M_y = M_{xy} = 0$ in every location of the specimen. With these given stress resultants, the stresses in each lamina of the laminates can be calculated by following the flowchart shown in Figure 2.8 for the laminate stress analysis. Since the stresses and strains are linear functions of the applied loads if the laminae exhibit linear elastic behavior, the stresses and strains may be uniformly scaled upward. Therefore, by multiplying the results obtained from the unit load by a load factor L for the applied load $N_x = L$, and comparing the stresses and strains in each lamina with suitable lamina failure criterion, the load $L^{(1)}$ at which the *first ply failure* (FPF) occurs may be predicted. After the FPF, the laminate response will deviate from its initial behavior and show discontinuities at the points of ply failure. Figure 2.10 shows a representative load-deformation curve for a hypothetical laminate in which each knee of the curve represents a ply failure. If the laminae are assumed to show a linear behavior up to fracture, the response between two discontinuities may be assumed to be linear. Therefore, the load–strain relationship, (2.35), for each segment of the curve may be written for incremental load and incremental strain, and the corresponding stiffness matrices **A**, **B**, and **D** should be recalculated by neglecting the stiffness of the failed plies. With this new relationship, the laminate stress analysis will then provide the incremental stresses and strains in each lamina after the FPF. By adding these incremental stresses and strains to the previous stresses

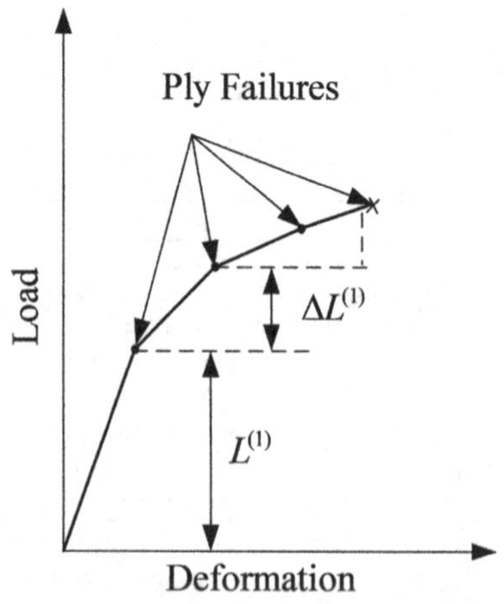

FIGURE 2.10 Load-deformation curve of a hypothetical laminate.

and strains, the stresses and strains in each lamina can be obtained with the unknown incremental load factor $\Delta L^{(1)}$ to be determined by the lamina failure criterion. Repeating this procedure until all plies of the laminate fail, the laminate may therefore not sustain any more loads. The laminate strength is then predicted by summation of all the incremental loads $\Delta L^{(i)}$ and initial load capacity $L^{(1)}$, that is, the laminate tensile strength in x-axis, N_{xu}, can be predicted as

$$N_{xu} = L^{(1)} + \sum_{i=1}^{n-1} \Delta L^{(i)}. \tag{2.43}$$

As shown in Figure 2.9 for the prediction of the other laminate strengths, same procedure as that described above can be applied.

REFERENCES

Agarwal, B.D. and Broutman, L.J., 1990, *Analysis and Performance of Fiber Composites*, 2nd Ed., John Wiley & Sons, Inc., New York.

Carlsson, L.A. and Pipes, R.B., 1987, *Experimental Characterization of Advanced Composite Materials*, Prentice-Hall Inc., New Jersey.

Halpin, J.C. and Tsai, S.W., 1969, "Effects of Environmental Factors on Composite Materials," AFML-TR-67-423.

Hill, R., 1950, *The Mathematical Theory of Plasticity*, Oxford University press, Oxford.

Hwu, C., 2010, *Anisotropic Elastic Plates*, Springer, New York.

Jones, R.M., 1974, *Mechanics of Composite Materials*, Scripta, Washington, D.C..

Mendelson, A., 1970, *Plasticity: Theory and Application*, Macmillan Company, New York.

Rosen, B. W., 1964, "Tensile Failure of Fibrous Composites," *AIAA Journal*, Vol. 2, No. 11, pp. 1985–1991.

Tsai, S.W., 1968, "Strength Theories of Filamentary Structures," in R.T. Schwartz and H.S. Schwartz, Eds., *Fundamental Aspects of Fiber Reinforced Plastic Composites*, Interscience, New York, Chapter 1.

Tsai, S.W. and Wu, E.M., 1971, "A General Theory of Strength for Anisotropic Materials," *J. of Composite Materials*, Vol. 5, No. 1, pp. 58–80.

CHAPTER 3

Laminated Composite Plates

In Chapter 2 we discussed the mechanical behavior of laminated composites and the elastic behavior of its constituents – laminae. From the discussion we know that if the resultant forces and moments of the laminated plates are known, the stresses and strains of each lamina can be determined through the laminate stress analysis. To find the resultant forces and moments as well as the related mechanical responses through the structural analysis, general formulations and methods of solution for the laminated composite structural members – plates, beams, and shells, will be introduced in this and the following two chapters.

Laminated plate is one of the simplest and most widespread structural members of composite laminates. In most practical applications, plates are usually designed in the configuration that the thickness of the plates is much smaller than the other two dimensions, which are commonly called *thin plates*. For simplicity, the general formulation described in the first section will mostly be restricted to the cases of thin plates with small deformations. The cases of *thick plates* will be briefly discussed in Section 3.1, followed by a more detailed presentation in Section 3.4. Section 3.2 then presents some classical solutions for the static, buckling, and free vibration of symmetric laminated thin plates. Section 3.3 further extends

DOI: 10.1201/9781003470465-3

these classical solutions to modern smart structures such as the *unsymmetric magento-electro-elastic* (MEE) laminated plates, for which the coupling effects occur not only from lamination sequence and anisotropic properties but also from mechanical, electric, and magnetic interaction. The last section presents the first-order shear deformation theory for thick laminated composite plates. For more general applications, *Green's functions* which are the key solutions for the development of boundary element methods will then be presented in Section 3.4.2.

3.1 GENERAL FORMULATION

3.1.1 Thin Plates

According to the classical lamination theory discussed in Section 2.4 for the thin laminated plates with small deformations, the mechanical behavior of laminated plates may be described by using the relations given in (2.30)–(2.37). For the convenience of the following derivation, these expressions are now repeated as follows:

Displacement Fields

$$
\begin{aligned}
u(x,y,z) &= u_0(x,y) - z\frac{\partial w(x,y)}{\partial x},\\
v(x,y,z) &= v_0(x,y) - z\frac{\partial w(x,y)}{\partial y},\\
w(x,y,z) &= w_0(x,y).
\end{aligned}
\tag{3.1}
$$

Strain–Displacement Relations

$$
\begin{aligned}
\varepsilon_x &= \frac{\partial u_0}{\partial x} - z\frac{\partial^2 w}{\partial x^2} = \varepsilon_x^0 + z\kappa_x,\\
\varepsilon_y &= \frac{\partial v_0}{\partial y} - z\frac{\partial^2 w}{\partial y^2} = \varepsilon_y^0 + z\kappa_y,\\
\gamma_{xy} &= \frac{\partial u_0}{\partial y} + \frac{\partial v_0}{\partial x} - 2z\frac{\partial^2 w}{\partial x \partial y} = \gamma_{xy}^0 + z\kappa_{xy}.
\end{aligned}
\tag{3.2}
$$

Constitutive Laws

$$\begin{Bmatrix} N_x \\ N_y \\ N_{xy} \\ M_x \\ M_y \\ M_{xy} \end{Bmatrix} = \begin{bmatrix} A_{11} & A_{12} & A_{16} & B_{11} & B_{12} & B_{16} \\ A_{12} & A_{22} & A_{26} & B_{12} & B_{22} & B_{26} \\ A_{16} & A_{26} & A_{66} & B_{16} & B_{26} & B_{66} \\ B_{11} & B_{12} & B_{16} & D_{11} & D_{12} & D_{16} \\ B_{12} & B_{22} & B_{26} & D_{12} & D_{22} & D_{26} \\ B_{16} & B_{26} & B_{66} & D_{16} & D_{26} & D_{66} \end{bmatrix} \begin{Bmatrix} \varepsilon_x^0 \\ \varepsilon_y^0 \\ \gamma_{xy}^0 \\ \kappa_x \\ \kappa_y \\ \kappa_{xy} \end{Bmatrix}. \tag{3.3}$$

Note that the equations for the displacement fields and strain–displacement relations given in (3.1) and (3.2) are independent of the material types. The expressions for the displacement fields given in (3.1) are due to the *Kirchhoff assumptions* for the deformations that straight lines initially normal to the middle surface remain straight lines and normal to the middle surface, and the normals have constant length so that the strain perpendicular to the middle surface is ignored. This assumption is appropriate for the thin plate whose thickness is much smaller compared to its other dimensions. The strain–displacement relations given in (3.2) are related to the assumptions of small deformation, that is, the deflections are much smaller compared to the plate thickness. As to the constitutive equations described in (3.3), we consider each lamina to be generally orthotropic and assume the laminate consists of perfectly bonded laminae and the bonds are infinitesimally thin as well as non-shear-deformable.

Besides Equations (3.1)–(3.3) which have been discussed in Section 2.4, to complete the structural analysis we need the equilibrium equations for the laminated plates.

Equilibrium Equations

Since the equilibrium equations concern only the balance of forces acting upon the structures, it should be independent of the material types. For thin plates, they are often developed by integrating the usual equilibrium equations of elasticity through the plate thickness. Neglecting the body forces, the equilibrium equations for every point inside the plates can be written as

$$\frac{\partial \sigma_x}{\partial x}+\frac{\partial \tau_{xy}}{\partial y}+\frac{\partial \tau_{xz}}{\partial z}=0,$$
$$\frac{\partial \tau_{xy}}{\partial x}+\frac{\partial \sigma_y}{\partial y}+\frac{\partial \tau_{yz}}{\partial z}=0,$$
$$\frac{\partial \tau_{xz}}{\partial x}+\frac{\partial \tau_{yz}}{\partial y}+\frac{\partial \sigma_z}{\partial z}=0. \tag{3.4}$$

With the resultant forces $\boldsymbol{N}$ and moments $\boldsymbol{M}$ defined in (2.34) and the transverse shear resultants defined by

$$\mathbf{Q}=\begin{Bmatrix} Q_x \\ Q_y \end{Bmatrix}=\int_{-h/2}^{h/2}\begin{Bmatrix} \tau_{xz} \\ \tau_{yz} \end{Bmatrix}dz, \tag{3.5}$$

integration of (3.4) with respect to z from the top surface ($z = -h/2$) to the bottom surface ($z = h/2$) of the plate will now give the force equilibrium equations of the plates as (please refer to Figure 2.7 for the sign convention)

$$\frac{\partial N_x}{\partial x}+\frac{\partial N_{xy}}{\partial y}+\tau_{1x}-\tau_{2x}=0,$$
$$\frac{\partial N_{xy}}{\partial x}+\frac{\partial N_y}{\partial y}+\tau_{1y}-\tau_{2y}=0,$$
$$\frac{\partial Q_x}{\partial x}+\frac{\partial Q_y}{\partial y}+q_1-q_2=0, \tag{3.6}$$

where $\tau_{1x} = \tau_{zx}(h/2)$, $\tau_{2x} = \tau_{zx}(-h/2)$, $\tau_{1y} = \tau_{zy}(h/2)$, $\tau_{2y} = \tau_{zy}(-h/2)$ and $q_1 = \sigma_z(h/2)$, $q_2 = \sigma_z(-h/2)$ are the tractions on the bottom and top surfaces of the plates. Since the stresses across the plate thickness are not uniformly distributed, for the entire plate not only the force equilibrium but also the moment equilibrium should be considered. Multiplying $(3.4)_{1,2}$ by z and then integrating through the thickness, the moment equilibrium equations can be derived as

$$\frac{\partial M_x}{\partial x}+\frac{\partial M_{xy}}{\partial y}-Q_x+\frac{h}{2}\left(\tau_{1x}+\tau_{2x}\right)=0,$$
$$\frac{\partial M_{xy}}{\partial x}+\frac{\partial M_y}{\partial y}-Q_y+\frac{h}{2}\left(\tau_{1y}+\tau_{2y}\right)=0. \tag{3.7}$$

For a laminated composite plate subjected to a lateral distributed load $q(x, y)$, the tractions on the top and bottom surfaces of the plate will then become

$$\tau_{1x} = \tau_{2x} = \tau_{1y} = \tau_{2y} = q_1 = 0, \quad q_2 = -q(x, y). \tag{3.8}$$

Consideration of (3.8), the force and moment equilibrium equations obtained in (3.6) and (3.7) now become

$$\begin{aligned}
&\frac{\partial N_x}{\partial x} + \frac{\partial N_{xy}}{\partial y} = 0, \quad \frac{\partial N_{xy}}{\partial x} + \frac{\partial N_y}{\partial y} = 0, \\
&\frac{\partial Q_x}{\partial x} + \frac{\partial Q_y}{\partial y} + q = 0, \quad \frac{\partial M_x}{\partial x} + \frac{\partial M_{xy}}{\partial y} - Q_x = 0, \\
&\frac{\partial M_{xy}}{\partial x} + \frac{\partial M_y}{\partial y} - Q_y = 0.
\end{aligned} \tag{3.9}$$

Since only three unknown displacement functions u_0, v_0, and w are used in the classical lamination theory, the five equilibrium equations shown in (3.9) may not be independent of each other. The three independent equilibrium equations should be those corresponding to the force balance in the x, y, and z directions. With this understanding, we now substitute $(3.9)_4$ and $(3.9)_5$ into $(3.9)_3$ to get the force equilibrium equation in the z-direction, which is

$$\frac{\partial^2 M_x}{\partial x^2} + 2\frac{\partial^2 M_{xy}}{\partial x \partial y} + \frac{\partial^2 M_y}{\partial y^2} + q = 0. \tag{3.10}$$

Governing Equations

The basic equations for the laminated plates are given in (3.1) for the displacement fields, (3.2) for the strain–displacement relations, (3.3) for the constitutive laws, and (3.9) for the equilibrium equations. Among these basic equations, only the constitutive laws depend on the material properties. All the other equations are exactly the same as those of the classical plate theory. To get governing equations satisfying all these basic equations, we first use (3.2) to express the midplane strains $\varepsilon_x^0, \varepsilon_y^0, \gamma_{xy}^0$ and curvatures κ_x, κ_y, κ_{xy} in terms of the midplane displacements u_0, v_0, and w, then use (3.3) to express the resultant forces N_x, N_y, N_{xy} and moments M_x, M_y, M_{xy} in terms of the midplane displacements. After these direct substitutions, the three equilibrium Equations $(3.9)_{1,2}$ and (3.10) can now be written in terms of three unknown displacement functions u_0, v_0, and w as

$$A_{11}\frac{\partial^2 u_0}{\partial x^2}+2A_{16}\frac{\partial^2 u_0}{\partial x\partial y}+A_{66}\frac{\partial^2 u_0}{\partial y^2}+A_{16}\frac{\partial^2 v_0}{\partial x^2}+\left(A_{12}+A_{66}\right)\frac{\partial^2 v_0}{\partial x\partial y}+A_{26}\frac{\partial^2 v_0}{\partial y^2}$$
$$-B_{11}\frac{\partial^3 w}{\partial x^3}-3B_{16}\frac{\partial^3 w}{\partial x^2\partial y}-\left(B_{12}+2B_{66}\right)\frac{\partial^3 w}{\partial x\partial y^2}-B_{26}\frac{\partial^3 w}{\partial y^3}=0, \tag{3.11a}$$

$$A_{16}\frac{\partial^2 u_0}{\partial x^2}+\left(A_{12}+A_{66}\right)\frac{\partial^2 u_0}{\partial x\partial y}+A_{26}\frac{\partial^2 u_0}{\partial y^2}+A_{66}\frac{\partial^2 v_0}{\partial x^2}+2A_{26}\frac{\partial^2 v_0}{\partial x\partial y}+A_{22}\frac{\partial^2 v_0}{\partial y^2}$$
$$-B_{16}\frac{\partial^3 w}{\partial x^3}-\left(B_{12}+2B_{66}\right)\frac{\partial^3 w}{\partial x^2\partial y}-3B_{26}\frac{\partial^3 w}{\partial x\partial y^2}-B_{22}\frac{\partial^3 w}{\partial y^3}=0, \tag{3.11b}$$

$$D_{11}\frac{\partial^4 w}{\partial x^4}+4D_{16}\frac{\partial^4 w}{\partial x^3\partial y}+2(D_{12}+2D_{66})\frac{\partial^4 w}{\partial x^2\partial y^2}+4D_{26}\frac{\partial^4 w}{\partial x\partial y^3}+D_{22}\frac{\partial^4 w}{\partial y^4}$$
$$-B_{11}\frac{\partial^3 u_0}{\partial x^3}-3B_{16}\frac{\partial^3 u_0}{\partial x^2\partial y}-(B_{12}+2B_{66})\frac{\partial^3 u_0}{\partial x\partial y^2}-B_{26}\frac{\partial^3 u_0}{\partial y^3}$$
$$-B_{16}\frac{\partial^3 v_0}{\partial x^3}-(B_{12}+2B_{66})\frac{\partial^3 v_0}{\partial x^2\partial y}-3B_{26}\frac{\partial^3 v_0}{\partial x\partial y^2}-B_{22}\frac{\partial^3 v_0}{\partial y^3}=q, \tag{3.11c}$$

which are the governing equations for the laminated plates.

The governing equations shown in (3.11) are system of partial differential equations with three unknown functions u_0, v_0, and w. Due to the mathematical complexity of these equations, it is not easy to get solutions by solving these partial differential equations. In practical engineering applications, it is common to have a symmetric laminate or to construct a balanced laminate. In those cases the coupling stiffness components like B_{ij} and/or A_{16}, A_{26} and/or D_{16}, D_{26} will be zero, and Equations (3.11a,b,c) will be drastically simplified. The problems of these special cases and their associated solution techniques will be discussed in the next section.

Boundary Conditions

For the general cases of laminated plates, the in-plane and plate bending problems will couple each other. Hence, every boundary of the plates should be described by four prescribed values. Two of them correspond to the in-plane problems and the other two correspond to the plate bending problems. Generally, they may be expressed as

$$\begin{aligned}
&u_n = \hat{u}_n \quad \text{or} \quad N_n = \hat{N}_n \quad \text{or} \quad N_n = k_n u_n, \\
&u_s = \hat{u}_s \quad \text{or} \quad N_{ns} = \hat{N}_{ns} \quad \text{or} \quad N_{ns} = k_s u_s, \\
&w_{,n} = \hat{w}_{,n} \quad \text{or} \quad M_n = \hat{M}_n \quad \text{or} \quad M_n = k_m w_{,n}, \\
&w = \hat{w} \quad \text{or} \quad V_n = \hat{V}_n \quad \text{or} \quad V_n = k_v w,
\end{aligned} \tag{3.12}$$

where V_n is the well-known *Kirchhoff force* of classical plate theory, or called *effective transverse shear force* defined by

$$V_n = Q_n + \frac{\partial M_{ns}}{\partial s}. \tag{3.13}$$

The subscripts n and s denote, respectively, the directions normal and tangent to the boundary. The overhat $(\hat{\bullet})$ denotes the prescribed value. k_n, k_s, k_m, and k_v are spring constants. If θ denotes the angle directed clockwise from the positive x-axis to the tangent direction of s, the values in the n–s coordinate can be calculated from the values in the x–y coordinate according to the following transformation laws:

$$\begin{aligned}
&u_s = \cos\theta u_0 + \sin\theta v_0, \quad u_n = -\sin\theta u_0 + \cos\theta v_0, \\
&\frac{\partial w}{\partial s} = \cos\theta \frac{\partial w}{\partial x} + \sin\theta \frac{\partial w}{\partial y}, \quad \frac{\partial w}{\partial n} = -\sin\theta \frac{\partial w}{\partial x} + \cos\theta \frac{\partial w}{\partial y},
\end{aligned} \tag{3.14a}$$

$$\begin{aligned}
&N_s = \cos^2\theta N_x + \sin^2\theta N_y + 2\sin\theta\cos\theta N_{xy}, \\
&N_n = \sin^2\theta N_x + \cos^2\theta N_y - 2\sin\theta\cos\theta N_{xy}, \\
&N_{ns} = \sin\theta\cos\theta\left(N_y - N_x\right) + \left(\cos^2\theta - \sin^2\theta\right)N_{xy},
\end{aligned} \tag{3.14b}$$

$$\begin{aligned}
&M_s = \cos^2\theta M_x + \sin^2\theta M_y + 2\sin\theta\cos\theta M_{xy}, \\
&M_n = \sin^2\theta M_x + \cos^2\theta M_y - 2\sin\theta\cos\theta M_{xy}, \\
&M_{ns} = \sin\theta\cos\theta\left(M_y - M_x\right) + \left(\cos^2\theta - \sin^2\theta\right)M_{xy},
\end{aligned} \tag{3.14c}$$

$$Q_s = \cos\theta Q_x + \sin\theta Q_y, \quad Q_n = -\sin\theta Q_x + \cos\theta Q_y. \tag{3.14d}$$

Due to the coupling of in-plane and plate bending problems, each of the commonly used boundary conditions such as, simply supported, clamped or free edge boundary conditions, may have four different possible types. Using the prefix *S* for the simply supported edges, prefix *C* for the clamped edges, and prefix *F* for the free edges, twelve commonly used boundary conditions are classified as

$$\begin{aligned}
&S1:\ w=0,\ M_n=0,\ u_n=\hat{u}_n,\ u_s=\hat{u}_s;\\
&S2:\ w=0,\ M_n=0,\ N_n=\hat{N}_n,\ u_s=\hat{u}_s;\\
&S3:\ w=0,\ M_n=0,\ u_n=\hat{u}_n,\ N_{ns}=\hat{N}_{ns};\\
&S4:\ w=0,\ M_n=0,\ N_n=\hat{N}_n,\ N_{ns}=\hat{N}_{ns};
\end{aligned} \tag{3.15a}$$

$$\begin{aligned}
&C1:\quad w=0,\quad w_{,n}=0,\quad u_n=\hat{u}_n,\quad u_s=\hat{u}_s;\\
&C2:\quad w=0,\quad w_{,n}=0,\quad N_n=\hat{N}_n,\quad u_s=\hat{u}_s;\\
&C3:\quad w=0,\quad w_{,n}=0,\quad u_n=\hat{u}_n,\quad N_{ns}=\hat{N}_{ns};\\
&C4:\quad w=0,\quad w_{,n}=0,\quad N_n=\hat{N}_n,\quad N_{ns}=\hat{N}_{ns};
\end{aligned} \tag{3.15b}$$

$$\begin{aligned}
&F1:\quad V_n=0,\quad M_n=0,\quad u_n=\hat{u}_n,\quad u_s=\hat{u}_s;\\
&F2:\quad V_n=0,\quad M_n=0,\quad N_n=\hat{N}_n,\quad u_s=\hat{u}_s;\\
&F3:\quad V_n=0,\quad M_n=0,\quad u_n=\hat{u}_n,\quad N_{ns}=\hat{N}_{ns};\\
&F4:\quad V_n=0,\quad M_n=0,\quad N_n=\hat{N}_n,\quad N_{ns}=\hat{N}_{ns}.
\end{aligned} \tag{3.15c}$$

In (3.12) and (3.15b), the subscript ",*n*" denotes the differentiation with respect to *n*, that is, $w_{,n} = \partial w/\partial n$.

3.1.2 Effects of Transverse Shear Deformation

In the previous subsection, although the transverse shear forces are included in the equilibrium equations the effects of transverse shear deformation are neglected, which can be observed from (3.1) that

$$\gamma_{xz}=\frac{\partial u}{\partial z}+\frac{\partial w}{\partial x}=0,\quad \text{and}\quad \gamma_{yz}=\frac{\partial v}{\partial z}+\frac{\partial w}{\partial y}=0. \tag{3.16}$$

If we want to consider the effects of transverse shear deformation, the assumption of displacement fields given in (3.1) should be modified, such as

$$u(x,y,z) = u_0(x,y) + z\left(\gamma_{xz}(x,y) - \frac{\partial w(x,y)}{\partial x}\right),$$
$$v(x,y,z) = v_0(x,y) + z\left(\gamma_{yz}(x,y) - \frac{\partial w(x,y)}{\partial y}\right),$$
$$w(x,y,z) = w_0(x,y). \tag{3.17}$$

All the following derivation should also be modified. The resulting theory based upon the displacement fields assumed in (3.17) is called *shear deformation theory*. It has been shown that shear deformations are much more important in composite material plates because of the much lower transverse shear stiffness relative to homogeneous materials. The shear deformation becomes relatively more important in thicker plates. A detailed discussion of the consideration of transverse shear deformation will be given in Section 3.4 for the thick laminated plates and Chapter 6 for the composite sandwich constructions.

3.1.3 High-Order Plate Theory

The classical lamination theory and the shear deformation theory are based upon the displacement fields assumed in (3.1) and (3.17), respectively. Both of these two theories assume that deformations are such that initially planar cross-sections will remain plane after deformation. In other words, the displacements are in linear order of the thickness coordinate z. Sometimes it is difficult to describe the deformation of thick laminates by the linear order theory. Considering warp of the cross-section that probably occurs in the thick laminates, the high-order theory is developed by assuming the displacement fields as follows (Lo, et al., 1977):

$$u(x,y,z) = u_0(x,y) + z\psi_x(x,y) + z^2\zeta_x(x,y) + z^3\phi_x(x,y),$$
$$v(x,y,z) = v_0(x,y) + z\psi_y(x,y) + z^2\zeta_y(x,y) + z^3\phi_y(x,y),$$
$$w(x,y,z) = w_0(x,y) + z\psi_z(x,y) + z^2\zeta_z(x,y). \tag{3.18}$$

One may refer to Christensen (1979) for further derivation based upon these assumed displacement fields.

3.2 SYMMETRIC LAMINATED PLATES

From the discussion in Section 2.5, we know that the coupling stiffness $\boldsymbol{B}$ will be identically zero for symmetric laminates. Substituting $B_{ij} = 0$, i,

$j = 1, 2, 6$, into (3.3) and using the relations shown in (3.2) for the curvatures, the resultant moments can now be expressed in terms of the lateral deflection as

$$
\begin{aligned}
M_x &= -\left(D_{11}\frac{\partial^2 w}{\partial x^2} + D_{12}\frac{\partial^2 w}{\partial y^2} + 2D_{16}\frac{\partial^2 w}{\partial x \partial y} \right),\\
M_y &= -\left(D_{12}\frac{\partial^2 w}{\partial x^2} + D_{22}\frac{\partial^2 w}{\partial y^2} + 2D_{26}\frac{\partial^2 w}{\partial x \partial y} \right),\\
M_{xy} &= -\left(D_{16}\frac{\partial^2 w}{\partial x^2} + D_{26}\frac{\partial^2 w}{\partial y^2} + 2D_{66}\frac{\partial^2 w}{\partial x \partial y} \right).
\end{aligned}
\tag{3.19}
$$

From $(3.9)_{4,5}$, we may further express the transverse shear forces in terms of the deflection as

$$
\begin{aligned}
Q_x &= -\left[D_{11}\frac{\partial^3 w}{\partial x^3} + 3D_{16}\frac{\partial^3 w}{\partial x^2 \partial y} + \left(D_{12} + 2D_{66}\right)\frac{\partial^3 w}{\partial x \partial y^2} + D_{26}\frac{\partial^3 w}{\partial y^3} \right],\\
Q_y &= -\left[D_{16}\frac{\partial^3 w}{\partial x^3} + \left(D_{12} + 2D_{66}\right)\frac{\partial^3 w}{\partial x^2 \partial y} + 3D_{26}\frac{\partial^3 w}{\partial x \partial y^2} + D_{22}\frac{\partial^3 w}{\partial y^3} \right].
\end{aligned}
\tag{3.20}
$$

Substituting $(3.19)_3$ and (3.20) into (3.13), the effective transverse shear force can be written in terms of the deflection as

$$
\begin{aligned}
V_x &= -\left[D_{11}\frac{\partial^3 w}{\partial x^3} + 4D_{16}\frac{\partial^3 w}{\partial x^2 \partial y} + \left(D_{12} + 4D_{66}\right)\frac{\partial^3 w}{\partial x \partial y^2} + 2D_{26}\frac{\partial^3 w}{\partial y^3} \right],\\
V_y &= -\left[2D_{16}\frac{\partial^3 w}{\partial x^3} + \left(D_{12} + 4D_{66}\right)\frac{\partial^3 w}{\partial x^2 \partial y} + 4D_{26}\frac{\partial^3 w}{\partial x \partial y^2} + D_{22}\frac{\partial^3 w}{\partial y^3} \right].
\end{aligned}
\tag{3.21}
$$

Substituting $B_{ij} = 0$, $i, j = 1, 2, 6$, into (3.11), we find that the governing equations for the in-plane and bending problems are uncoupled and the one corresponding to the plate bending problems is

$$
D_{11}\frac{\partial^4 w}{\partial x^4} + 4D_{16}\frac{\partial^4 w}{\partial x^3 \partial y} + 2\left(D_{12} + 2D_{66}\right)\frac{\partial^4 w}{\partial x^2 \partial y^2} + 4D_{26}\frac{\partial^4 w}{\partial x \partial y^3} + D_{22}\frac{\partial^4 w}{\partial y^4} = q.
\tag{3.22}
$$

Its associated boundary conditions can be simplified as

$$\begin{aligned} w_{,n} &= \hat{w}_{,n} \quad \text{or} \quad M_n = \hat{M}_n \quad \text{or} \quad M_n = k_m w_{,n}, \\ w &= \hat{w} \quad \text{or} \quad V_n = \hat{V}_n \quad \text{or} \quad V_n = k_v w. \end{aligned} \tag{3.23}$$

Among all the possible boundary conditions, the commonly used conditions like simply supported, clamped, and free edge boundary conditions can then be simplified as

$$\begin{aligned} &\text{Simply Supported:} \quad w = 0, \quad M_n = 0; \\ &\text{Clamped:} \quad w = 0, \quad w_{,n} = 0; \\ &\text{Free Edge:} \quad V_n = 0, \quad M_n = 0. \end{aligned} \tag{3.24}$$

3.2.1 Static Analysis

Although the governing Equation (3.22) has been drastically simplified from (3.11) due to the zero coupling stiffness, it is still not so easy to get analytical solutions for any practical plate structures even the simplest rectangular plates. The main difficulty encountered is due to the existence of bending-twisting coupling stiffnesses D_{16} and D_{26}. To handle this difficulty, several approximation methods have been proposed such as the perturbation solutions (Vinson and Sierakowski, 1986) and the Rayleigh–Ritz method (Ashton, 1969; Ashton and Whitney, 1970). For simplicity, here we only consider the cases whose $B_{ij} = 0$, $i, j = 1, 2, 6$, and $D_{16} = D_{26} = 0$.

Symmetric Cross-Ply Laminates

A simple example of the laminates with $B_{ij} = 0$ and $D_{16} = D_{26} = 0$ is a symmetric cross-ply laminate. With this consideration, the governing Equation (3.22) for the plate bending problems can be further simplified as

$$D_{11}\frac{\partial^4 w}{\partial x^4} + 2\left(D_{12} + 2D_{66}\right)\frac{\partial^4 w}{\partial x^2 \partial y^2} + D_{22}\frac{\partial^4 w}{\partial y^4} = q. \tag{3.25}$$

Without the bending-twisting coupling stiffnesses D_{16} and D_{26}, most of the problems can be solved by following the same techniques as that of the isotropic plates. For example, the *Navier solutions* for the case of rectangular plate being simply supported on all four edges, and the *Levy solutions* for the case of rectangular plate being simply supported on two opposite edges.

EXAMPLE 1: NAVIER SOLUTION

Consider a rectangular symmetric cross-ply laminated plate of sides a and b, simply supported on all edges and subjected to a distributed load $q(x, y)$. The origin of coordinates is placed at the upper left corner of the plate as shown in Figure 3.1.

Solution: By $(3.24)_1$ and the relations given in (3.19) with $D_{16} = D_{26} = 0$, the simply supported boundary conditions on all edges can be expressed as

$$w = 0,\ M_x = -D_{11}\frac{\partial^2 w}{\partial x^2} - D_{12}\frac{\partial^2 w}{\partial y^2} = 0, \quad \text{on } x = 0, a,$$
$$w = 0,\ M_y = -D_{12}\frac{\partial^2 w}{\partial x^2} - D_{22}\frac{\partial^2 w}{\partial y^2} = 0, \quad \text{on } y = 0, b, \tag{3.26}$$

which can be further simplified as

$$w = \frac{\partial^2 w}{\partial x^2} = 0, \quad \text{on } x = 0, a,$$
$$w = \frac{\partial^2 w}{\partial y^2} = 0, \quad \text{on } y = 0, b. \tag{3.27}$$

To find a solution satisfying the governing Equation (3.25) and the boundary conditions $(3.24)_1$, in Navier approach one simply expands the lateral deflection $w(x, y)$ and the applied load $q(x, y)$ into a double Fourier series as

$$w(x,y) = \sum_{m=1}^{\infty}\sum_{n=1}^{\infty} w_{mn} \sin\frac{m\pi x}{a} \sin\frac{n\pi y}{b},$$
$$q(x,y) = \sum_{m=1}^{\infty}\sum_{n=1}^{\infty} q_{mn} \sin\frac{m\pi x}{a} \sin\frac{n\pi y}{b}. \tag{3.28}$$

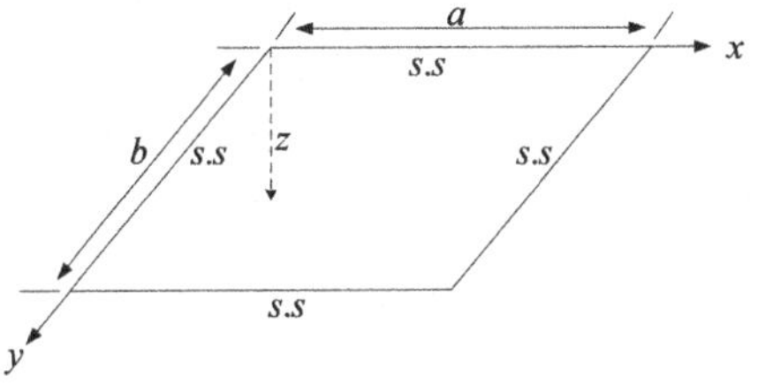

FIGURE 3.1 A rectangular plate simply supported on all four edges.

The deflection $w(x, y)$ assumed in $(3.28)_1$ satisfies the simply supported boundary conditions (3.27) exactly. The coefficients q_{mn}, $m, n = 1, \ldots, \infty$, related to the applied load $q(x, y)$ can be determined by

$$q_{mn} = \frac{4}{ab}\int_0^b\int_0^a q(x,y)\sin\frac{m\pi x}{a}\sin\frac{n\pi y}{b}\,dxdy. \tag{3.29}$$

Substituting (3.28) into (3.25), we get

$$w_{mn} = \frac{q_{mn}}{\pi^4\left\{D_{11}\left(\frac{m}{a}\right)^4 + 2(D_{12}+2D_{66})\left(\frac{m}{a}\right)^2\left(\frac{n}{b}\right)^2 + D_{22}\left(\frac{n}{b}\right)^4\right\}}. \tag{3.30}$$

In particular, for a uniform lateral load $q(x, y) = q_0$, (3.29) leads to

$$q_{mn} = \frac{16q_0}{\pi^2 mn}, \quad m,n = 1,3,\ldots\ldots.. \tag{3.31}$$

By substituting (3.31) into (3.30) and $(3.28)_1$, the solution for the deflection of the laminated plates can be written explicitly as

$$w = \frac{16q_0}{\pi^6}\sum_m^\infty\sum_n^\infty \frac{\sin\frac{m\pi x}{a}\sin\frac{n\pi y}{b}}{mn\left\{D_{11}\left(\frac{m}{a}\right)^4 + 2(D_{12}+2D_{66})\left(\frac{m}{a}\right)^2\left(\frac{n}{b}\right)^2 + D_{22}\left(\frac{n}{b}\right)^4\right\}},$$
$$m,n = 1,3,5,\ldots \tag{3.32}$$

Once the deflections are known, the strains can be straightforwardly obtained by substitution in the strain–displacement relation (3.2). The resultant moments can then be obtained by using the constitutive laws given in (3.3) with the coupling stiffness $\boldsymbol{B}$ being zero. Moreover, the stresses can be calculated by using the stress–strain relation (2.33) for each lamina.

EXAMPLE 2: LEVY SOLUTION

Consider a rectangular symmetric cross-ply laminated plate of sides a and b. The plate is simply supported on two opposite edges $x = 0$ and $x = a$, and is free on the third edge $y = b$, and the fourth edge $y = 0$ is clamped. A distributed load $q(x, y)$ is applied on the plate upper surface.

Solution: The Navier solution presented in the previous example contains a double series which is not very satisfactory for the slow convergence of the series. Another approach overcoming this non-satisfaction is the Levy solution that deals with a single series. While the Navier solution is only applicable for the plates simply supported on all four edges, the Levy solution is applicable to the bending of rectangular plates whose two opposite edges are simply supported and the remaining two edges can have arbitrarily supporting conditions.

With the assistance of (3.19), (3.21), (3.24), and (3.27), the boundary conditions of this example can be expressed in terms of deflection as

$$w = \frac{\partial^2 w}{\partial x^2} = 0, \qquad \text{on } x = 0, a,$$

$$w = \frac{\partial w}{\partial y} = 0, \qquad \text{on } y = 0,$$

$$D_{11}\frac{\partial^2 w}{\partial x^2} + D_{22}\frac{\partial^2 w}{\partial y^2} = 0, \ \text{on } y = b,$$

$$\left(D_{12} + 4D_{66}\right)\frac{\partial^3 w}{\partial x^2 \partial y} + D_{22}\frac{\partial^3 w}{\partial y^3} = 0, \quad \text{on } y = b. \tag{3.33}$$

To find a solution satisfying the governing Equation (3.25) and the boundary conditions (3.33), in Levy approach one expands the lateral deflection $w(x, y)$ and the applied load $q(x, y)$ into a single Fourier series instead of a double series as

$$w(x,y) = \sum_{m=1}^{\infty} w_m(y)\sin\frac{m\pi x}{a},$$

$$q(x,y) = \sum_{m=1}^{\infty} q_m(y)\sin\frac{m\pi x}{a}, \tag{3.34a}$$

where

$$q_m(y) = \frac{2}{a}\int_0^a q(x,y)\sin\frac{m\pi x}{a}dx. \tag{3.34b}$$

The deflection $w(x, y)$ assumed in $(3.34a)_1$ satisfies the simply supported boundary conditions $(3.33)_{1,2}$ on two opposite edges $x = 0$ and $x = a$ exactly. Substituting (3.34a) into (3.25), we get

$$w_m''''(y) - 2r_2\lambda_m^2 w_m''(y) + r_1\lambda_m^4 w_m(y) = q_m(y)/D_{22}, \tag{3.35a}$$

where

$$r_1 = \frac{D_{11}}{D_{22}}, \quad r_2 = \frac{D_{12} + 2D_{66}}{D_{22}}, \quad \lambda_m = \frac{m\pi}{a}. \tag{3.35b}$$

The general solution of the fourth-order ordinary differential Equation (3.35a) is

$$w_m(y) = w_m^{(h)}(y) + w_m^{(p)}(y), \tag{3.36}$$

where $w_m^{(p)}$ is the particular solution of (3.35a), and its determination depends on the function of $q_m(y)$; $w_m^{(h)}$ is the homogeneous solution of (3.35a), which has the following three different forms depending on whether r_2^2 is greater than, equal to, or less than r_1:

When $r_2^2 > r_1$,

$$\begin{aligned} w_m^{(h)}(y) = {} & c_1\cosh(\lambda_m s_1 y) + c_2\sinh(\lambda_m s_1 y) \\ & + c_3\cosh(\lambda_m s_2 y) + c_4\sinh(\lambda_m s_2 y). \end{aligned} \tag{3.37a}$$

When $r_2^2 = r_1$,

$$w_m^{(h)}(y) = (c_1 + c_2 y)\cosh(\lambda_m s_3 y) + (c_3 + c_4 y)\sinh(\lambda_m s_3 y). \tag{3.37b}$$

When $r_2^2 < r_1$,

$$\begin{aligned} w_m^{(h)}(y) = {} & \big(c_1\cos(\lambda_m s_4 y) + c_2\sin(\lambda_m s_4 y)\big)\cosh(\lambda_m s_5 y) \\ & + \big(c_3\cos(\lambda_m s_4 y) + c_4\sin(\lambda_m s_4 y)\big)\sinh(\lambda_m s_5 y). \end{aligned} \tag{3.37c}$$

In the above,

$$\begin{aligned} & s_1 = \sqrt{r_2 + \sqrt{r_2^2 - r_1}}, \quad s_2 = \sqrt{r_2 - \sqrt{r_2^2 - r_1}}, \quad s_3 = \sqrt{r_2}, \\ & s_4 = \sqrt{\left(\sqrt{r_1} - r_2\right)/2}, \quad s_5 = \sqrt{\left(\sqrt{r_1} + r_2\right)/2}. \end{aligned} \tag{3.38}$$

The unknown coefficients c_1, c_2, c_3, and c_4 are then determined through the satisfaction of the remaining boundary conditions shown in $(3.33)_{3,4}$ for $y = 0$ and $y = b$.

3.2.2 Buckling and Free Vibration

The general formulations presented in Section 3.1.1 are valid for the thin laminated plates with small deformations. To consider the buckling and free vibration problems, the effects of large deformation and time variation should be added. Moreover, these two kinds of problems involve solutions of eigenvalue problems as opposed to the boundary value problems of equilibrium analysis discussed in Section 3.2.1. Similar to the complexity encountered previously, only the case of *symmetric cross-ply laminates* will be discussed in this section. For the other cases, one may refer to the books such as Jones (1974) and Vinson and Sierakowski (1986).

Buckling

Since the buckling is an unstable state that the plate will transit from zero lateral deflection to possibly large lateral deflection, the equilibrium Equations (3.9) based upon the undeformed states should be modified to include the effect of lateral deflection. Consideration of a deformed plate element (deflected in lateral direction) leads to the results that the force and moment equilibrium equations in the plate surface directions (x- and y-directions) will keep the same as those shown in (3.9) but the force equilibrium in the lateral direction (z-direction) should be modified as (Szilard, 1974):

$$\frac{\partial^2 M_x}{\partial x^2}+2\frac{\partial^2 M_{xy}}{\partial x\partial y}+\frac{\partial^2 M_y}{\partial y^2}+q+N_x\frac{\partial^2 w}{\partial x^2}+2N_{xy}\frac{\partial^2 w}{\partial x\partial y}+N_y\frac{\partial^2 w}{\partial y^2}=0. \quad (3.39)$$

With this result, the governing differential Equation (3.22) for the laminates with $B_{ij}=0$ and $A_{16}=A_{26}=D_{16}=D_{26}=0$ becomes

$$D_{11}\frac{\partial^4 w}{\partial x^4}+2\left(D_{12}+2D_{66}\right)\frac{\partial^4 w}{\partial x^2\partial y^2}+D_{22}\frac{\partial^4 w}{\partial y^4}$$
$$=q+N_x\frac{\partial^2 w}{\partial x^2}+2N_{xy}\frac{\partial^2 w}{\partial x\partial y}+N_y\frac{\partial^2 w}{\partial y^2}. \quad (3.40)$$

Consider now a rectangular symmetric cross-ply laminated plate simply supported on all edges and subjected to a uniaxial in-plane compressive force N. For a homogeneous plate subjected to a uniform compressive load, the stress resultants in the plates will be uniformly distributed, that is, $N_x=-N=$ constant, $N_y=N_{xy}=0$. The governing Equation (3.40) becomes

$$D_{11}\frac{\partial^4 w}{\partial x^4}+2\left(D_{12}+2D_{66}\right)\frac{\partial^4 w}{\partial x^2\partial y^2}+D_{22}\frac{\partial^4 w}{\partial y^4}+N\frac{\partial^2 w}{\partial x^2}=0. \quad (3.41)$$

Again for the plates simply supported on all edges, the boundary conditions will be satisfied automatically if we assume the solution in the form of Navier solutions. That is,

$$w\left(x,y\right)=\sum_{m=1}^{\infty}\sum_{n=1}^{\infty}w_{mn}\sin\frac{m\pi x}{a}\sin\frac{n\pi y}{b} \quad (3.42)$$

Substituting (3.42) into (3.41), we obtain

$$\sum_{m=1}^{\infty}\sum_{n=1}^{\infty}\left\{\pi^4\left[D_{11}\left(\frac{m}{a}\right)^4+2\left(D_{12}+2D_{66}\right)\left(\frac{mn}{ab}\right)^2+D_{22}\left(\frac{n}{b}\right)^4\right]\right.$$
$$\left.-N\left(\frac{m}{a}\right)^2\pi^2\right\}w_{mn}\sin\frac{m\pi x}{a}\sin\frac{n\pi y}{b}=0. \quad (3.43)$$

A nontrivial solution ($w_{mn} \neq 0$) to (3.43) can be obtained by letting the coefficients of the double sine series be identical to zero, from which the critical buckling load N_{cr} is obtained as

$$N_{cr}=\left(\frac{\pi a}{m}\right)^2\left[D_{11}\left(\frac{m}{a}\right)^4+2\left(D_{12}+2D_{66}\right)\left(\frac{mn}{ab}\right)^2+D_{22}\left(\frac{n}{b}\right)^4\right] \quad (3.44)$$

It is observed from (3.44) that the minimum value of N_{cr} occurs when $n = 1$. However, since m appears in both of the numerator and denominator, it is not clear which value of m results in the lowest critical buckling load. For a given laminated plate it can be easily determined computationally.

Free Vibration

Any continuous structure mathematically has an infinite number of natural frequencies and mode shapes. In designing a structure, we are not only insuring that the structure will not be over-deflected or become overstressed, but also caring that resonance will not occur (i.e., imposed loads will not have the same frequency as one or more natural frequencies).

To illustrate how to find the natural frequencies and mode shapes, we consider the easiest example of symmetric cross-ply laminates.

In writing the governing differential equations of motion for free vibration, we may apply d'Alembert's dynamic equilibrium principle by adding the inertia force to the right hand side of (3.25) and neglecting the external force q, that is,

$$D_{11}\frac{\partial^4 w}{\partial x^4}+2\left(D_{12}+2D_{66}\right)\frac{\partial^4 w}{\partial x^2\partial y^2}+D_{22}\frac{\partial^4 w}{\partial y^4}=-\rho h\frac{\partial^2 w}{\partial t^2}, \tag{3.45}$$

where ρ is the mass density of the plate and h is the plate thickness.

Again, we consider a rectangular symmetric cross-ply laminated plate of sides a and b, simply supported on all edges (Figure 3.1). The vibration mode shapes satisfying the boundary conditions (3.27) can be assumed to be

$$w\left(x,y\right)=\sum_{m=1}^{\infty}\sum_{n=1}^{\infty}w_{mn}\sin\frac{m\pi x}{a}\sin\frac{n\pi y}{b}\cos\omega_{mn}t \tag{3.46}$$

Substituting these mode shapes into (3.45) provides the corresponding natural frequencies

$$\omega_{mn}=\frac{\pi^2}{\sqrt{\rho h}}\sqrt{D_{11}\left(\frac{m}{a}\right)^4+2\left(D_{12}+2D_{66}\right)\left(\frac{mn}{ab}\right)^2+D_{22}\left(\frac{n}{b}\right)^4} \tag{3.47}$$

The fundamental frequency occurs with $m = n = 1$, which is for one half sine wave in each direction.

3.3 MAGNETO-ELECTRO-ELASTIC LAMINATED PLATES

For unsymmetric magento-electro-elastic (MEE) laminated composite thin plates, the coupling effects occur not only from mechanical, electric, and magnetic interaction but also from lamination sequence and anisotropic properties. Stretching and bending deformation as well as electric and magnetic potentials will all be coupled with each other. Thus, even a simple rectangular MEE laminated thin plate with simply supported edges under generalized loads, their analytical solutions are much more complicated than those shown in Section 3.2.1 for symmetric laminated plates. In this section to simplify the possible complicated mathematical expressions, the governing partial differential equations are written in matrix form. With simple matrix form expressions, we present the explicit solutions for two different boundary conditions corresponding to Navier's solution and Lévy's solution in Sections 3.3.2 and 3.3.3, respectively. The former is valid for a rectangular plate with all edges simply supported, and the latter is valid for the one with two opposite edges simply supported and no restriction is set on the other two edges. To avoid the ill-conditioned matrices induced by the wide-ranged MEE material constants and the matrix exponential operation, some remarks on scaling and numerical calculation are provided. Parts of this section are adapted from our recent work (Hsu and Hwu, 2022a,b).

3.3.1 Governing Equations

If a multilayered MEE composite is made up of different layers such as a fiber-reinforced composite layer and a composite layer consisting of the piezoelectric materials and/or piezomagnetic materials, each layer can be treated as plane stress condition and its constitutive relation can be expressed as (Hwu, 2021)

$$\begin{Bmatrix} \sigma_1 \\ \sigma_2 \\ \sigma_6 \\ D_1 \\ D_2 \\ B_1 \\ B_2 \end{Bmatrix} = \begin{bmatrix} C_{11} & C_{12} & C_{16} & e_{11} & e_{21} & q_{11} & q_{21} \\ C_{12} & C_{22} & C_{26} & e_{12} & e_{22} & q_{12} & q_{22} \\ C_{16} & C_{26} & C_{66} & e_{16} & e_{26} & q_{16} & q_{26} \\ e_{11} & e_{12} & e_{16} & -\omega_{11} & -\omega_{12} & -m_{11} & -m_{21} \\ e_{21} & e_{22} & e_{26} & -\omega_{12} & -\omega_{22} & -m_{12} & -m_{22} \\ q_{11} & q_{12} & q_{16} & -m_{11} & -m_{12} & -\xi_{11} & -\xi_{12} \\ q_{21} & q_{22} & q_{26} & -m_{21} & -m_{22} & -\xi_{12} & -\xi_{22} \end{bmatrix} \begin{Bmatrix} \varepsilon_1 \\ \varepsilon_2 \\ \varepsilon_6 \\ -E_1 \\ -E_2 \\ -H_1 \\ -H_2 \end{Bmatrix}, \tag{3.48}$$

where $\sigma_i, \varepsilon_i, D_k, E_k, B_k, H_k, i = 1, 2, 6, k = 1, 2$, denote, respectively, the in-plane components of stress, strain, electric displacement, electric field, magnetic flux, and magnetic field; $C_{ij}, e_{kj}, q_{kj}, m_{kl}, \omega_{kl}, \xi_{kl}, i, j = 1, 2, 6, k, l = 1, 2$, are, respectively, the in-plane components of elastic, piezoelectric, piezomagnetic, magneto-electric, permittivity, and permeability coefficients, whose definitions are specifically presented in Hwu (2021). By letting

$$\begin{aligned}
&\sigma_p = D_{\bar{p}}, \quad \varepsilon_p = -E_{\bar{p}}, \quad p = 7,8,\\
&\sigma_r = B_{\tilde{r}}, \quad \varepsilon_r = -H_{\tilde{r}}, \quad r = 9,10,\\
&C_{ip} = C_{pi} = e_{\bar{p}i}, \qquad i = 1,2,6, \quad p = 7,8,\\
&C_{ir} = C_{ri} = q_{\tilde{r}i}, \qquad i = 1,2,6, \quad r = 9,10,\\
&C_{pq} = C_{qp} = -\omega_{\bar{p}\bar{q}}, \quad p,q = 7,8,\\
&C_{pr} = C_{rp} = -m_{\tilde{r}\bar{p}}, \quad p = 7,8, \; r = 9,10,\\
&C_{rs} = C_{sr} = -\xi_{\tilde{r}\tilde{s}}, \quad r,s = 9,10,\\
&\bar{p} = p-6, \; \bar{q} = q-6, \; \tilde{r} = r-8, \; \tilde{s} = s-8,
\end{aligned} \tag{3.49}$$

the relation (3.48) can be written in an expanded matrix form as

$$\boldsymbol{\sigma} = \mathbf{C}\boldsymbol{\varepsilon}, \tag{3.50}$$

where $\boldsymbol{\sigma}$ and $\boldsymbol{\varepsilon}$ are, respectively, the 7 × 1 vector of generalized stresses and strains, and $\mathbf{C}$ is the 7 × 7 generalfized elastic matrix. To see clearly the indices used in (3.50), we may re-write (3.50) as

$$\sigma_p = \sum_q C_{pq}\varepsilon_q, \quad p,q = 1,2,6,7,8,9,10. \tag{3.51}$$

If the MEE laminates are the plate structure with thickness much smaller than the plane dimensions, the Kirchhoff assumption is employed such that the displacements, electric and magnetic fields are assumed to be linear variations along the laminate thickness and the transverse shear deformation is ignored. That is,

$$\begin{aligned}
&u_i(x_1,x_2,x_3) = u_i^0(x_1,x_2) + x_3\beta_i(x_1,x_2),\\
&E_i(x_1,x_2,x_3) = E_i^0(x_1,x_2) + x_3E_i^*(x_1,x_2),\\
&H_i(x_1,x_2,x_3) = H_i^0(x_1,x_2) + x_3H_i^*(x_1,x_2), \quad i = 1,2,
\end{aligned} \tag{3.52a}$$

and

$$\beta_1 = -\partial w / \partial x_1, \beta_2 = -\partial w / \partial x_2. \tag{3.52b}$$

In (3.52), $u_i^0, E_i^0, H_i^0, i = 1,2$, are the associated mid-plane quantities; w is the out-of-plane displacement, or generally called plate deflection; β_1 and β_2 are the plate slopes; and $E_i^*, H_i^*, i = 1,2$, are the associated rate changes of electric fields and magnetic fields along thickness direction.

Based upon the Kirchhoff assumption and the constitutive relation (3.48) for each layer, by following the derivation of classical lamination theory stated in Section 2.4.1 we can write down the constitutive laws for MEE laminates as

$$\begin{Bmatrix} \mathbf{N} \\ \mathbf{M} \end{Bmatrix} = \begin{bmatrix} \mathbf{A} & \mathbf{B} \\ \mathbf{B} & \mathbf{D} \end{bmatrix} \begin{Bmatrix} \boldsymbol{\varepsilon}_0 \\ \boldsymbol{\kappa} \end{Bmatrix}, \tag{3.53a}$$

where

$$\mathbf{N} = \begin{Bmatrix} N_{11} \\ N_{22} \\ N_{12} \\ N_{41} \\ N_{42} \\ N_{51} \\ N_{52} \end{Bmatrix}, \quad \mathbf{M} = \begin{Bmatrix} M_{11} \\ M_{22} \\ M_{12} \\ M_{41} \\ M_{42} \\ M_{51} \\ M_{52} \end{Bmatrix}, \quad \boldsymbol{\varepsilon}_0 = \begin{Bmatrix} \varepsilon_{11}^0 \\ \varepsilon_{22}^0 \\ 2\varepsilon_{12}^0 \\ 2\varepsilon_{41}^0 \\ 2\varepsilon_{42}^0 \\ 2\varepsilon_{51}^0 \\ 2\varepsilon_{52}^0 \end{Bmatrix}, \quad \boldsymbol{\kappa} = \begin{Bmatrix} \kappa_{11} \\ \kappa_{22} \\ 2\kappa_{12} \\ 2\kappa_{41} \\ 2\kappa_{42} \\ 2\kappa_{51} \\ 2\kappa_{52} \end{Bmatrix}, \tag{3.53b}$$

and

$$\begin{aligned} &\varepsilon_{Ji}^0 = (u_{J,i}^0 + u_{i,J}^0)/2, \quad \kappa_{Ji} = (\beta_{J,i} + \beta_{i,J})/2, \\ &N_{Ji} = \int_{-h/2}^{h/2} \sigma_{Ji} dx_3, \quad M_{Ji} = \int_{-h/2}^{h/2} \sigma_{Ji} x_3 dx_3, \quad J = 1,2,4,5, \quad i = 1,2, \\ &A_{pq} = \sum_{k=1}^{n} C_{pq}^{(k)} \left(h_k - h_{k-1}\right), \quad B_{pq} = \frac{1}{2} \sum_{k=1}^{n} C_{pq}^{(k)} \left(h_k^2 - h_{k-1}^2\right), \\ &D_{pq} = \frac{1}{3} \sum_{k=1}^{n} C_{pq}^{(k)} \left(h_k^3 - h_{k-1}^3\right), \quad p,q = 1,2,6,7,8,9,10. \end{aligned} \tag{3.53c}$$

In the above, $\varepsilon_{Ji}^0, \kappa_{Ji}$ are the generalized mid-plane strains and plate curvatures; N_{Ji}, M_{Ji} are the generalized stress resultants and bending moments; A_{pq}, B_{pq}, D_{pq} are the generalized extensional, coupling, and bending stiffness constants; the subscript comma followed by an index denotes partial differentiation with respect to the coordinate of the subsequent index, for example, $u_{J,i} = \partial u_J / \partial x_i$; the superscript 0 denotes the value of the mid-plane surface; h is the laminate thickness, h_{k-1} and h_k denote the locations (x_3-coordinates) of top and bottom surfaces of the kth layer; the superscript

(k) denotes the value of the kth layer; n is the number of layers in laminate, and $u^0_{j,4} = u^0_{j,5} = \beta_{j,4} = \beta_{j,5} = 0$ is assumed.

To have a better understanding for some symbols appeared in (3.53c), alternative notations of $x_i, u^0_J, \beta_J, \sigma_{Ji}$ are given and explained as follows:

$$
\begin{aligned}
&x_1 = x,\ x_2 = y,\ x_3 = z, \\
&u^0_1 = u_0,\ u^0_2 = v_0,\ u^0_4 = u_E,\ u^0_5 = u_H, \\
&\beta_1 = \beta_x,\ \beta_2 = \beta_y,\ \beta_4 = \beta_E,\ \beta_5 = \beta_H, \\
&\sigma_{11} = \sigma_1,\ \sigma_{22} = \sigma_2,\ \sigma_{12} = \sigma_6, \\
&\sigma_{41} = \sigma_7 = D_1,\ \sigma_{42} = \sigma_8 = D_2, \\
&\sigma_{51} = \sigma_9 = B_1,\ \sigma_{52} = \sigma_{10} = B_2,
\end{aligned}
\tag{3.54}
$$

where (x, y, z) is an alternative symbol used for the Cartesian coordinate; u_0, v_0 are the mid-plane displacements along x_1 and x_2 directions; β_x, β_y are the plate slopes related to the deflection w by (3.52b); u_E, u_H are the mid-plane electric and magnetic potentials; β_E, β_H are the rate change of electric and magnetic potentials along thickness direction. Their relations to electric and magnetic fields are

$$
\begin{aligned}
&u_E = -\int E^0_1 dx_1 = -\int E^0_2 dx_2,\ \ u_H = -\int H^0_1 dx_1 = -\int H^0_2 dx_2, \\
&\beta_E = -\int E^*_1 dx_1 = -\int E^*_2 dx_2,\ \ \beta_H = -\int H^*_1 dx_1 = -\int H^*_2 dx_2.
\end{aligned}
\tag{3.55}
$$

To express the complete set of equilibrium equations for MEE laminated thin plates, besides the mechanical equilibrium (3.9) additional electro-magneto static conditions should be considered such as

$$
\begin{aligned}
&\frac{\partial N_{41}}{\partial x_1} + \frac{\partial N_{42}}{\partial x_2} = q_D,\quad \frac{\partial M_{41}}{\partial x_1} + \frac{\partial M_{42}}{\partial x_2} = m_D, \\
&\frac{\partial N_{51}}{\partial x_1} + \frac{\partial N_{52}}{\partial x_2} = q_B,\quad \frac{\partial M_{51}}{\partial x_1} + \frac{\partial M_{52}}{\partial x_2} = m_B,
\end{aligned}
\tag{3.56}
$$

where q_D, q_B are the negatives of distributed electric and magnetic charges applied on the laminate surfaces; m_D, m_B are the distributed electric and magnetic moments, which may be induced by the eccentric loading of q_D, q_B from mid-plane.

With the field expressions assumed in (3.52), the kinematic relations $(3.53c)_{1,2}$, the constitutive laws (3.53a) and the equilibrium Equations (3.9) and (3.56) for MEE laminated thin plates can all be written in matrix form as

$$\begin{Bmatrix} \boldsymbol{\varepsilon}_0 \\ \boldsymbol{\kappa} \end{Bmatrix} = \begin{bmatrix} \mathcal{L}_s & \mathbf{0} \\ \mathbf{0} & \mathcal{L}_b \end{bmatrix} \begin{Bmatrix} \mathbf{u}_s \\ \mathbf{u}_b \end{Bmatrix},$$
$$\begin{Bmatrix} N \\ M \end{Bmatrix} = \begin{bmatrix} A & B \\ B & D \end{bmatrix} \begin{Bmatrix} \boldsymbol{\varepsilon}_0 \\ \boldsymbol{\kappa} \end{Bmatrix},$$
$$\begin{bmatrix} \mathcal{L}_s^T & \mathbf{0} \\ \mathbf{0} & \mathcal{L}_b^T \end{bmatrix} \begin{Bmatrix} N \\ M \end{Bmatrix} + \begin{Bmatrix} \mathbf{q}_s \\ \mathbf{q}_b \end{Bmatrix} = \mathbf{0}, \tag{3.57}$$

where the superscript T denotes matrix transpose, the subscripts s and b represent the in-plane stretching and out-of-plane bending parts,

$$\mathbf{u}_s = \begin{Bmatrix} u_0 \\ v_0 \\ u_E \\ u_H \end{Bmatrix}, \quad \mathbf{u}_b = \begin{Bmatrix} -w \\ \beta_E \\ \beta_H \end{Bmatrix}, \quad \mathbf{q}_s = \begin{Bmatrix} q_x \\ q_y \\ q_D \\ q_B \end{Bmatrix}, \quad \mathbf{q}_b = \begin{Bmatrix} q_z \\ m_D \\ m_B \end{Bmatrix}, \tag{3.58a}$$

and $\mathcal{L}_s$, $\mathcal{L}_b$ are two partial differential operators defined by

$$\mathcal{L}_s = \begin{bmatrix} \partial_x & 0 & 0 & 0 \\ 0 & \partial_y & 0 & 0 \\ \partial_y & \partial_x & 0 & 0 \\ 0 & 0 & \partial_x & 0 \\ 0 & 0 & \partial_y & 0 \\ 0 & 0 & 0 & \partial_x \\ 0 & 0 & 0 & \partial_y \end{bmatrix}, \quad \mathcal{L}_b = \begin{bmatrix} \partial_x^2 & 0 & 0 \\ \partial_y^2 & 0 & 0 \\ 2\partial_x\partial_y & 0 & 0 \\ 0 & \partial_x & 0 \\ 0 & \partial_y & 0 \\ 0 & 0 & \partial_x \\ 0 & 0 & \partial_y \end{bmatrix},$$
$$\partial_x = \frac{\partial}{\partial x}, \quad \partial_y = \frac{\partial}{\partial y}, \quad \partial_x^2 = \frac{\partial^2}{\partial x^2}, \quad \partial_y^2 = \frac{\partial^2}{\partial y^2}. \tag{3.58b}$$

By choosing the generalized displacements u_0, v_0, u_E, u_H, w, β_E, β_H in vectors $\mathbf{u}_s$ and $\mathbf{u}_b$ as basic variables, the constitutive relations between generalized stress resultants/bending moments $\boldsymbol{N}$, $\boldsymbol{M}$ and generalized displacements $\mathbf{u}_s$, $\mathbf{u}_b$ can be obtained through the substitution of $(3.57)_1$ into $(3.57)_2$ as

$$\begin{Bmatrix} \boldsymbol{N} \\ \boldsymbol{M} \end{Bmatrix} = \begin{bmatrix} \boldsymbol{A}\boldsymbol{\mathcal{L}}_s & \boldsymbol{B}\boldsymbol{\mathcal{L}}_b \\ \boldsymbol{B}\boldsymbol{\mathcal{L}}_s & \boldsymbol{D}\boldsymbol{\mathcal{L}}_b \end{bmatrix} \begin{Bmatrix} \mathbf{u}_s \\ \mathbf{u}_b \end{Bmatrix}. \tag{3.59}$$

By further substitution of (3.59) into $(3.57)_3$, the governing equations for MEE composite laminated plates can be obtained as

$$\begin{bmatrix} \boldsymbol{\mathcal{L}}_s^T\boldsymbol{A}\boldsymbol{\mathcal{L}}_s & \boldsymbol{\mathcal{L}}_s^T\boldsymbol{B}\boldsymbol{\mathcal{L}}_b \\ \boldsymbol{\mathcal{L}}_b^T\boldsymbol{B}\boldsymbol{\mathcal{L}}_s & \boldsymbol{\mathcal{L}}_b^T\boldsymbol{D}\boldsymbol{\mathcal{L}}_b \end{bmatrix} \begin{Bmatrix} \mathbf{u}_s \\ \mathbf{u}_b \end{Bmatrix} + \begin{Bmatrix} \mathbf{q}_s \\ \mathbf{q}_b \end{Bmatrix} = \mathbf{0}, \tag{3.60}$$

in which seven partial differential equations are constructed to solve the seven basic variables in $\mathbf{u}_s$ and $\mathbf{u}_b$.

Note that the basic Equations (3.57) and associated governing equations of (3.60) are valid for the general MEE laminated plates with any form of generalized extensional, coupling and bending stiffness matrices, $\boldsymbol{A}$, $\boldsymbol{B}$, and $\boldsymbol{D}$. If the laminated plate is made up of specially orthotropic MEE layers whose material principal axes coincide with global coordinate, the stiffness matrices $\boldsymbol{A}$, $\boldsymbol{B}$, and $\boldsymbol{D}$ for the *orthotropic MEE laminated plates* can be reduced to the form of

$$\boldsymbol{X} = \begin{bmatrix} X_{11} & & & & & & \text{symm.} \\ X_{12} & X_{22} & & & & & \\ 0 & 0 & X_{66} & & & & \\ 0 & 0 & X_{67} & X_{77} & & & \\ X_{18} & X_{28} & 0 & 0 & X_{88} & & \\ 0 & 0 & X_{69} & X_{79} & 0 & X_{99} & \\ X_{1,10} & X_{2,10} & 0 & 0 & X_{8,10} & 0 & X_{10,10} \end{bmatrix}, \quad X = A, B, D. \tag{3.61}$$

With the definition of (3.58b) for differential operators and the form of generalized extensional, coupling, and bending stiffness matrices (3.61) for orthotropic MEE laminates, the explicit expression of differential operators in the constitutive relations of (3.59) and in governing equations of (3.60), which will be used later, can be obtained after the associated matrix multiplication, as follows. In (3.59),

$$\boldsymbol{X}\boldsymbol{\mathcal{L}}_s = \begin{bmatrix} X_{11}\partial_x & X_{12}\partial_y & X_{18}\partial_y & X_{1,10}\partial_y \\ X_{12}\partial_x & X_{22}\partial_y & X_{28}\partial_y & X_{2,10}\partial_y \\ X_{66}\partial_y & X_{66}\partial_x & X_{67}\partial_x & X_{69}\partial_x \\ X_{67}\partial_y & X_{67}\partial_x & X_{77}\partial_x & X_{79}\partial_x \\ X_{18}\partial_x & X_{28}\partial_y & X_{88}\partial_y & X_{8,10}\partial_y \\ X_{69}\partial_y & X_{69}\partial_x & X_{79}\partial_x & X_{99}\partial_x \\ X_{1,10}\partial_x & X_{2,10}\partial_y & X_{8,10}\partial_y & X_{10,10}\partial_y \end{bmatrix}, \quad X = A, B, \tag{3.62a}$$

$$\boldsymbol{X}\boldsymbol{\mathcal{L}}_b = \begin{bmatrix} X_{11}\partial_x^2 + X_{12}\partial_y^2 & X_{18}\partial_y & X_{1,10}\partial_y \\ X_{12}\partial_x^2 + X_{22}\partial_y^2 & X_{28}\partial_y & X_{2,10}\partial_y \\ 2X_{66}\partial_x\partial_y & X_{67}\partial_x & X_{69}\partial_x \\ 2X_{67}\partial_x\partial_y & X_{77}\partial_x & X_{79}\partial_x \\ X_{18}\partial_x^2 + X_{28}\partial_y^2 & X_{88}\partial_y & X_{8,10}\partial_y \\ 2X_{69}\partial_x\partial_y & X_{79}\partial_x & X_{99}\partial_x \\ X_{1,10}\partial_x^2 + X_{2,10}\partial_y^2 & X_{8,10}\partial_y & X_{10,10}\partial_y \end{bmatrix}, \quad X = B, D, \tag{3.62b}$$

and in (3.60),

$$\boldsymbol{\mathcal{L}}_s^T\boldsymbol{A}\boldsymbol{\mathcal{L}}_s = \begin{bmatrix} A_{11}\partial_x^2 + A_{66}\partial_y^2 & & & \text{symm.} \\ (A_{12} + A_{66})\partial_x\partial_y & A_{66}\partial_x^2 + A_{22}\partial_y^2 & & \\ (A_{18} + A_{67})\partial_x\partial_y & A_{67}\partial_x^2 + A_{28}\partial_y^2 & A_{77}\partial_x^2 + A_{88}\partial_y^2 & \\ (A_{1,10} + A_{69})\partial_x\partial_y & A_{69}\partial_x^2 + A_{2,10}\partial_y^2 & A_{79}\partial_x^2 + A_{8,10}\partial_y^2 & A_{99}\partial_x^2 + A_{10,10}\partial_y^2 \end{bmatrix}, \tag{3.63a}$$

$$\boldsymbol{\mathcal{L}}_s^T\boldsymbol{B}\boldsymbol{\mathcal{L}}_b = \begin{bmatrix} B_{11}\partial_x^3 + (B_{12} + 2B_{66})\partial_x\partial_y^2 & (B_{18} + B_{67})\partial_x\partial_y & (B_{1,10} + B_{69})\partial_x\partial_y \\ (B_{12} + 2B_{66})\partial_x^2\partial_y + B_{22}\partial_y^3 & B_{67}\partial_x^2 + B_{28}\partial_y^2 & B_{69}\partial_x^2 + B_{2,10}\partial_y^2 \\ (B_{18} + 2B_{67})\partial_x^2\partial_y + B_{28}\partial_y^3 & B_{77}\partial_x^2 + B_{88}\partial_y^2 & B_{79}\partial_x^2 + B_{8,10}\partial_y^2 \\ (B_{1,10} + 2B_{69})\partial_x^2\partial_y + B_{2,10}\partial_y^3 & B_{79}\partial_x^2 + B_{8,10}\partial_y^2 & B_{99}\partial_x^2 + B_{10,10}\partial_y^2 \end{bmatrix}, \tag{3.63b}$$

$$\mathcal{L}_b^T \boldsymbol{D} \mathcal{L}_b = \begin{bmatrix} D_{11}\partial_x^4 + 2(D_{12}+2D_{66})\partial_x^2\partial_y^2 + D_{22}\partial_y^4 & & \text{symm.} \\ (D_{18}+2D_{67})\partial_x^2\partial_y + D_{28}\partial_y^3 & D_{77}\partial_x^2 + D_{88}\partial_y^2 & \\ (D_{1,10}+2D_{69})\partial_x^2\partial_y + D_{2,10}\partial_y^3 & D_{79}\partial_x^2 + D_{8,10}\partial_y^2 & D_{99}\partial_x^2 + D_{10,10}\partial_y^2 \end{bmatrix}. \tag{3.63c}$$

3.3.2 Navier's Solution

Consider a rectangular orthotropic MEE laminated thin plate whose length and width are L and W. This plate is subjected to the generalized distributed forces q_x, q_y, q_z, q_D, q_B and moments m_D, m_B, with the two sets of simply supported edges described by

$$\begin{aligned} &\text{S1:} \begin{cases} T_x = v_0 = u_E = u_H = 0, \\ w = M_{nn} = \beta_E = \beta_H = 0, \end{cases} \quad \text{along } x = 0, L, \\ &\text{S2:} \begin{cases} u_0 = T_y = N_D = N_B = 0, \\ w = M_{nn} = M_D = M_B = 0, \end{cases} \quad \text{along } y = 0, W, \end{aligned} \tag{3.64}$$

as shown in Figure 3.2. In (3.64),

$$\begin{aligned} &T_x = N_{11}n_1 + N_{12}n_2, \quad T_y = N_{12}n_1 + N_{22}n_2, \\ &N_D = N_{41}n_1 + N_{42}n_2, \; N_B = N_{51}n_1 + N_{52}n_2, \\ &M_{nn} = M_{11}n_1^2 + 2M_{12}n_1n_2 + M_{22}n_2^2, \\ &M_D = M_{41}n_1 + M_{42}n_2, \; M_B = M_{51}n_1 + M_{52}n_2, \end{aligned} \tag{3.65}$$

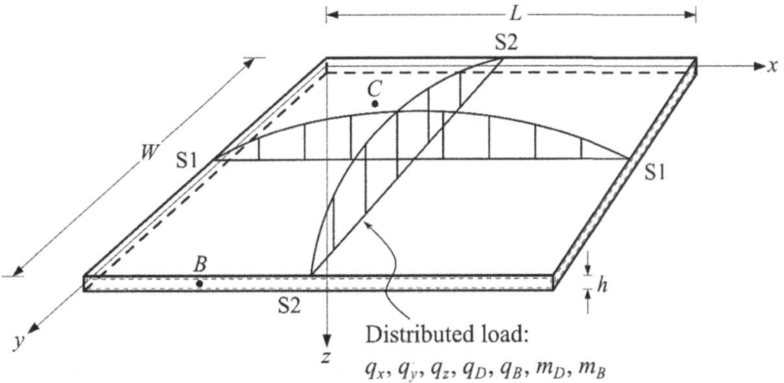

FIGURE 3.2 A rectangular MEE laminated plate simply supported on all edges subjected to the distributed forces/moments on plate surface. (Hsu and Hwu, 2022b)

where n_1, n_2 are the x- and y-components of unit outward normal vector $\mathbf{n}$. Knowing that $\mathbf{n}$ is parallel to x or y axis for the four edges of rectangular plates, with the constitutive relations of (3.59) and (3.62), the boundary conditions of (3.64) can be rewritten in terms of the generalized displacements as

$$\text{S1:} \begin{cases} \partial_x(u_0) = v_0 = u_E = u_H = 0, \\ w = \partial_x^2(w) = \beta_E = \beta_H = 0, \end{cases} \quad \text{along } x = 0, L,$$

$$\text{S2:} \begin{cases} u_0 = \partial_y(v_0) = \partial_y(u_E) = \partial_y(u_H) = 0, \\ w = \partial_y^2(w) = \partial_y(\beta_E) = \partial_y(\beta_H) = 0, \end{cases} \quad \text{along } y = 0, W. \tag{3.66}$$

Express the functions of generalized displacements in the double Fourier series form as

$$u_0(x,y) = \sum_{i=0}^{\infty}\sum_{j=1}^{\infty} \upsilon_{ij}^u \cos(\lambda_i x)\sin(\omega_j y),$$

$$v_0(x,y) = \sum_{i=1}^{\infty}\sum_{j=0}^{\infty} \upsilon_{ij}^v \sin(\lambda_i x)\cos(\omega_j y),$$

$$u_E(x,y) = \sum_{i=1}^{\infty}\sum_{j=0}^{\infty} \upsilon_{ij}^E \sin(\lambda_i x)\cos(\omega_j y),$$

$$u_H(x,y) = \sum_{i=1}^{\infty}\sum_{j=0}^{\infty} \upsilon_{ij}^H \sin(\lambda_i x)\cos(\omega_j y),$$

$$w(x,y) = \sum_{i=1}^{\infty}\sum_{j=1}^{\infty} \upsilon_{ij}^w \sin(\lambda_i x)\sin(\omega_j y),$$

$$\beta_E(x,y) = \sum_{i=1}^{\infty}\sum_{j=0}^{\infty} \beta_{ij}^E \sin(\lambda_i x)\cos(\omega_j y),$$

$$\beta_H(x,y) = \sum_{i=1}^{\infty}\sum_{j=0}^{\infty} \beta_{ij}^H \sin(\lambda_i x)\cos(\omega_j y), \tag{3.67}$$

where $\lambda_i = i\pi/L$, $\omega_j = j\pi/W$, and the constant multipliers of trigonometric functions are to be solved through the satisfaction of governing Equations (3.60) with (3.63). It can be readily found that all the boundary conditions of (3.66) can be automatically satisfied by the expressions (3.67) because $\sin(j\pi)$ is zero for any integer j. Similar to the generalized displacements, the functions of applied distributed loads can also be expressed in the double Fourier series form as

$$
\begin{aligned}
q_x(x,y) &= \sum_{i=0}^{\infty}\sum_{j=1}^{\infty} q_{ij}^{x}\cos(\lambda_i x)\sin(\omega_j y),\\
q_y(x,y) &= \sum_{i=1}^{\infty}\sum_{j=0}^{\infty} q_{ij}^{y}\sin(\lambda_i x)\cos(\omega_j y),\\
q_D(x,y) &= \sum_{i=1}^{\infty}\sum_{j=0}^{\infty} q_{ij}^{D}\sin(\lambda_i x)\cos(\omega_j y),\\
q_B(x,y) &= \sum_{i=1}^{\infty}\sum_{j=0}^{\infty} q_{ij}^{B}\sin(\lambda_i x)\cos(\omega_j y),\\
q_z(x,y) &= \sum_{i=1}^{\infty}\sum_{j=1}^{\infty} q_{ij}^{z}\sin(\lambda_i x)\sin(\omega_j y),\\
m_D(x,y) &= \sum_{i=1}^{\infty}\sum_{j=0}^{\infty} m_{ij}^{D}\sin(\lambda_i x)\cos(\omega_j y),\\
m_B(x,y) &= \sum_{i=1}^{\infty}\sum_{j=0}^{\infty} m_{ij}^{B}\sin(\lambda_i x)\cos(\omega_j y).
\end{aligned}
\tag{3.68}
$$

Unlike the unknown constants in generalized displacements of (3.67), the constants $q_{ij}^{x}, q_{ij}^{y}, q_{ij}^{D}, q_{ij}^{B}, q_{ij}^{z}, m_{ij}^{D}, m_{ij}^{B}$ in (3.68) can be evaluated through the orthogonality of trigonometric functions with the use of prescribed loading functions, for example,

$$
q_{ij}^{x} = \begin{cases} \dfrac{2}{LW}\displaystyle\int_0^W\int_0^L q_x(x,y)\sin(\omega_j y)\,dxdy, & \text{when } i=0,\\[2ex] \dfrac{4}{LW}\displaystyle\int_0^W\int_0^L q_x(x,y)\cos(\lambda_i x)\sin(\omega_j y)\,dxdy, & \text{when } i\neq 0, \end{cases}
\tag{3.69a}
$$

$$
q_{ij}^{y} = \begin{cases} \dfrac{2}{LW}\displaystyle\int_0^W\int_0^L q_y(x,y)\sin(\lambda_i x)\,dxdy, & \text{when } j=0,\\[2ex] \dfrac{4}{LW}\displaystyle\int_0^W\int_0^L q_y(x,y)\sin(\lambda_i x)\cos(\omega_j y)\,dxdy, & \text{when } j\neq 0, \end{cases}
\tag{3.69b}
$$

$$q_{ij}^{z}=\frac{4}{LW}\int_{0}^{W}\int_{0}^{L}q_{z}\left(x,y\right)\sin\left(\lambda_{i}x\right)\sin\left(\omega_{j}y\right)dxdy, \qquad (3.69c)$$

and $q_{ij}^{D}, q_{ij}^{B}, m_{ij}^{D}, m_{ij}^{B}$ can be evaluated in a similar way to (3.69b). Note that since sin(0) = 0, the double series of (3.67) and (3.68) starting from index $i = 1$ (and/or $j = 1$) can be extended to those from $i = 0$ (and/or $j = 0$) with the inclusion of the trivial zero terms. Thus, all the summations can be unified to be from zero to infinity, which is convenient for the later derivation of field solutions.

With the expressions of generalized displacements and applied distributed loads in the double Fourier series form, the system of linear equations for the unknown constants of $\upsilon_{ij}^{u}, \upsilon_{ij}^{v}, \upsilon_{ij}^{E}, \upsilon_{ij}^{H}, \upsilon_{ij}^{w}, \beta_{ij}^{E}, \beta_{ij}^{H}$ can be constructed by substituting (3.63), (3.67), and (3.68) into the governing Equations (3.60), which leads to

$$\begin{bmatrix} \mathbf{A}_{ij}^{*} & \mathbf{B}_{ij}^{*} \\ \left(\mathbf{B}_{ij}^{*}\right)^{T} & \mathbf{D}_{ij}^{*} \end{bmatrix} \begin{Bmatrix} \boldsymbol{\upsilon}_{ij} \\ \boldsymbol{\beta}_{ij} \end{Bmatrix} = \begin{Bmatrix} \mathbf{q}_{ij} \\ \mathbf{m}_{ij} \end{Bmatrix}, \quad i,j=0,1,2,\ldots, \qquad (3.70a)$$

where

$$\mathbf{A}_{ij}^{*}=\begin{bmatrix} A_{11}\lambda_i^2+A_{66}\omega_j^2 & & & \text{symm.} \\ (A_{12}+A_{66})\lambda_i\omega_j & A_{66}\lambda_i^2+A_{22}\omega_j^2 & & \\ (A_{18}+A_{67})\lambda_i\omega_j & A_{67}\lambda_i^2+A_{28}\omega_j^2 & A_{77}\lambda_i^2+A_{88}\omega_j^2 & \\ (A_{1,10}+A_{69})\lambda_i\omega_j & A_{69}\lambda_i^2+A_{2,10}\omega_j^2 & A_{79}\lambda_i^2+A_{8,10}\omega_j^2 & A_{99}\lambda_i^2+A_{10,10}\omega_j^2 \end{bmatrix},$$

$$\mathbf{B}_{ij}^{*}=\begin{bmatrix} B_{11}\lambda_i^3+(B_{12}+2B_{66})\lambda_i\omega_j^2 & (B_{18}+B_{67})\lambda_i\omega_j & (B_{1,10}+B_{69})\lambda_i\omega_j \\ (B_{12}+2B_{66})\lambda_i^2\omega_j+B_{22}\omega_j^3 & B_{67}\lambda_i^2+B_{28}\omega_j^2 & B_{69}\lambda_i^2+B_{2,10}\omega_j^2 \\ (B_{18}+2B_{67})\lambda_i^2\omega_j+B_{28}\omega_j^3 & B_{77}\lambda_i^2+B_{88}\omega_j^2 & B_{79}\lambda_i^2+B_{8,10}\omega_j^2 \\ (B_{1,10}+2B_{69})\lambda_i^2\omega_j+B_{2,10}\omega_j^3 & B_{79}\lambda_i^2+B_{8,10}\omega_j^2 & B_{99}\lambda_i^2+B_{10,10}\omega_j^2 \end{bmatrix},$$

$$\mathbf{D}_{ij}^{*}=\begin{bmatrix} D_{11}\lambda_i^4+2(D_{12}+2D_{66})\lambda_i^2\omega_j^2+D_{22}\omega_j^4 & & \text{symm.} \\ (D_{18}+2D_{67})\lambda_i^2\omega_j+D_{28}\omega_j^3 & D_{77}\lambda_i^2+D_{88}\omega_j^2 & \\ (D_{1,10}+2D_{69})\lambda_i^2\omega_j+D_{2,10}\omega_j^3 & D_{79}\lambda_i^2+D_{8,10}\omega_j^2 & D_{99}\lambda_i^2+D_{10,10}\omega_j^2 \end{bmatrix},$$

(3.70b)

and

$$\boldsymbol{\upsilon}_{ij}=\begin{Bmatrix}\upsilon_{ij}^{u}\\ \upsilon_{ij}^{v}\\ \upsilon_{ij}^{E}\\ \upsilon_{ij}^{H}\end{Bmatrix},\ \boldsymbol{\beta}_{ij}=\begin{Bmatrix}-\upsilon_{ij}^{w}\\ \beta_{ij}^{E}\\ \beta_{ij}^{H}\end{Bmatrix},\ \mathbf{q}_{ij}=\begin{Bmatrix}q_{ij}^{x}\\ q_{ij}^{y}\\ q_{ij}^{D}\\ q_{ij}^{B}\end{Bmatrix},\ \mathbf{m}_{ij}=\begin{Bmatrix}-q_{ij}^{z}\\ m_{ij}^{D}\\ m_{ij}^{B}\end{Bmatrix}. \tag{3.70c}$$

Once the constants in $\boldsymbol{\upsilon}_{ij}$ and $\boldsymbol{\beta}_{ij}$ are solved, the generalized displacement fields can be calculated by (3.67), and the associated generalized mid-plane strains, plate curvatures, stress resultants, and bending moments can then be obtained through the relations of $(3.57)_1$ and (3.59) with the use of (3.62) and differentiations of (3.67).

Remarks on Numerical Calculation

For the general MEE materials, it is common that some elastic stiffness constants are above the order of 10^9, while some magneto-electric coefficients are below the order of 10^{-9}. The thickness of MEE laminates considered is in the order of 10^{-3} m. Hence, it is very possible that the associated generalized extensional stiffness matrix $\boldsymbol{A}$ in SI unit contains the components ranging from 10^{-12} to 10^6, which is numerically ill-conditioned. Similar situation occurs for the generalized coupling and bending stiffness matrices $\boldsymbol{B}$, $\boldsymbol{D}$. Accordingly, the numerical scaling in components of $\boldsymbol{A}$, $\boldsymbol{B}$, $\boldsymbol{D}$ is suggested to avoid ill-conditioned coefficient matrix and to obtain accurate solutions of linear Equations (3.70).

To have a clear explanation of numerical scaling procedure in $\boldsymbol{A}$, $\boldsymbol{B}$, $\boldsymbol{D}$, we split all the quantities according to mechanical and the other (electric and magnetic) parts denoted by the subscripts m, e, respectively. Then, the constitutive relations of $(3.57)_2$ can also be expressed as

$$\begin{Bmatrix}\boldsymbol{N}_m\\ \boldsymbol{N}_e\\ \boldsymbol{M}_m\\ \boldsymbol{M}_e\end{Bmatrix}=\begin{bmatrix}\boldsymbol{A}_{mm} & \boldsymbol{A}_{me} & \boldsymbol{B}_{mm} & \boldsymbol{B}_{me}\\ \boldsymbol{A}_{me}^{T} & \boldsymbol{A}_{ee} & \boldsymbol{B}_{me}^{T} & \boldsymbol{B}_{ee}\\ \boldsymbol{B}_{mm} & \boldsymbol{B}_{me} & \boldsymbol{D}_{mm} & \boldsymbol{D}_{me}\\ \boldsymbol{B}_{me}^{T} & \boldsymbol{B}_{ee} & \boldsymbol{D}_{me}^{T} & \boldsymbol{D}_{ee}\end{bmatrix}\begin{Bmatrix}\boldsymbol{\varepsilon}_m\\ \boldsymbol{\varepsilon}_e\\ \boldsymbol{\kappa}_m\\ \boldsymbol{\kappa}_e\end{Bmatrix}, \tag{3.71a}$$

where

$$\mathbf{N}_m = \begin{Bmatrix} N_{11} \\ N_{22} \\ N_{12} \end{Bmatrix}, \ \mathbf{M}_m = \begin{Bmatrix} M_{11} \\ M_{22} \\ M_{12} \end{Bmatrix}, \ \boldsymbol{\varepsilon}_m = \begin{Bmatrix} \varepsilon_{11}^0 \\ \varepsilon_{22}^0 \\ 2\varepsilon_{12}^0 \end{Bmatrix}, \ \boldsymbol{\kappa}_m = \begin{Bmatrix} \kappa_{11} \\ \kappa_{22} \\ 2\kappa_{12} \end{Bmatrix},$$

$$\mathbf{N}_e = \begin{Bmatrix} N_{41} \\ N_{42} \\ N_{51} \\ N_{52} \end{Bmatrix}, \ \mathbf{M}_e = \begin{Bmatrix} M_{41} \\ M_{42} \\ M_{51} \\ M_{52} \end{Bmatrix}, \ \boldsymbol{\varepsilon}_e = \begin{Bmatrix} 2\varepsilon_{41}^0 \\ 2\varepsilon_{42}^0 \\ 2\varepsilon_{51}^0 \\ 2\varepsilon_{52}^0 \end{Bmatrix}, \ \boldsymbol{\kappa}_e = \begin{Bmatrix} 2\kappa_{41} \\ 2\kappa_{42} \\ 2\kappa_{51} \\ 2\kappa_{52} \end{Bmatrix}, \tag{3.71b}$$

and $\boldsymbol{X}_{mm}$, $\boldsymbol{X}_{me}$, $\boldsymbol{X}_{ee}$, $\boldsymbol{X} = \boldsymbol{A}, \boldsymbol{B}, \boldsymbol{D}$, are the corresponding submatrices of $\boldsymbol{A}$, $\boldsymbol{B}$, $\boldsymbol{D}$.

Based upon the associated numerical scaling procedure of $\boldsymbol{A}$, $\boldsymbol{B}$, $\boldsymbol{D}$ for elastic and electro-elastic laminates presented in Hwu (2021) we now extend such procedure to MEE laminates. That is, we apply the following scaling procedure to (3.71a) and obtain

$$\begin{Bmatrix} \mathbf{N}_m/(E_0h_0) \\ \mathbf{N}_e/h_0 \\ \mathbf{M}_m/(E_0h_0^2) \\ \mathbf{M}_e/h_0^2 \end{Bmatrix} = \begin{bmatrix} \mathbf{A}_{mm}/(E_0h_0) & \mathbf{A}_{me}/h_0 & \mathbf{B}_{mm}/(E_0h_0^2) & \mathbf{B}_{me}/h_0^2 \\ \mathbf{A}_{me}^T/h_0 & \mathbf{A}_{ee}E_0/h_0 & \mathbf{B}_{me}^T/h_0^2 & \mathbf{B}_{ee}E_0/h_0^2 \\ \mathbf{B}_{mm}/(E_0h_0^2) & \mathbf{B}_{me}/h_0^2 & \mathbf{D}_{mm}/(E_0h_0^3) & \mathbf{D}_{me}/h_0^3 \\ \mathbf{B}_{me}^T/h_0^2 & \mathbf{B}_{ee}E_0/h_0^2 & \mathbf{D}_{me}^T/h_0^3 & \mathbf{D}_{ee}E_0/h_0^3 \end{bmatrix} \begin{Bmatrix} \boldsymbol{\varepsilon}_m \\ \boldsymbol{\varepsilon}_e/E_0 \\ \boldsymbol{\kappa}_m h_0 \\ \boldsymbol{\kappa}_e h_0/E_0 \end{Bmatrix}, \tag{3.72}$$

where E_0 and h_0 are scaling factors, and their values are suggested to be

$$E_0 = 10^9, \ \ h_0 = 10^{-3}. \tag{3.73}$$

With the scaled stiffnesses of (3.72), the associated system of linear Equations (3.70) can be rewritten as

$$\begin{bmatrix} \overset{\leftrightarrow}{\mathbf{A}}{}_{ij}^* & \overset{\leftrightarrow}{\mathbf{B}}{}_{ij}^* \\ \left(\overset{\leftrightarrow}{\mathbf{B}}{}_{ij}^*\right)^T & \overset{\leftrightarrow}{\mathbf{D}}{}_{ij}^* \end{bmatrix} \begin{Bmatrix} \overset{\leftrightarrow}{\boldsymbol{\upsilon}}_{ij} \\ \overset{\leftrightarrow}{\boldsymbol{\beta}}_{ij} \end{Bmatrix} = \begin{Bmatrix} \overset{\leftrightarrow}{\mathbf{q}}_{ij} \\ \overset{\leftrightarrow}{\mathbf{m}}_{ij} \end{Bmatrix}, \ \ i, j = 0, 1, 2, \cdots, \tag{3.74a}$$

where

$$\overset{\leftrightarrow}{\mathbf{A}}{}_{ij}^* = \left(\mathbf{P}_{ij}^{(s)}\right)^T \overset{\leftrightarrow}{\mathbf{A}}\mathbf{P}_{ij}^{(s)}, \ \overset{\leftrightarrow}{\mathbf{B}}{}_{ij}^* = \left(\mathbf{P}_{ij}^{(s)}\right)^T \overset{\leftrightarrow}{\mathbf{B}}\mathbf{P}_{ij}^{(b)}, \ \overset{\leftrightarrow}{\mathbf{D}}{}_{ij}^* = \left(\mathbf{P}_{ij}^{(b)}\right)^T \overset{\leftrightarrow}{\mathbf{D}}\mathbf{P}_{ij}^{(b)}, \tag{3.74b}$$

$$\vec{\boldsymbol{\upsilon}}_{ij}=\begin{Bmatrix}\upsilon_{ij}^{u}\\ \upsilon_{ij}^{v}\\ \upsilon_{ij}^{E}/E_0\\ \upsilon_{ij}^{H}/E_0\end{Bmatrix},\quad \vec{\boldsymbol{\beta}}_{ij}=\begin{Bmatrix}-\upsilon_{ij}^{w}h_0\\ \beta_{ij}^{E}h_0/E_0\\ \beta_{ij}^{H}h_0/E_0\end{Bmatrix},$$

$$\vec{\mathbf{q}}_{ij}=\begin{Bmatrix}q_{ij}^{x}/(E_0h_0)\\ q_{ij}^{y}/(E_0h_0)\\ q_{ij}^{D}/h_0\\ q_{ij}^{B}/h_0\end{Bmatrix},\quad \vec{\mathbf{m}}_{ij}=\begin{Bmatrix}-q_{ij}^{z}/(E_0h_0^2)\\ m_{ij}^{D}/h_0^2\\ m_{ij}^{B}/h_0^2\end{Bmatrix}. \tag{3.74c}$$

In (3.74b), $\vec{\boldsymbol{A}},\vec{\boldsymbol{B}},\vec{\boldsymbol{D}}$ are the scaled $\boldsymbol{A}$, $\boldsymbol{B}$, $\boldsymbol{D}$ as shown in (3.72), and the matrices $\mathbf{P}_{ij}^{(s)}$ and $\mathbf{P}_{ij}^{(b)}$ of (3.74b) are the corresponding matrices of the operators $\boldsymbol{\mathcal{L}}_s$ and $\boldsymbol{\mathcal{L}}_b$ of (3.58b) and are

$$\mathbf{P}_{ij}^{(s)}=\begin{bmatrix}\lambda_i & 0 & 0 & 0\\ 0 & \omega_j & 0 & 0\\ \omega_j & \lambda_i & 0 & 0\\ 0 & 0 & \lambda_i & 0\\ 0 & 0 & \omega_j & 0\\ 0 & 0 & 0 & \lambda_i\\ 0 & 0 & 0 & \omega_j\end{bmatrix},\quad \mathbf{P}_{ij}^{(b)}=\begin{bmatrix}\lambda_i^2 & 0 & 0\\ \omega_j^2 & 0 & 0\\ 2\lambda_i\omega_j & 0 & 0\\ 0 & \lambda_i & 0\\ 0 & \omega_j & 0\\ 0 & 0 & \lambda_i\\ 0 & 0 & \omega_j\end{bmatrix}. \tag{3.75}$$

By solving the scaled linear Equations (3.74a), the trouble of ill-conditioned matrix can be avoided, and through (3.74c) the original numerical values of constants $\boldsymbol{\upsilon}_{ij}$ and $\boldsymbol{\beta}_{ij}$ in (3.70c) can be returned from the scaled ones.

3.3.3 Levy's Solution

Consider a rectangular orthotropic MEE laminated thin plate whose length and width are L and W. This plate is subjected to the generalized distributed forces q_x, q_y, q_z, q_D, q_B and moments m_D, m_B, with the edges $x = 0$ and $x = L$ being simply supported as described in $(3.66)_1$. Unlike Navier's solution in previous section where the top and bottom edges are restricted to be simply supported (S2 conditions defined in (3.64)), here the boundary conditions along the top edge $y = -W/2$ and the bottom edge $y = W/2$ can be any combination of following conditions (Note that here we have moved the x-axis from the top edge to the center line of the rectangular plate of Figure 3.2):

$$\begin{aligned}
u_0 &= 0, \quad \text{or} \quad T_x = 0, \\
v_0 &= 0, \quad \text{or} \quad T_y = 0, \\
u_E &= 0, \quad \text{or} \quad N_D = 0, \\
u_H &= 0, \quad \text{or} \quad N_B = 0, \\
w &= 0, \quad \text{or} \quad V_n = 0, \\
\beta_n &= 0, \quad \text{or} \quad M_{nn} = 0, \\
\beta_E &= 0, \quad \text{or} \quad M_D = 0, \\
\beta_H &= 0, \quad \text{or} \quad M_B = 0.
\end{aligned} \tag{3.76}$$

In the above, $\beta_n = -\partial w/\partial n$ is the negative plate slope along normal direction, and V_n is the effective transverse shear force related to the bending moments and transverse shear force by (Hwu, 2010)

$$\begin{aligned}
V_n &= Q_n + \frac{\partial M_{sn}}{\partial s}, \quad Q_n = Q_1 n_1 + Q_2 n_2, \\
M_{sn} &= n_1 n_2 \left(M_{11} - M_{22}\right) + \left(n_2^2 - n_1^2\right) M_{12},
\end{aligned} \tag{3.77a}$$

where s is the tangential direction, n_1, n_2 have been explained after (3.65), and

$$Q_1 = \frac{\partial M_{11}}{\partial x} + \frac{\partial M_{12}}{\partial y}, \quad Q_2 = \frac{\partial M_{12}}{\partial x} + \frac{\partial M_{22}}{\partial y}. \tag{3.77b}$$

Express the functions of generalized displacements in the Fourier series as

$$\begin{aligned}
&u_0(x,y) = \sum_{i=0}^{\infty} \upsilon_i^u(y)\cos(\lambda_i x), \quad v_0(x,y) = \sum_{i=1}^{\infty} \upsilon_i^v(y)\sin(\lambda_i x), \\
&u_E(x,y) = \sum_{i=1}^{\infty} \upsilon_i^E(y)\sin(\lambda_i x), \quad u_H(x,y) = \sum_{i=1}^{\infty} \upsilon_i^H(y)\sin(\lambda_i x), \\
&w(x,y) = \sum_{i=1}^{\infty} \upsilon_i^w(y)\sin(\lambda_i x), \\
&\beta_E(x,y) = \sum_{i=1}^{\infty} \beta_i^E(y)\sin(\lambda_i x), \quad \beta_H(x,y) = \sum_{i=1}^{\infty} \beta_i^H(y)\sin(\lambda_i x),
\end{aligned} \tag{3.78}$$

where $\lambda_i = i\pi/L$ and $\upsilon_i^u, \upsilon_i^v, \upsilon_i^E, \upsilon_i^H, \upsilon_i^w, \beta_i^E, \beta_i^H$ are unknown functions to be solved through the satisfaction of governing Equations (3.60). With (3.78), the boundary conditions of $(3.66)_1$ along $x = 0$ and $x = L$ can be satisfied automatically. Similarly, the loading can be expressed in the consistent series form as

$$
\begin{aligned}
&q_x(x,y) = \sum_{i=0}^{\infty} q_i^x(y)\cos(\lambda_i x), \quad q_y(x,y) = \sum_{i=1}^{\infty} q_i^y(y)\sin(\lambda_i x), \\
&q_D(x,y) = \sum_{i=1}^{\infty} q_i^D(y)\sin(\lambda_i x), \quad q_B(x,y) = \sum_{i=1}^{\infty} q_i^B(y)\sin(\lambda_i x), \\
&q_z(x,y) = \sum_{i=1}^{\infty} q_i^z(y)\sin(\lambda_i x), \\
&m_D(x,y) = \sum_{i=1}^{\infty} m_i^D(y)\sin(\lambda_i x), \quad m_B(x,y) = \sum_{i=1}^{\infty} m_i^B(y)\sin(\lambda_i x),
\end{aligned}
\tag{3.79a}
$$

where

$$
q_i^x(y) = \begin{cases} \dfrac{1}{L}\displaystyle\int_0^L q_x(x,y)\,dx, & \text{when } i = 0, \\[2ex] \dfrac{2}{L}\displaystyle\int_0^L q_x(x,y)\cos(\lambda_i x)\,dx, & \text{when } i \neq 0, \end{cases}
\tag{3.79b}
$$

$$
q_i^y(y) = \frac{2}{L}\int_0^L q_y(x,y)\sin(\lambda_i x)\,dx,
\tag{3.79c}
$$

and $q_i^D, q_i^B, q_i^z, m_i^D, m_i^B$ can be evaluated in a similar way to (3.79c).

After expressing the generalized displacements and the applied distributed loads in series form, substitution of (3.78) and (3.79) into the governing Equations (3.60) with (3.63) gives us the following ordinary differential equations (ODEs) for the unknown functions $\upsilon_i^u, \upsilon_i^v, \upsilon_i^E, \upsilon_i^H, \upsilon_i^w, \beta_i^E, \beta_i^H$

$$
\begin{bmatrix} \mathcal{L}_i^{(A)} & \mathcal{L}_i^{(B)} \\ \left(\mathcal{L}_i^{(B)}\right)^T & \mathcal{L}_i^{(D)} \end{bmatrix}
\begin{Bmatrix} \boldsymbol{\upsilon}_i \\ \boldsymbol{\beta}_i \end{Bmatrix} = \begin{Bmatrix} \mathbf{q}_i \\ \mathbf{m}_i \end{Bmatrix}, \quad i = 0,1,2,\cdots,
\tag{3.80a}
$$

in which

$$\boldsymbol{\upsilon}_i = \begin{Bmatrix} \upsilon_i^u \\ \upsilon_i^v \\ \upsilon_i^E \\ \upsilon_i^H \end{Bmatrix}, \quad \boldsymbol{\beta}_i = \begin{Bmatrix} -\upsilon_i^w \\ \beta_i^E \\ \beta_i^H \end{Bmatrix}, \quad \mathbf{q}_i = \begin{Bmatrix} -q_i^x \\ q_i^y \\ q_i^D \\ q_i^B \end{Bmatrix}, \quad \mathbf{m}_i = \begin{Bmatrix} q_i^z \\ m_i^D \\ m_i^B \end{Bmatrix}, \tag{3.80b}$$

are function vectors of variable y, and $\mathcal{L}_i^{(A)}, \mathcal{L}_i^{(B)}, \mathcal{L}_i^{(D)}$ are differential operators defined by

$$\mathcal{L}_i^{(A)} = \begin{bmatrix} -A_{11}\lambda_i^2 + A_{66}\partial_y^2 & & & \text{symm.} \\ (A_{12} + A_{66})\lambda_i\partial_y & A_{66}\lambda_i^2 - A_{22}\partial_y^2 & & \\ (A_{18} + A_{67})\lambda_i\partial_y & A_{67}\lambda_i^2 - A_{28}\partial_y^2 & A_{77}\lambda_i^2 - A_{88}\partial_y^2 & \\ (A_{1,10} + A_{69})\lambda_i\partial_y & A_{69}\lambda_i^2 - A_{2,10}\partial_y^2 & A_{79}\lambda_i^2 - A_{8,10}\partial_y^2 & A_{99}\lambda_i^2 - A_{10,10}\partial_y^2 \end{bmatrix}, \tag{3.81a}$$

$$\mathcal{L}_i^{(B)} = \begin{bmatrix} -B_{11}\lambda_i^3 + (B_{12} + 2B_{66})\lambda_i\partial_y^2 & (B_{18} + B_{67})\lambda_i\partial_y & (B_{1,10} + B_{69})\lambda_i\partial_y \\ (B_{12} + 2B_{66})\lambda_i^2\partial_y - B_{22}\partial_y^3 & B_{67}\lambda_i^2 - B_{28}\partial_y^2 & B_{69}\lambda_i^2 - B_{2,10}\partial_y^2 \\ (B_{18} + 2B_{67})\lambda_i^2\partial_y - B_{28}\partial_y^3 & B_{77}\lambda_i^2 - B_{88}\partial_y^2 & B_{79}\lambda_i^2 - B_{8,10}\partial_y^2 \\ (B_{1,10} + 2B_{69})\lambda_i^2\partial_y - B_{2,10}\partial_y^3 & B_{79}\lambda_i^2 - B_{8,10}\partial_y^2 & B_{99}\lambda_i^2 - B_{10,10}\partial_y^2 \end{bmatrix}, \tag{3.81b}$$

$$\mathcal{L}_i^{(D)} = \begin{bmatrix} -D_{11}\lambda_i^4 + 2(D_{12} + 2D_{66})\lambda_i^2\partial_y^2 - D_{22}\partial_y^4 & & \text{symm.} \\ (D_{18} + 2D_{67})\lambda_i^2\partial_y - D_{28}\partial_y^3 & D_{77}\lambda_i^2 - D_{88}\partial_y^2 & \\ (D_{1,10} + 2D_{69})\lambda_i^2\partial_y - D_{2,10}\partial_y^3 & D_{79}\lambda_i^2 - D_{8,10}\partial_y^2 & D_{99}\lambda_i^2 - D_{10,10}\partial_y^2 \end{bmatrix}. \tag{3.81c}$$

Equations (3.80) are the system of fourth-order ODE, which can be reorganized into the following first-order ODE,

$$\mathbf{f}_i' = \mathbf{J}_i\mathbf{f}_i + \mathbf{r}_i \quad (\text{no sum on } i), \quad i = 0, 1, 2, \cdots, \tag{3.82}$$

where the prime •′ denotes the derivative with respect to y, and $\mathbf{f}_i$ are the 16×1 unknown function vectors defined by

$$\mathbf{f}_i = \begin{Bmatrix} \mathbf{w}_i \\ \mathbf{w}_i' \end{Bmatrix}, \quad \mathbf{w}_i = \left[\upsilon_i^u, \upsilon_i^v, \upsilon_i^E, \upsilon_i^H, -\upsilon_i^w, -\left(\upsilon_i^w\right)'', \beta_i^E, \beta_i^H \right]^T. \tag{3.83}$$

$\mathbf{J}_i$ are 16×16 constant matrices independent of y, and $\mathbf{r}_i = \mathbf{r}_i(y)$ are the 16×1 function vectors related to $\mathbf{q}_i$ and $\mathbf{m}_i$ of (3.80b). The derivation and details of $\mathbf{J}_i$ and $\mathbf{r}_i$ can be found in Hsu and Hwu (2022b). The solution to (3.82) can be written in matrix form as (Franklin, 1968)

$$\mathbf{f}_i(y) = e^{\mathbf{J}_i y}\left\{ \mathbf{d}_i + \int e^{-\mathbf{J}_i y}\mathbf{r}_i(y)dy \right\}, \tag{3.84}$$

where $\mathbf{d}_i$ is a constant vector to be determined through the satisfaction of boundary conditions set in (3.76) along $y = \pm W/2$. Note that in (3.84) an indefinite integral is used, which can be evaluated numerically by using standard Gaussian quadrature rule with an assigned lower limit. This limit can be assigned arbitrarily, which serves as an integration constant and will influence the result of $\mathbf{d}_i$ but not affect the final result of $\mathbf{f}_i$. For simple loading conditions such as constant loads or sinusoidal loads, the integral in (3.84) can be evaluated analytically and no numerical integration is involved.

Remarks on Numerical Calculation

When we perform numerical calculation by using the solutions obtained in (3.84), we found that the coefficient matrix $\mathbf{J}_i$ will become ill-conditioned when the index i increases to a large number such as $i = 10$. Therefore, its corresponding matrix exponential $e^{\mathbf{J}_i y}$ evaluated directly by the built-in function *expm* of Matlab usually provides inaccurate solutions. To avoid such trouble induced by the inaccurate results of matrix exponential, we suggest the evaluation by matrix decomposition as

$$\mathbf{J}_i = \mathbf{P}_i\mathbf{\Lambda}_i\mathbf{P}_i^{-1}, \quad e^{\mathbf{J}_i y} = \mathbf{P}_i e^{\mathbf{\Lambda}_i y}\mathbf{P}_i^{-1}, \tag{3.85}$$

in which $\mathbf{\Lambda}_i = diag\begin{bmatrix} \eta_1 & \eta_2 & \cdots & \eta_{16} \end{bmatrix}$ is the diagonal matrix whose components η_k, $k = 1, 2, \cdots, 16$, are eigenvalues of $\mathbf{J}_i$, and $\mathbf{P}_i$ is the corresponding eigenvector matrix. Note that the eigenvalues and eigenvectors of real matrix $\mathbf{J}_i$ may be complex. Substituting (3.85) into (3.84), we get a solution form which is better for numerical calculation, that is,

$$\mathbf{f}_i\left(y\right)=\mathbf{P}_i e^{\Lambda_i y}\left\{\mathbf{d}_i^*+\int e^{-\Lambda_i y}\mathbf{r}_i^*\left(y\right)dy\right\}, \tag{3.86a}$$

where

$$\mathbf{d}_i^*=\mathbf{P}_i^{-1}\mathbf{d}_i, \quad \mathbf{r}_i^*=\mathbf{P}_i^{-1}\mathbf{r}_i. \tag{3.86b}$$

To solve $\mathbf{d}_i^*$ of (3.86a), we apply the boundary conditions set in (3.76) for $y = \pm W/2$, which will lead to 16 linear algebraic equations for each index i (the ith term of the Fourier series of (3.78)). Based upon the series of (3.78), with the aid of the constitutive laws (3.59) and (3.62) and the relation (3.77), the quantities in (3.76) can be expressed in the associated Fourier series related to the functions covered in $\mathbf{f}_i$ and $\mathbf{f}_i'$. The collection of 16 boundary conditions (3.76) along $y = \pm W/2$ (8 conditions along each edge) would lead to the boundary-valued relations of $\mathbf{f}_i$ and $\mathbf{f}_i'$ for each index i, which can be symbolically expressed as

$$\mathbf{G}_i^+\mathbf{f}_i'^+ + \mathbf{H}_i^+\mathbf{f}_i^+ = \mathbf{0}, \quad \mathbf{G}_i^-\mathbf{f}_i'^- + \mathbf{H}_i^-\mathbf{f}_i^- = \mathbf{0}, \tag{3.87}$$

where the superscripts + and – denote, respectively, the values related to the boundaries $y = W/2$ and $y = -W/2$, and $\mathbf{G}_i$ and $\mathbf{H}_i$ are two 8×16 coefficient matrices dependent on the prescribed conditions of (3.76), laminate stiffness constants $\boldsymbol{A}$, $\boldsymbol{B}$, $\boldsymbol{D}$, and values of λ_i. By substituting (3.82) and (3.86a) into (3.87), the 16 linear algebraic equations for $\mathbf{d}_i^*$ can then be obtained as

$$\mathbf{K}_i^+ e^{\Lambda_i W/2}\mathbf{d}_i^* = \boldsymbol{\rho}_i^+, \quad \mathbf{K}_i^- e^{-\Lambda_i W/2}\mathbf{d}_i^* = \boldsymbol{\rho}_i^-, \tag{3.88a}$$

where

$$\begin{aligned}&\mathbf{K}_i^\pm = [\mathbf{G}_i^\pm\mathbf{J}_i + \mathbf{H}_i^\pm]\mathbf{P}_i, \quad \boldsymbol{\rho}_i^\pm = -\mathbf{K}_i^\pm e^{\pm\Lambda_i W/2}\tilde{\mathbf{r}}_i^\pm - \mathbf{G}_i^\pm\mathbf{r}_i^\pm,\\ &\tilde{\mathbf{r}}_i(y) = \int e^{-\Lambda_i y}\mathbf{r}_i^*(y)dy.\end{aligned} \tag{3.88b}$$

After performing the matrix multiplication for $\mathbf{K}_i^\pm e^{\pm\Lambda_i W/2}$, the linear Equations (3.88a) can be organized as

$$\mathbf{K}_i^e\mathbf{d}_i^* = \boldsymbol{\rho}_i, \tag{3.89a}$$

where

$$\mathbf{K}_i^e = \begin{bmatrix} k_{11}^+ e^{\eta_1 W/2} & k_{12}^+ e^{\eta_2 W/2} & \cdots & k_{1\ell}^+ e^{\eta_\ell W/2} & \cdots & k_{1,16}^+ e^{\eta_{16} W/2} \\ k_{21}^+ e^{\eta_1 W/2} & k_{22}^+ e^{\eta_2 W/2} & \cdots & k_{2\ell}^+ e^{\eta_\ell W/2} & \cdots & k_{2,16}^+ e^{\eta_{16} W/2} \\ \vdots & \vdots & \cdots & \vdots & \cdots & \vdots \\ k_{81}^+ e^{\eta_1 W/2} & k_{82}^+ e^{\eta_2 W/2} & \cdots & k_{8\ell}^+ e^{\eta_\ell W/2} & \cdots & k_{8,16}^+ e^{\eta_{16} W/2} \\ k_{11}^- e^{-\eta_1 W/2} & k_{12}^- e^{-\eta_2 W/2} & \cdots & k_{1\ell}^- e^{-\eta_\ell W/2} & \cdots & k_{1,16}^- e^{-\eta_{16} W/2} \\ \vdots & \vdots & \cdots & \vdots & \cdots & \vdots \\ k_{81}^- e^{-\eta_1 W/2} & k_{82}^- e^{-\eta_2 W/2} & \cdots & k_{8\ell}^- e^{-\eta_\ell W/2} & \cdots & k_{8,16}^- e^{-\eta_{16} W/2} \end{bmatrix}_i,$$

$$\mathbf{d}_i^* = \begin{Bmatrix} d_1^* \\ d_2^* \\ \vdots \\ d_8^* \\ d_9^* \\ \vdots \\ d_{16}^* \end{Bmatrix}_i, \quad \boldsymbol{\rho}_i = \begin{Bmatrix} \boldsymbol{\rho}_i^+ \\ \boldsymbol{\rho}_i^- \end{Bmatrix}. \tag{3.89b}$$

In (3.89b), $k_{j\ell}^\pm, j = 1,2,\cdots,8, \ell = 1,2\cdots,16$, of $\mathbf{K}_i^e$ are the components of $\mathbf{K}_i^\pm$ given in (3.88b). As the index i increases, due to the matrix exponential operation, the ill-conditioned matrix $\mathbf{K}_i^e$ is unavoidable even if the scaling procedure for generalized extensional, coupling, and bending matrices $\boldsymbol{A}$, $\boldsymbol{B}$, $\boldsymbol{D}$ stated at the end of Section 3.3.2 is performed. As suggested in Nosier and Reddy (1992) for the buckling and free vibration analyses of elastic laminated shell, to avoid the ill-condition here the scaling of $\mathbf{K}_i^e$ is made as follows:

$$\text{If } N_\ell > N, \text{ then} \begin{cases} \overleftrightarrow{k}_{j\ell}^+ = k_{j\ell}^+ / e^{N_\ell}, \\ \overleftrightarrow{k}_{j\ell}^- = 0, \end{cases} \text{ and } \overleftrightarrow{d}_\ell^* = d_\ell^* e^{N_\ell};$$

$$\text{If } N_\ell < -N, \text{then} \begin{cases} \overleftrightarrow{k}_{j\ell}^+ = 0, \\ \overleftrightarrow{k}_{j\ell}^- = k_{j\ell}^- / e^{-N_\ell}, \end{cases} \text{ and } \overleftrightarrow{d}_\ell^* = d_\ell^* e^{-N_\ell}. \tag{3.90}$$

In (3.90), $N_\ell = \text{Re}\{\eta_\ell W/2\}$, and N is a large positive number, which is set to be 20 in the study of Hsu and Hwu (2022b).

Once the function vector $\mathbf{f}_i(y)$ is obtained through (3.86a), the generalized displacement fields can be calculated by (3.78), and the associated generalized mid-plane strains, plate curvatures, stress resultants, and bending moments can then be obtained through the relations of $(3.57)_1$ and (3.59).

3.3.4 Numerical Illustration

For Navier's solution (NS) and Levy's solution (LS) presented in Sections 3.3.2 and 3.3.3, the MEE laminated plates are restricted to rectangular shape and are made up of specially orthotropic layers only, and the boundary conditions are limited to simply supported on all edges or on two opposite edges. To solve the problems of MEE laminated plates with generally anisotropic layers, arbitrary geometries, loadings, and supporting conditions, we can only count on the numerical methods such as finite element method (FEM) and boundary element method (BEM), both of which will be presented later in Chapter 9. Due to the coupling among magnetic, electro and elastic behaviors as well as the coupling between stretching and bending deformation, till now many available FEM and BEM still have some restrictions on their applications. For the MEE laminated plates under coupled stretching–bending (CSB) deformation, to develop a proper BEM we may employ the associated boundary integral equations for the general laminated plate presented later in (9.43) simply by extending the range of their subscripts to include the contents of electro-magnetic quantities (Hsu and Hwu, 2022a,b).

In the following examples, all the laminated plates are six-layered, and each layer is 1 mm thick and is made up of the following three materials: (i) graphite/epoxy fiber-reinforced composite lamina, (ii) piezoelectric material PZT-4, and (iii) MEE material made by mixing 50% $BaTiO_3$ and 50% $CoFe_2O_4$. For these materials, the nonzero material constants in SI unit are listed as follows:

i. Graphite/epoxy composite lamina

$$
\begin{aligned}
&E_L = 172.37\,\text{GPa},\ E_T = E_z = 6.895\,\text{GPa},\\
&G_{LT} = G_{Lz} = 3.447\,\text{GPa},\ G_{Tz} = 1.379\,\text{GPa},\\
&\nu_{LT} = \nu_{Lz} = 0.25.
\end{aligned}
\tag{3.91}
$$

ii. Piezoelectric material PZT-4

$$\begin{aligned}
&C_{11} = C_{33} = 139\,\text{GPa},\ C_{22} = 115\,\text{GPa},\\
&C_{13} = 77.8\,\text{GPa},\ C_{12} = C_{23} = 74.3\,\text{GPa},\\
&C_{44} = C_{66} = 25.6\,\text{GPa},\ C_{55} = 30.6\,\text{GPa},\\
&e_{21} = e_{23} = -5.2\,\text{C/m}^2,\ e_{22} = 15.1\,\text{C/m}^2,\ e_{16} = e_{34} = 12.7\,\text{C/m}^2,\\
&\omega_{11} = \omega_{33} = 6.46\times 10^{-9}\,\text{C}^2/\text{Nm}^2,\ \omega_{22} = 5.62\times 10^{-9}\,\text{C}^2/\text{Nm}^2.
\end{aligned} \tag{3.92}$$

iii. Magneto-electro-elastic material $BaTiO_3$-$CoFe_2O_4$

$$\begin{aligned}
&C_{11} = C_{33} = 226\ \text{GPa},\ C_{22} = 216\ \text{GPa},\\
&C_{13} = 125\ \text{GPa},\ C_{12} = C_{23} = 124\ \text{GPa},\\
&C_{44} = C_{66} = 44.2\ \text{GPa},\ C_{55} = 50.5\ \text{GPa},\\
&e_{21} = e_{23} = -2.2\ \text{C/m}^2,\ e_{22} = 9.3\ \text{C/m}^2,\ e_{16} = e_{34} = 5.8\ \text{C/m}^2,\\
&\omega_{11} = \omega_{33} = 5.64\times 10^{-9}\ \text{C}^2/\text{Nm}^2,\ \omega_{22} = 6.35\times 10^{-9}\ \text{C}^2/\text{Nm}^2,\\
&q_{21} = q_{23} = 290.2\ \text{N/Am},\ q_{22} = 350\ \text{N/Am},\ q_{16} = q_{34} = 275\ \text{N/Am},\\
&m_{11} = m_{33} = 5.37\times 10^{-12}\ \text{Ns/VC},\ m_{22} = 2.74\times 10^{-9}\ \text{Ns/VC},\\
&\xi_{11} = \xi_{33} = 297\times 10^{-6}\ \text{Ns}^2/\text{C}^2,\ \xi_{22} = 83.5\times 10^{-6}\ \text{Ns}^2/\text{C}^2.
\end{aligned} \tag{3.93}$$

In (3.91), E, G, and v stand for Young's modulus, shear modulus, and Poisson's ratio, and the subscripts L and T represent the longitudinal and transverse directions with respect to fiber orientation. In (3.92) and (3.93), C_{ij}, ω_{pq}, ξ_{pq}, e_{pj}, q_{pj}, m_{pq}, $i, j = 1, 2, \cdots, 6$, $p, q = 1, 2, 3$, are, respectively, elastic stiffness, permittivity, permeability, piezoelectric, piezomagnetic, and magneto-electric coefficients. With these constants, the generalized extensional, coupling, and bending stiffness matrices $\boldsymbol{A}$, $\boldsymbol{B}$, $\boldsymbol{D}$ can be evaluated as shown in (3.53).

To identify these materials in the expression of stacking sequence of laminated plate, the fiber orientation is used to specify the layers made up of Material (i), while the symbols "P" and "M" are used to represent Material (ii) and (iii) with the material principal direction aligned with the global coordinate. For example, $[M/0_2/90_2/P]$ represents the laminated plate whose layers from top to bottom are Material (iii), Material (i) with fiber orientation being 0°, 0°, 90°, and 90°, and Material (ii).

The accuracy of NS and LS will depend on the truncated series we select, and that of BEM will depend on the meshes we discretize and the integration we approximate. Therefore, before we present the numerical solutions, the standard convergence test has been done for NS, LS, and BEM. Through the convergence test, we select the first 35 and 20 terms for the truncated infinite series of NS and LS, respectively. If the same fully simply supported conditions are considered, usually LS converges faster than NS. Hence, less series terms are needed for LS. As to BEM, the integration is performed by using Gaussian quadrature rule with 64 Gaussian points, and its discretization is made with 10 equidistant elements on each edge (totally 40 boundary elements on the entire outer boundary).

EXAMPLE 1: A MEE LAMINATED PLATE WITH ALL EDGES SIMPLY SUPPORTED UNDER UNIFORMLY DISTRIBUTED IN-PLANE AND OUT-OF-PLANE FORCES

Consider a $[M/0_2/90_2/P]$ MEE unsymmetric laminated plate with all edges simply supported under two different uniformly distributed loads separately, $q_x = 1\,\text{kPa}$ and $q_z = 1\,\text{kPa}$. The length and width of the plate are $L = 1\,\text{m}$ and $W = 0.8L$. Table 3.1 shows the results of displacements, generalized

TABLE 3.1 Displacements, Generalized Stress Resultants and Bending Moments at Boundary Point B and Internal Point C of a $[M/0_2/90_2/P]$ Laminated Plate with All Edges Simply Supported Subjected to a Uniformly Distributed Force

	Point *B*			Point *C*		
	NS	**LS**	**BEM**	**NS**	**LS**	**BEM**
In-plane force q_x = 1kPa						
u_0(nm)	0	0	0	717.8	717.8	705.1
$N_{12}h$(mN)	−2373	−2400	−2336	1200	1200	1206
M_{12}(mN)	220	222.5	216.3	−111.3	−111.3	−112.8
$N_{41}h$(pC)	−525.2	−531.2	−516.8	265.6	265.6	267
M_{41}(pC)	−81.62	−82.55	−80.31	41.28	41.27	41.48
$N_{51}h$(nV/s)	−7.807	−7.896	−7.669	3.948	3.948	3.971
M_{51}(nV/s)	3.253	3.290	3.196	−1.645	−1.645	−1.654
Out-of-plane force q_z = 1kPa						
w(mm)	0	0	0	0.678	0.678	0.676
$N_{12}h$(N)	1.638	1.638	1.644	−1	−1	−0.999
M_{12}(N)	13.05	13.05	12.94	−9.036	−9.036	−9.016
$N_{42}h$(pC)	0	0	0	−923.6	−923.7	−917.9
M_{42}(pC)	0	0	0	−369.2	−369.3	−366.7
$N_{52}h$(nV/s)	0	0	0	−16.24	−16.24	−16.16
M_{52}(nV/s)	0	0	0	6.845	6.848	6.815

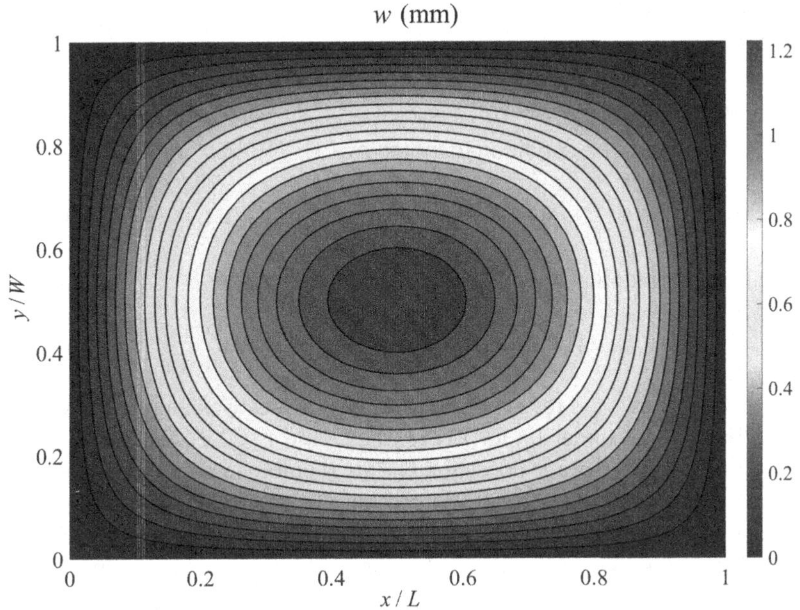

FIGURE 3.3 Deflection contour of [M/0_2/90_2/P] laminated plates under uniform transverse load q_z = 1kPa.

stress resultants, and bending moments at the boundary point B and internal point C, where B and C are located at (L/4, W) and (L/4, W/4), respectively (see Figure 3.2). From this table, we see that the results of NS, LS, and BEM are almost the same. The S2 boundary conditions described in (3.64) at boundary point B are satisfied exactly. Moreover, the coupling effects can be observed by the nonzero values of M_{12}, M_{41}, and M_{51} under in-plane load, and the nonzero values of N_{12}, N_{42}, and N_{52} under transverse load. Figure 3.3 shows the contour plots of deflection w under the transverse load q_z = 1kPa obtained by NS. In this figure, the contour curves of deflections encircle the plate center, which agrees with our engineering intuition.

EXAMPLE 2: A MEE LAMINATED PLATE SIMPLY SUPPORTED ON TWO OPPOSITE EDGES UNDER SINUSOIDAL ELECTRIC LOADS

Consider a square [M/0_2/90_2/P] unsymmetric MEE laminated plate under the following sinusoidal electric load,

$$q_D(x,y) = \sin(2\pi x) \ \mu C/m^2. \tag{3.94}$$

The length and width of the plate are $L = W = 1\text{m}$. The edges along $x = 0$ and $x = L$ are simply supported as described in (3.64) (see Figure 3.2), whereas the following two kinds of boundary conditions are enforced on the other two edges $y = \pm W/2$:

i. First-kind free condition (denoted by F1):

$$T_x = T_y = N_D = N_B = V_n = M_{nn} = M_D = M_B = 0; \qquad (3.95a)$$

ii. Second-kind clamped condition (denoted by C2):

$$u_0 = v_0 = u_E = u_H = w = \beta_n = \beta_E = \beta_H = 0. \qquad (3.95b)$$

Figure 3.4 shows the results of M_{51} and N_{41}, respectively, along the vertical line $x = L/2$ under q_D for F1F1, C2C2, and F1C2 conditions, where F1C2 means that the free condition (3.95a) is enforced on $y = W/2$ and the clamped condition (3.95b) is enforced on $y = -W/2$, etc. From Figure 3.4 we see that the solutions obtained by LS and BEM well agree with each other for these three different combinations of boundary conditions.

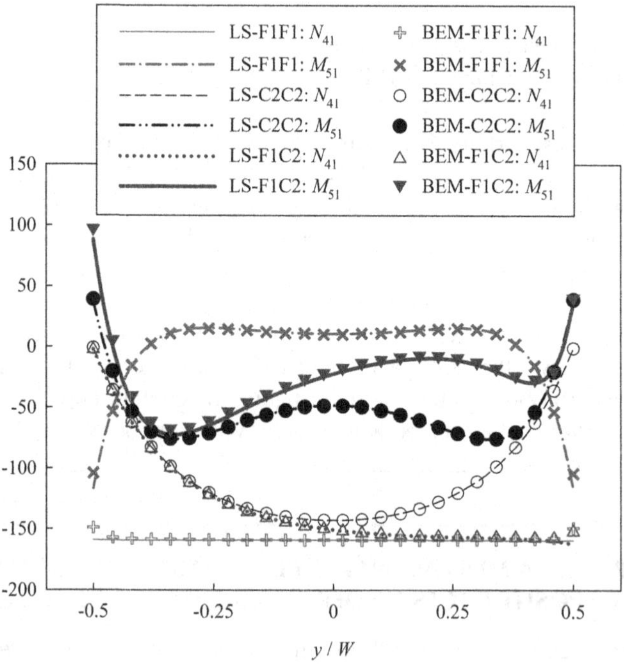

FIGURE 3.4 Electric resultant and magnetic moment along $x = L/2$ of $[M/0_2/90_2/P]$ laminated plate under electric charge $q_D = \sin(2\pi x)(\mu C/m^2)$ (unit: nC/m for N_{41}; pV/s for M_{51}).

3.4 THICK LAMINATED PLATES

As mentioned in Section 3.1.2 if a thick plate is considered, the transverse shear deformation cannot be ignored. In this situation, the first-order shear deformation theory (FSDT) whose displacement fields are assumed in the form of (3.17) is an important theory for the description of the mechanical behavior of thick laminated plates. Based upon this theory, recently we formulated the governing equation of thick laminated plates in matrix form and solved the Green's function explicitly (Hsu and Hwu, 2023). Portions of this section are therefore extracted from our recent work (Hsu and Hwu, 2023).

3.4.1 General Formulation

Under the assumption of FSDT, the displacement field of composite laminated plates can be expressed as

$$\begin{aligned} u(x,y,z) &= u_0(x,y) + z\beta_x(x,y), \\ v(x,y,z) &= v_0(x,y) + z\beta_y(x,y), \\ w(x,y,z) &= w_0(x,y), \end{aligned} \tag{3.96a}$$

where u, v, and w denote, respectively, the displacements in the x, y, and z directions, and u_0, v_0, and w_0 are their associated middle surface displacements; β_x and β_y are the rotation with respect to the x and y axes, and are related to the transverse shear strains γ_{xz} and γ_{yz} by

$$\beta_x = \gamma_{xz} - \frac{\partial w}{\partial x}, \quad \beta_y = \gamma_{yz} - \frac{\partial w}{\partial y}. \tag{3.96b}$$

If small deformations are considered, the laminate strains can be written in terms of the middle surface displacements and rotations as

$$\varepsilon_x = \varepsilon_x^0 + z\kappa_x, \ \varepsilon_y = \varepsilon_y^0 + z\kappa_y, \ \gamma_{xy} = \gamma_{xy}^0 + z\kappa_{xy}, \tag{3.97a}$$

where $\varepsilon_x^0, \varepsilon_y^0, \gamma_{xy}^0$ are the mid-surface strains and κ_x, κ_y, κ_{xy} are the plate curvatures, and are related to the mid-surface displacements and rotations by

$$\begin{aligned} &\varepsilon_x^0 = \frac{\partial u_0}{\partial x}, \ \varepsilon_y^0 = \frac{\partial v_0}{\partial y}, \ \gamma_{xy}^0 = \frac{\partial u_0}{\partial y} + \frac{\partial v_0}{\partial x}, \\ &\kappa_x = \frac{\partial \beta_x}{\partial x}, \ \kappa_y = \frac{\partial \beta_y}{\partial y}, \ \kappa_{xy} = \frac{\partial \beta_x}{\partial y} + \frac{\partial \beta_y}{\partial x}. \end{aligned} \tag{3.97b}$$

If the thickness of the plate is small compared to its other dimensions even a thick plate is considered, instead of dealing with the stress distribution across the laminate thickness, usually an integral equivalent system of forces and moments acting on the laminate cross-section is used. With the integral equivalent system, the constitutive laws of laminated plates can be expressed as

$$\begin{Bmatrix} N_x \\ N_y \\ N_{xy} \\ M_x \\ M_y \\ M_{xy} \end{Bmatrix} = \begin{bmatrix} A_{11} & A_{12} & A_{16} & B_{11} & B_{12} & B_{16} \\ A_{12} & A_{22} & A_{26} & B_{12} & B_{22} & B_{26} \\ A_{16} & A_{26} & A_{66} & B_{16} & B_{26} & B_{66} \\ B_{11} & B_{12} & B_{16} & D_{11} & D_{12} & D_{16} \\ B_{12} & B_{22} & B_{26} & D_{12} & D_{22} & D_{26} \\ B_{16} & B_{26} & B_{66} & D_{16} & D_{26} & D_{66} \end{bmatrix} \begin{Bmatrix} \varepsilon_x^0 \\ \varepsilon_y^0 \\ \gamma_{xy}^0 \\ \kappa_x \\ \kappa_y \\ \kappa_{xy} \end{Bmatrix}, \quad (3.98a)$$

$$\begin{Bmatrix} Q_x \\ Q_y \end{Bmatrix} = \begin{bmatrix} A_{55} & A_{45} \\ A_{45} & A_{44} \end{bmatrix} \begin{Bmatrix} \gamma_{xz} \\ \gamma_{yz} \end{Bmatrix}, \quad (3.98b)$$

where N_x, N_y, N_{xy} are the stress resultants, M_x, M_y, M_{xy} are the bending moments, and Q_x, Q_y are the transverse shear forces defined in (2.34) and (3.5), and repeated here as

$$\begin{Bmatrix} N_x \\ N_y \\ N_{xy} \end{Bmatrix} = \int_{-h/2}^{h/2} \begin{Bmatrix} \sigma_x \\ \sigma_y \\ \tau_{xy} \end{Bmatrix} dz, \quad \begin{Bmatrix} M_x \\ M_y \\ M_{xy} \end{Bmatrix} = \int_{-h/2}^{h/2} \begin{Bmatrix} \sigma_x \\ \sigma_y \\ \tau_{xy} \end{Bmatrix} z dz, \quad \begin{Bmatrix} Q_x \\ Q_y \end{Bmatrix} = \int_{-h/2}^{h/2} \begin{Bmatrix} \tau_{xz} \\ \tau_{yz} \end{Bmatrix} dz. \quad (3.98c)$$

A_{ij}, B_{ij}, D_{ij}, $i, j = 1, 2, 6$, are, respectively, the *extensional*, *coupling*, and *bending stiffnesses* defined in (2.36) and repeated here as

$$A_{ij} = \sum_{k=1}^{n} Q_{ij}^{(k)} \left(h_k - h_{k-1}\right), \; B_{ij} = \frac{1}{2} \sum_{k=1}^{n} Q_{ij}^{(k)} \left(h_k^2 - h_{k-1}^2\right),$$

$$D_{ij} = \frac{1}{3} \sum_{k=1}^{n} Q_{ij}^{(k)} \left(h_k^3 - h_{k-1}^3\right), \quad i, j = 1,2,6. \quad (3.98d)$$

in which n is the total laminae number of the laminated plate; h_k and h_{k-1} denote, respectively, the position of bottom and top surface of the kth lamina; $Q_{ij}^{(k)}$ is the elastic constants of the kth lamina. In (3.98b), A_{44}, A_{45}, and

A_{55} are the transverse shear stiffnesses and can be evaluated by the same way as A_{ij} of (3.98d) with $Q_{ij}^{(k)}$ being replaced by $Q_{44}^{(k)}$, $Q_{45}^{(k)}$, and $Q_{55}^{(k)}$, and further multiplied by a shear correction factor α. If a composite sandwich plate with an orthotropic core whose transverse shear moduli in x–z and y–z planes are G_{xz} and G_{yz}, we have

$$A_{44} = \alpha h G_{yz}, \; A_{55} = \alpha h G_{xz}, \; A_{45} = 0, \tag{3.98e}$$

where h is the plate thickness.

Note that in the first-order shear deformation theory the transverse shear strains γ_{xz} and γ_{yz} are assumed to be independent of z. If a linear elastic material is considered, through the use of linear stress–strain relation their associated transverse shear stresses should also be independent of z. However, if we consider the satisfaction of equilibrium equations and traction-free boundary conditions on the top and bottom surfaces of the plate, we will obtain a parabolic distribution along the thickness direction for the transverse shear stresses. Thus, in this theory, the distribution of transverse shear stresses derived by constitutive laws is different from that by equilibrium equations. To compromise the inconsistency between constant and parabolic distributions, the assumed constants of transverse shear strains can be taken to be the average of transverse shear strains across the thickness. By taking the average sense for the transverse shear strains and resultant transverse shear force, usually a *shear correction factor* α appears in the relation of (3.98e). Although this factor may depend on lamination properties that will be stated later in (4.49), for simplicity we take $\alpha = 5/6$, which is usually set for isotropic elastic plates with rectangular cross-section (Cowper, 1966; Reddy, 2003).

Besides Equations (3.96)–(3.98), to complete the structural analysis we need the equilibrium equations for the laminated plates, which should be independent of the material types and can be expressed as

$$\begin{aligned}
&\frac{\partial N_x}{\partial x} + \frac{\partial N_{xy}}{\partial y} + q_x = 0, \quad \frac{\partial N_{xy}}{\partial x} + \frac{\partial N_y}{\partial y} + q_y = 0, \\
&\frac{\partial Q_x}{\partial x} + \frac{\partial Q_y}{\partial y} + q_z = 0, \frac{\partial M_x}{\partial x} + \frac{\partial M_{xy}}{\partial y} - Q_x + m_x = 0, \\
&\frac{\partial M_{xy}}{\partial x} + \frac{\partial M_y}{\partial y} - Q_y + m_y = 0,
\end{aligned} \tag{3.99}$$

where q_x, q_y, q_z and m_x, m_y are the distributed forces and moments acting on the plate surfaces.

Governing Equations

As described previously, the basic equations for the laminated composite thick plates are given in (3.96) for the displacement fields, (3.97) for the strain–displacement relations, (3.98) for the constitutive laws, and (3.99) for the equilibrium equations. To get a simple expression of governing equations satisfying all these basic equations, we now rewrite these equations in matrix form as follows:

$$\begin{Bmatrix} \boldsymbol{\varepsilon} \\ \boldsymbol{\kappa} \\ \boldsymbol{\gamma} \end{Bmatrix} = \begin{bmatrix} \boldsymbol{\mathcal{L}} & \mathbf{0} & \mathbf{0} \\ \mathbf{0} & \boldsymbol{\mathcal{L}} & \mathbf{0} \\ \mathbf{0} & \mathbf{I} & \nabla \end{bmatrix} \begin{Bmatrix} \mathbf{u} \\ \boldsymbol{\beta} \\ w \end{Bmatrix}, \quad \begin{Bmatrix} \boldsymbol{N} \\ \boldsymbol{M} \\ \boldsymbol{Q} \end{Bmatrix} = \begin{bmatrix} \boldsymbol{A} & \boldsymbol{B} & \mathbf{0} \\ \boldsymbol{B} & \boldsymbol{D} & \mathbf{0} \\ \mathbf{0} & \mathbf{0} & \boldsymbol{A}_t \end{bmatrix} \begin{Bmatrix} \boldsymbol{\varepsilon} \\ \boldsymbol{\kappa} \\ \boldsymbol{\gamma} \end{Bmatrix},$$

$$\begin{bmatrix} \boldsymbol{\mathcal{L}}^T & \mathbf{0} & \mathbf{0} \\ \mathbf{0} & \boldsymbol{\mathcal{L}}^T & -\mathbf{I} \\ \mathbf{0} & \mathbf{0} & \nabla^T \end{bmatrix} \begin{Bmatrix} \boldsymbol{N} \\ \boldsymbol{M} \\ \boldsymbol{Q} \end{Bmatrix} + \begin{Bmatrix} \mathbf{q} \\ \mathbf{m} \\ q_z \end{Bmatrix} = \mathbf{0}, \tag{3.100a}$$

where

$$\mathbf{u} = \begin{Bmatrix} u_0 \\ v_0 \end{Bmatrix}, \quad \boldsymbol{\beta} = \begin{Bmatrix} \beta_x \\ \beta_y \end{Bmatrix}, \quad \boldsymbol{\varepsilon} = \begin{Bmatrix} \varepsilon_x^0 \\ \varepsilon_y^0 \\ \gamma_{xy}^0 \end{Bmatrix}, \quad \boldsymbol{\kappa} = \begin{Bmatrix} \kappa_x \\ \kappa_y \\ \kappa_{xy} \end{Bmatrix}, \quad \boldsymbol{\gamma} = \begin{Bmatrix} \gamma_{xz} \\ \gamma_{yz} \end{Bmatrix},$$

$$\boldsymbol{N} = \begin{Bmatrix} N_x \\ N_y \\ N_{xy} \end{Bmatrix}, \quad \boldsymbol{M} = \begin{Bmatrix} M_x \\ M_y \\ M_{xy} \end{Bmatrix}, \quad \boldsymbol{Q} = \begin{Bmatrix} Q_x \\ Q_y \end{Bmatrix}, \quad \mathbf{q} = \begin{Bmatrix} q_x \\ q_y \end{Bmatrix}, \quad \mathbf{m} = \begin{Bmatrix} m_x \\ m_y \end{Bmatrix},$$

$$\boldsymbol{A} = \begin{bmatrix} A_{11} & A_{12} & A_{16} \\ A_{12} & A_{22} & A_{26} \\ A_{16} & A_{26} & A_{66} \end{bmatrix}, \quad \boldsymbol{B} = \begin{bmatrix} B_{11} & B_{12} & B_{16} \\ B_{12} & B_{22} & B_{26} \\ B_{16} & B_{26} & B_{66} \end{bmatrix},$$

$$\boldsymbol{D} = \begin{bmatrix} D_{11} & D_{12} & D_{16} \\ D_{12} & D_{22} & D_{26} \\ D_{16} & D_{26} & D_{66} \end{bmatrix}, \quad \boldsymbol{A}_t = \begin{bmatrix} A_{55} & A_{45} \\ A_{45} & A_{44} \end{bmatrix}, \tag{3.100b}$$

and $\boldsymbol{\mathcal{L}}$, ∇ are two differential operators defined by

$$\boldsymbol{\mathcal{L}} = \begin{bmatrix} \partial_x & 0 \\ 0 & \partial_y \\ \partial_y & \partial_x \end{bmatrix}, \quad \nabla = \begin{Bmatrix} \partial_x \\ \partial_y \end{Bmatrix}, \quad \partial_x = \frac{\partial}{\partial x}, \quad \partial_y = \frac{\partial}{\partial y}. \tag{3.100c}$$

Substituting $(3.100\text{a})_1$ and $(3.100\text{a})_2$ into $(3.100\text{a})_3$, we get

$$\mathcal{L}^*\mathbf{v} + \mathbf{p} = \mathbf{0}, \tag{3.101a}$$

where

$$\mathcal{L}^* = \begin{bmatrix} \mathcal{L}^T \mathbf{A}\mathcal{L} & \mathcal{L}^T \mathbf{B}\mathcal{L} & \mathbf{0} \\ \mathcal{L}^T \mathbf{B}\mathcal{L} & \mathcal{L}^T \mathbf{D}\mathcal{L} - \mathbf{A}_t & -\mathbf{A}_t\nabla \\ \mathbf{0} & \nabla^T \mathbf{A}_t & \nabla^T \mathbf{A}_t \nabla \end{bmatrix}, \quad \mathbf{v} = \begin{Bmatrix} \mathbf{u} \\ \boldsymbol{\beta} \\ w \end{Bmatrix}, \quad \mathbf{p} = \begin{Bmatrix} \mathbf{q} \\ \mathbf{m} \\ q_z \end{Bmatrix}. \tag{3.101b}$$

After performing the matrix multiplication, we have

$$\mathcal{L}^* = \begin{bmatrix} L_{11}^{(A)} & L_{12}^{(A)} & L_{11}^{(B)} & L_{12}^{(B)} & 0 \\ L_{21}^{(A)} & L_{22}^{(A)} & L_{21}^{(B)} & L_{22}^{(B)} & 0 \\ L_{11}^{(B)} & L_{12}^{(B)} & L_{11}^{(D)} - A_{55} & L_{12}^{(D)} - A_{45} & -L_1^{(G)} \\ L_{21}^{(B)} & L_{22}^{(B)} & L_{21}^{(D)} - A_{45} & L_{22}^{(D)} - A_{44} & -L_2^{(G)} \\ 0 & 0 & L_{1*}^{(G)} & L_{2*}^{(G)} & L_{11}^{(G)} \end{bmatrix}, \tag{3.102a}$$

where

$$\begin{aligned} L_{11}^{(X)} &= \frac{\partial}{\partial x}\left(X_{11}\frac{\partial}{\partial x} + X_{16}\frac{\partial}{\partial y} \right) + \frac{\partial}{\partial y}\left(X_{16}\frac{\partial}{\partial x} + X_{66}\frac{\partial}{\partial y} \right), \\ L_{12}^{(X)} &= \frac{\partial}{\partial x}\left(X_{16}\frac{\partial}{\partial x} + X_{12}\frac{\partial}{\partial y} \right) + \frac{\partial}{\partial y}\left(X_{66}\frac{\partial}{\partial x} + X_{26}\frac{\partial}{\partial y} \right), \\ L_{21}^{(X)} &= \frac{\partial}{\partial y}\left(X_{12}\frac{\partial}{\partial x} + X_{26}\frac{\partial}{\partial y} \right) + \frac{\partial}{\partial x}\left(X_{16}\frac{\partial}{\partial x} + X_{66}\frac{\partial}{\partial y} \right), \\ L_{22}^{(X)} &= \frac{\partial}{\partial y}\left(X_{26}\frac{\partial}{\partial x} + X_{22}\frac{\partial}{\partial y} \right) + \frac{\partial}{\partial x}\left(X_{66}\frac{\partial}{\partial x} + X_{26}\frac{\partial}{\partial y} \right), \qquad X = A, B, D, \end{aligned} \tag{3.102b}$$

$$\begin{aligned} & L_1^{(G)} = A_{55}\frac{\partial}{\partial x} + A_{45}\frac{\partial}{\partial y}, \quad L_2^{(G)} = A_{45}\frac{\partial}{\partial x} + A_{44}\frac{\partial}{\partial y}, \\ & L_{1*}^{(G)} = \frac{\partial}{\partial x}(A_{55}) + \frac{\partial}{\partial y}(A_{45}), \quad L_{2*}^{(G)} = \frac{\partial}{\partial x}(A_{45}) + \frac{\partial}{\partial y}(A_{44}), \\ & L_{11}^{(G)} = \frac{\partial}{\partial x}\left(A_{55}\frac{\partial}{\partial x} + A_{45}\frac{\partial}{\partial y} \right) + \frac{\partial}{\partial y}\left(A_{45}\frac{\partial}{\partial x} + A_{44}\frac{\partial}{\partial y} \right). \end{aligned} \tag{3.102c}$$

By (3.102b), we see that the governing Equation (3.101) is a set of five second-order partial differential equations with five unknown functions $u_0, v_0, w, \beta_x, \beta_y$.

Note that in the above expressions, the thickness of the plate is not required to be uniform. In other words, it is allowed to be a function of variables x and y. Thus, in (3.98c) and (3.98d),

$$h_k = h_k(x,y),\ h = h(x,y). \tag{3.103}$$

And hence,

$$\begin{aligned} &A_{ij} = A_{ij}(x,y),\ B_{ij} = B_{ij}(x,y), D_{ij} = D_{ij}(x,y),\ i,j = 1,2,6, \\ &A_{44} = A_{44}(x,y),\ A_{45} = A_{45}(x,y), A_{55} = A_{55}(x,y). \end{aligned} \tag{3.104}$$

When a thick plate with uniform thickness is considered, that is, h = constant, we have

$$\begin{aligned} &L_{11}^{(X)} = X_{11}\frac{\partial^2}{\partial x^2} + 2X_{16}\frac{\partial^2}{\partial x \partial y} + X_{66}\frac{\partial^2}{\partial y^2}, \\ &L_{12}^{(X)} = L_{21}^{(X)} = X_{16}\frac{\partial^2}{\partial x^2} + (X_{12} + X_{66})\frac{\partial^2}{\partial x \partial y} + X_{26}\frac{\partial^2}{\partial y^2}, \\ &L_{22}^{(X)} = X_{66}\frac{\partial^2}{\partial x^2} + 2X_{26}\frac{\partial^2}{\partial x \partial y} + X_{22}\frac{\partial^2}{\partial y^2}, \qquad X = A, B, D, \end{aligned} \tag{3.105a}$$

$$\begin{aligned} &L_1^{(G)} = L_{1*}^{(G)} = A_{55}\frac{\partial}{\partial x} + A_{45}\frac{\partial}{\partial y},\quad L_2^{(G)} = L_{2*}^{(G)} = A_{45}\frac{\partial}{\partial x} + A_{44}\frac{\partial}{\partial y}, \\ &L_{11}^{(G)} = A_{55}\frac{\partial^2}{\partial x^2} + 2A_{45}\frac{\partial^2}{\partial x \partial y} + A_{44}\frac{\partial^2}{\partial y^2}. \end{aligned} \tag{3.105b}$$

Governing Equations Derived from Hamilton's Principle

Hamilton's principle states that the motion of a continuum acted on by conservative forces between two arbitrary instants of time t_1 and t_2 is such that the line integral over the Lagrangian function is an extremum for the path motion (Reddy, 1993). The Lagrangian function is the difference between kinetic and total potential energies, and the total potential energy is the sum of the strain energy and potential energy of external forces. Thus, the Hamilton's principle can be expressed by

$$\delta\int_{t_1}^{t_2}\left(\Pi - T\right)dt = 0, \tag{3.106}$$

where δ is the variational operator, t_1 and t_2 are the integration limits of time. Π is the total potential energy, and T is the kinetic energy, which can be written as

$$\Pi = \int_V \left(W - f_i u_i\right)dV - \int_{S_\sigma} \hat{t}_i u_i dS, \quad T = \frac{1}{2}\int_V \rho \dot{u}_i \dot{u}_i dV, \tag{3.107}$$

where W, f_i, u_i, $\hat{t}_i$, and ρ are, respectively, the strain energy density, body forces, displacements, prescribed surface tractions, and density; the overdot $\dot{\bullet}$ denotes the time derivative; repeated indices imply summation through 1 to 3; V and S_σ are the regions for volume and traction-prescribed surface integrals, respectively. If a linear elastic body is considered, $W = \sigma_{ij}\varepsilon_{ij}/2$.

Using the relations (3.97a), (3.98c), and (3.100a)$_2$, we have

$$\begin{aligned}\int_V W dV &= \frac{1}{2}\int_V \sigma_{ij}\varepsilon_{ij} dV = \frac{1}{2}\int_A \left(\boldsymbol{N}^T \boldsymbol{\varepsilon} + \boldsymbol{M}^T \boldsymbol{\kappa} + \boldsymbol{Q}^T \boldsymbol{\gamma}\right) dA \\ &= \frac{1}{2}\int_A \left(\boldsymbol{\varepsilon}^T \boldsymbol{A}\boldsymbol{\varepsilon} + 2\boldsymbol{\varepsilon}^T \boldsymbol{B}\boldsymbol{\kappa} + \boldsymbol{\kappa}^T \boldsymbol{D}\boldsymbol{\kappa} + \boldsymbol{\gamma}^T \boldsymbol{A}_t \boldsymbol{\gamma}\right) dA,\end{aligned} \tag{3.108a}$$

$$\int_V f_i u_i dV = \int_A \left(\mathbf{u}^T \mathbf{q} + \boldsymbol{\beta}^T \mathbf{m} + w q_z\right) dA, \tag{3.108b}$$

$$\int_{S_\sigma} \hat{t}_i u_i dS = \int_{\Gamma_\sigma} \left(\mathbf{u}^T \hat{\mathbf{t}}_n + \boldsymbol{\beta}^T \hat{\mathbf{h}}_n + w\hat{Q}_n\right) d\Gamma, \tag{3.108c}$$

$$\frac{1}{2}\int_V \rho \dot{u}_i \dot{u}_i dV = \frac{1}{2}\int_A \left\{g_0\left(\dot{\mathbf{u}}^T \dot{\mathbf{u}} + \dot{w}^2\right) + 2g_1 \dot{\mathbf{u}}^T \dot{\boldsymbol{\beta}} + g_2 \dot{\boldsymbol{\beta}}^T \dot{\boldsymbol{\beta}}\right\} dA, \tag{3.108d}$$

where

$$\hat{\mathbf{t}}_n = \begin{Bmatrix} \hat{T}_x \\ \hat{T}_y \end{Bmatrix}, \ \hat{T}_x = \hat{N}_x n_1 + \hat{N}_{xy} n_2, \ \ \hat{T}_y = \hat{N}_{xy} n_1 + \hat{N}_y n_2,$$

$$\hat{\mathbf{h}}_n = \begin{Bmatrix} \hat{H}_x \\ \hat{H}_y \end{Bmatrix}, \ \hat{H}_x = \hat{M}_x n_1 + \hat{M}_{xy} n_2, \ \ \hat{H}_y = \hat{M}_{xy} n_1 + \hat{M}_y n_2,$$

$$\hat{Q}_n = \hat{Q}_x n_1 + \hat{Q}_y n_2,$$

$$g_0 = \int_{-h/2}^{h/2} \rho dz, \quad g_1 = \int_{-h/2}^{h/2} \rho z dz, \quad g_2 = \int_{-h/2}^{h/2} \rho z^2 dz. \tag{3.108e}$$

Taking the variation for each equation of (3.108a)–(3.108d), we have

$$\begin{aligned}
&\delta\int_V WdV = \int_A \left\{(\delta\boldsymbol{\varepsilon})^T \boldsymbol{A}\boldsymbol{\varepsilon} + (\delta\boldsymbol{\varepsilon})^T \boldsymbol{B}\boldsymbol{\kappa} + (\delta\boldsymbol{\kappa})^T \boldsymbol{B}\boldsymbol{\varepsilon} + (\delta\boldsymbol{\kappa})^T \boldsymbol{D}\boldsymbol{\kappa} + (\delta\boldsymbol{\gamma})^T \boldsymbol{A}_t\boldsymbol{\gamma}\right\} dA,\\
&\delta\int_V f_i u_i dV = \int_A \left\{(\delta\mathbf{u})^T \mathbf{q} + (\delta\boldsymbol{\beta})^T \mathbf{m} + (\delta w) q_z\right\} dA,\\
&\delta\int_{S_\sigma} \hat{t}_i u_i dS = \int_{\Gamma_\sigma} \left\{(\delta\mathbf{u})^T \hat{\mathbf{t}}_n + (\delta\boldsymbol{\beta})^T \hat{\mathbf{h}}_n + (\delta w)\hat{Q}_n\right\} \Gamma,\\
&\delta\int_V \frac{1}{2}\rho \dot{u}_i \dot{u}_i dV = \int_A \left\{g_0\left[(\delta\dot{\mathbf{u}})^T \dot{\mathbf{u}} + (\delta\dot{w})\dot{w}\right]\right.\\
&\qquad \left. + g_1\left[(\delta\dot{\mathbf{u}})^T \dot{\boldsymbol{\beta}} + (\delta\dot{\boldsymbol{\beta}})^T \dot{\mathbf{u}}\right] + g_2 (\delta\dot{\boldsymbol{\beta}})^T \dot{\boldsymbol{\beta}}\right\} dA.
\end{aligned} \tag{3.109}$$

Using the symbols for matrices and vectors defined in (3.100b) and (3.100c), from (3.97b) and (3.96b) we have

$$\boldsymbol{\varepsilon} = \boldsymbol{\mathcal{L}}\mathbf{u}, \quad \boldsymbol{\kappa} = \boldsymbol{\mathcal{L}}\boldsymbol{\beta}, \quad \boldsymbol{\gamma} = \boldsymbol{\beta} + \nabla w. \tag{3.110}$$

Substituting (3.110) into $(3.109)_1$, and using integration by parts and the conditions that either displacements or tractions are prescribed on the boundary, we get

$$\begin{aligned}
\delta\int_V WdV = \int_A \Big\{ &-(\delta\mathbf{u})^T \boldsymbol{\mathcal{L}}_A \mathbf{u} - (\delta\mathbf{u})^T \boldsymbol{\mathcal{L}}_B \boldsymbol{\beta} - (\delta\boldsymbol{\beta})^T \boldsymbol{\mathcal{L}}_B \mathbf{u} - (\delta\boldsymbol{\beta})^T \boldsymbol{\mathcal{L}}_B \boldsymbol{\beta}\\
&+ (\delta\boldsymbol{\beta})^T \boldsymbol{A}_t \boldsymbol{\beta} + (\delta\boldsymbol{\beta})^T \boldsymbol{A}_t \nabla w - (\delta w)\nabla^T \boldsymbol{A}_t \boldsymbol{\beta} - (\delta w)\nabla^T \boldsymbol{A}_t \nabla w \Big\} dA,
\end{aligned} \tag{3.111a}$$

where

$$\boldsymbol{\mathcal{L}}_A = \boldsymbol{\mathcal{L}}^T \boldsymbol{A}\boldsymbol{\mathcal{L}}, \quad \boldsymbol{\mathcal{L}}_B = \boldsymbol{\mathcal{L}}^T \boldsymbol{B}\boldsymbol{\mathcal{L}}, \quad \boldsymbol{\mathcal{L}}_D = \boldsymbol{\mathcal{L}}^T \boldsymbol{D}\boldsymbol{\mathcal{L}}. \tag{3.111b}$$

Similarly, by using the integration by parts and the initial conditions set on the boundary, in $(3.109)_4$ the variation with respect to the velocities, $\delta\dot{\mathbf{u}}$, $\delta\dot{\boldsymbol{\beta}}$, and $\delta\dot{w}$, can also be replaced by $\delta\mathbf{u}$, $\delta\boldsymbol{\beta}$, and δw, and $(3.109)_4$ becomes

$$\delta\int_V \frac{1}{2}\rho\dot{u}_i\dot{u}_i dV = -\int_A \Big\{ g_0\Big[(\delta\mathbf{u})^T\ddot{\mathbf{u}} + (\delta w)\ddot{w}\Big] + g_1\Big[(\delta\mathbf{u})^T\ddot{\boldsymbol{\beta}} + (\delta\boldsymbol{\beta})^T\ddot{\mathbf{u}}\Big] + g_2(\delta\boldsymbol{\beta})^T\ddot{\boldsymbol{\beta}}\Big\} dA. \quad (3.112)$$

With the results of $(3.109)_{2,3}$, (3.111a), and (3.112), the governing equations as well as the boundary conditions are obtained as follows:

$$\left.\begin{aligned} &\mathcal{L}_A\mathbf{u} + \mathcal{L}_B\boldsymbol{\beta} + \mathbf{q} = g_0\ddot{\mathbf{u}} + g_1\ddot{\boldsymbol{\beta}} \\ &\mathcal{L}_B\mathbf{u} + (\mathcal{L}_D - \mathbf{A}_t)\boldsymbol{\beta} - \mathbf{A}_t\nabla w + \mathbf{m} = g_1\ddot{\mathbf{u}} + g_2\ddot{\boldsymbol{\beta}} \\ &\nabla^T\mathbf{A}_t\boldsymbol{\beta} + \nabla^T\mathbf{A}_t\nabla w + q_z = g_0\ddot{w} \end{aligned}\right\}, \quad \mathbf{x}\in A,$$
$$\mathbf{t}_n = \hat{\mathbf{t}}_n,\ \mathbf{h}_n = \hat{\mathbf{h}}_n,\ Q_n = \hat{Q}_n, \quad \mathbf{x}\in\Gamma_\sigma. \quad (3.113)$$

The static part of $(3.113)_{1-3}$ can be proved to be exactly the same as that of (3.101), which is derived directly from the basic equations.

3.4.2 Green's Functions

Green's Functions in Terms of Matrix Exponential

Consider a laminated composite thick plate with an infinite plane dimension subjected to concentrated forces $\hat{q}_x, \hat{q}_y, \hat{q}_z$ and moments $\hat{m}_x, \hat{m}_y$ at the source point $\hat{\mathbf{x}} = (\hat{x}, \hat{y})$ as shown in Figure 3.5.

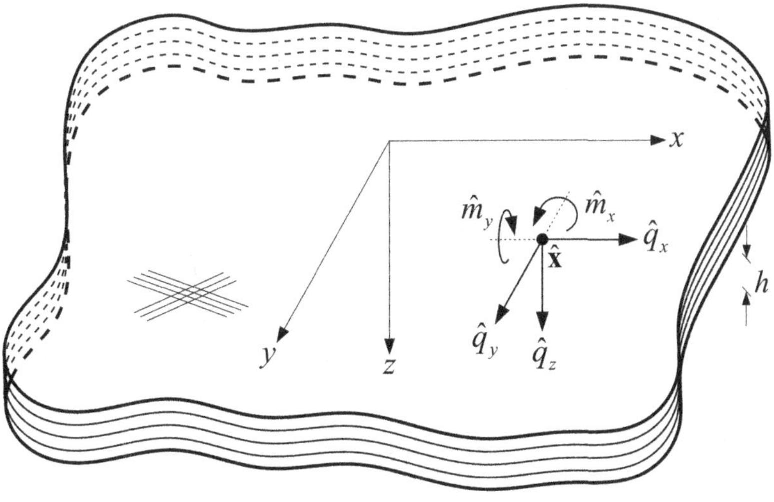

FIGURE 3.5 An infinite laminated composite plate subjected to concentrated forces and moments.

The concentrated forces/moments can be expressed as

$$\mathbf{p} = \hat{\mathbf{p}}\delta(\hat{\mathbf{x}},\mathbf{x}), \text{ where } \hat{\mathbf{p}} = (\hat{q}_x, \hat{q}_y, \hat{m}_x, \hat{m}_y, \hat{q}_z)^T, \tag{3.114}$$

where $\delta(\hat{\mathbf{x}},\mathbf{x})$ is the Dirac delta function with impulse at $\mathbf{x} = \hat{\mathbf{x}}$. Substituting (3.114) into (3.101a), we have

$$\mathcal{L}^*\mathbf{v} + \hat{\mathbf{p}}\delta(\hat{\mathbf{x}},\mathbf{x}) = \mathbf{0}. \tag{3.115}$$

According to the plane wave decomposition method (Gel'fand and Shilov, 1964), the Dirac delta function $\delta(\hat{\mathbf{x}},\mathbf{x})$ in two-dimensional domain can be represented in the transform integral form as

$$\delta(\hat{\mathbf{x}},\mathbf{x}) = -\frac{1}{4\pi^2}\int_0^{2\pi} \rho^{-2} d\theta, \tag{3.116a}$$

where

$$\rho = \rho(\hat{\mathbf{x}},\mathbf{x},\theta) = (\mathbf{x} - \hat{\mathbf{x}}) \cdot \boldsymbol{\omega}, \quad \boldsymbol{\omega} = (\cos\theta, \sin\theta)^T, \tag{3.116b}$$

and the dot denotes the inner product. The variable θ is the orientation angle of a moving point on the unit circle of two-dimensional domain, which is independent of the locations of $\hat{\mathbf{x}}$ and $\mathbf{x}$. The illustration of transformed variable ρ and angle θ is shown in Figure 3.6. Similarly, we can

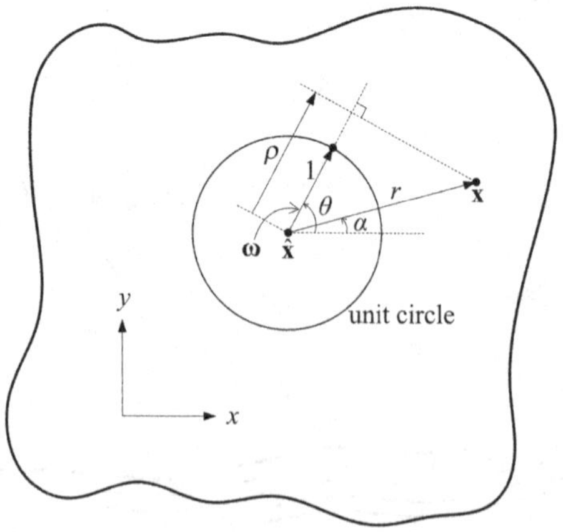

FIGURE 3.6 Illustration of transformed variable ρ and angle θ.

also transform the displacement vector $\mathbf{v}(\hat{\mathbf{x}},\mathbf{x})$ through its corresponding one $\breve{\mathbf{v}}(\rho)$ in ρ-domain by

$$\mathbf{v}(\hat{\mathbf{x}},\mathbf{x}) = \int_0^{2\pi} \breve{\mathbf{v}}(\rho)\, d\theta, \tag{3.117}$$

With (3.116) and (3.117) and knowing that

$$\frac{\partial \mathbf{v}}{\partial x} = \int_0^{2\pi} \frac{\partial \breve{\mathbf{v}}}{\partial \rho} \cos\theta\, d\theta, \quad \frac{\partial \mathbf{v}}{\partial y} = \int_0^{2\pi} \frac{\partial \breve{\mathbf{v}}}{\partial \rho} \sin\theta\, d\theta, \tag{3.118}$$

the governing partial differential Equations (3.115) can be rewritten in the integral form of $\int_0^{2\pi}\{\cdots\}d\theta = 0$. By taking the integrand as zero function we can obtain a system of five ordinary differential equations in matrix form as

$$\mathbf{L}_2 \breve{\mathbf{v}}'' + \mathbf{L}_1 \breve{\mathbf{v}}' + \mathbf{L}_0 \breve{\mathbf{v}} - \frac{1}{4\pi^2 \rho^2} \hat{\mathbf{p}} = \mathbf{0}, \tag{3.119}$$

where the symbol prime •′ denotes the derivative with respect to ρ, that is, $\breve{\mathbf{v}}' = \partial \breve{\mathbf{v}} / \partial \rho$, $\breve{\mathbf{v}}'' = \partial^2 \breve{\mathbf{v}} / \partial \rho^2$, and

$$\mathbf{L}_2 = \begin{bmatrix} \tilde{A} & \tilde{B} & \mathbf{0} \\ \tilde{B} & \tilde{D} & \mathbf{0} \\ \mathbf{0} & \mathbf{0} & \tilde{A}_t \end{bmatrix}, \; \mathbf{L}_1 = \begin{bmatrix} \mathbf{0} & \mathbf{0} & \mathbf{0} \\ \mathbf{0} & \mathbf{0} & -A_t \boldsymbol{\omega} \\ \mathbf{0} & \boldsymbol{\omega}^T A_t & 0 \end{bmatrix}, \; \mathbf{L}_0 = \begin{bmatrix} \mathbf{0} & \mathbf{0} & \mathbf{0} \\ \mathbf{0} & -A_t & \mathbf{0} \\ \mathbf{0} & \mathbf{0} & 0 \end{bmatrix}, \tag{3.120a}$$

in which

$$\tilde{A} = \boldsymbol{\Omega}^T A \boldsymbol{\Omega}, \; \tilde{B} = \boldsymbol{\Omega}^T B \boldsymbol{\Omega}, \; \tilde{D} = \boldsymbol{\Omega}^T D \boldsymbol{\Omega}, \; \tilde{A}_t = \boldsymbol{\omega}^T A_t \boldsymbol{\omega},$$
$$\boldsymbol{\Omega} = \begin{bmatrix} \cos\theta & 0 \\ 0 & \sin\theta \\ \sin\theta & \cos\theta \end{bmatrix}, \; \boldsymbol{\omega} = \begin{Bmatrix} \cos\theta \\ \sin\theta \end{Bmatrix}. \tag{3.120b}$$

It can be proved that $\mathbf{L}_2$ is a positive-definite matrix, and its inverse can be obtained as

$$\mathbf{L}_2^{-1} = \begin{bmatrix} \tilde{A}^* & \tilde{B}^* & \mathbf{0} \\ \tilde{B}^{*T} & \tilde{D}^* & \mathbf{0} \\ \mathbf{0} & \mathbf{0} & \tilde{A}_t^{-1} \end{bmatrix}, \quad \begin{aligned} &\tilde{A}^* = \tilde{A}^{-1} + \tilde{A}^{-1}\tilde{B}\tilde{D}^*\tilde{B}\tilde{A}^{-1}, \\ &\tilde{B}^* = -\tilde{A}^{-1}\tilde{B}\tilde{D}^*, \\ &\tilde{D}^* = \left(\tilde{D} - \tilde{B}\tilde{A}^{-1}\tilde{B}\right)^{-1}. \end{aligned} \tag{3.121}$$

Combining (3.119) with the trivial identities $\breve{\mathbf{v}}' = \breve{\mathbf{v}}'$, we have

$$\begin{bmatrix} \mathbf{I} & \mathbf{0} \\ \mathbf{0} & \mathbf{L}_2 \end{bmatrix} \begin{Bmatrix} \breve{\mathbf{v}}' \\ \breve{\mathbf{v}}'' \end{Bmatrix} = \begin{bmatrix} \mathbf{0} & \mathbf{I} \\ -\mathbf{L}_0 & -\mathbf{L}_1 \end{bmatrix} \begin{Bmatrix} \breve{\mathbf{v}} \\ \breve{\mathbf{v}}' \end{Bmatrix} + \frac{1}{4\pi^2\rho^2} \begin{Bmatrix} \mathbf{0} \\ \hat{\mathbf{p}} \end{Bmatrix}, \tag{3.122}$$

or simply as

$$\boldsymbol{\upsilon}' = \mathbf{K}\boldsymbol{\upsilon} + \rho^{-2}\hat{\mathbf{f}}, \tag{3.123a}$$

where

$$\boldsymbol{\upsilon} = \begin{Bmatrix} \breve{\mathbf{v}} \\ \breve{\mathbf{v}}' \end{Bmatrix}, \ \mathbf{K} = \begin{bmatrix} \mathbf{0} & \mathbf{I} \\ \mathbf{K}_0 & \mathbf{K}_1 \end{bmatrix}, \ \hat{\mathbf{f}} = \frac{1}{4\pi^2} \begin{Bmatrix} \mathbf{0} \\ \mathbf{L}_2^{-1}\hat{\mathbf{p}} \end{Bmatrix}, \tag{3.123b}$$

and

$$\mathbf{K}_0 = -\mathbf{L}_2^{-1}\mathbf{L}_0 = \begin{bmatrix} \mathbf{0} & \tilde{B}^* A_t & \mathbf{0} \\ \mathbf{0} & \tilde{D}^* A_t & \mathbf{0} \\ \mathbf{0} & \mathbf{0} & 0 \end{bmatrix},$$

$$\mathbf{K}_1 = -\mathbf{L}_2^{-1}\mathbf{L}_1 = \begin{bmatrix} \mathbf{0} & \mathbf{0} & \tilde{B}^* A_t\boldsymbol{\omega} \\ \mathbf{0} & \mathbf{0} & \tilde{D}^* A_t\boldsymbol{\omega} \\ \mathbf{0} & -\tilde{A}_t^{-1}\boldsymbol{\omega}^T A_t & 0 \end{bmatrix}. \tag{3.123c}$$

Equation (3.123a) is in the standard state-space form whose solution can be expressed as (Franklin, 1968)

$$\boldsymbol{\upsilon} = \boldsymbol{\upsilon}(\rho) = e^{\mathbf{K}\rho}\left\{\mathbf{h} + \int \rho^{-2} e^{-\mathbf{K}\rho}\hat{\mathbf{f}}d\rho\right\}, \tag{3.124}$$

where the constant vector $\mathbf{h}$ is to be determined through the satisfaction of boundary conditions and belongs to the homogeneous solution, while the indefinite integral related to $\hat{\mathbf{f}}$ is associated with the particular solution.

Note that the integral constant induced by the indefinite integral can be absorbed by $\mathbf{h}$ and would have no influence on the complete solution $\boldsymbol{\upsilon}(\rho)$. Since the Green's function of the present problem does not involve any other boundary conditions, the homogeneous solution can be ignored, and hence,

$$\boldsymbol{\upsilon}(\rho)=e^{\mathbf{K}\rho}\left\{\int \rho^{-2}e^{-\mathbf{K}\rho}d\rho\right\}\hat{\mathbf{f}}, \tag{3.125}$$

Once the function $\boldsymbol{\upsilon}(\rho)$ is solved, the transform function $\breve{\mathbf{v}}(\rho)$ can be obtained through (3.123b)$_1$. The Green's functions of generalized displacements $\mathbf{v}(\hat{\mathbf{x}},\mathbf{x})$ can then be evaluated by the transform integral (3.117).

For the other physical quantities such as mid-plane strains, plate curvatures, stress resultants, bending moments, transverse shear strains and forces, their associated field functions can all be expressed in the transform integral form similar to (3.117). That is,

$$\mathbf{Y}(\hat{\mathbf{x}},\mathbf{x})=\int_0^{2\pi}\breve{\mathbf{Y}}(\rho)d\theta,\quad \mathbf{Y}=\boldsymbol{\varepsilon},\boldsymbol{\kappa},\boldsymbol{\gamma},\boldsymbol{N},\boldsymbol{M},\mathbf{Q}, \tag{3.126}$$

where $\breve{\mathbf{Y}}(\rho)$ are the corresponding transformed functions. By using the strain–displacement relations (3.100a)$_1$ and constitutive laws (3.100a)$_2$, these transformed functions can be related to $\breve{\mathbf{v}}(\rho)$ by

$$\begin{Bmatrix}\breve{\boldsymbol{\varepsilon}}(\rho)\\ \breve{\boldsymbol{\kappa}}(\rho)\\ \breve{\boldsymbol{\gamma}}(\rho)\end{Bmatrix}=\begin{Bmatrix}\boldsymbol{\Omega}\mathbf{I}_u\breve{\mathbf{v}}'(\rho)\\ \grave{\mathbf{U}}\mathbf{I}_\beta\breve{\mathbf{v}}'(\rho)\\ \mathbf{I}_\beta\breve{\mathbf{v}}(\rho)+\boldsymbol{\omega}\mathbf{i}_5^T\breve{\mathbf{v}}'(\rho)\end{Bmatrix},\quad \begin{Bmatrix}\breve{\boldsymbol{N}}(\rho)\\ \breve{\boldsymbol{M}}(\rho)\\ \breve{\mathbf{Q}}(\rho)\end{Bmatrix}=\begin{bmatrix}\mathbf{A} & \boldsymbol{B} & \mathbf{0}\\ \mathbf{B} & \boldsymbol{D} & \mathbf{0}\\ \mathbf{0} & \mathbf{0} & \mathbf{A}_t\end{bmatrix}\begin{Bmatrix}\breve{\boldsymbol{\varepsilon}}(\rho)\\ \breve{\boldsymbol{\kappa}}(\rho)\\ \breve{\boldsymbol{\gamma}}(\rho)\end{Bmatrix}, \tag{3.127a}$$

where

$$\mathbf{I}_u=\begin{bmatrix}\mathbf{i}_1^T\\ \mathbf{i}_2^T\end{bmatrix},\quad \mathbf{I}_\beta=\begin{bmatrix}\mathbf{i}_3^T\\ \mathbf{i}_4^T\end{bmatrix}, \tag{3.127b}$$

and $\mathbf{i}_n$, $n = 1, 2, \cdots, 5$ is the 5×1 base vector whose n^{th} component is one and the others are zero, for example, $\mathbf{i}_2=(0\quad 1\quad 0\quad 0\quad 0)^T$.

At this stage, the transformed function $\boldsymbol{\upsilon}(\rho)$ has been solved in (3.125), which involves a matrix exponential $e^{\mathbf{K}\rho}$ and an indefinite integral containing $\rho^{-2}e^{-\mathbf{K}\rho}$. Since some computer software such as Matlab and Mathematica have the capability of computing matrix exponential and indefinite integral, it seems no problem for the evaluation of $\boldsymbol{\upsilon}(\rho)$. However, the indefinite integral in (3.125) is related to the matrix function of exponential integral Ei($\mathbf{K}\rho$) (Ei(x) will be stated in the next section) and is nonelementary. Although Ei of scalar x can be also computed through the software, the corresponding matrix function is not available. If one tries to directly evaluate the integral in (3.125) through a numerical quadrature rule, the double numerical integration, which is extremely inefficient, would be required to obtain the real-domain solutions: one is at this stage; the other is for the transform integrals (3.117) and (3.126). Therefore, in the next subsection, we like to provide an accurate and efficient way to evaluate $\boldsymbol{\upsilon}(\rho)$ by deriving the explicit solutions for the matrix exponential and its related integral. When the transformed function $\boldsymbol{\upsilon}(\rho)$ can be evaluated accurately and efficiently, we may encounter the problem of singular integral appeared in (3.117) and (3.126) for getting the Green's function of physical quantities in real domain. To solve this problem, a special treatment of numerical integration will then be provided later.

Explicit Solutions of Green's Functions in Transformed Domain

The Green's function $\breve{\mathbf{v}}(\rho)$ in transformed domain is related to the Green's function $\mathbf{v}(\hat{\mathbf{x}},\mathbf{x})$ by (3.117) and is contained in the vector $\boldsymbol{\upsilon}$ of (3.123b)$_1$. Thus, to derive the explicit solutions of $\breve{\mathbf{v}}(\rho)$, firstly we need to derive the explicit solution of $\boldsymbol{\upsilon}(\rho)$. The solution obtained in (3.125) for $\boldsymbol{\upsilon}(\rho)$ shows that it contains two important functions to be solved explicitly. One is the matrix exponential $e^{\mathbf{K}\rho}$, and the other is the integral $\int\rho^{-2}e^{-\mathbf{K}\rho}d\rho$. To solve them explicitly, we use the method of eigen-decomposition (Bronson, 1969). From (3.123b)$_2$ and its related equations, we see that $\mathbf{K}$ is a 10×10 matrix. Let λ and $\mathbf{z}$ be the eigenvalue and eigenvector of $\mathbf{K}$, we have

$$(\mathbf{K}-\lambda\mathbf{I})\mathbf{z}=\mathbf{0}. \tag{3.128}$$

By successively using the formula for block matrix determinant (Horn and Johnson, 2013), we obtain (Hsu and Hwu, 2023)

$$\det(\mathbf{K}-\lambda\mathbf{I})=\lambda^8\left(\lambda^2-\lambda_d^2\right), \tag{3.129a}$$

where

$$\lambda_d^2 = \mathrm{tr}\left(\mathbf{D}_0\right) = d_{11} + d_{22} > 0,$$
$$\mathbf{D}_0 = \begin{bmatrix} d_{11} & d_{12} \\ d_{21} & d_{22} \end{bmatrix} = \tilde{\boldsymbol{D}}^* \boldsymbol{A}_t \left(\mathbf{I} - \tilde{\boldsymbol{A}}_t^{-1} \boldsymbol{\omega}\boldsymbol{\omega}^T \boldsymbol{A}_t\right), \tag{3.129b}$$

and tr denotes the trace of a matrix.

Without loss of generality we choose $\lambda_d > 0$, and the ten eigenvalues of $\mathbf{K}$ can be determined as

$$\lambda_1 = \lambda_2 = \ldots. = \lambda_8 = 0, \quad \lambda_9 = \lambda_d, \ \lambda_{10} = -\lambda_d. \tag{3.130}$$

Substituting (3.130) into (3.128), only five independent eigenvectors can be found and three of them are obtained from the repeated root of $\lambda = 0$. To get a complete set of eigenvectors, we need to consider the generalized eigenvectors (Bronson, 1969). With this consideration, we assign

$$\begin{aligned}
&\mathbf{K}\mathbf{z}_1 = \mathbf{0}, \ \mathbf{K}\mathbf{z}_2 = \mathbf{z}_1, \\
&\mathbf{K}\mathbf{z}_3 = \mathbf{0}, \ \mathbf{K}\mathbf{z}_4 = \mathbf{z}_3, \\
&\mathbf{K}\mathbf{z}_5 = \mathbf{0}, \ \mathbf{K}\mathbf{z}_6 = \mathbf{z}_5, \ \mathbf{K}\mathbf{z}_7 = \mathbf{z}_6, \ \mathbf{K}\mathbf{z}_8 = \mathbf{z}_7, \\
&\mathbf{K}\mathbf{z}_9 = \lambda_d \mathbf{z}_9, \ \mathbf{K}\mathbf{z}_{10} = -\lambda_d \mathbf{z}_{10},
\end{aligned} \tag{3.131}$$

in which $\mathbf{z}_1$, $\mathbf{z}_3$, and $\mathbf{z}_5$ are the three independent eigenvectors associated with $\lambda = 0$; $\mathbf{z}_9$ and $\mathbf{z}_{10}$ are the eigenvectors associated with $\lambda = \pm\lambda_d$; and $\mathbf{z}_2$, $\mathbf{z}_4$, $\mathbf{z}_6$, $\mathbf{z}_7$, $\mathbf{z}_8$ are the generalized eigenvectors of $\lambda = 0$. From (3.131) with $\mathbf{K}$ given in $(3.123\text{b})_2$, we obtain

$$\begin{aligned}
&\mathbf{z}_1 = \begin{Bmatrix} \mathbf{i}_1 \\ \mathbf{0} \end{Bmatrix}, \ \mathbf{z}_2 = \begin{Bmatrix} \mathbf{i}_1 \\ \mathbf{i}_1 \end{Bmatrix}, \ \mathbf{z}_3 = \begin{Bmatrix} \mathbf{i}_2 \\ \mathbf{0} \end{Bmatrix}, \ \mathbf{z}_4 = \begin{Bmatrix} \mathbf{i}_2 \\ \mathbf{i}_2 \end{Bmatrix}, \\
&\mathbf{z}_5 = \begin{Bmatrix} \mathbf{i}_5 \\ \mathbf{0} \end{Bmatrix}, \ \mathbf{z}_6 = \begin{Bmatrix} \boldsymbol{\omega}_6 \\ \mathbf{i}_5 \end{Bmatrix}, \ \mathbf{z}_7 = \begin{Bmatrix} \boldsymbol{\omega}_7 \\ \boldsymbol{\omega}_6 \end{Bmatrix}, \ \mathbf{z}_8 = \begin{Bmatrix} \boldsymbol{\omega}_8 \\ \boldsymbol{\omega}_7 \end{Bmatrix}, \\
&\mathbf{z}_9 = \begin{Bmatrix} \mathbf{g}^+ \\ \lambda_d \mathbf{g}^+ \end{Bmatrix}, \ \mathbf{z}_{10} = \begin{Bmatrix} \mathbf{g}^- \\ -\lambda_d \mathbf{g}^- \end{Bmatrix},
\end{aligned} \tag{3.132a}$$

where

$$\mathbf{g}^{\pm}=\begin{Bmatrix} -\tilde{A}^{-1}\tilde{\boldsymbol{B}}\mathbf{d} \\ \mathbf{d} \\ \mp\lambda_d^{-1}\tilde{A}_t^{-1}\boldsymbol{\omega}^T\boldsymbol{A}_t\mathbf{d} \end{Bmatrix},\quad \mathbf{d}=\begin{cases} \{1 \quad d_{21}/d_{11}\}^T, & \text{when } d_{11}\neq 0, \\ \{0 \quad 1\}^T, & \text{when } d_{11}=0, \end{cases}$$

$$\boldsymbol{\omega}_6=\begin{Bmatrix} \tilde{A}^{-1}\tilde{\boldsymbol{B}}\boldsymbol{\omega} \\ -\boldsymbol{\omega} \\ 1 \end{Bmatrix},\quad \boldsymbol{\omega}_7=\begin{Bmatrix} \mathbf{0} \\ -\boldsymbol{\omega} \\ 1 \end{Bmatrix},\quad \boldsymbol{\omega}_8=\begin{Bmatrix} \mathbf{0} \\ -\left(\mathbf{I}+A_t^{-1}\tilde{\boldsymbol{D}}^{*-1}\right)\boldsymbol{\omega} \\ 0 \end{Bmatrix}. \tag{3.132b}$$

With the results of (3.130)–(3.132), we can now decompose **K** through the Jordan canonical form as (Bronson, 1969)

$$\mathbf{KZ}=\mathbf{ZJ},\ \text{ or }\ \mathbf{K}=\mathbf{ZJZ}^{-1}, \tag{3.133a}$$

where

$$\mathbf{Z}=\begin{bmatrix}\mathbf{z}_1 & \mathbf{z}_2 & \cdots & \mathbf{z}_{10}\end{bmatrix},\ \mathbf{J}=\text{diag}\begin{bmatrix}\mathbf{J}_a, \mathbf{J}_b, \mathbf{J}_c, \mathbf{J}_d\end{bmatrix}, \tag{3.133b}$$

and

$$\mathbf{J}_a=\mathbf{J}_b=\begin{bmatrix} 0 & 1 \\ 0 & 0 \end{bmatrix},\ \mathbf{J}_c=\begin{bmatrix} 0 & 1 & 0 & 0 \\ 0 & 0 & 1 & 0 \\ 0 & 0 & 0 & 1 \\ 0 & 0 & 0 & 0 \end{bmatrix},\ \mathbf{J}_d=\begin{bmatrix} \lambda_d & 0 \\ 0 & -\lambda_d \end{bmatrix}. \tag{3.133c}$$

The "diag" of (3.133b) denotes a *block-diagonal matrix* whose components outside the main diagonal blocks are all zero matrices.

By using $(3.133\text{a})_2$, the function $f(\mathbf{K}\rho)$ of matrix **K** can be decomposed as

$$f(\mathbf{K}\rho)=\mathbf{Z}f(\mathbf{J}\rho)\mathbf{Z}^{-1}, \tag{3.134a}$$

where

$$f(\mathbf{J}\rho)=\text{diag}\left[f(\mathbf{J}_a\rho), f(\mathbf{J}_a\rho), f(\mathbf{J}_c\rho), f(\mathbf{J}_d\rho)\right],$$

$$f(\mathbf{J}_a\rho)=\begin{bmatrix} f(\lambda\rho) & f^{(1)}(\lambda\rho) \\ 0 & f(\lambda\rho) \end{bmatrix}_{\lambda=0},$$

$$f(\mathbf{J}_c\rho)=\begin{bmatrix} f(\lambda\rho) & f^{(1)}(\lambda\rho) & f^{(2)}(\lambda\rho) & f^{(3)}(\lambda\rho) \\ 0 & f(\lambda\rho) & f^{(1)}(\lambda\rho) & f^{(2)}(\lambda\rho) \\ 0 & 0 & f(\lambda\rho) & f^{(1)}(\lambda\rho) \\ 0 & 0 & 0 & f(\lambda\rho) \end{bmatrix}_{\lambda=0},$$

$$f(\mathbf{J}_d\rho)=\begin{bmatrix} f(\lambda\rho) & 0 \\ 0 & f(-\lambda\rho) \end{bmatrix}_{\lambda=\lambda_d}, \tag{3.134b}$$

and

$$f^{(n)}(\lambda\rho)=\frac{1}{n!}\frac{d^n f(\lambda\rho)}{d\lambda^n},\quad n=1,2,3. \tag{3.134c}$$

Taking the function as the matrix exponential, we have

$$f(\mathbf{K}\rho)=e^{\mathbf{K}\rho},\ \ f(\lambda\rho)=e^{\lambda\rho},\ \ f^{(n)}(\lambda\rho)=\frac{1}{n!}\rho^n e^{\lambda\rho}. \tag{3.135}$$

Substituting (3.135) into (3.134), we obtain

$$e^{\mathbf{K}\rho}=\mathbf{Z}\mathbf{E}(\rho)\mathbf{Z}^{-1}, \tag{3.136a}$$

where

$$\mathbf{E}(\rho)=\mathrm{diag}\left[\mathbf{E}_a(\rho),\mathbf{E}_a(\rho),\mathbf{E}_c(\rho),\mathbf{E}_d(\rho)\right],$$

$$\mathbf{E}_a(\rho)=\begin{bmatrix} 1 & \rho \\ 0 & 1 \end{bmatrix},\ \ \mathbf{E}_c(\rho)=\begin{bmatrix} 1 & \rho & \frac{1}{2}\rho^2 & \frac{1}{6}\rho^3 \\ 0 & 1 & \rho & \frac{1}{2}\rho^2 \\ 0 & 0 & 1 & \rho \\ 0 & 0 & 0 & 1 \end{bmatrix},$$

$$\mathbf{E}_d(\rho)=\begin{bmatrix} e^{\lambda_d\rho} & 0 \\ 0 & e^{-\lambda_d\rho} \end{bmatrix}. \tag{3.136b}$$

With the expression of (3.136a), the solution of $\boldsymbol{\upsilon}(\rho)$ obtained in (3.125) can be rewritten as

$$\boldsymbol{\upsilon}(\rho)=\mathbf{Z}\mathbf{F}(\rho)\mathbf{Z}^{-1}\hat{\mathbf{f}}, \tag{3.137a}$$

where

$$\mathbf{F}(\rho)=\mathbf{E}(\rho)\left\{\int \rho^{-2}\mathbf{E}(-\rho)d\rho\right\}. \tag{3.137b}$$

Using the results of (3.136b) and performing the related integrals of (3.137), we get

$$\begin{aligned}
&\mathbf{F}(\rho)=\mathrm{diag}\left[\mathbf{F}_a(\rho),\mathbf{F}_a(\rho),\mathbf{F}_c(\rho),\mathbf{F}_d(\rho)\right],\\
&\mathbf{F}_a(\rho)=\begin{bmatrix}-\rho^{-1} & -\ln\rho-1\\ 0 & -\rho^{-1}\end{bmatrix},\\
&\mathbf{F}_c(\rho)=\begin{bmatrix}-\rho^{-1} & -\ln\rho-1 & -\rho\ln\rho & \rho^2(1-2\ln\rho)/4\\ 0 & -\rho^{-1} & -\ln\rho-1 & -\rho\ln\rho\\ 0 & 0 & -\rho^{-1} & -\ln\rho-1\\ 0 & 0 & 0 & -\rho^{-1}\end{bmatrix},\\
&\mathbf{F}_d(\rho)=\begin{bmatrix}F_1(\rho) & 0\\ 0 & F_2(\rho)\end{bmatrix},
\end{aligned} \tag{3.138a}$$

where

$$F_1(\rho)=e^{\lambda_d\rho}\int\rho^{-2}e^{-\lambda_d\rho}d\rho,\quad F_2(\rho)=e^{-\lambda_d\rho}\int\rho^{-2}e^{\lambda_d\rho}d\rho. \tag{3.138b}$$

Note that in (3.138) only the real part of $\ln\rho$ is taken when ρ is negative, that is, $\ln\rho=\ln|\rho|$, and $F_1(\rho)$ and $F_2(\rho)$ can be evaluated with the help of exponential integral $\mathrm{Ei}(\rho)$. Taking the integration by parts and using the change of variables for the integrals of (3.138b), we find that

$$F_1(\rho)=-\rho^{-1}-\lambda_d E^*(-\lambda_d\rho),\quad F_2(\rho)=-\rho^{-1}+\lambda_d E^*(\lambda_d\rho), \tag{3.139}$$

in which the function $E^*(x)$ whose $x = \pm\lambda_d\rho$ is related to the exponential integral Ei(x) by

$$E^*(x) = e^{-x}\text{Ei}(x), \tag{3.140a}$$

and (Gautschi and Cahill, 1964)

$$\text{Ei}(x) = \int_{-\infty}^{x} \eta^{-1} e^{\eta} d\eta = -\int_{-x}^{\infty} \eta^{-1} e^{-\eta} d\eta = -\text{Re}\{E_1(-x)\}. \tag{3.140b}$$

In (3.140b), E_1 is the exponential integral function defined in the complex domain by another way.

Note that Ei(x) is defined for nonzero real variable x. For positive values of x, the integral is interpreted as the Cauchy principal value due to the singularity of the integrand at zero. The efficient numerical approximations of $E_1(x)$ and Ei(x) for positive x can be found in Cody and Thacher (1968, 1969). Note that the coefficients in Table II of Cody and Thacher (1969) should be corrected by multiplying 0.5 to the values presented there. With the relation (3.140b) and by combining the approximations in Cody and Thacher (1968, 1969), the function value Ei(x) for any nonzero real x can be efficiently evaluated. And by further multiplying e^{-x} with Ei(x), we construct the efficient approximation for $E^*(x)$. The approximation of $E^*(x)$ with accuracy at least to 6 significant digits can be found in Hsu and Hwu (2023). Note that if one tries to evaluate the values of Ei(x) and e^{-x} separately, and then multiply them together to obtain $E^*(x)$, the trouble of numerical overflow and underflow may occur when $|x|$ is large. Therefore, the evaluation presented in Hsu and Hwu (2023), which avoids such troubles by evaluating $E^*(x)$ at once, is suggested.

With the explicit solution (3.137) for $\boldsymbol{\upsilon}(\rho)$, the transformed function $\breve{\mathbf{v}}(\rho)$ and its derivatives can be readily obtained from $(3.123\text{b})_1$ as

$$\breve{\mathbf{v}}(\rho) = \mathbf{I}_\ell \mathbf{Z}\mathbf{F}(\rho)\mathbf{Z}^{-1}\hat{\mathbf{f}}, \quad \breve{\mathbf{v}}'(\rho) = \mathbf{I}_r \mathbf{Z}\mathbf{F}(\rho)\mathbf{Z}^{-1}\hat{\mathbf{f}}, \tag{3.141}$$

where $\mathbf{I}_\ell = [\mathbf{I}, \mathbf{0}]$ and $\mathbf{I}_r = [\mathbf{0}, \mathbf{I}]$ are two 5×10 constant matrices. Green's functions for the other physical quantities in the transformed domain can then be obtained by substituting (3.141) into (3.127a).

Numerical Integration for Transform Integrals

After we obtain the explicit solutions of Green's functions $\breve{\mathbf{v}}(\rho)$ in ρ-domain, the Green's functions $\mathbf{v}(\hat{\mathbf{x}},\mathbf{x})$ can be evaluated via the transform integral shown in (3.117). It seems that the standard Gaussian quadrature rule can be applied directly for numerical integration. However, since $\rho = 0$ when $\theta = \theta^*$ or $\theta^*+\pi$ ($\theta^* = \alpha+\pi/2$, see Figure 3.6) and $\breve{\mathbf{v}}(\rho)$ of (3.137) contains ρ^{-1}, $\ln\rho$, and $E^*(\pm\lambda_d\rho)$ as shown in (3.138) and (3.139), singularity may occur at these locations. To find an accurate and efficient way to evaluate the singular integral of $\breve{\mathbf{v}}(\rho)$, we consider

$$I = \int_0^{2\pi} g(\rho)h(\theta)d\theta, \tag{3.142}$$

where $g(\rho) = \rho^{-1}$, $\ln\rho$, $E^*(\pm\lambda_d\rho)$, $\cdots$.

To have more integral points located in the region around the singular points, which may improve the accuracy of numerical integration, Wang and Huang (1991) suggested the evaluation of I by cutting the entire integral region into four parts as

$$I = \sum_{n=1}^{4} \int_{\theta_{n-1}}^{\theta_n} g(\rho)h(\theta)d\theta, \quad \text{where } \theta_n = \theta^* + n\pi/2, \quad n = 0,1,\cdots,4. \tag{3.143}$$

In each sub-region the standard Gaussian quadrature rule is employed directly. Since all the related functions of $\breve{\mathbf{v}}(\rho)$ can be proved to be periodic about the argument θ with periodicity π, (3.143) can be further written as

$$I = 2\sum_{n=1}^{2} \int_{\theta_{n-1}}^{\theta_n} g(\rho)h(\theta)d\theta. \tag{3.144}$$

When we apply (3.144) to evaluate the Green's function, accurate results can be obtained but the convergence rate is too slow (see Figure 3 of Hsu and Hwu (2023)). To improve the convergence of numerical integration, we make the following change of variables (Telles, 1987; dos Reis, et al., 2013):

$$\text{First by } \theta(\varsigma) = \theta^* + (2n+\varsigma-1)\pi/4, \text{ and then } \varsigma(t) = c_3t^3 + c_2t^2 + c_1t + c_0. \tag{3.145}$$

With (3.145), the integral I becomes

$$I = \frac{\pi}{2}\sum_{n=1}^{2}\int_{-1}^{1} g(\rho(\varsigma))h(\theta(\varsigma))d\varsigma,$$
$$= \frac{\pi}{2}\sum_{n=1}^{2}\int_{-1}^{1} g(\rho(\varsigma(t)))h(\theta(\varsigma(t)))J(t)dt, \quad J(t) = \frac{d\varsigma}{dt}. \tag{3.146}$$

In order to keep the interval of integration be [−1, 1] and eliminate the singularity of the integrand, the transformation $\varsigma(t)$ should satisfy

$$\varsigma(1) = 1, \quad \varsigma(-1) = -1, \quad \left.\frac{d\varsigma}{dt}\right|_{t=t^*} = 0, \quad \left.\frac{d^2\varsigma}{dt^2}\right|_{t=t^*} = 0, \tag{3.147}$$

where t^* is the singular point corresponding to θ^* when $n = 1$, and to $\theta^* + \pi$ when $n = 2$, and hence

$$t^* = -1, \quad \text{when } n = 1;$$
$$t^* = 1, \quad \text{when } n = 2. \tag{3.148}$$

Substituting $(3.145)_2$ into (3.147), we get

$$c_3 = \frac{1}{1+3(t^*)^2}, \quad c_2 = \frac{-3t^*}{1+3(t^*)^2}, \quad c_1 = \frac{3(t^*)^2}{1+3(t^*)^2}, \quad c_0 = \frac{3t^*}{1+3(t^*)^2}. \tag{3.149}$$

With the results of (3.148) and (3.149), the transformation $\varsigma(t)$ assumed in $(3.145)_2$ and its derivative (Jacobian $J(t)$) become

$$\varsigma(t) = \frac{1}{4}(t+1)^3 - 1, \quad J(t) = \frac{3}{4}(t+1)^2 \quad \text{when } n = 1;$$
$$\varsigma(t) = \frac{1}{4}(t-1)^3 + 1, \quad J(t) = \frac{3}{4}(t-1)^2 \quad \text{when } n = 2. \tag{3.150}$$

By employing the Gaussian quadrature rule to $(3.146)_2$, the integral (3.142) can now be evaluated numerically by

$$I \approx \frac{\pi}{2}\sum_{n=1}^{2}\sum_{i=1}^{n_G} w_i^G g\left(\rho\left(\varsigma\left(t_i^G\right)\right)\right) h\left(\theta\left(\varsigma\left(t_i^G\right)\right)\right) J\left(t_i^G\right), \tag{3.151}$$

where t_i^G, w_i^G are the abscissae and weights of Gaussian points, n_G is the number of Gaussian points.

Note that in boundary element analysis not only the Green's function but also its derivative is required to be evaluated. In this situation, the function $g(\rho)$ of (3.142) may contain the strong singularity induced by ρ^{-2}. To eliminate this kind of singularity, we suggest (Wang and Huang, 1991)

$$I = \int_0^{2\pi} \rho^{-2} h(\theta) d\theta = \int_0^{2\pi} \rho^{-2}\left[h(\theta) - h(\theta^*)\right] d\theta + h(\theta^*) \int_0^{2\pi} \rho^{-2} d\theta. \tag{3.152}$$

On the right-hand side of (3.152), the first term can be evaluated by the same way as (3.151) with $h(\theta)$ replaced by $h(\theta) - h(\theta^*)$. Using the relation (3.116a), the second term can be proved to be

$$h(\theta^*) \int_0^{2\pi} \rho^{-2} d\theta = 0, \quad \text{when } \mathbf{x} \neq \hat{\mathbf{x}}. \tag{3.153}$$

Verification through Numerical Examples

Since no suitable reference solution is available for Green's function of laminated composite thick plates, the correctness of the obtained solutions was verified by Hsu and Hwu (2023) through the rationality of the results and its application to BEM, which will be shown later in Chapter 9.

REFERENCES

Ashton, J.E., 1969, "An Analogy for Certain Anisotropic Plates," *Journal of Composite Materials*, Vol. 3, No. 2, pp. 355–358.

Ashton, J.E. and Whitney, J.M., 1970, *Theory of Laminated Plates*, Technomic Pub., Stamford.

Bronson, R., 1969, *Matrix Methods: An Introduction*, Academic Press, New York.

Christensen, R.M., 1979, *Mechanics of Composite Materials*, Wiley-Interscience, New York.

Cody, W.J. and Thacher, H.C., 1968, "Rational Chebyshev approximations for the exponential integral E1(x)," *Mathematics of Computation*, Vol. 22, pp. 641–649.

Cody, W.J. and Thacher, H.C., 1969, "Chebyshev approximations for the exponential integral Ei(x)," *Mathematics of Computation*, Vol. 23, pp. 289–303.

Cowper, G. R., 1966, "The Shear Coefficient in Timoshenko's Beam Theory," *Journal of Applied Mechanics*, Vol. 3, No. 2, pp. 335–340.

dos Reis A., Lima Albuquerque É. and Palermo Júnior, L., 2013, "The Boundary Element Method Applied to Orthotropic Shear Deformable Plates," *Engineering Analysis with Boundary Elements*, Vol. 37, pp. 38–46.

Franklin, J.N., 1968, *Matrix Theory*, Prentice-Hall, New Jersey.

Gautschi, W. and Cahill, W.F., 1964, "Exponential Integral and Related Functions," in M. Abramowitz and I.A. Stegun, Eds., *Handbook of Mathematical Functions*, pp. 227–251. Dover, New York.

Gel'fand I.M. and Shilov G.E., 1964, *Generalized Functions (Volume I)*, Academic Press, New York.

Horn R.A. and Johnson C.R., 2013, *Matrix Analysis*, 2nd ed. Cambridge University Press, New York.

Hsu, C.W. and Hwu, C., 2022a, "Coupled Stretching-Bending Boundary Element Analysis for Unsymmetric Magneto-Electro-Elastic Laminates with Multiple Holes, Cracks and Inclusions," *Engineering Analysis with Boundary Elements*, Vol. 139, pp. 137–151.

Hsu, C.W. and Hwu, C., 2022b, "Classical Solutions for Coupling Analysis of Unsymmetric Magneto-Electro-Elastic Composite Laminated Thin Plates," *Thin-Walled Structures*, Vol. 181, 110112.

Hsu, C.W. and Hwu, C., 2023, "Green's Functions for Thick Laminated Composite Plates with Coupled Stretching-Bending and Transverse Shear Deformation," *Composite Structures*, Vol. 320, 117179.

Hwu, C., 2010, *Anisotropic Elastic Plates*, Springer, New York.

Hwu, C., 2021, *Anisotropic Elasticity with Matlab*, Springer, Cham.

Jones, R.M., 1974, *Mechanics of Composite Materials*, Scripta, Washington, D.C..

Lo, K.H., Christensen, R.M., and Wu, E.M., 1977, "A High Order Theory of Plate Deformation – Part I: Homogeneous Plates," *Journal of Applied Mechanics*, Vol. 44, pp. 663–668.

Nosier, A. and Reddy, J.N., 1992, "Vibration and Stability Analyses of Cross-Ply Laminated Circular Cylindrical Shells," *Journal of Sound and Vibration*, Vol. 157, pp. 139–159.

Reddy, J. N., 1993, *Applied Functional Analysis and Variational Methods in Engineering*, McGraw-Hill, New York.

Reddy J. N., 2003, *Mechanics of Laminated Composite Plates and Shells*. 2nd ed. CRC Press, Boca Raton.

Szilard, R., 1974, *Theory and Analysis of Plates – Classical and Numerical Methods*, Prentice-Hall, Inc., Englewood Cliffs.

Telles, J.C.F., 1987, "A Self-Adaptive Co-Ordinate Transformation for Efficient Numerical Evaluation of General Boundary Element Integrals," *International Journal for Numerical Methods in Engineering*, Vol. 24, pp. 959–973.

Vinson, J.R. and Sierakowski, R.L., 1986, *The Behavior of Structure Composed of Composite Materials*, Martinus Nijhoff Pub., Dordrecht.
Wang, J. and Huang, M., 1991, "Boundary Element Method for Orthotropic Thick Plates," *Acta Mechanica Sinica*, Vol. 7, pp. 258–266.

CHAPTER 4

Laminated Composite Beams

Beam is a long thin structural component and is a one-dimensional counterpart of the plate (Figure 4.1). Along with the name of beam, there are also other names denoting the long thin structural components like rod, column, shaft, bar, etc. These different names are distinguished not by the structural geometries but by the loading conditions. The term *beam* is used when the structure is subjected to lateral loads such that bending occurs. The term *rod* is used when the structure is subjected to tensile axial loads such that stretching occurs. The term *column* is used when the structure is subjected to compressive axial loads, which may compress and/or buckle the structures. *Shaft* is a cylindrical member subjected to torques and transmits power through rotation. *Bar* is a general term denoting one-dimensional structural members which are capable of carrying and transmitting bending, shearing, torsional, and axial loads or a combination of all four. In this chapter, both the lateral loads and axial loads will be considered. In addition to the thin laminated composite beams presented in the first two sections, the thick laminated beams with transverse shear deformation will be presented in the last two sections of this chapter. Sections 4.1.1 and 4.1.2 discuss the modification of stiffness for narrow and wide beams. Sections 4.2.1 and 4.2.2 present the static, buckling, and free vibration of symmetric laminated beams. Based upon the Timoshenko beam theory, Section 4.3.1 presents the governing equations of thick laminated composite beams in matrix form, which are

DOI: 10.1201/9781003470465-4

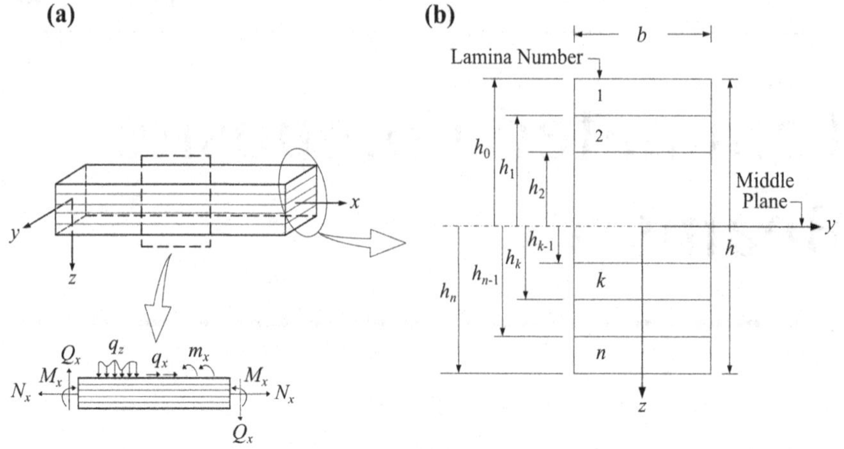

FIGURE 4.1 Geometry and sign convention of a laminated composite beam.

solved in terms of matrix exponential and can be applicable to any possible loading and boundary conditions. With the obtained general solutions, several analytical solutions are derived explicitly such as the Green's functions for an infinite laminated composite thick beam, and the solutions for some typical beam problems, which will then be presented in Sections 4.3.2, 4.4.1, and 4.4.2.

4.1 GENERAL FORMULATION OF THIN LAMINATED BEAMS

Since the beams are one-dimensional counterparts of the plates, it seems that all the mathematical formulations used in the thin plate theory may be employed in beam theory just by neglecting all the terms related to y-coordinate. In this way, the displacement fields assumed in (3.1), strain–displacement relations for small deformation (3.2), constitutive laws for laminated plates (3.3), equilibrium equations (3.9), governing equations (3.11), and boundary conditions (3.12) can all be reduced to those for the thin laminated composite beams as follows:

Displacement fields

$$u(x,z)=u_0(x)-z\frac{\partial w(x)}{\partial x},\quad w(x,z)=w_0(x). \tag{4.1}$$

Strain–displacement relations

$$\varepsilon_x = \frac{\partial u_0}{\partial x} - z\frac{\partial^2 w}{\partial x^2} = \varepsilon_x^0 + z\kappa_x. \tag{4.2}$$

Constitutive laws

$$\begin{Bmatrix} N_x \\ M_x \end{Bmatrix} = \begin{bmatrix} A_{11} & B_{11} \\ B_{11} & D_{11} \end{bmatrix} \begin{Bmatrix} \varepsilon_x^0 \\ k_x \end{Bmatrix}. \tag{4.3}$$

Equilibrium equations

$$\frac{\partial N_x}{\partial x} = 0, \ \frac{\partial Q_x}{\partial x} + q = 0, \ \frac{\partial M_x}{\partial x} - Q_x = 0, \ \frac{\partial^2 M_x}{\partial x^2} + q = 0. \tag{4.4}$$

Governing equations

$$A_{11}\frac{\partial^2 u_0}{\partial x^2} - B_{11}\frac{\partial^3 w}{\partial x^3} = 0, \ -B_{11}\frac{\partial^3 u_0}{\partial x^3} + D_{11}\frac{\partial^4 w}{\partial x^4} = q, \tag{4.5}$$

which can be decoupled into

$$\tilde{B}_{11}\frac{\partial^3 u_0}{\partial x^3} = q, \ \tilde{D}_{11}\frac{\partial^4 w}{\partial x^4} = q, \tag{4.6a}$$

where

$$\tilde{B}_{11} = \frac{A_{11}D_{11} - B_{11}^2}{B_{11}}, \ \tilde{D}_{11} = \frac{A_{11}D_{11} - B_{11}^2}{A_{11}}. \tag{4.6b}$$

Boundary conditions

$$\begin{aligned} &u_0 = \bar{u}_0 && \text{or} && N_x = \bar{N}_x && \text{or} && N_x = k_x u_0, \\ &w_{,x} = \bar{w}_{,x} && \text{or} && M_x = \bar{M}_x && \text{or} && M_x = k_m w_{,x}, \\ &w = \bar{w} && \text{or} && Q_x = \bar{Q}_x && \text{or} && Q_x = k_v w. \end{aligned} \tag{4.7}$$

Note that in the above the resultant forces (N_x, Q_x) and moment M_x are defined for the cross-sectional area of the beam instead of the thickness of the plate. The definitions and units of the associated stiffnesses A_{11}, B_{11}, D_{11} of the beam theory are therefore all different from those of the plate. To avoid misusing these symbols, the definitions of the plate, (2.34), (3.5) and (2.36), should now be replaced by

$$N_x = b\sum_{k=1}^{n}\int_{h_{k-1}}^{h_k}\sigma_x^{(k)}dz,\quad M_x = b\sum_{k=1}^{n}\int_{h_{k-1}}^{h_k}\sigma_x^{(k)}zdz,\quad Q_x = b\sum_{k=1}^{n}\int_{h_{k-1}}^{h_k}\tau_{xz}^{(k)}dz,$$
$$A_{11} = b\sum_{k=1}^{n}Q_{11}^{(k)}\left(h_k - h_{k-1}\right),\quad B_{11} = \frac{b}{2}\sum_{k=1}^{n}Q_{11}^{(k)}\left(h_k^2 - h_{k-1}^2\right),$$
$$D_{11} = \frac{b}{3}\sum_{k=1}^{n}Q_{11}^{(k)}\left(h_k^3 - h_{k-1}^3\right), \tag{4.8}$$

where b is the beam width

By careful thinking of the aforementioned reduction results, several points should be noted. For example, when we write down the constitutive laws (4.3) from (3.3), the Poisson's ratio effects have been ignored, that is, we assume

$$\varepsilon_y^0 = \gamma_{xy}^0 = \kappa_y = \kappa_{xy} = 0 \tag{4.9}$$

When we write down the equilibrium equations (4.4) from (3.9), the resultant forces and moments in y-direction have been ignored, that is, we assume

$$N_y = N_{xy} = Q_y = M_y = M_{xy} = 0. \tag{4.10}$$

However, assumptions (4.9) and (4.10) are not consistent, and hence some modifications are necessary for the beam theory.

4.1.1 Narrow Beams

The applied loads generally considered in a beam structure are the lateral load q, the bending moment M_x, and the axial force N_x. Along the boundaries parallel to the x-axis, no transverse forces and moments, N_y, N_{xy}, M_y, M_{xy}, are applied. When a beam is narrow, it is reasonable to assume that the transverse resultant forces and moments are zero in the entire beams, that is, $N_y = N_{xy} = M_y = M_{xy} = 0$ in the entire narrow beams. Under this assumption, it is better to express the constitutive laws by the inversion of (3.3) instead of (3.3) itself. This inversion has been shown in (2.37). Substituting $N_y = N_{xy} = M_y = M_{xy} = 0$ into (2.37a), we obtain

$$\begin{Bmatrix} \varepsilon_x^0 \\ k_x \end{Bmatrix} = \begin{bmatrix} \hat{A}_{11} & \hat{B}_{11} \\ \hat{B}_{11} & \hat{D}_{11} \end{bmatrix} \begin{Bmatrix} N_x \\ M_x \end{Bmatrix}. \tag{4.11}$$

In (4.11), the relations for the transverse strain, shear strain, and curvatures, $\varepsilon_y^0, \gamma_{xy}^0, \kappa_y, \kappa_{xy}$, are not shown, which does not mean that they vanish. They can be calculated through (2.37), for example, $\varepsilon_y^0 = \hat{A}_{12}N_x + \hat{B}_{12}M_x$, etc. Actually, their existence means that the Poisson's ratio effects are not ignored.

From (4.11), we know that for the narrow beams the constitutive laws (4.3) shown in the general formulation should be corrected as

$$\begin{Bmatrix} N_x \\ M_x \end{Bmatrix} = \begin{bmatrix} \hat{A}_{11} & \hat{B}_{11} \\ \hat{B}_{11} & \hat{D}_{11} \end{bmatrix}^{-1} \begin{Bmatrix} \varepsilon_{x0} \\ \kappa_x \end{Bmatrix} = \frac{1}{\Delta} \begin{bmatrix} \hat{D}_{11} & -\hat{B}_{11} \\ -\hat{B}_{11} & \hat{A}_{11} \end{bmatrix} \begin{Bmatrix} \varepsilon_{x0} \\ \kappa_x \end{Bmatrix}, \tag{4.12a}$$

where

$$\Delta = \hat{A}_{11}\hat{D}_{11} - \hat{B}_{11}^2. \tag{4.12b}$$

This also means that for the narrow beams the stiffnesses used in (4.3), (4.5), and (4.6) should be replaced by the following designation:

$$A_{11} \to \hat{D}_{11}/\Delta, \quad D_{11} \to \hat{A}_{11}/\Delta, \quad B_{11} \to -\hat{B}_{11}/\Delta. \tag{4.13}$$

If the laminates are *symmetric*, the coupling stiffness will be zero, that is, $B_{ij} = 0$. With this result, (2.37b) now leads to

$$\hat{\boldsymbol{A}} = \boldsymbol{A}^{-1}, \quad \hat{\boldsymbol{B}} = \boldsymbol{0}, \quad \hat{\boldsymbol{D}} = \boldsymbol{D}^{-1}. \tag{4.14}$$

The replacement shown in (4.13) may then be simplified as

$$A_{11} \to 1/\hat{A}_{11}, \quad D_{11} \to 1/\hat{D}_{11}, \quad B_{11} \to 0. \tag{4.15}$$

If the laminates are made such that the coupling terms, A_{16}, A_{26}, D_{16}, D_{26}, all vanish, for example, the *symmetric cross-ply laminates*, (4.15) can then be further simplified as

$$A_{11} \rightarrow \frac{A_{11}A_{22} - A_{12}^2}{A_{22}}, \quad D_{11} \rightarrow \frac{D_{11}D_{22} - D_{12}^2}{D_{22}}, \quad B_{11} \rightarrow 0. \qquad (4.16)$$

If the laminates are not only symmetric but also *unidirectional* in x-axis, according to (2.36) the extensional stiffness A_{ij} and bending stiffness D_{ij} may have a very simple relation with the stiffness Q_{ij} as

$$A_{ij} = Q_{ij}h, \quad D_{ij} = Q_{ij}h^3 / 12. \qquad (4.17)$$

With the expressions given in (2.2) for the stiffness Q_{ij} and the relation obtained in (4.17), the replacement (4.16) may be further simplified as

$$A_{11} \rightarrow E_L h, \quad D_{11} \rightarrow E_L h^3 / 12, \quad B_{11} \rightarrow 0. \qquad (4.18)$$

Before making the replacement (4.18), A_{11}, D_{11} from (2.2) and (4.17) are

$$A_{11} = \frac{E_L h}{1 - \nu_{LT}\nu_{TL}}, \quad D_{11} = \frac{E_L h^3}{12\left(1 - \nu_{LT}\nu_{TL}\right)}. \qquad (4.19)$$

By comparison of (4.18) and (4.19), we see that the difference comes from the Poisson's effects.

Note that when we consider the constitutive laws (4.12) instead of (4.3) for the narrow beams, its associated assumption is consistent with the force condition assumed in (4.10).

4.1.2 Wide Beams

A wide beam acts essentially as a plate. Unlike the narrow beams, N_y, N_{xy}, M_y, M_{xy} are not necessary to be zero in the entire beams if no transverse forces and moments are applied along the boundaries parallel to the x-axis. On the contrary, for the wide beams, it is customary to assume that (Swanson, 1997)

$$N_y = N_{xy} = \kappa_y = \kappa_{xy} = 0, \text{ in the entire wide beam.} \qquad (4.20)$$

In order to find the constitutive laws based upon the above assumption, firstly we re-write (2.35) into

$$\begin{Bmatrix} \boldsymbol{\varepsilon}_0 \\ \boldsymbol{M} \end{Bmatrix} = \begin{bmatrix} \boldsymbol{A}^v & \boldsymbol{B}^v \\ -\boldsymbol{B}^{vT} & \boldsymbol{D}^v \end{bmatrix} \begin{Bmatrix} \boldsymbol{N} \\ \boldsymbol{\kappa} \end{Bmatrix}, \qquad (4.21a)$$

where the superscript v is just a symbol to denote that the matrices come from the process of semi-inverse, and

$$\boldsymbol{A}^v = \boldsymbol{A}^{-1}, \quad \boldsymbol{B}^v = -\boldsymbol{A}^{-1}\boldsymbol{B}, \quad \boldsymbol{D}^v = \boldsymbol{D} - \boldsymbol{B}\boldsymbol{A}^{-1}\boldsymbol{B}. \tag{4.21b}$$

Substituting (4.20) into (4.21), we obtain

$$\begin{Bmatrix} \varepsilon_x^0 \\ M_x \end{Bmatrix} = \begin{bmatrix} A_{11}^v & B_{11}^v \\ -B_{11}^v & D_{11}^v \end{bmatrix} \begin{Bmatrix} N_x \\ \kappa_x \end{Bmatrix}, \tag{4.22}$$

which can be re-organized as

$$\begin{Bmatrix} N_x \\ M_x \end{Bmatrix} = \begin{bmatrix} \dfrac{1}{A_{11}^v} & -\dfrac{B_{11}^v}{A_{11}^v} \\ -\dfrac{B_{11}^v}{A_{11}^v} & D_{11}^v + \dfrac{B_{11}^{v2}}{A_{11}^v} \end{bmatrix} \begin{Bmatrix} \varepsilon_x^0 \\ \kappa_x \end{Bmatrix}. \tag{4.23}$$

Like the discussions given for the narrow beams, (4.23) tells us the following replacements are necessary if the assumptions (4.20) are considered for the wide beams.

Non-symmetric laminates:

$$A_{11} \to \frac{1}{A_{11}^v}, \quad D_{11} \to D_{11}^v + \frac{B_{11}^{v2}}{A_{11}^v}, \quad B_{11} \to -\frac{B_{11}^v}{A_{11}^v}. \tag{4.24}$$

Symmetric laminates:

$$A_{11} \to 1/A_{11}^v, \quad D_{11} \to D_{11}, \quad B_{11} \to 0. \tag{4.25}$$

Symmetric cross-ply laminates:

$$A_{11} \to \frac{A_{11}A_{22} - A_{12}^2}{A_{22}}, \quad D_{11} \to D_{11}, \quad B_{11} \to 0. \tag{4.26}$$

Specially orthotropic lamina:

$$A_{11} \to E_L h, \quad D_{11} \to \frac{E_L h^3}{12(1 - \nu_{LT}\nu_{TL})}, \quad B_{11} \to 0. \tag{4.27}$$

Based upon the assumptions (4.20), the reduction of equilibrium equations (3.9) and governing equations (3.11) also leads to the results of (4.4)–(4.6) with the replacements shown in (4.24)–(4.27).

4.2 SYMMETRIC LAMINATED BEAMS

4.2.1 Static Analysis

If a narrow symmetric laminated composite beam is considered, the governing equations (4.5) together with the correction (4.15) become

$$\frac{d^2u_0}{dx^2}=0,\quad \frac{d^4w}{dx^4}=q\widehat{D}_{11}. \tag{4.28}$$

EXAMPLE: CANTILEVER BEAMS

Consider a cantilever symmetric laminated composite beam subjected to a uniformly distributed load q_0 (Figure 4.2). The cross-section of the beam is rectangular with width b and thickness h, and the width-to-thickness ratio is identified to be narrow. Determine the maximum deflection, bending moment, and stress in the beam.

Solution: Refer to (4.7), the boundary condition of this example which is a cantilever beam can be expressed as

$$u_0(0)=w(0)=w'(0)=N_x(\ell)=M_x(\ell)=Q_x(\ell)=0. \tag{4.29}$$

By using the relations (4.2), (4.3), and (4.5) with the replacement (4.15), we have

$$N_x=\frac{1}{\widehat{A}_{11}}\frac{du_0}{dx},\quad M_x=-\frac{1}{\widehat{D}_{11}}\frac{d^2w}{dx^2},\quad Q_x=-\frac{1}{\widehat{D}_{11}}\frac{d^3w}{dx^3}. \tag{4.30}$$

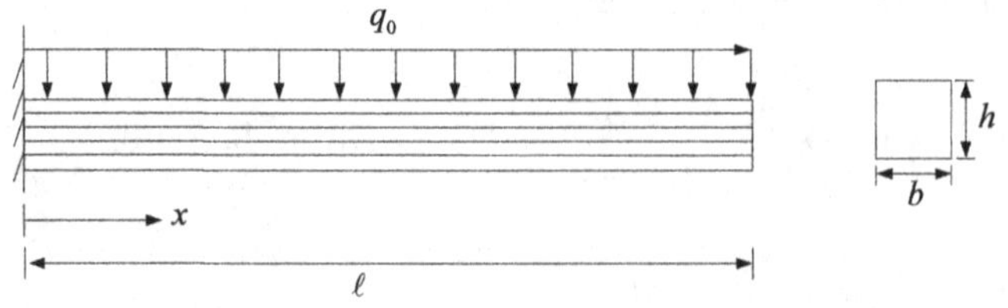

FIGURE 4.2 A cantilever laminated beam subjected to a uniform load.

The general solutions to the governing equations (4.28) are

$$u_0(x) = a_1 x + a_2,$$
$$w(x) = \frac{q_0 \widehat{D}_{11}}{24b} x^4 + \frac{c_1}{6} x^3 + \frac{c_2}{2} x^2 + c_3 x + c_4. \tag{4.31}$$

The unknown coefficients of (4.31) can be determined by satisfying the boundary conditions (4.29) with the assistance of the relations (4.30). The results are

$$a_1 = a_2 = c_3 = c_4 = 0, \quad c_1 = -q_0 \widehat{D}_{11} \ell, \quad c_2 = \frac{1}{2} q_0 \widehat{D}_{11} \ell^2. \tag{4.32}$$

Thus,

$$u_0(x) = 0, \quad w(x) = \frac{q_0 \widehat{D}_{11}}{24} \left(x^4 - 4\ell x^3 + 6\ell^2 x^2 \right). \tag{4.33}$$

Substituting (4.33) into (4.30), we have

$$N_x = 0, \quad M_x = -\frac{q_0}{2} (x - \ell)^2, \quad Q_x = -q_0 (x - \ell). \tag{4.34}$$

The strains and curvatures can be calculated by (2.37) with N_x, M_x obtained in (4.34) and $B_{ij} = 0$, $N_y = N_{xy} = M_y = M_{xy} = 0$, which gives us

$$\varepsilon_x^0 = \varepsilon_y^0 = \gamma_{xy}^0 = 0, \quad \kappa_x = \widehat{D}_{11} M_x, \quad \kappa_y = \widehat{D}_{12} M_x, \kappa_{xy} = \widehat{D}_{16} M_x. \tag{4.35}$$

The stresses in each layer of the laminated composite beams can then be obtained by the stress–strain relation (2.33) in which $\boldsymbol{\varepsilon}_0 = \mathbf{0}$ and $\boldsymbol{\kappa}$ is found in (4.35). With all these results, it can easily be seen that the maximum deflection occurs at the free end and the maximum bending moment occurs at the root of the cantilever beam. They are

$$w_{\max} = w(l) = \frac{q_0 \widehat{D}_{11} \ell^4}{8}, \quad M_{\max} = M_x(0) = -\frac{q_0 \ell^2}{2}. \tag{4.36}$$

As to the maximum stress, no exact solution form can be provided here because it depends on the lamina composition which is not given in this example.

From the formulations given in Section 4.1 and discussions given in Sections 4.1.1 and 4.1.2, and the simple example discussed above, we know that the mathematical formulations of the laminated composite beams are almost the same as those of the isotropic beams. They can even be made exactly the same by the replacement of the bending rigidity. Therefore, almost all the solution techniques discussed in the isotropic beams can be used in the laminated composite beams. For further examples, one may follow the basic fundamentals usually presented in the textbooks of mechanics of materials for isotropic beams such as Gere and Timoshenko (1984).

4.2.2 Buckling and Free Vibration

Similar to the plate buckling and free vibration discussed in Section 3.2.2, to consider these problems for laminated composite beams, the effects of large deformation and time variation should be added.

Buckling

Similar to the derivation of the plate buckling equation (3.40), the buckling equation for symmetric laminated composite beams can be obtained from (3.40) by adding a term related to the axial compressive load N and disregarding the lateral load q, as

$$\tilde{D}_{11}\frac{d^4w}{dx^4}+N\frac{d^2w}{dx^2}=0, \tag{4.37}$$

where $\tilde{D}_{11}$ defined in $(4.6b)_2$ can be calculated by using the replacement (4.13) for the narrow beams or (4.24) for the wide beams, that is,

$$\tilde{D}_{11}=\begin{cases}\dfrac{1}{\hat{D}_{11}}, & \text{for narrow beam,}\\ D_{11}^{v}, & \text{for wide beam.}\end{cases} \tag{4.38}$$

For *simply supported beams*, the boundary conditions ($w(0) = M_x(0) = w(\ell) = M_x(\ell) = 0$) will be satisfied automatically if we assume the deflection as

$$w(x)=\sum_{n=1}^{\infty}w_n\sin\frac{n\pi x}{\ell}. \tag{4.39}$$

Substituting (4.39) into (4.37) yields

$$\sum_{n=1}^{\infty} w_n \left[\tilde{D}_{11} \frac{n^4 \pi^4}{\ell^4} - N \frac{n^2 \pi^2}{\ell^2} \right] \sin \frac{n\pi x}{\ell} = 0. \tag{4.40}$$

A nontrivial solution to (4.40) can be obtained by letting the coefficients of the sine series be zero, from which the critical buckling load N_{cr} is obtained as

$$N_{cr} = \frac{\tilde{D}_{11} n^2 \pi^2}{\ell^2}. \tag{4.41}$$

It is observed that the minimum value of N_{cr} occurs when $n = 1$.

Free Vibration

In writing the governing differential equations of motion for free vibration, we apply d'Alembert's dynamic equilibrium principle by adding the inertia force to the right hand side of $(4.6a)_2$ and neglecting the external force q, that is,

$$\tilde{D}_{11} \frac{\partial^4 w}{\partial x^4} + \rho A \frac{\partial^2 w}{\partial t^2} = 0, \tag{4.42}$$

where $\tilde{D}_{11}$ is given in (4.38); ρ is the mass density of the beam materials and A is the cross-sectional area of the beam.

Consider a *simply supported laminated composite beam*. The vibration mode shapes satisfying the boundary conditions can be assumed to be

$$w(x,t) = \sum_{n=1}^{\infty} w_n \sin \frac{n\pi x}{\ell} \cos \omega_n t. \tag{4.43}$$

Substituting these mode shapes into (4.42) provides the corresponding natural frequencies

$$\omega_n = \frac{n^2 \pi^2}{\ell^2} \sqrt{\frac{\tilde{D}_{11}}{\rho A}}. \tag{4.44}$$

For each n there is a different natural frequency and a different mode shape. The lowest natural frequency, corresponding to $n = 1$, is termed the *fundamental frequency.*

4.3 THICK LAMINATED BEAMS

In the general formulation of thin laminated composite beams, under the assumption of the displacement field given in (4.1), $\gamma_{xz} = \partial u/\partial z + \partial w/\partial x = 0$. In other words, the effects of transverse shear deformation are neglected in the thin laminated composite beams, which may not be true for the thick laminated composite beams. If we consider the first-order shear deformation theory, the assumption of displacement fields given in (4.1) should be modified as

$$u(x,z) = u_0(x) + z\beta_x(x), \quad w(x,z) = w_0(x), \tag{4.45}$$

where β_x is the slope of the deflection and is related to the transverse shear deformation by

$$\beta_x = \gamma_{xz} - \frac{\partial w}{\partial x}. \tag{4.46}$$

Based upon the first-order shear deformation theory (generally called *Timoshenko beam theory* for beam problems), recently we expressed the governing equations in matrix form and solved them in terms of matrix exponential, which is applicable for any possible loading and boundary conditions (Huang et al., 2024). Expanding the matrix exponential through the Taylor series, an explicit analytical solution is obtained for the arbitrarily laminated composite beams under general loading conditions. With the obtained general solutions, several analytical solutions are derived explicitly such as the Green's functions for an infinite laminated composite thick beam, and the solutions for some typical beam problems, which will then be presented in the following two sections. Most of the contents of these two sections are therefore extracted from our recent work (Huang et al., 2024).

4.3.1 General Formulation

By the assumption of the displacement fields (4.45), the strain–displacement relations, the constitutive laws, and the equilibrium equations of the thick laminated composite beams can now be modified as follows:

$$\varepsilon_x = \varepsilon_x^0 + z\kappa_x,\quad \varepsilon_x^0 = \frac{\partial u_0}{\partial x},\quad \kappa_x = \frac{\partial \beta_x}{\partial x},$$
$$N_x = A_{11}\varepsilon_x^0 + B_{11}\kappa_x,\qquad M_x = B_{11}\varepsilon_x^0 + D_{11}\kappa_x,\qquad Q_x = A_{55}\gamma_{xz},$$
$$\frac{\partial N_x}{\partial x} + q_x = 0,\quad \frac{\partial M_x}{\partial x} - Q_x + m_x = 0,\quad \frac{\partial Q_x}{\partial x} + q_z = 0. \tag{4.47}$$

The extensional stiffness A_{11}, coupling stiffness B_{11}, and bending stiffness D_{11} have been defined in (4.8). The additional constant A_{55} counting for the transverse shear resistance is called the transverse shear stiffness and can be evaluated by

$$A_{55} = \alpha A_{55}^0,\quad A_{55}^0 = b\sum_{k=1}^{n} G_{xz}^{(k)}\left(h_k - h_{k-1}\right), \tag{4.48}$$

where $G_{xz}^{(k)}$ is the shear modulus in x-z plane of the kth layer; α is the shear correction factor depending upon the geometry of cross-section and layer information, which should be considered to compromise the inconsistency raised by the assumption of constant transverse shear strain (constant in thickness direction, but may vary in axial direction) made in the Timoshenko beam theory. By taking the average sense for transverse shear strain/stress, and considering the equivalence of strain energy for the assumed constant and the derived non-uniform strain/stress distributions, several formulae for shear correction factor have been presented in the literature, for example, Cowper (1966), Chow (1971), Whitney (1973), Vlachoutsis (1992), Laitinen et al. (1995). To improve the accuracy of Timoshenko beams theory, this factor is required and included in A_{55} of (4.48). By referring to the formula listed in Chow (1971), Vlachoutsis (1992), and Laitinen et al. (1995), we re-express the shear correction factor as (Huang et al., 2024)

$$\alpha = \frac{\Delta^2}{bA_{55}^0 A_{11}^2}\left\{\int_{-h/2}^{h/2} \frac{g^2(z)}{G(z)}dz\right\}^{-1},\quad \Delta = A_{11}D_{11} - B_{11}^2, \tag{4.49a}$$

where

$$g(z) = \int_{-h/2}^{z} Q_{11}(z)\left[z - \frac{B_{11}}{A_{11}}\right]dz,$$
$$G(z) = G_{xz}^{(k)},\quad Q_{11}(z) = Q_{11}^{(k)},\quad \text{when } h_{k-1} \le z < h_k. \tag{4.49b}$$

When a homogeneous isotropic rectangular cross-section is considered, we have $Q_{11}(z) = E/(1 - \nu^2)$, where E and ν are the Young's modulus and Poisson's ratio. With the definitions given in (4.48) and (4.49b), we have

$$A_{11} = \frac{Ebh}{1-\nu^2}, \quad B_{11} = 0, \quad D_{11} = \frac{Ebh^3}{12\left(1-\nu^2\right)},$$

$$A_{55}^0 = \frac{Ebh}{2(1+\nu)}, \quad G(z) = \frac{E}{2(1+\nu)}, \quad g(z) = \frac{E}{2\left(1-\nu^2\right)}\left(z^2 - \frac{h^2}{4}\right). \quad (4.50)$$

Substituting (4.50) into (4.49a), we obtain $\alpha = 5/6$, which is a well-known shear correction factor used for the homogeneous isotropic rectangular cross-section (Reissner, 1947).

Although slight changes may result from the selection of shear correction factor, the correctness of our derived solutions will not be influenced by the selection of its value because when the solutions are compared for a specified laminated cross-section same shear correction factor should be used for the solutions obtained from different methods. In other words, if the comparison is correct for the selection of $\alpha = 5/6$, it will also be correct for the other selected values. Thus, for simplicity of our presentation, without further notification, the commonly selected factor $\alpha = 5/6$ will be used for all the laminated composite beams. If a composite sandwich beam with an orthotropic core whose effective transverse shear modulus in x-z plane is G_{xz}, we have $A_{55} = \alpha bhG_{xz}$, where h is the thickness of the beam.

To get governing equations satisfying all the basic equations of (4.47), we first use $(4.47)_{2,3}$ and (4.46) to express the midplane axial strain ε_x^0, curvature κ_x, and transverse shear strain γ_{xz} in terms of the midplane displacement u_0, slope β_x, and deflection w, then use $(4.46)_{4-6}$ to express the resultant forces N_x, Q_x and moment M_x in terms of the midplane displacements and slope. After these direct substitutions, the three equilibrium equations $(4.46)_{7-9}$ can now be written in terms of three unknown displacement functions u_0, β_x and w as

$$A_{11}\frac{\partial^2 u_0}{\partial x^2} + B_{11}\frac{\partial^2 \beta_x}{\partial x^2} + q_x = 0,$$

$$B_{11}\frac{\partial^2 u_0}{\partial x^2} + D_{11}\frac{\partial^2 \beta_x}{\partial x^2} - A_{55}\left(\beta_x + \frac{\partial w}{\partial x}\right) + m_x = 0,$$

$$A_{55}\left(\frac{\partial \beta_x}{\partial x} + \frac{\partial^2 w}{\partial x^2}\right) + q_z = 0. \quad (4.51)$$

The governing equations obtained in (4.51) are a system of 3 × 3 second-order differential equations, which can be written in matrix form as

$$\mathbf{K}_2\boldsymbol{\delta}'' + \mathbf{K}_1\boldsymbol{\delta}' + \mathbf{K}_0\boldsymbol{\delta} + \mathbf{p} = \mathbf{0}, \tag{4.52a}$$

where $\boldsymbol{\delta}' = \partial\boldsymbol{\delta}/\partial x$, $\boldsymbol{\delta}'' = \partial^2\boldsymbol{\delta}/\partial x^2$, and

$$\mathbf{K}_2 = \begin{bmatrix} A_{11} & B_{11} & 0 \\ B_{11} & D_{11} & 0 \\ 0 & 0 & A_{55} \end{bmatrix}, \ \mathbf{K}_1 = \begin{bmatrix} 0 & 0 & 0 \\ 0 & 0 & -A_{55} \\ 0 & A_{55} & 0 \end{bmatrix},$$
$$\mathbf{K}_0 = \begin{bmatrix} 0 & 0 & 0 \\ 0 & -A_{55} & 0 \\ 0 & 0 & 0 \end{bmatrix}, \ \boldsymbol{\delta} = \begin{Bmatrix} u_0 \\ \beta_x \\ w \end{Bmatrix}, \ \mathbf{p} = \begin{Bmatrix} q_x \\ m_x \\ q_z \end{Bmatrix}. \tag{4.52b}$$

By using the state-space representation, equation (4.52) can be further reduced to a system of 6 × 6 first-order differential equations as

$$\mathbf{K}_2^*\boldsymbol{\Delta}' + \mathbf{K}_1^*\boldsymbol{\Delta} + \mathbf{p}^* = \mathbf{0}, \tag{4.53a}$$

where

$$\mathbf{K}_2^* = \begin{bmatrix} \mathbf{I} & \mathbf{0} \\ \mathbf{0} & \mathbf{K}_2 \end{bmatrix}, \ \mathbf{K}_1^* = \begin{bmatrix} \mathbf{0} & -\mathbf{I} \\ \mathbf{K}_0 & \mathbf{K}_1 \end{bmatrix}, \ \boldsymbol{\Delta} = \begin{Bmatrix} \boldsymbol{\delta} \\ \boldsymbol{\delta}' \end{Bmatrix}, \ \mathbf{p}^* = \begin{Bmatrix} \mathbf{0} \\ \mathbf{p} \end{Bmatrix}. \tag{4.53b}$$

Multiplied by the inverse of $\mathbf{K}_2^*$, equation (4.53a) can now be written in the following standard matrix form

$$\boldsymbol{\Delta}' = \mathbf{L}\boldsymbol{\Delta} + \mathbf{f}, \tag{4.54a}$$

where

$$\mathbf{L} = -\left(\mathbf{K}_2^*\right)^{-1}\mathbf{K}_1^* = \begin{bmatrix} \mathbf{0} & \mathbf{I} \\ -\mathbf{K}_2^{-1}\mathbf{K}_0 & -\mathbf{K}_2^{-1}\mathbf{K}_1 \end{bmatrix},$$
$$\mathbf{f} = -\left(\mathbf{K}_2^*\right)^{-1}\mathbf{p}^* = \begin{Bmatrix} \mathbf{0} \\ -\mathbf{K}_2^{-1}\mathbf{p} \end{Bmatrix}. \tag{4.54b}$$

With the standard matrix form of first-order differential equation, the general solution of (4.54a) can be expressed in terms of *matrix exponential* $e^{\mathbf{L}x}$ as

$$\Delta(x) = e^{\mathbf{L}x}\left\{\mathbf{b} + \int e^{-\mathbf{L}x}\mathbf{f}(x)dx\right\}, \tag{4.55}$$

where **b** stands for the homogeneous solution and is an arbitrary constant vector to be determined through the satisfaction of boundary conditions.

Note that the integral in (4.55) is indefinite and may be analytically calculated for simple form of loading functions such as uniform, concentrated, or polynomial function loads, or numerically for any possible loading conditions. No matter whether integration is performed analytically or numerically, the lower integral limit can be arbitrarily assigned with the upper integral limit being x, and the selection of lower limit will not influence the final solution because they will be absorbed by the homogenous solution. Thus, if we select the lower limit to be zero, (4.55) can be re-written as

$$\Delta(x) = e^{\mathbf{L}x}\mathbf{b} + \int_0^x e^{\mathbf{L}(x-t)}\mathbf{f}(t)dt, \tag{4.56}$$

To know more details about the matrix exponential $e^{\mathbf{L}x}$, we can now calculate it by the following infinite matrix power series (expansion of Taylor series of $e^{\mathbf{L}x}$)

$$e^{\mathbf{L}x} = \sum_{k=0}^{\infty}\frac{(\mathbf{L}x)^k}{k!} = \mathbf{I} + \mathbf{L}x + \frac{(\mathbf{L}x)^2}{2!} + \frac{(\mathbf{L}x)^3}{3!} + \frac{(\mathbf{L}x)^4}{4!} + \cdots. \tag{4.57}$$

In (4.57), the matrix **L** and its kth power $\mathbf{L}^k$ can be calculated by substituting $(4.52\text{b})_{1\text{-}3}$ to $(4.54\text{b})_1$, which leads to

$$\mathbf{L} = \begin{bmatrix} \mathbf{0} & \mathbf{I} \\ \mathbf{L}_0 & \mathbf{L}_1 \end{bmatrix}, \quad \mathbf{L}^2 = \begin{bmatrix} \mathbf{L}_0 & \mathbf{L}_1 \\ \mathbf{L}_2 & \mathbf{L}_3 \end{bmatrix}, \quad \mathbf{L}^3 = \begin{bmatrix} \mathbf{L}_2 & \mathbf{L}_3 \\ \mathbf{0} & \mathbf{0} \end{bmatrix},$$
$$\mathbf{L}^k = \mathbf{0}, \quad \text{when } k \geq 4, \tag{4.58a}$$

where

$$\mathbf{L}_0 = \begin{bmatrix} 0 & -B_0 & 0 \\ 0 & A_0 & 0 \\ 0 & 0 & 0 \end{bmatrix}, \ \mathbf{L}_1 = \begin{bmatrix} 0 & 0 & -B_0 \\ 0 & 0 & A_0 \\ 0 & -1 & 0 \end{bmatrix},$$

$$\mathbf{L}_2 = \begin{bmatrix} 0 & 0 & 0 \\ 0 & 0 & 0 \\ 0 & -A_0 & 0 \end{bmatrix}, \ \mathbf{L}_3 = \begin{bmatrix} 0 & 0 & 0 \\ 0 & 0 & 0 \\ 0 & 0 & -A_0 \end{bmatrix},$$

$$A_0 = A_{55}A_{11}^*, \quad B_0 = A_{55}B_{11}^*. \tag{4.58b}$$

Substituting (4.58) into (4.57), we have

$$e^{\mathbf{L}x} = \begin{bmatrix} 1 & -\frac{B_0}{2}x^2 & 0 & x & 0 & -\frac{B_0}{2}x^2 \\ 0 & 1+\frac{A_0}{2}x^2 & 0 & 0 & x & \frac{A_0}{2}x^2 \\ 0 & -\frac{A_0}{6}x^3 & 1 & 0 & -\frac{1}{2}x^2 & x-\frac{A_0}{6}x^3 \\ 0 & -B_0x & 0 & 1 & 0 & -B_0x \\ 0 & A_0x & 0 & 0 & 1 & A_0x \\ 0 & -\frac{A_0}{2}x^2 & 0 & 0 & -x & 1-\frac{A_0}{2}x^2 \end{bmatrix}. \tag{4.59}$$

Also, substituting $(4.52\text{b})_{1,5}$ into $(4.54\text{b})_2$, we get

$$\mathbf{f} = \begin{Bmatrix} \mathbf{0} \\ \mathbf{f}_0 \end{Bmatrix}, \ \mathbf{f}_0 = \begin{Bmatrix} f_1 \\ f_2 \\ f_3 \end{Bmatrix},$$

$$f_1 = -D_{11}^*q_x + B_{11}^*m_x, \ f_2 = B_{11}^*q_x - A_{11}^*m_x, \ f_3 = -q_z / A_{55}. \tag{4.60}$$

In (4.58b) and (4.60), A_{11}^*, B_{11}^*, and D_{11}^* are defined by

$$A_{11}^* = \frac{A_{11}}{A_{11}D_{11} - B_{11}^2}, \quad B_{11}^* = \frac{B_{11}}{A_{11}D_{11} - B_{11}^2}, \quad D_{11}^* = \frac{D_{11}}{A_{11}D_{11} - B_{11}^2}. \tag{4.61}$$

With the results of (4.59) and (4.60), the solutions of each component of the vector $\mathbf{\Delta}(x)$ of (4.56) can now be written as

$$
\begin{aligned}
u_0(x) &= \phi_u(x) + b_1 + b_4 x - \frac{B_0}{2}(b_2 + b_6)x^2, \\
\beta_x(x) &= \phi_\beta(x) + b_2 + b_5 x + \frac{A_0}{2}(b_2 + b_6)x^2, \\
w(x) &= \phi_w(x) + b_3 + b_6 x - \frac{1}{2}b_5 x^2 - \frac{A_0}{6}(b_2 + b_6)x^3,
\end{aligned} \tag{4.62}
$$

where b_1, b_2, …, b_6 are the components of the vector $\mathbf{b}$, and ϕ_u, ϕ_β, ϕ_w are the first three components of the integral $\int_0^x e^{\mathbf{L}(x-t)}\mathbf{f}(t)dt$, which can be written explicitly as

$$
\begin{aligned}
\phi_u(x) &= \int_0^x \left\{ f_1(t)(x-t) - \frac{B_0}{2} f_3(t)(x-t)^2 \right\} dt, \\
\phi_\beta(x) &= \int_0^x \left\{ f_2(t)(x-t) + \frac{A_0}{2} f_3(t)(x-t)^2 \right\} dt, \\
\phi_w(x) &= \int_0^x \left\{ f_3(t)(x-t) - \frac{1}{2} f_2(t)(x-t)^2 - \frac{A_0}{6} f_3(t)(x-t)^3 \right\} dt.
\end{aligned} \tag{4.63}
$$

Since most of the distributed and concentrated loads can be expressed by using polynomial, sinusoidal, or delta function, we now use the following expressions to represent the loads,

$$
\begin{aligned}
f_i(x) &= f_i^0 x^k, \quad \text{polynomial function}, \\
f_i(x) &= f_i^0 \sin(kx + \alpha), \quad \text{sinusoidal function}, \\
f_i(x) &= f_i^0 \delta(x - \hat{x}), \quad \text{delta function}.
\end{aligned} \tag{4.64}
$$

Substituting (4.64) into (4.63), we have

$$
\begin{aligned}
\phi_u(x) &= f_1^0 T_1(x) - B_0 f_3^0 T_2(x), \\
\phi_\beta(x) &= f_2^0 T_1(x) + A_0 f_3^0 T_2(x), \\
\phi_w(x) &= f_3^0 \{T_1(x) - A_0 T_3(x)\} - f_2^0 T_2(x),
\end{aligned} \tag{4.65a}
$$

where

$$T_1(x)=\frac{x^{k+2}}{(k+2)(k+1)},\ T_2(x)=\frac{xT_1(x)}{(k+3)},\ T_3(x)=\frac{xT_2(x)}{(k+4)},\ \text{polynomial},$$
$$T_n(x)=\frac{1}{6k^4}\frac{d^{3-n}}{dx^{3-n}}\left\{ckx\left(k^2x^2-6\right)+3s\left(k^2x^2-2\right)+\sin(kx+\alpha)\right\},$$
$$n=1,2,3,\ \text{sinusoidal},$$
$$T_1(x)=\bar{x}H(\bar{x}),\ T_2(x)=\frac{\bar{x}^2}{2}H(\bar{x}),\ T_3(x)=\frac{\bar{x}^3}{6}H(\bar{x}),\ \text{delta}. \tag{4.65b}$$

In (4.65b), $c=\cos\alpha, s=\sin\alpha, \bar{x}=x-\hat{x}$, and $H(\bar{x})$ is the Heaviside step function defined by $H(\bar{x})=1$ when $\bar{x}\geq 0$ and $H(\bar{x})=0$ when $\bar{x}<0$.

With the closed-form solutions obtained in (4.62), the midplane axial strain ε_x^0, curvature κ_x, and transverse shear strain γ_{xz} can be calculated directly by

$$\varepsilon_x^0=\frac{\partial u_0}{\partial x},\quad \kappa_x=\frac{\partial \beta_x}{\partial x},\quad \gamma_{xz}=\beta_x+\frac{\partial w}{\partial x}, \tag{4.66}$$

whereas the resultant forces N_x, transverse shear force Q_x, and bending moment M_x are calculated by

$$N_x=A_{11}\frac{\partial u_0}{\partial x}+B_{11}\frac{\partial \beta_x}{\partial x},\quad M_x=B_{11}\frac{\partial u_0}{\partial x}+D_{11}\frac{\partial \beta_x}{\partial x},\quad Q_x=A_{55}\left(\beta_x+\frac{\partial w}{\partial x}\right). \tag{4.67}$$

4.3.2 Green's Functions

Consider an infinite laminated composite beam subjected to a concentrated load $\hat{\mathbf{p}}=\left(\hat{q}_x,\hat{m}_x,\hat{q}_z\right)$ at the point $\hat{x}$. The solution of this problem can be used as a fundamental solution of boundary element method and is generally called *Green's function*. For this problem, the load vector $\mathbf{p}$ of $(4.52\text{b})_5$ can be represented by

$$\mathbf{p}=\delta(x-\hat{x})\hat{\mathbf{p}}, \tag{4.68}$$

where $\delta(x-\hat{x})$ is the Dirac delta function that is zero everywhere except at the point $\hat{x}$ where it is infinite. Combining the solutions obtained in (4.62), (4.65a), and (4.65b)$_3$, we have

$$u_0(x)=b_1+b_4x-\frac{B_0}{2}(b_2+b_6)x^2+\left(\hat{f}_1\bar{x}-\frac{B_0}{2}\hat{f}_3\bar{x}^2\right)H(\bar{x}),$$
$$\beta_x(x)=b_2+b_5x+\frac{A_0}{2}(b_2+b_6)x^2+\left(\hat{f}_2\bar{x}+\frac{A_0}{2}\hat{f}_3\bar{x}^2\right)H(\bar{x}),$$
$$w(x)=b_3+b_6x-\frac{b_5}{2}x^2-\frac{A_0}{6}(b_2+b_6)x^3+\left(\hat{f}_3\bar{x}-\frac{1}{2}\hat{f}_2\bar{x}^2-\frac{A_0}{6}\hat{f}_3\bar{x}^3\right)H(\bar{x}),$$
(4.69a)

where

$$\hat{f}_1=-D^*_{11}\hat{q}_x+B^*_{11}\hat{m}_x,\ \hat{f}_2=B^*_{11}\hat{q}_x-A^*_{11}\hat{m}_x,\ \hat{f}_3=-\hat{q}_z/A_{55},\ \bar{x}=x-\hat{x}. \quad (4.69b)$$

From the viewpoint of mathematical expressions, the solutions shown in (4.69a) consist of two parts: homogeneous solution and particular solution. The former is a solution consisting of arbitrary constants b_i, and the latter is the one related to the concentrated loads. In (4.69a), these two parts are expressed in terms of x and $x-\hat{x}$ separately. Since the constants b_i can be chosen arbitrarily, to have a consistent expression we may also write down the homogenous solution in terms of the polynomial of $x-\hat{x}$. Simply by comparison of coefficients with (4.69a), the homogenous solution noted by the superscript (h) can also be expressed as

$$u_0^{(h)}=c_0+c_1\bar{x}-\frac{B_0}{2}\gamma_0\bar{x}^2,\quad \beta_x^{(h)}=d_0+d_1\bar{x}+\frac{A_0}{2}\gamma_0\bar{x}^2,$$
$$w^{(h)}=e_0+(\gamma_0-d_0)\bar{x}-\frac{d_1}{2}\bar{x}^2-\frac{A_0}{6}\gamma_0\bar{x}^3, \quad (4.70)$$

and c_0, c_1, d_0, d_1, e_0, γ_0 are newly set arbitrary constants.

If we consider to express the particular solution in terms of the distance between x and $\hat{x}$, we may let $r=|x-\hat{x}|=|\bar{x}|$ which leads to

$$\bar{x}=rr',\ H(\bar{x})=(1+r')/2,\ (r')^2=1,$$
$$\bar{x}H(\bar{x})=(r+\bar{x})/2,\ \bar{x}^2H(\bar{x})=(r^2r'+\bar{x}^2)/2,\ \bar{x}^3H(\bar{x})=(r^3+\bar{x}^3)/2. \quad (4.71)$$

Here, $r' = \partial r/\partial x$, and $r' = 1$ when $x \geq \hat{x}$ and $r' = -1$ when $x < \hat{x}$. With the relations given in (4.71), the particular solution of (4.69a) noted by the superscript (p) can be expressed as

$$\begin{aligned} u_0^{(p)} &= \frac{1}{4}\left\{2\hat{f}_1 r - B_0\hat{f}_3 r^2 r' + 2\hat{f}_1\bar{x} - B_0\hat{f}_3\bar{x}^2\right\}, \\ \beta_x^{(p)} &= \frac{1}{4}\left\{2\hat{f}_2 r + A_0\hat{f}_3 r^2 r' + 2\hat{f}_2\bar{x} + A_0\hat{f}_3\bar{x}^2\right\}, \\ w^{(p)} &= \frac{1}{12}\left\{6\hat{f}_3 r - 3\hat{f}_2 r^2 r' - A_0\hat{f}_3 r^3 + 6\hat{f}_3\bar{x} - 3\hat{f}_2\bar{x}^2 - A_0\hat{f}_3\bar{x}^3\right\}. \end{aligned} \tag{4.72}$$

In (4.72), we see that the terms with $\bar{x}$ are parts of the homogeneous solutions (4.70) with $c_1 = \hat{f}_1/2, d_1 = \hat{f}_2/2, \gamma_0 = \hat{f}_3/2, c_0 = d_0 = e_0 = 0$. Therefore, to employ a simpler Green's function in practical engineering applications, we may choose

$$\begin{aligned} u_0 &= \frac{1}{4}\left\{2\hat{f}_1 r - B_0\hat{f}_3 r^2 r'\right\},\ \beta_x = \frac{1}{4}\left\{2\hat{f}_2 r + A_0\hat{f}_3 r^2 r'\right\}, \\ w &= \frac{1}{12}\left\{6\hat{f}_3 r - 3\hat{f}_2 r^2 r' - A_0\hat{f}_3 r^3\right\}. \end{aligned} \tag{4.73}$$

With the selected Green's function (4.73), all the other physical quantities can be evaluated by (4.66) and (4.67), and are obtained as

$$\begin{aligned} \varepsilon_{x0} &= \frac{1}{2}\left\{\hat{f}_1 r' - B_0\hat{f}_3 r\right\},\ \kappa_x = \frac{1}{2}\left\{\hat{f}_2 r' + A_0\hat{f}_3 r\right\},\ \gamma_{xz} = \frac{1}{2}\hat{f}_3 r', \\ N_x &= -\frac{1}{2}\hat{q}_x r',\ M_x = -\frac{1}{2}\left\{\hat{m}_x r' + \hat{q}_z r\right\},\ Q_x = -\frac{1}{2}\hat{q}_z r'. \end{aligned} \tag{4.74}$$

If a Timoshenko beam made by isotropic elastic materials is considered, the Green's function obtained in (4.73) can be reduced to the one presented in (Antes, 2003).

With the Green's function obtained in (4.73) and (4.74), the boundary element method (BEM) for laminated composite beams can be developed and will be presented later in Section 9.2.1. By using the technique of boundary-based finite element method (Hwu et al., 2017), the BEM can be further extended to more complicated beam structures such as curved laminated composite beams with non-uniform cross-sectional properties. To illustrate its applicability, some examples of curved laminated composite beam with uniform, stepped, or smoothly varying thickness subjected to combined loads will be solved by BEM and presented later in Chapter 9.

4.4 TYPICAL THICK BEAM PROBLEMS

4.4.1 Explicit Solutions

As stated in Section 4.3.1, the solutions of $u_0(x)$, $\beta_x(x)$, $w(x)$ have been shown explicitly in (4.62) and (4.63) for the general loading conditions. If a specific loading type is considered such as polynomial and sinusoidal distributed loads, or a concentrated load, the integrals of (4.63) have been obtained analytically in (4.65). With these solutions, all the other physical quantities can be obtained by using the relations given in (4.66) and (4.67). Thus, only six unknown constants $b_1, b_2, \ldots, b_6$ remain to be solved through the boundary conditions. To demonstrate how to get the complete solutions, some typical beams such as cantilever beam, simply supported beam, and fixed-end beam are solved explicitly in this section.

A Cantilever Beam Subjected to a Uniformly Distributed Load

Consider a cantilever laminated composite beam of length ℓ subjected to a uniformly distributed transverse load q_0 (see Figure 4.3 without F_0 and m_0). Under this loading condition, $q_x = m_x = 0$ and $q_z = q_0$ = constant, which leads the components of vector **f** in (4.60) to

$$f_1 = f_2 = 0, \; f_3 = -q_0 / A_{55}. \tag{4.75}$$

Substituting (4.75) into (4.65a) and $(4.65\text{b})_1$ with $k = 0$, we have

$$\phi_u(x) = \frac{B_{11}^*}{6} q_0 x^3, \; \phi_\beta(x) = -\frac{A_{11}^*}{6} q_0 x^3, \; \phi_w(x) = -\frac{q_0}{2A_{55}} x^2 + \frac{A_{11}^*}{24} q_0 x^4. \tag{4.76}$$

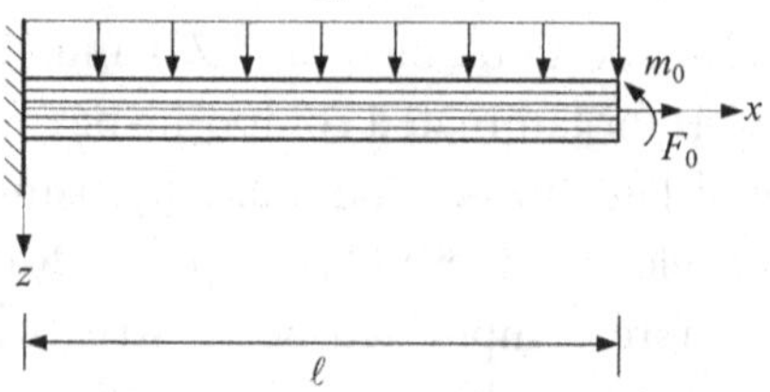

FIGURE 4.3 A cantilever beam subjected to a uniformly distributed load, concentrated axial force, and bending moment.

For a cantilever beam, the boundary conditions can be described by

$$u_0(0)=\beta_x(0)=w(0)=N_x(\ell)=M_x(\ell)=Q_x(\ell)=0. \tag{4.77}$$

Substituting (4.76) into (4.62), and using (4.66) and (4.67) to derive ε_{x0}, κ_x, γ_{xz}, N_x, Q_x, and M_x, and employing the boundary conditions (4.77), we get

$$b_1=b_2=b_3=0,\ \ b_4=\frac{B_{11}^*}{2}q_0\ell^2,\ b_5=-\frac{A_{11}^*}{2}q_0\ell^2,\ b_6=\frac{q_0\ell}{A_{55}}. \tag{4.78}$$

With the results of (4.76) and (4.78), the explicit analytical solutions of this problem can be written as

$$\begin{aligned}
&u_0(x)=\frac{B_{11}^*q_0x}{6}\left(3\ell^2-3\ell x+x^2\right),\ \ \beta_x(x)=-\frac{A_{11}^*q_0x}{6}\left(3\ell^2-3\ell x+x^2\right),\\
&w(x)=q_0x\left[\frac{A_{11}^*}{24}x\left(6\ell^2-4\ell x+x^2\right)+\frac{2\ell-x}{2A_{55}}\right],
\end{aligned} \tag{4.79a}$$

$$\begin{aligned}
&\varepsilon_{x0}(x)=\frac{B_{11}^*q_0}{2}(\ell-x)^2,\ \kappa_x(x)=-\frac{A_{11}^*q_0}{2}(\ell-x)^2,\ \gamma_{xz}(x)=\frac{q_0}{A_{55}}(\ell-x),\\
&N_x(x)=0,\ \ M_x(x)=-\frac{q_0}{2}(\ell-x)^2,\ Q_x(x)=q_0(\ell-x).
\end{aligned} \tag{4.79b}$$

A Cantilever Beam Subjected to a Concentrated Axial Force and Moment

A cantilever laminated composite beam of length ℓ subjected to an axial concentrated force $q_x=F_0\delta(x-\hat{x})$ and a bending moment $m_x=m_0\delta(x-\hat{x})$ at $\hat{x}=\ell$ is considered in this case (see Figure 4.3 without q_0). Through the same derivation procedure described in the previous case, we get

$$\begin{aligned}
&\phi_u(x)=\phi_\beta(x)=\phi_w(x)=0,\\
&b_1=b_2=b_3=b_6=0,\ \ b_4=D_{11}^*F_0-B_{11}^*m_0,\ \ b_5=-B_{11}^*F_0+A_{11}^*m_0.
\end{aligned} \tag{4.80}$$

With the results of (4.80), the explicit solutions of this problem can be written as

$$u_0(x) = \left(D_{11}^* F_0 - B_{11}^* m_0\right)x, \ \beta_x(x) = \left(-B_{11}^* F_0 + A_{11}^* m_0\right)x,$$
$$w(x) = \frac{1}{2}x^2\left(B_{11}^* F_0 - A_{11}^* m_0\right), \tag{4.81a}$$

$$\varepsilon_{x0}(x) = D_{11}^* F_0 - B_{11}^* m_0, \ \kappa_x(x) = -B_{11}^* F_0 + A_{11}^* m_0, \ \gamma_{xz}(x) = 0,$$
$$N_x(x) = F_0, \ M_x(x) = m_0, \ Q_x(x) = 0. \tag{4.81b}$$

A Simply Supported Beam Subjected to Sinusoidal Load

Consider a simply supported laminated composite beam of length ℓ subjected to a sinusoidal distributed transverse load q_0 (see Figure 4.4). The loading and the boundary condition can be expressed as

$$q_x = m_x = 0, \ q_z = q_0 \sin(\pi x/\ell);$$
$$u_0(0) = w(0) = M_x(0) = N_x(\ell) = w(\ell) = M_x(\ell) = 0. \tag{4.82}$$

By the same process as that described previously, we can obtain

$$\phi_u(x) = \frac{B_{11}^* q_0}{2k^3}\left\{k^2x^2 - 2 + 2\cos kx\right\}, \ \phi_\beta(x) = \frac{-A_{11}^* q_0}{2k^3}\left\{k^2x^2 - 2 + 2\cos kx\right\},$$
$$\phi_w(x) = \frac{A_{11}^* q_0}{6k^4}\left\{k^3x^3 - 6kx + 6\sin kx\right\} - \frac{q_0}{k^2 A_{55}}\left\{kx - \sin kx\right\},$$
$$b_1 = b_3 = b_4 = b_5 = 0, \ b_2 = -\frac{A_{11}^* q_0}{k^3}, \ b_6 = \left\{\frac{A_{11}^*}{k^3} + \frac{1}{kA_{55}}\right\}q_0, \ k = \pi/\ell. \tag{4.83}$$

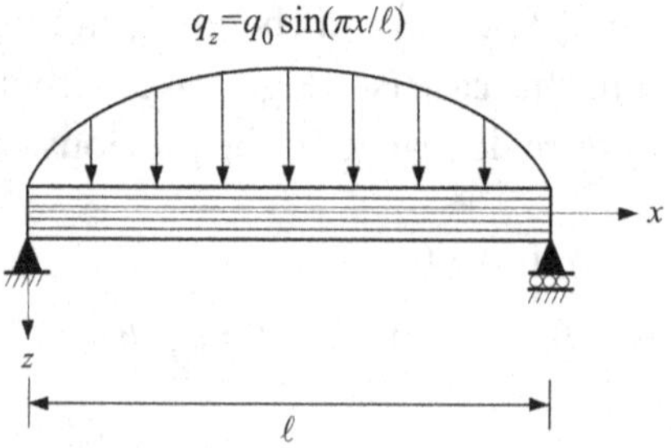

FIGURE 4.4 A simply supported beam subjected to a sinusoidal distributed load.

With the results of (4.83), the explicit solutions of this problem can be written as

$$u_0(x)=\frac{B_{11}^*q_0}{k^3}(\cos kx-1),\ \beta_x(x)=-\frac{A_{11}^*q_0}{k^3}\cos kx,$$
$$w(x)=\frac{q_0}{k^2}\left(\frac{A_{11}^*}{k^2}+\frac{1}{A_{55}}\right)\sin kx, \tag{4.84a}$$

$$\varepsilon_{x0}(x)=-\frac{B_{11}^*q_0}{k^2}\sin kx,\ \kappa_x(x)=\frac{A_{11}^*q_0}{k^2}\sin kx,\ \gamma_{xz}(x)=\frac{q_0}{A_{55}k}\cos kx,$$
$$N_x(x)=0,\ M_x(x)=\frac{q_0}{k^2}\sin kx,\ Q_x(x)=\frac{q_0}{k}\cos kx. \tag{4.84b}$$

A Fixed-End Beam Subjected to a Concentrated Force

Consider a fixed-end beam of length ℓ loaded by a concentrated force P_0 acting at a distance a from the left-hand support (see Figure 4.5). The loading and the boundary condition can be expressed as

$$q_x=m_x=0,\ q_z=P_0\delta(x-a);$$
$$u_0(0)=\beta_x(0)=w(0)=u_0(\ell)=\beta_x(\ell)=w(\ell)=0. \tag{4.85}$$

By the same process as that described in the previous three cases, we obtain the explicit solutions for $0\le x\le a$ as

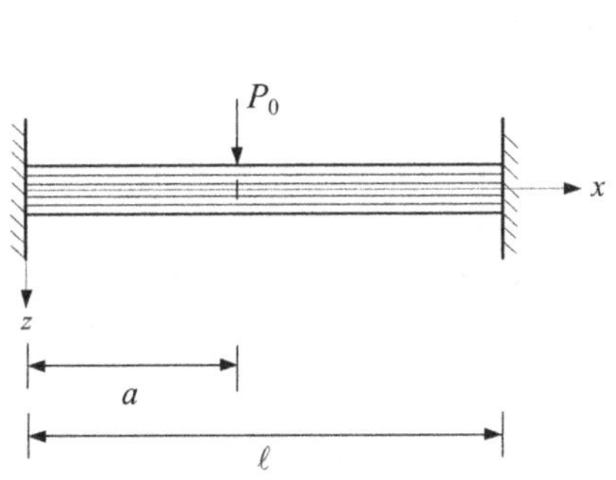

FIGURE 4.5 A fixed-end beam subjected to a concentrated force.

$$u_0(x) = \frac{B_{11}^* P_0}{2\ell h_0} x(a-\ell)\left[12(x-a) + A_0(a-\ell)(2a\ell - 2ax - \ell x)\right],$$
$$\beta_x(x) = -\frac{A_{11}^* P_0}{2\ell h_0} x(a-\ell)\left[12(x-a) + A_0(a-\ell)(2a\ell - 2ax - \ell x)\right],$$
$$w(x) = \frac{P_0}{6A_{55}\ell h_0} x(a-\ell)\left\{-72 + A_0\left\{6\left[2a^2 - \ell^2 + 2x^2 - a(\ell + 3x)\right]\right.\right.$$
$$\left.\left. + A_0 x(a-\ell)(3a\ell - 2ax - \ell x)\right\}\right\},$$
(4.86a)

$$\varepsilon_{x0} = \frac{B_{11}^* P_0}{\ell h_0}(a-\ell)\left\{A_0(a-\ell)\left[a\ell - (2a+\ell)x\right] - 6(a-2x)\right\},$$
$$\kappa_x = \frac{A_{11}^* P_0}{\ell h_0}(a-\ell)\left\{6(a-2x) + A_0(\ell - a)\left[\ell a - (2a+\ell)x\right]\right\},$$
$$\gamma_{xz} = \frac{P_0}{A_{55}\ell h_0}(a-\ell)\left\{A_0 a(2a-\ell) - h_0\right\}, \tag{4.86b}$$

$$N_x = 0,$$
$$M_x = \frac{P_0}{\ell h_0}(a-\ell)\left\{6(a-2x) + A_0(\ell - a)\left[\ell a - (2a+\ell)x\right]\right\},$$
$$Q_x = \frac{P_0}{\ell h_0}(a-\ell)\left\{A_0 a(2a-\ell) - h_0\right\}. \tag{4.86c}$$

where

$$h_0 = 12 + A_0\ell^2. \tag{4.86d}$$

The solutions for $a \le x \le \ell$ can be obtained by using (4.86a–4.86c) with the following replacement

$$a \to \ell - a, \; x \to \ell - x, \tag{4.87}$$

and the sign change for u_0, β_x, γ_{xz}, and Q_x. Note that the values obtained by using the replacement made in (4.87) are viewed from the right end of the beam, which is opposite to the coordinate set at the left end of the beam. And hence, we have the opposite sign for u_0, β_x, γ_{xz}, and Q_x, and the identical sign for w, ε_{x0}, κ_x, N_x, and M_x.

Some Other Typical Beam Problems

In the above, we present the analytical solutions for four representative cases of laminated composite beams. Each of them is supported in a different way and is subjected to different loading types, but is solved in the same way using the results obtained in (4.62)–(4.67) for the general loading and supporting conditions. By following the same approach, without showing the detailed derivation we now list the explicit solutions for some typical problems of laminated composite beams with transverse shear deformation in Tables 4.1 through 4.3, respectively, for the cases of cantilever beams, simply supported beams, and fixed-end beams.

TABLE 4.1 Axial Displacement, Slope, and Deflection of Cantilever Beams (Huang et al., 2024)

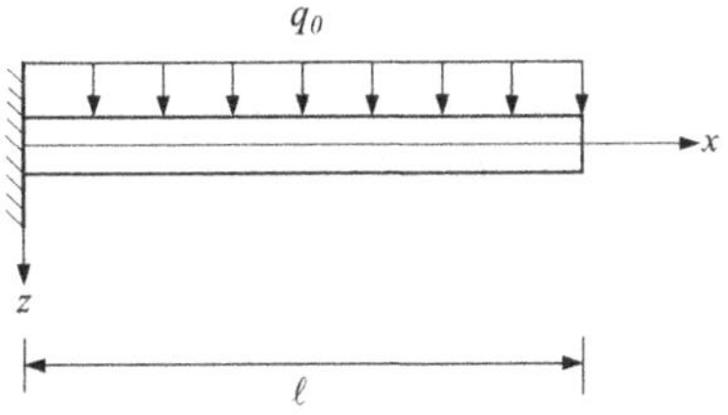	$u_0(x)=\frac{B_{11}^* q_0 x}{6}\left(3\ell^2-3\ell x+x^2\right)$, $\beta_x(x)=-\frac{A_{11}^* q_0 x}{6}\left(3\ell^2-3\ell x+x^2\right)$, $w(x)=q_0 x\left[\frac{A_{11}^*}{24}x\left(6\ell^2-4\ell x+x^2\right)+\frac{2\ell-x}{2A_{55}}\right]$.
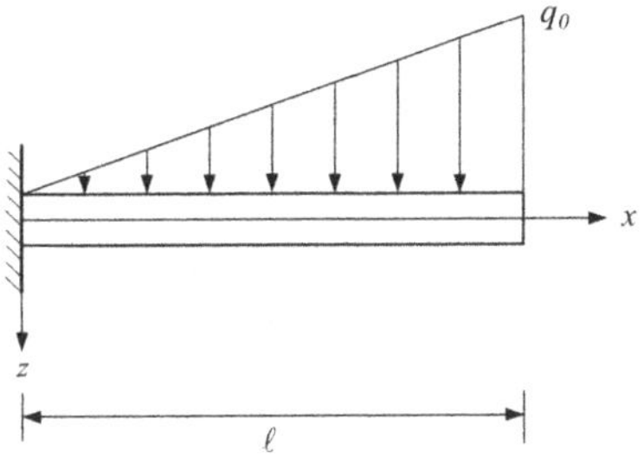	$u_0(x)=\frac{B_{11}^* q_0}{24\ell}x\left(8\ell^3-6\ell^2 x+x^3\right)$, $\beta_x(x)=-\frac{A_{11}^* q_0}{24\ell}x\left(8\ell^3-6\ell^2 x+x^3\right)$, $w(x)=\frac{q_0 x}{\ell}\left[\frac{A_{11}^*}{120}x\left(20\ell^3-10\ell^2 x+x^3\right)+\frac{3\ell^2-x^2}{6A_{55}}\right]$.
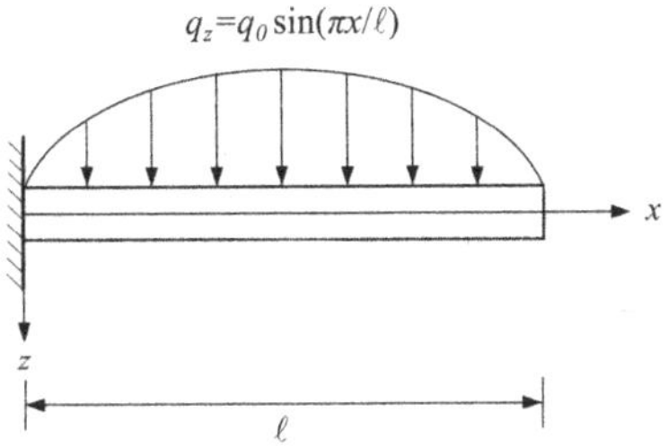	$u_0(x)=\frac{B_{11}^* q_0 \ell^3}{2\pi^3}\left[-2+\left(\frac{\pi}{\ell}\right)^2(2\ell-x)x+2\cos\frac{\pi x}{\ell}\right]$, $\beta_x(x)=\frac{A_{11}^* q_0 \ell^3}{2\pi^3}\left[2+\left(\frac{\pi}{\ell}\right)^2(x-2\ell)x-2\cos\frac{\pi x}{\ell}\right]$, $w(x)=\frac{q_0\ell^2}{\pi^2}\left\{\frac{A_{11}^*\ell^2}{6\pi^2}\left[-\frac{6\pi x}{\ell}-\frac{\pi^3 x^2}{\ell^3}(x-3\ell)+6\sin\frac{\pi x}{\ell}\right]+\frac{1}{A_{55}}\left(\frac{\pi x}{\ell}+\sin\frac{\pi x}{\ell}\right)\right\}$.

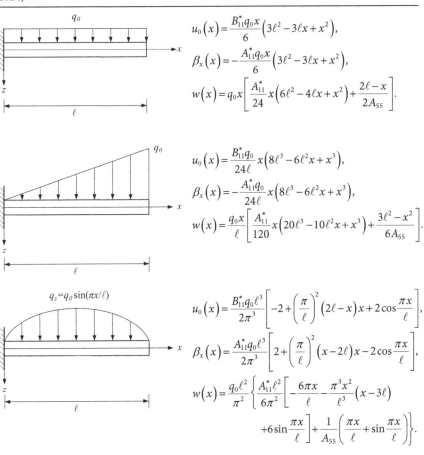

(Continued)

TABLE 4.1 (Continued)

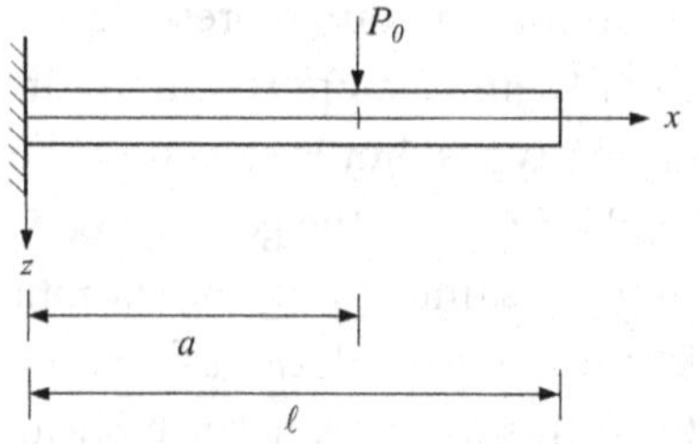	(i) $0 \le x \le a$ $u_0(x) = \frac{B_{11}^* P_0}{2} x(2a - x)$, $\beta_x(x) = \frac{A_{11}^* P_0}{2} x(x - 2a)$, $w(x) = \frac{P_0 x}{6}\left[A_{11}^* x(3a - x) + \frac{6}{A_{55}}\right]$. (ii) $a \le x \le \ell$ $u_0(x) = \frac{B_{11}^* P_0}{2} a^2,\ \beta_x(x) = -\frac{A_{11}^* P_0}{2} a^2$, $w(x) = \frac{P_0 a}{6}\left[A_{11}^* a(3x - a) + \frac{6}{A_{55}}\right]$.
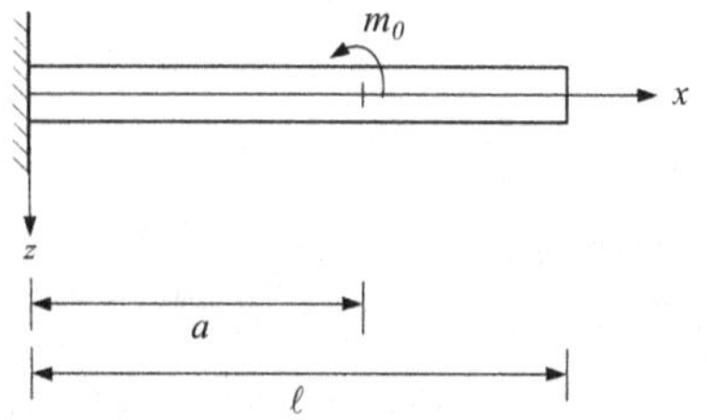	(i) $0 \le x \le a$ $u_0(x) = -B_{11}^* m_0 x,\ \beta_x(x) = A_{11}^* m_0 x$, $w(x) = -\frac{A_{11}^* m_0}{2} x^2$. (ii) $a \le x \le \ell$ $u_0(x) = -B_{11}^* m_0 a,\ \beta_x(x) = A_{11}^* m_0 a$, $w(x) = \frac{A_{11}^* m_0}{2}\left(a^2 - 2ax\right)$.
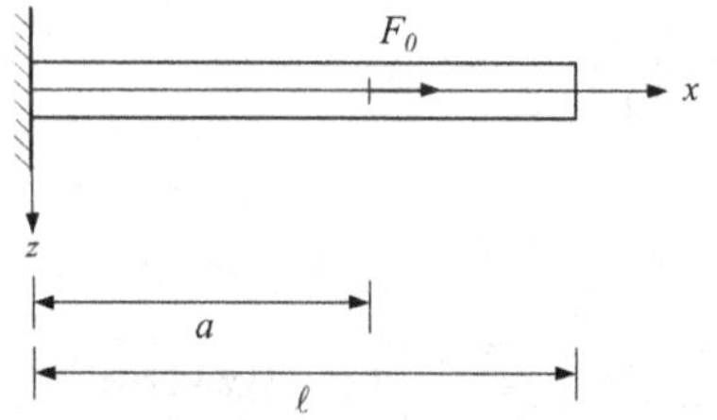	(i) $0 \le x \le a$ $u_0(x) = D_{11}^* F_0 x,\ \ \beta_x(x) = -B_{11}^* F_0 x,\ w(x)$ $= \frac{B_{11}^* F_0}{2} x^2$, (ii) $a \le x \le \ell$ $u_0(x) = D_{11}^* F_0 a,\ \ \beta_x(x) = -B_{11}^* F_0 a$, $w(x) = \frac{B_{11}^* F_0}{2}\left(2xa - a^2\right)$,

TABLE 4.2 Axial Displacement, Slope, and Deflection of Simply Supported Beams (Huang et al., 2024)

Sinusoidal Load (see Figure 4.4) **(4.84)**

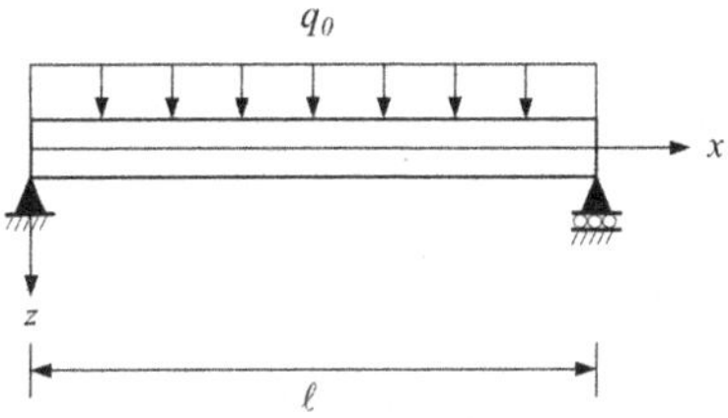

$$u_0(x) = -\frac{B_{11}^* q_0}{12}\left(3\ell x^2 - 2x^3\right),$$

$$\beta_x(x) = -\frac{A_{11}^* q_0}{24}\left(\ell^3 - 6\ell x^2 + 4x^3\right),$$

$$w(x) = \frac{A_{11}^* q_0}{24}\left(\ell^3 x - 2\ell x^3 + x^4\right) + \frac{q_0}{2A_{55}}\left(\ell x - x^2\right).$$

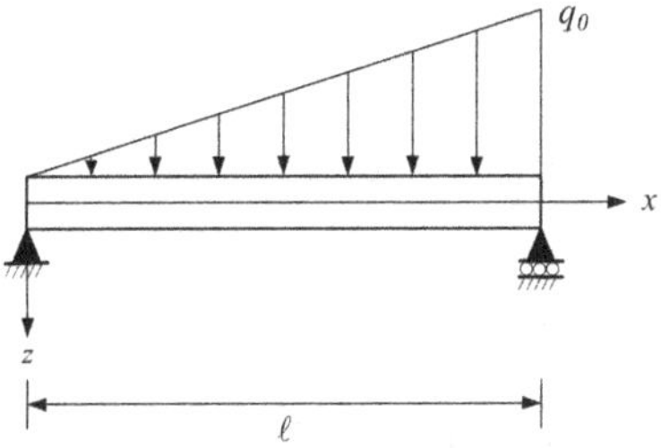

$$u_0(x) = \frac{B_{11}^* q_0}{24\ell} x^2\left(x^2 - 2\ell^2\right),$$

$$\beta_x(x) = -\frac{A_{11}^* q_0}{360\ell}\left(7\ell^4 - 30\ell^2 x^2 + 15x^4\right),$$

$$w(x) = \frac{q_0 x}{\ell}\left(\ell^2 - x^2\right)\left[\frac{A_{11}^*}{360}\left(7\ell^2 - 3x^2\right) + \frac{1}{6A_{55}}\right].$$

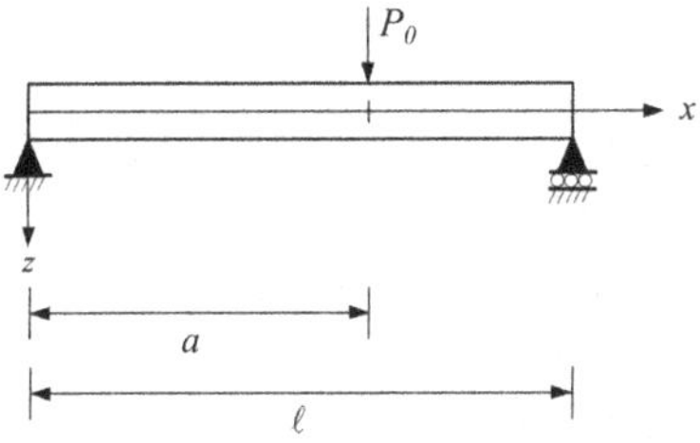

(i) $0 \le x \le a$

$$u_0(x) = \frac{B_{11}^* P_0}{2\ell}(a - \ell)x^2,$$

$$\beta_x(x) = -\frac{A_{11}^* P_0}{6\ell}(a - \ell)\left(a^2 - 2a\ell + 3x^2\right),$$

$$w(x) = \frac{P_0}{\ell}(a - \ell)x\left[\frac{A_{11}^*}{6}\left(a^2 - 2a\ell + x^2\right) - \frac{1}{A_{55}}\right].$$

(ii) $a \le x \le \ell$

$$u_0(x) = \frac{B_{11}^* a P_0}{2\ell}\left[a\ell - 2\ell x + x^2\right],$$

$$\beta_x(x) = \frac{A_{11}^* a P_0}{6\ell}\left(a^2 + 2\ell^2 - 6\ell x + 3x^2\right),$$

$$w(x) = \frac{a P_0}{\ell}(\ell - x)\left[-\frac{A_{11}^*}{6}\left(a^2 - 2x\ell + x^2\right) + \frac{1}{A_{55}}\right].$$

(*Continued*)

TABLE 4.2 (Continued)

Sinusoidal Load (see Figure 4.4) **(4.84)**

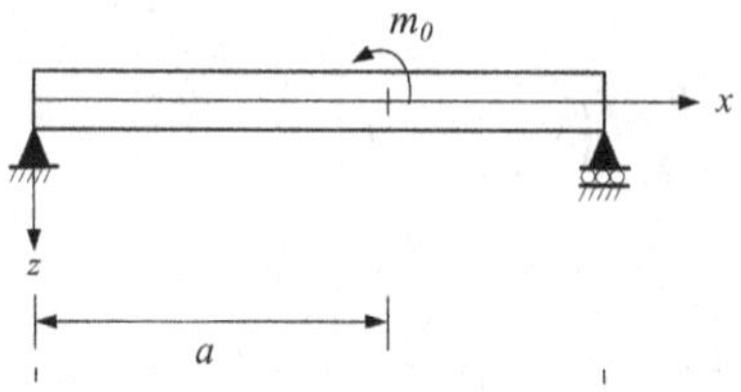

(i) $0 \le x \le a$

$$u_0(x) = \frac{-B_{11}^* m_0}{2\ell} x^2,$$

$$\beta_x(x) = \frac{m_0}{\ell}\left[\frac{A_{11}^*}{6}\left(3a^2 - 6a\ell + 2\ell^2 + 3x^2\right) + \frac{1}{A_{55}}\right],$$

$$w(x) = -\frac{A_{11}^* m_0 x}{6\ell}\left(3a^2 - 6a\ell + 2\ell^2 + x^2\right).$$

(ii) $a \le x \le \ell$

$$u_0(x) = -\frac{B_{11}^* m_0}{2\ell}\left(2a\ell - 2x\ell + x^2\right),$$

$$\beta_x(x) = \frac{m_0}{\ell}\left[\frac{A_{11}^*}{6}\left(3a^2 + 2\ell^2 - 6\ell x + 3x^2\right) + \frac{1}{A_{55}}\right],$$

$$w(x) = \frac{A_{11}^* m_0}{6\ell}(\ell - x)\left(3a^2 - 2\ell x + x^2\right).$$

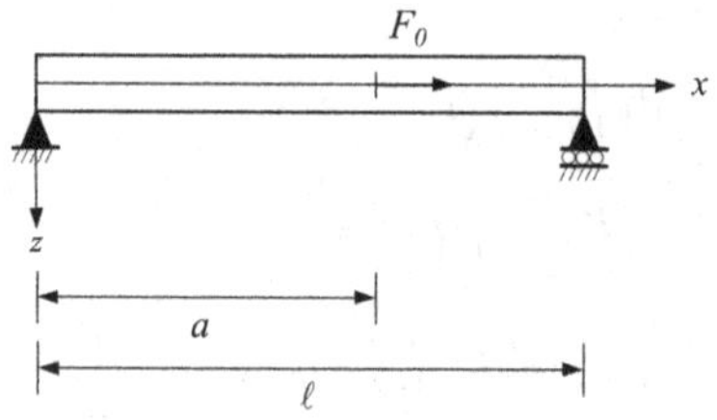

(i) $0 \le x \le a$

$$u_0(x) = D_{11}^* F_0 x, \quad \beta_x(x) = -\frac{B_{11}^* F_0}{2\ell}\left(a^2 - 2a\ell + 2\ell x\right),$$

$$w(x) = \frac{B_{11}^* F_0}{2\ell}\left(a^2 - 2a\ell + \ell x\right)x.$$

(ii) $a \le x \le \ell$

$$u_0(x) = D_{11}^* F_0 a, \quad \beta_x(x) = -\frac{B_{11}^* F_0}{2\ell} a^2,$$

$$w(x) = \frac{B_{11}^* F_0}{2\ell} a^2 (x - \ell).$$

From the explicit solutions shown in Tables 4.1–4.3, we see that all the solutions can be reduced to those of Timoshenko beams made by isotropic elastic materials (Carrer et al., 2014) in which $B_{11} = 0$, $D_{11} = EI$, $A_{55} = 5bhG/6$, where EI and G are the bending rigidity and shear modulus of the beams, and by (4.61) we have $A_{11}^* = 1/EI$ and $B_{11}^* = 0$. Further reduction to the Euler–Bernoulli isotropic beams can be made by dropping the terms associated with A_{55} contributed by transverse shear deformation, and the reduced solutions are identical to those presented in (Goodno and Gere,

TABLE 4.3 Axial Displacement, Slope, and Deflection of Fixed-End Beams (Huang et al., 2024)

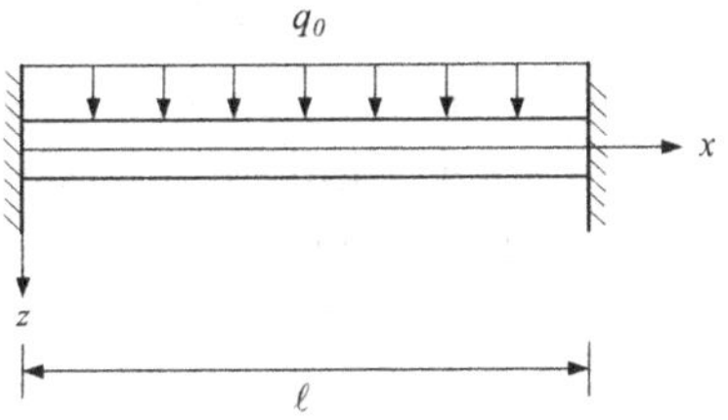

$$u_0(x)=\frac{B_{11}^{*}q_0x}{12}(\ell-2x)(\ell-x),$$

$$\beta_x(x)=-\frac{A_{11}^{*}q_0x}{12}(\ell-2x)(\ell-x),$$

$$w(x)=\frac{q_0x}{24}(\ell-x)\left[A_{11}^{*}x(\ell-x)+\frac{12}{A_{55}}\right].$$

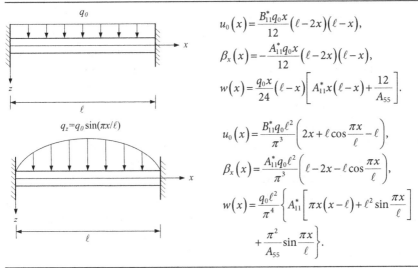

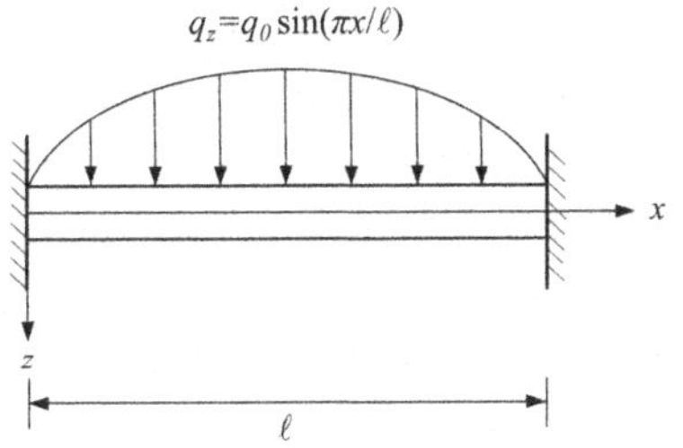

$$u_0(x)=\frac{B_{11}^{*}q_0\ell^2}{\pi^3}\left(2x+\ell\cos\frac{\pi x}{\ell}-\ell\right),$$

$$\beta_x(x)=\frac{A_{11}^{*}q_0\ell^2}{\pi^3}\left(\ell-2x-\ell\cos\frac{\pi x}{\ell}\right),$$

$$w(x)=\frac{q_0\ell^2}{\pi^4}\left\{A_{11}^{*}\left[\pi x(x-\ell)+\ell^2\sin\frac{\pi x}{\ell}\right]+\frac{\pi^2}{A_{55}}\sin\frac{\pi x}{\ell}\right\}.$$

2016). Reversely, without further derivation we may get parts of the explicit solutions of laminated composite beams in the following way:

i. If we have the existing solutions obtained for the isotropic Timoshenko beams, simply by replacing $1/EI$ and $5bhG/6$, respectively, with A_{11}^{*} and A_{55}, we can get the corresponding solutions for the out-of-plane bending and transverse shear deformation of laminated composite beams. However, the replacement would also exclude the terms related to B_{11}^{*} of laminated beams. Hence, through such replacement the solutions obtained reversely from isotropic Timoshenko beams would be valid only for the laminated composite beams with $B_{11}^{*}=0$.

ii. If we have the existing solutions obtained for the isotropic Euler–Bernoulli beams, simply by replacing $1/EI$ with A_{11}^{*}, we can get the corresponding solutions for the laminated composite beams with out-of-plane bending deformation. However, the contributions induced from B_{11}^{*} and A_{55} have been ignored. In other words, in this way the results contributed by the bending–stretching coupling and transverse shear deformation are not included.

iii. If we have the existing solutions obtained for an axially loaded isotropic bar, simply by replacing $1/EA$ with D_{11}^*, we can get the corresponding solutions for the laminated composite beams with lengthen or shorten deformation. However, the contributions induced from B_{11}^* have been ignored. In other words, in this way the results contributed by the bending–stretching coupling are not included.

From the above discussions, we know that if both the coupled stretching–bending and transverse shear effects for arbitrarily laminated composite beams are considered, the complete solutions cannot be obtained just by knowing their associated solutions of isotropic Euler–Bernoulli beams or Timoshenko beams. Thus, the general solution obtained in (4.56) and its related explicit expressions given in (4.62)–(4.67) become very useful for deriving the explicit solutions for general problems of laminated composite beams.

4.4.2 Numerical Results

The analytical solutions of Green's functions and some typical beam problems have been verified through their reduction to the isotropic Timoshenko beams. In order to investigate how lamination sequence, thickness, and width affect laminated composite beams, several numerical examples have been provided in Huang et al. (2024). Based upon the numerical data presented in that paper, we can draw the following conclusion: (1) In the case of symmetric laminated composite beams, stretching and bending deformation decouple, that is, zero axial displacement occurs for the bending cases and zero deflection/slope for the stretching case; (2) When considering an unsymmetric laminated beam, coupling occurs for all different loading types and the magnitude of coupling depends on the values of B_{11}; (3) The stacking sequence significantly influences the deformations, regardless of the types of applied load; (4) As the thickness of the beam increases, the deflection decreases; (5) When the thickness/length ratio increases, the error caused by neglecting the transverse shear effect becomes noticeable. In other words, when the beam is thick the transverse shear effect cannot be ignored; (6) When the beam is narrow or wide, proper stiffness replacement by (4.13) or (4.24) is required for getting more accurate solutions from beam models; (7) For beams with a width between narrow and wide, it is advisable to use a mixture linear relation obtained by (4.13) and (4.24) for appropriate stiffness replacement. This approach can lead to more accurate solutions.

REFERENCES

Antes, H., 2003, "Fundamental solution and integral equations for Timoshenko beams," *Compos Struct.*, Vol. 81, pp. 383–396.

Carrer, J.A.M., Mansur, W.J., Scuciato, R.F., Fliischfresser, S.A., 2014, "Analysis of Euler-Bernoulli and Timoshenko beams by the boundary element method," in: *The 10th World Congress on Computational Mechanics*, São Paulo.

Chow, T.S., 1971, "On the propagation of flexural waves in an orthotropic laminated plate and its response to an impulsive load," *J. Compos. Mater.*, Vol. 5, pp. 306–318.

Cowper, G.R., 1966, "The shear coefficient in Timoshenko's beam theory," *ASME J Mech Appl.*, Vol. 3, pp. 335–340.

Gere, J.M. and Timoshenko, S.P., 1984, *Mechanics of Materials*, PWS Publishers, Boston.

Goodno, B.J. and Gere, J.M., 2016, *Mechanics of Materials*, 9th ed., Cengage Learning, Boston.

Huang, W.Y., Hwu, C. and Hsu, C.W., 2024, "Explicit analytical solutions for some typical problems of arbitrarily laminated composite beams with transverse shear deformation," *Eur. J. Mech. A Solids*, Vol. 103, 105147.

Hwu, C., Huang, S.T., and Li, C.C., 2017, "Boundary-based finite element method for two-dimensional anisotropic elastic solids with multiple holes and cracks," *Eng. Anal. Bound. Elem.*, Vol. 79, pp. 13–22.

Laitinen, M., Lahtinen, H., and Sjölind, S.G., 1995, "Transverse shear correction factors for laminates in cylindrical bending," *Commun. Numer. Meth. Eng.*, Vol. 11, pp. 41–47.

Reissner, E., 1947, "On bending of elastic plates," *Q. Appl. Math.*, Vol. 5, pp. 55–68.

Swanson, S.R., 1997, *Introduction to Design and Analysis with Advanced Composite Materials*, Prentice-Hall, Inc., Upper Saddle River, New Jersey.

Vlachoutsis, S., 1992, "Shear correction factors for plates and shells," *Int. J. Numer. Meth. Eng.* Vol. 33, pp. 1537–1552.

Whitney, J.M., 1973, "Shear correction factors for orthotropic laminates under static load," *ASME. J. Appl. Mech.*, Vol. 40, pp. 302–304.

CHAPTER 5

Laminated Composite Shells

A THIN SHELL IS A body bounded by two closely curved surfaces and can be considered as the materialization of a curved surface. When composed of laminated composites, it becomes a laminated composite shell. To describe the mechanical behavior of laminated composite shells, in Section 5.1 we introduce the basic equations of elasticity such as the strain–displacement relations, constitutive laws, and equilibrium equations for general three-dimensional bodies in terms of curvilinear coordinates, and reduce them to the two-dimensional curvilinear surfaces in shell coordinates. By following the concept of classical lamination theory and utilizing the shell coordinate, mathematical formulation for general laminated composite shells is presented in Section 5.2. These equations form the basis of most of shell analysis. Based upon these general formulations, the special and commonly used shells such as shells of revolution are presented in Section 5.3. Further simplification is made for some special types of shells of revolution such as conical shells, circular cylindrical shells, spherical shells, and shallow spherical shells, etc. To have a clear picture of the solution procedure, a simple case for the shells of revolution with bending resistance is presented in the last part of this section. Special consideration of membrane shells together with their

DOI: 10.1201/9781003470465-5

associated reductions is presented in Section 5.4. Brief discussions on the vibration analysis of general shells will then be presented in the last section of this chapter.

5.1 COORDINATES

The basic equations which describe the behavior of a thin elastic shell were originally derived more than one hundred years ago (Love, 1888). After that, Reissner's version (1941) of the Love theory was derived and described in Kraus's book (1972). Most of the works in the theory of thin elastic shells restrict their attention to shells composed of a single isotropic layer. By generalizing the constitutive laws to the anisotropic laminates, Ambartsumyan (1961) wrote a book devoted to the theory of anisotropic shells. Because all these theories are well established and their basic assumptions are almost the same as the classical plate/lamination theory, no detailed derivation will be given. Only the final results of these equations will be provided in this section.

5.1.1 Curvilinear Coordinates

Because a thin shell is the materialization of a curved surface, in order to describe the mechanical behavior of the shells it is better to use the curvilinear coordinates. Consider a set of three independent functions of the Cartesian variables x, y, z,

$$\alpha_i = \alpha_i\left(x, y, z\right), \quad i = 1,2,3. \tag{5.1}$$

The intersections of the surfaces $\alpha_i(x, y, z) =$ constant, $i = 1, 2, 3$, pair by pair determine the coordinate lines of the curvilinear coordinate system. Any point in the shell is completely defined when α_i, $i = 1, 2, 3$, are given and its associated rectangular Cartesian coordinates x, y, z can also be determined by the inversion of (5.1), that is,

$$x = x\left(\alpha_1,\alpha_2,\alpha_3\right), \quad y = y\left(\alpha_1,\alpha_2,\alpha_3\right), \quad z = z\left(\alpha_1,\alpha_2,\alpha_3\right). \tag{5.2}$$

If the curvilinear coordinate system $(\alpha_1, \alpha_2, \alpha_3)$ is *orthogonal*, the element of arc ds has the form

$$ds^2 = H_1^2 d\alpha_1^2 + H_2^2 d\alpha_2^2 + H_3^2 d\alpha_3^2, \tag{5.3a}$$

where H_1, H_2, and H_3 are the *Lame coefficients* and are determined by

$$H_1^2 = \left(\frac{\partial x}{\partial \alpha_1}\right)^2 + \left(\frac{\partial y}{\partial \alpha_1}\right)^2 + \left(\frac{\partial z}{\partial \alpha_1}\right)^2,$$
$$H_2^2 = \left(\frac{\partial x}{\partial \alpha_2}\right)^2 + \left(\frac{\partial y}{\partial \alpha_2}\right)^2 + \left(\frac{\partial z}{\partial \alpha_2}\right)^2,$$
$$H_3^2 = \left(\frac{\partial x}{\partial \alpha_3}\right)^2 + \left(\frac{\partial y}{\partial \alpha_3}\right)^2 + \left(\frac{\partial z}{\partial \alpha_3}\right)^2. \tag{5.3b}$$

For example, in cylindrical coordinates with $\alpha_1 = r, \alpha_2 = \theta, \alpha_3 = z$ and the relations (5.2) of the form

$$x = r\cos\theta, \quad y = r\sin\theta, \quad z = z, \tag{5.4a}$$

from (5.3b) we obtain the Lame coefficients as

$$H_1 = 1, \quad H_2 = r, \quad H_3 = 1. \tag{5.4b}$$

In spherical coordinates with $\alpha_1 = r, \alpha_2 = \varphi, \alpha_3 = \theta$ and

$$x = r\sin\varphi\cos\theta, \quad y = r\sin\varphi\sin\theta, \quad z = r\cos\varphi, \tag{5.5a}$$

we have

$$H_1 = 1, \quad H_2 = r, \quad H_3 = r\sin\varphi. \tag{5.5b}$$

The Lame coefficients H_1, H_2, and H_3 considered as functions of α_i, $i = 1, 2, 3$ are not independent but are related by the six conditions which secure that the quadratic differential form (5.3a) may be transformable into

$$ds^2 = dx^2 + dy^2 + dz^2. \tag{5.6}$$

These six conditions are

$$\frac{\partial}{\partial \alpha_1}\left(\frac{\partial H_2}{H_1 \partial \alpha_1}\right) + \frac{\partial}{\partial \alpha_2}\left(\frac{\partial H_1}{H_2 \partial \alpha_2}\right) + \frac{1}{H_3^2}\frac{\partial H_1}{\partial \alpha_3}\frac{\partial H_2}{\partial \alpha_3} = 0,$$
$$\frac{\partial}{\partial \alpha_2}\left(\frac{\partial H_3}{H_2 \partial \alpha_2}\right) + \frac{\partial}{\partial \alpha_3}\left(\frac{\partial H_2}{H_3 \partial \alpha_3}\right) + \frac{1}{H_1^2}\frac{\partial H_2}{\partial \alpha_1}\frac{\partial H_3}{\partial \alpha_1} = 0,$$
$$\frac{\partial}{\partial \alpha_3}\left(\frac{\partial H_1}{H_3 \partial \alpha_3}\right) + \frac{\partial}{\partial \alpha_1}\left(\frac{\partial H_3}{H_1 \partial \alpha_1}\right) + \frac{1}{H_2^2}\frac{\partial H_3}{\partial \alpha_2}\frac{\partial H_1}{\partial \alpha_2} = 0, \tag{5.7a}$$

$$\frac{\partial^2 H_1}{\partial\alpha_2\partial\alpha_3}-\frac{1}{H_2}\frac{\partial H_2}{\partial\alpha_3}\frac{\partial H_1}{\partial\alpha_2}-\frac{1}{H_3}\frac{\partial H_3}{\partial\alpha_2}\frac{\partial H_1}{\partial\alpha_3}=0,$$
$$\frac{\partial^2 H_2}{\partial\alpha_1\partial\alpha_3}-\frac{1}{H_3}\frac{\partial H_3}{\partial\alpha_1}\frac{\partial H_2}{\partial\alpha_3}-\frac{1}{H_1}\frac{\partial H_1}{\partial\alpha_3}\frac{\partial H_2}{\partial\alpha_1}=0,$$
$$\frac{\partial^2 H_3}{\partial\alpha_1\partial\alpha_2}-\frac{1}{H_1}\frac{\partial H_1}{\partial\alpha_2}\frac{\partial H_3}{\partial\alpha_1}-\frac{1}{H_2}\frac{\partial H_2}{\partial\alpha_1}\frac{\partial H_3}{\partial\alpha_2}=0. \tag{5.7b}$$

Let u_i, $i = 1, 2, 3$ denote the displacements in the directions tangent to the coordinate lines α_1, α_2 and α_3, and let ε_{ij} and σ_{ij} denote, respectively, the strains and stresses in $\alpha_1 - \alpha_2 - \alpha_3$ coordinates. The strain–displacement relations, constitutive laws, and equilibrium equations based upon the curvilinear coordinate system can be written as follows (Love, 1944; Sokolnikoff, 1956; Ambartsumyan, 1961):

Strain–displacement relations

$$\varepsilon_{11}=\frac{1}{H_1}\frac{\partial u_1}{\partial\alpha_1}+\frac{u_2}{H_1H_2}\frac{\partial H_1}{\partial\alpha_2}+\frac{u_3}{H_1H_3}\frac{\partial H_1}{\partial\alpha_3},$$
$$\varepsilon_{22}=\frac{1}{H_2}\frac{\partial u_2}{\partial\alpha_2}+\frac{u_3}{H_2H_3}\frac{\partial H_2}{\partial\alpha_3}+\frac{u_1}{H_1H_2}\frac{\partial H_2}{\partial\alpha_1},$$
$$\varepsilon_{33}=\frac{1}{H_3}\frac{\partial u_3}{\partial\alpha_3}+\frac{u_1}{H_1H_3}\frac{\partial H_3}{\partial\alpha_1}+\frac{u_2}{H_2H_3}\frac{\partial H_3}{\partial\alpha_2}, \tag{5.8a}$$

$$2\varepsilon_{12}=\gamma_{12}=\frac{H_1}{H_2}\frac{\partial}{\partial\alpha_2}\left(\frac{u_1}{H_1}\right)+\frac{H_2}{H_1}\frac{\partial}{\partial\alpha_1}\left(\frac{u_2}{H_2}\right),$$
$$2\varepsilon_{23}=\gamma_{23}=\frac{H_2}{H_3}\frac{\partial}{\partial\alpha_3}\left(\frac{u_2}{H_2}\right)+\frac{H_3}{H_2}\frac{\partial}{\partial\alpha_2}\left(\frac{u_3}{H_3}\right),$$
$$2\varepsilon_{13}=\gamma_{13}=\frac{H_3}{H_1}\frac{\partial}{\partial\alpha_1}\left(\frac{u_3}{H_3}\right)+\frac{H_1}{H_3}\frac{\partial}{\partial\alpha_3}\left(\frac{u_1}{H_1}\right). \tag{5.8b}$$

Constitutive laws (homogeneous curvilinearly anisotropic solids)

$$\begin{Bmatrix}\sigma_{11}\\ \sigma_{22}\\ \sigma_{33}\\ \sigma_{23}\\ \sigma_{13}\\ \sigma_{12}\end{Bmatrix}=\begin{bmatrix}C_{11} & C_{12} & C_{13} & C_{14} & C_{15} & C_{16}\\ C_{12} & C_{22} & C_{23} & C_{24} & C_{25} & C_{26}\\ C_{13} & C_{23} & C_{33} & C_{34} & C_{35} & C_{36}\\ C_{14} & C_{24} & C_{34} & C_{44} & C_{45} & C_{46}\\ C_{15} & C_{25} & C_{35} & C_{45} & C_{55} & C_{56}\\ C_{16} & C_{26} & C_{36} & C_{46} & C_{56} & C_{66}\end{bmatrix}\begin{Bmatrix}\varepsilon_{11}\\ \varepsilon_{22}\\ \varepsilon_{33}\\ \gamma_{23}\\ \gamma_{13}\\ \gamma_{12}\end{Bmatrix}. \tag{5.9}$$

Equilibrium equations

$$\frac{\partial}{\partial\alpha_1}\left(H_2H_3\sigma_{11}\right)+\frac{\partial}{\partial\alpha_2}\left(H_1H_3\sigma_{12}\right)+\frac{\partial}{\partial\alpha_3}\left(H_1H_2\sigma_{13}\right)$$
$$-\sigma_{22}H_3\frac{\partial H_2}{\partial\alpha_1}-\sigma_{33}H_2\frac{\partial H_3}{\partial\alpha_1}+\sigma_{21}H_3\frac{\partial H_1}{\partial\alpha_2}+\sigma_{31}H_2\frac{\partial H_1}{\partial\alpha_3}+f_1H_1H_2H_3=0, \tag{5.10a}$$

$$\frac{\partial}{\partial\alpha_2}\left(H_3H_1\sigma_{22}\right)+\frac{\partial}{\partial\alpha_3}\left(H_2H_1\sigma_{23}\right)+\frac{\partial}{\partial\alpha_1}\left(H_2H_3\sigma_{21}\right)$$
$$-\sigma_{33}H_1\frac{\partial H_3}{\partial\alpha_2}-\sigma_{11}H_3\frac{\partial H_1}{\partial\alpha_2}+\sigma_{32}H_1\frac{\partial H_2}{\partial\alpha_3}+\sigma_{12}H_3\frac{\partial H_2}{\partial\alpha_1}+f_2H_1H_2H_3=0, \tag{5.10b}$$

$$\frac{\partial}{\partial\alpha_3}\left(H_1H_2\sigma_{33}\right)+\frac{\partial}{\partial\alpha_1}\left(H_2H_3\sigma_{31}\right)+\frac{\partial}{\partial\alpha_2}\left(H_3H_1\sigma_{32}\right)$$
$$-\sigma_{11}H_2\frac{\partial H_1}{\partial\alpha_3}-\sigma_{22}H_1\frac{\partial H_2}{\partial\alpha_3}+\sigma_{13}H_2\frac{\partial H_3}{\partial\alpha_1}+\sigma_{23}H_1\frac{\partial H_3}{\partial\alpha_2}+f_3H_1H_2H_3=0, \tag{5.10c}$$

in which f_i, $i = 1, 2, 3$ are body forces.

5.1.2 Shell Coordinates

As to the plate bending problems, the behavior of the shell is governed by the behavior of its reference surface, and the middle surface of the shell is generally selected as the reference surface. It is therefore appropriate to have an understanding of the representation of the reference surface before embarking upon the derivation of the theory of thin elastic shells. If the curvilinear coordinates of the reference surface are taken to be α_1 and α_2 which are orthogonal, *the first fundamental form of the reference surface* can be represented by

$$ds^2 = A_1^2 d\alpha_1^2 + A_2^2 d\alpha_2^2, \tag{5.11}$$

where A_1 and A_2 are the coefficients of the first fundamental form of the reference surface. With this reference surface, the space occupied by a thin shell can be represented by (Kraus, 1972)

$$ds^2 = A_1^2\left(1+\zeta/R_1\right)^2 d\alpha_1^2 + A_2^2\left(1+\zeta/R_2\right)^2 d\alpha_2^2 + d\zeta^2, \tag{5.12}$$

where ζ is the coordinate along the normal of the reference surface (Figure 5.1); and R_1 and R_2 are the *principal radii of curvature*. By comparing (5.12) with (5.3a), the Lame coefficients of the shells are

$$H_1 = A_1\left(1+\zeta / R_1\right), \quad H_2 = A_2\left(1+\zeta / R_2\right), \quad H_3 = 1. \tag{5.13}$$

As stated in the previous section, these three Lame coefficients are not independent but are related by six conditions shown in (5.7). Substituting (5.13) into (5.7) with $\zeta = 0$, we obtain the *Gauss–Codazzi relations* for the reference surface of shell

$$\begin{aligned} &\frac{\partial}{\partial \alpha_1}\left(\frac{1}{A_1}\frac{\partial A_2}{\partial \alpha_1}\right)+\frac{\partial}{\partial \alpha_2}\left(\frac{1}{A_2}\frac{\partial A_1}{\partial \alpha_2}\right)+\frac{A_1 A_2}{R_1 R_2}=0,\\ &\frac{\partial}{\partial \alpha_2}\left(\frac{A_1}{R_1}\right)-\frac{1}{R_2}\frac{\partial A_1}{\partial \alpha_2}=0, \quad \frac{\partial}{\partial \alpha_1}\left(\frac{A_2}{R_2}\right)-\frac{1}{R_1}\frac{\partial A_2}{\partial \alpha_1}=0. \end{aligned} \tag{5.14}$$

By using $(5.14)_2$ and $(5.14)_3$, we can prove that

$$\frac{1}{H_2}\frac{\partial H_1}{\partial \alpha_2}=\frac{1}{A_2}\frac{\partial A_1}{\partial \alpha_2}, \quad \frac{1}{H_1}\frac{\partial H_2}{\partial \alpha_1}=\frac{1}{A_1}\frac{\partial A_2}{\partial \alpha_1}, \tag{5.15}$$

which will be used in the subsequent discussions.

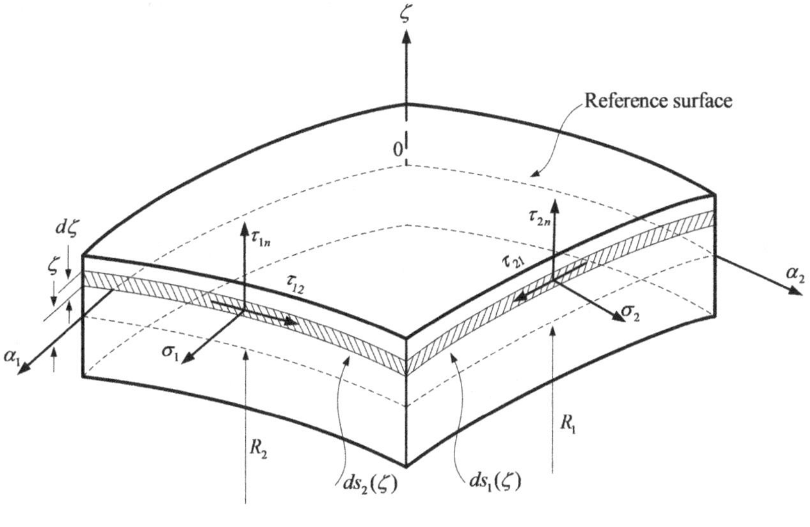

FIGURE 5.1 Differential element of a shell.

To have a clear picture of the shell representation, we now list below some special shell reference surfaces which are usually used in practical applications.

Flat plates:

$$ds^2 = dx^2 + dy^2, \tag{5.16a}$$

$$\alpha_1 = x,\ \ \alpha_2 = y,\ \ A_1 = A_2 = 1,\ \ R_1, R_2 \to \infty. \tag{5.16b}$$

Shells of revolution (Figure 5.2):

$$ds^2 = r_\varphi^2 d\varphi^2 + r^2 d\theta^2, \tag{5.17a}$$

$$\alpha_1 = \varphi,\ \ \alpha_2 = \theta,\ \ A_1 = R_1 = r_\varphi,\ \ A_2 = r,\ \ R_2 = r/\sin\varphi. \tag{5.17b}$$

Examples of shells of revolution are:

Conical shells (Figure 5.3(a)): $r_\varphi \to \infty$, $\varphi = \varphi_c$ = constant and can no longer serve as a coordinate on the meridian of shell, so φ is usually replaced by x and $dx = r_\varphi d\varphi$. Thus,

$$ds^2 = dx^2 + r^2 d\theta^2, \tag{5.18a}$$

$$\alpha_1 = x,\ \alpha_2 = \theta,\ A_1 = 1,\ R_1 \to \infty,\ A_2 = r,\ R_2 = r/\sin\varphi_c. \tag{5.18b}$$

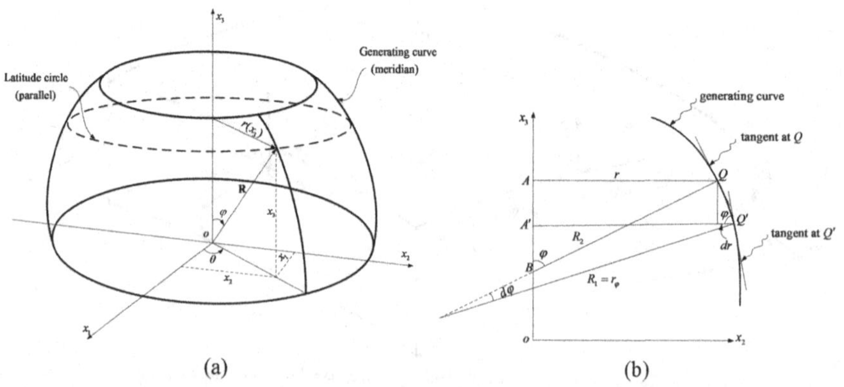

FIGURE 5.2 (a) Geometry of a surface of revolution; (b) generating curve of a surface of revolution.

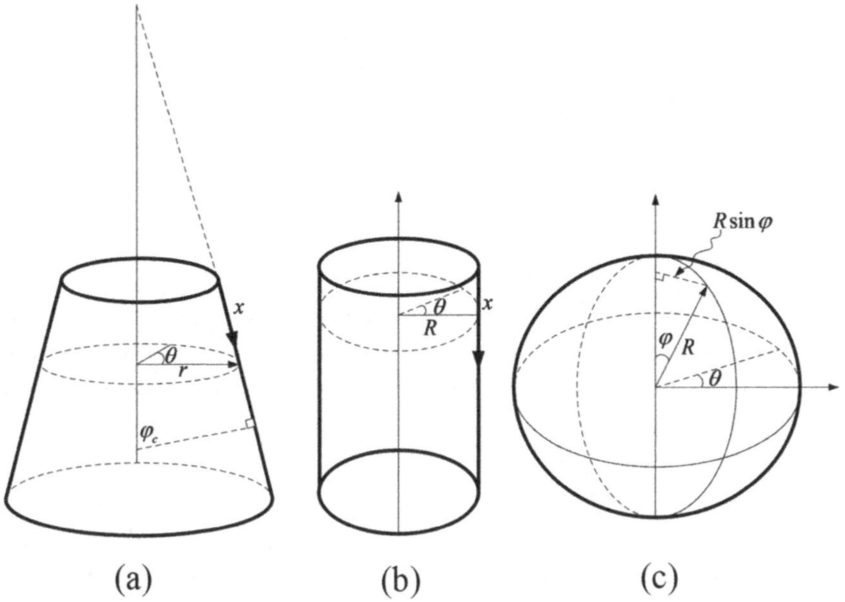

FIGURE 5.3 (a) Conical shells; (b) Circular cylindrical shells; (c) Spherical shells.

Circular cylindrical shells (Figure 5.3(b)): $r_\varphi \to \infty, \varphi = \pi/2, r = R =$ constant. Similar to the conical shells, φ is not suitable to be a coordinate and is replaced by x through $dx = r_\varphi d\varphi$. Thus,

$$ds^2 = dx^2 + R^2 d\theta^2, \tag{5.19a}$$

$$\alpha_1 = x,\ \alpha_2 = \theta,\ A_1 = 1,\ R_1 \to \infty,\ A_2 = R_2 = R. \tag{5.19b}$$

Spherical shells (Figure 5.3(c)): $r_\varphi = R$ and $r = R \sin \varphi$ if the radius of the sphere is R. Thus,

$$ds^2 = R^2 d\varphi^2 + \left(R \sin \varphi\right)^2 d\theta^2, \tag{5.20a}$$

$$\alpha_1 = \varphi,\quad \alpha_2 = \theta,\quad A_1 = R_1 = R_2 = R,\quad A_2 = R \sin \varphi. \tag{5.20b}$$

Ellipsoidal shells:

$$ds^2 = r_\varphi^2 d\varphi^2 + r^2 d\theta^2, \tag{5.21a}$$

$$\alpha_1=\varphi,\ \alpha_2=\theta,\ A_1=R_1=r_\varphi=\frac{a^2b^2}{\left(a^2\sin^2\varphi+b^2\cos^2\varphi\right)^{3/2}},$$

$$A_2=r,\quad R_2=r/\sin\varphi=\frac{a^2}{\left(a^2\sin^2\varphi+b^2\cos^2\varphi\right)^{1/2}}. \tag{5.21b}$$

5.2 GENERAL FORMULATION

By following the development of laminated composite plates described in Section 3.1 and using the shell coordinates given in the previous section, we may write down the general formulation of laminated composite shells as in the following sections (Kraus, 1972; Ambartsumyan, 1961).

5.2.1 Displacement Fields

Because the only difference between the laminated plates and laminated shells is the curvilinear characteristics of shells, we may follow the basic assumptions given in Section 2.4.1 for the laminated plates to describe the laminated shells. According to those assumptions, the displacement fields for the laminated shells can be expressed as

$$\begin{aligned}u_1(\alpha_1,\alpha_2,\zeta)&=u_1^0(\alpha_1,\alpha_2)+\zeta\beta_1(\alpha_1,\alpha_2),\\u_2(\alpha_1,\alpha_2,\zeta)&=u_2^0(\alpha_1,\alpha_2)+\zeta\beta_2(\alpha_1,\alpha_2),\\u_3(\alpha_1,\alpha_2,\zeta)&=w(\alpha_1,\alpha_2),\end{aligned} \tag{5.22}$$

where u_1^0, u_2^0 and w represent the displacements of a point on the reference surface, and β_1 and β_2 represent the rotations of tangents to the reference surface oriented along the parametric lines α_1 and α_2, respectively. If the Kirchhoff assumptions for the deformations of straight lines normal to the reference surface still apply to the laminated shells, that is, $\gamma_{13}=\gamma_{23}=0$, we have

$$\beta_1=\frac{u_1^0}{R_1}-\frac{1}{A_1}\frac{\partial w}{\partial\alpha_1},\quad \beta_2=\frac{u_2^0}{R_2}-\frac{1}{A_2}\frac{\partial w}{\partial\alpha_2}. \tag{5.23}$$

5.2.2 Strain–Displacement Relations

Through the use of (5.8), (5.13)–(5.15), (5.22), (5.23) and the thin shell assumption $\zeta/R_i << 1$, we have

$$\varepsilon_{11}=\varepsilon_{11}^0+\zeta\kappa_{11},\ \varepsilon_{22}=\varepsilon_{22}^0+\zeta\kappa_{22},\ \gamma_{12}=\gamma_{12}^0+\zeta\kappa_{12}, \tag{5.24}$$

where $\varepsilon_{11}^0, \varepsilon_{22}^0$, and γ_{12}^0 represent the normal and shearing strains of the reference surface, and are related to the displacements of reference surface by

$$\varepsilon_{11}^0 = \frac{1}{A_1}\frac{\partial u_1^0}{\partial \alpha_1} + \frac{u_2^0}{A_1 A_2}\frac{\partial A_1}{\partial \alpha_2} + \frac{w}{R_1},$$

$$\varepsilon_{22}^0 = \frac{1}{A_2}\frac{\partial u_2^0}{\partial \alpha_2} + \frac{u_1^0}{A_1 A_2}\frac{\partial A_2}{\partial \alpha_1} + \frac{w}{R_2},$$

$$\gamma_{12}^0 = \frac{A_1}{A_2}\frac{\partial}{\partial \alpha_2}\left(\frac{u_1^0}{A_1}\right) + \frac{A_2}{A_1}\frac{\partial}{\partial \alpha_1}\left(\frac{u_2^0}{A_2}\right), \tag{5.25a}$$

and κ_{11}, κ_{22}, and κ_{12} represent linear distributed components of strains, which are analogous to curvatures and related to the rotations by

$$\kappa_{11} = \frac{1}{A_1}\left(\frac{\partial \beta_1}{\partial \alpha_1} + \frac{\beta_2}{A_2}\frac{\partial A_1}{\partial \alpha_2}\right),$$

$$\kappa_{22} = \frac{1}{A_2}\left(\frac{\partial \beta_2}{\partial \alpha_2} + \frac{\beta_1}{A_1}\frac{\partial A_2}{\partial \alpha_1}\right),$$

$$\kappa_{12} = \frac{A_1}{A_2}\frac{\partial}{\partial \alpha_2}\left(\frac{\beta_1}{A_1}\right) + \frac{A_2}{A_1}\frac{\partial}{\partial \alpha_1}\left(\frac{\beta_2}{A_2}\right). \tag{5.25b}$$

5.2.3 Stress Resultants and Stress Couples

Define the stress resultants and stress couples as the resultants across the thickness of the shell per unit arc length on the reference surface. For example, the stress resultant and stress couple of the stress σ_{11} distributed over an α_1 = constant face of the fundamental element of the shell (see Figure 5.1 and Figure 5.4) are defined as

$$N_1 = \int_{\zeta} \frac{\sigma_{11} ds_2(\zeta) d\zeta}{ds_2(0)} = \int_{\zeta} \sigma_{11}(1+\zeta/R_2) d\zeta = \sum_{k=1}^{n} \int_{\zeta_{k-1}}^{\zeta_k} \sigma_{11}^{(k)}(1+\zeta/R_2) d\zeta,$$

$$M_1 = \int_{\zeta} \frac{\zeta \sigma_{11} ds_2(\zeta) d\zeta}{ds_2(0)} = \int_{\zeta} \sigma_{11}(1+\zeta/R_2)\zeta d\zeta = \sum_{k=1}^{n} \int_{\zeta_{k-1}}^{\zeta_k} \sigma_{11}^{(k)}(1+\zeta/R_2)\zeta d\zeta. \tag{5.26}$$

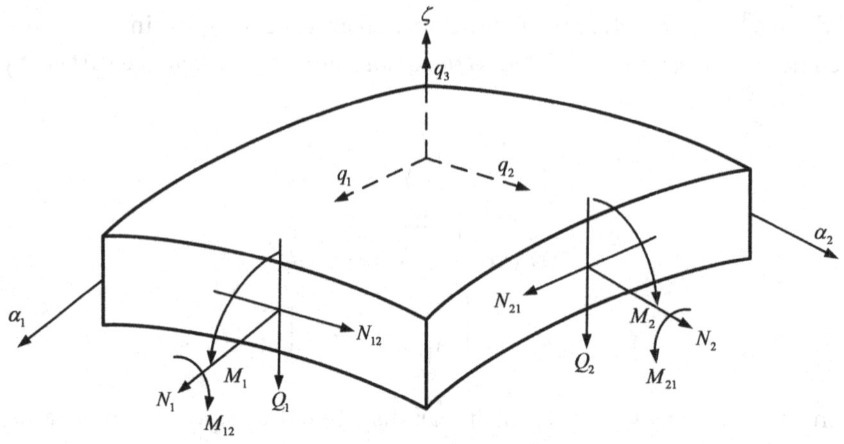

FIGURE 5.4 Stress resultants, stress couples, and surface loads acting on a differential element.

By extension of (5.26), the stress resultants and stress couples are defined by

$$\begin{Bmatrix} N_1 \\ N_{12} \\ Q_1 \end{Bmatrix} = \sum_{k=1}^{n} \int_{\zeta_{k-1}}^{\zeta_k} \begin{Bmatrix} \sigma_{11} \\ \sigma_{12} \\ \sigma_{13} \end{Bmatrix}_k (1+\zeta/R_2)d\zeta,$$

$$\begin{Bmatrix} N_2 \\ N_{21} \\ Q_2 \end{Bmatrix} = \sum_{k=1}^{n} \int_{\zeta_{k-1}}^{\zeta_k} \begin{Bmatrix} \sigma_{22} \\ \sigma_{21} \\ \sigma_{23} \end{Bmatrix}_k (1+\zeta/R_1)d\zeta, \qquad (5.27a)$$

$$\begin{Bmatrix} M_1 \\ M_{12} \end{Bmatrix} = \sum_{k=1}^{n} \int_{\zeta_{k-1}}^{\zeta_k} \begin{Bmatrix} \sigma_{11} \\ \sigma_{12} \end{Bmatrix}_k (1+\zeta/R_2)\zeta d\zeta,$$

$$\begin{Bmatrix} M_2 \\ M_{21} \end{Bmatrix} = \sum_{k=1}^{n} \int_{\zeta_{k-1}}^{\zeta_k} \begin{Bmatrix} \sigma_{22} \\ \sigma_{21} \end{Bmatrix}_k (1+\zeta/R_1)\zeta d\zeta. \qquad (5.27b)$$

It should be noted that by the above definitions the symmetry of the stresses does not necessarily imply that $N_{12} = N_{21}$ or $M_{12} = M_{21}$ except the principal radii of curvature R_1 and R_2 are equal or the shell is very thin (i.e., $\zeta/R_i << 1$).

5.2.4 Constitutive Laws

If each lamina of the laminated shell is a specially orthotropic lamina, the constitutive laws (5.9) for each lamina can be decoupled into two parts. One is in-plane stress–strain relation, and the other is out-of-plane stress–strain relation. Like the laminated plates discussed in Section 2.4, we use $\left(Q_{ij}^*\right)_k$ to stand for the in-plane transformed stiffness of the kth lamina. Thus, for the kth lamina of the shell

$$\begin{Bmatrix} \sigma_{11} \\ \sigma_{22} \\ \sigma_{12} \end{Bmatrix}_k = \begin{bmatrix} Q_{11}^* & Q_{12}^* & Q_{16}^* \\ Q_{12}^* & Q_{22}^* & Q_{26}^* \\ Q_{16}^* & Q_{26}^* & Q_{66}^* \end{bmatrix}_k \begin{Bmatrix} \varepsilon_{11} \\ \varepsilon_{22} \\ \gamma_{12} \end{Bmatrix}. \tag{5.28}$$

Substituting (5.24) into (5.28), and then employing its results in the definitions of stress resultants and stress couples, we obtain

$$\begin{Bmatrix} \mathbf{N} \\ \mathbf{M} \end{Bmatrix} = \begin{bmatrix} \mathbf{A} + \dfrac{1}{R_i}\mathbf{B} & \mathbf{B} + \dfrac{1}{R_i}\mathbf{D} \\ \mathbf{B} + \dfrac{1}{R_i}\mathbf{D} & \mathbf{D} + \dfrac{1}{R_i}\mathbf{E} \end{bmatrix} \begin{Bmatrix} \boldsymbol{\varepsilon}_0 \\ \boldsymbol{\kappa} \end{Bmatrix}, \tag{5.29a}$$

where

$$\mathbf{N} = \begin{Bmatrix} N_1 \\ N_2 \\ N_{12} \text{ or } N_{21} \end{Bmatrix}, \quad \mathbf{M} = \begin{Bmatrix} M_1 \\ M_2 \\ M_{12} \text{ or } M_{21} \end{Bmatrix}, \quad \boldsymbol{\varepsilon}_0 = \begin{Bmatrix} \varepsilon_{11}^0 \\ \varepsilon_{22}^0 \\ \gamma_{12}^0 \end{Bmatrix}, \quad \boldsymbol{\kappa} = \begin{Bmatrix} \kappa_{11} \\ \kappa_{22} \\ \kappa_{12} \end{Bmatrix}, \tag{5.29b}$$

and

$$\begin{aligned} R_i &= R_1 \quad \text{for the equations associated with } N_2, N_{21}, M_2, \text{and } M_{21}, \\ &= R_2 \quad \text{for the equations associated with } N_1, N_{12}, M_1, \text{and } M_{12}. \end{aligned} \tag{5.29c}$$

Same as the definitions given in (2.36) for the laminated plates, **A**, **B**, **D**, and **E** are the stiffnesses determined by

$$\boldsymbol{A}=\sum_{k=1}^{n}\mathbf{Q}_k^*\left(\zeta_k-\zeta_{k-1}\right),\quad \boldsymbol{B}=\frac{1}{2}\sum_{k=1}^{n}\mathbf{Q}_k^*\left(\zeta_k^2-\zeta_{k-1}^2\right),$$

$$\mathbf{D}=\frac{1}{3}\sum_{k=1}^{n}\mathbf{Q}_k^*\left(\zeta_k^3-\zeta_{k-1}^3\right),\quad \mathbf{E}=\frac{1}{4}\sum_{k=1}^{n}\mathbf{Q}_k^*\left(\zeta_k^4-\zeta_{k-1}^4\right). \tag{5.30}$$

If ζ/R_i is very small compared to unity, (5.29a) will be reduced to

$$\begin{Bmatrix}\boldsymbol{N}\\ \boldsymbol{M}\end{Bmatrix}=\begin{bmatrix}\boldsymbol{A} & \boldsymbol{B}\\ \boldsymbol{B} & \boldsymbol{D}\end{bmatrix}\begin{Bmatrix}\boldsymbol{\varepsilon}_0\\ \boldsymbol{\kappa}\end{Bmatrix}, \tag{5.31}$$

in which $N_{12}=N_{21}$ and $M_{12}=M_{21}$. It is also convenient to write (5.31) in component form as

$$\begin{Bmatrix}N_1\\ N_2\\ N_{12}\\ M_1\\ M_2\\ M_{12}\end{Bmatrix}=\begin{bmatrix}A_{11} & A_{12} & A_{16} & B_{11} & B_{12} & B_{16}\\ A_{12} & A_{22} & A_{26} & B_{12} & B_{22} & B_{26}\\ A_{16} & A_{26} & A_{66} & B_{16} & B_{26} & B_{66}\\ B_{11} & B_{12} & B_{16} & D_{11} & D_{12} & D_{16}\\ B_{12} & B_{22} & B_{26} & D_{12} & D_{22} & D_{26}\\ B_{16} & B_{26} & B_{66} & D_{16} & D_{26} & D_{66}\end{bmatrix}\begin{Bmatrix}\varepsilon_1^0\\ \varepsilon_2^0\\ \gamma_{12}^0\\ k_1\\ k_2\\ k_{12}\end{Bmatrix}. \tag{5.32}$$

5.2.5 Equilibrium Equations

To establish the relationships which express the equilibrium of the fundamental element of a laminated shell, we may use the results for isotropic thin shells because the equilibrium equations concern only the balance of forces acting upon the structures and they should be independent of the material types. Or, we may derive the equilibrium equations directly by integrating (5.10) with respect to ζ in which the Lame coefficients are given in (5.13) and assuming ζ/R_i infinitesimally small compared to unity. By either way, we can obtain six equilibrium equations for the shells as (Kraus, 1972).

$$\frac{\partial(N_1A_2)}{\partial\alpha_1}+\frac{\partial(N_{21}A_1)}{\partial\alpha_2}+N_{12}\frac{\partial A_1}{\partial\alpha_2}-N_2\frac{\partial A_2}{\partial\alpha_1}+A_1A_2\left(\frac{Q_1}{R_1}+q_1\right)=0,$$
$$\frac{\partial(N_{12}A_2)}{\partial\alpha_1}+\frac{\partial(N_2A_1)}{\partial\alpha_2}+N_{21}\frac{\partial A_2}{\partial\alpha_1}-N_1\frac{\partial A_1}{\partial\alpha_2}+A_1A_2\left(\frac{Q_2}{R_2}+q_2\right)=0,$$
$$\frac{\partial(Q_1A_2)}{\partial\alpha_1}+\frac{\partial(Q_2A_1)}{\partial\alpha_2}-A_1A_2\left(\frac{N_1}{R_1}+\frac{N_2}{R_2}+q\right)=0, \tag{5.33a}$$

$$\frac{\partial(M_1A_2)}{\partial\alpha_1}+\frac{\partial(M_{21}A_1)}{\partial\alpha_2}+M_{12}\frac{\partial A_1}{\partial\alpha_2}-M_2\frac{\partial A_2}{\partial\alpha_1}+A_1A_2(m_1-Q_1)=0,$$
$$\frac{\partial(M_{12}A_2)}{\partial\alpha_1}+\frac{\partial(M_2A_1)}{\partial\alpha_2}+M_{21}\frac{\partial A_2}{\partial\alpha_1}-M_1\frac{\partial A_1}{\partial\alpha_2}+A_1A_2(m_2-Q_2)=0,$$
$$\frac{M_{21}}{R_2}-\frac{M_{12}}{R_1}+N_{21}-N_{12}=0, \tag{5.33b}$$

where q_1, q_2 and $q(=-q_3)$ are the components of external force vector along α_1, α_2 and the opposite normal of the reference surface of the shell, and m_1, m_2 are the components of external moment vector. These force and moment vectors may include all possible types of body and surface loadings acting on a unit area of the reference surface. If f_i, $i = 1, 2, 3$ stand for the body forces and $\sigma_{i3}(\zeta^+)$, $\sigma_{i3}(\zeta^-)$ denote the surface tractions on the top and bottom surfaces of the shell, we have

$$q_i=\int_{\zeta^-}^{\zeta^+}f_i(1+\zeta/R_1)(1+\zeta/R_2)d\zeta+\left[(1+\zeta/R_1)(1+\zeta/R_2)\sigma_{i3}\right]_{\zeta^-}^{\zeta^+},\ i=1,2,3,$$
$$m_i=\int_{\zeta^-}^{\zeta^+}f_i(1+\zeta/R_1)(1+\zeta/R_2)\zeta d\zeta+\left[(1+\zeta/R_1)(1+\zeta/R_2)\sigma_{i3}\right]_{\zeta^-}^{\zeta^+}$$
$$+\int_{\zeta^-}^{\zeta^+}\sigma_{i3}\left(\frac{\zeta}{R_i}+\frac{\zeta^2}{R_1R_2}\right)d\zeta. \tag{5.34}$$

5.2.6 Governing Equations

The basic equations for the laminated shells are given in (5.22) and (5.23) for the displacement fields, (5.24) and (5.25) for the strain–displacement relations, (5.29) for the constitutive laws, and (5.33) for the equilibrium equations. Among these basic equations, only the constitutive laws depend on the material properties. All the other equations are exactly the same as those of the thin shell theory for the homogeneous isotropic elastic materials. To get governing equations satisfying all these basic equations, we first use (5.25) to express the reference surface strains $\varepsilon_{11}^0, \varepsilon_{22}^0, \gamma_{12}^0$ and curvatures κ_{11}, κ_{22}, κ_{12} in terms of the reference surface displacements u_1^0, u_2^0 and w, then use (5.29) to express the stress resultants N_1, N_2, N_{12}, N_{21} and stress couples M_1, M_2, M_{12}, M_{21} in terms of the reference surface displacements. With these direct substitutions, the equilibrium equations (5.33) can then be reorganized into three partial differential equations in terms of three unknown displacement functions u_1^0, u_2^0, and w. However, due to the mathematical complexity, no systematic approach has been proposed to solve the governing equations for the general laminated shells. Therefore, unlike the governing equations (3.11) for the laminated composite plates, it is meaningless for us to show the complicated form of the governing equations for the laminated composite shells.

5.2.7 Boundary Conditions

As is known, the arbitrary functions solved from the governing equations of the shells must be determined from the boundary conditions. When the laminated shell is treated as a whole by assuming perfect bond between adjacent laminae, its boundary conditions will not differ from the corresponding boundary conditions for the corresponding isotropic shell. Therefore, the boundary conditions for the laminated shells can be expressed in the same way as those of the homogeneous isotropic shells like (Kraus, 1972)

$$
\begin{aligned}
&u_n = \bar{u}_n \quad \text{or} \quad N_n = \bar{N}_n \quad \text{or} \quad N_n = k_n u_n, \\
&u_t = \bar{u}_t \quad \text{or} \quad T_{nt} = \bar{T}_{nt} \quad \text{or} \quad T_{nt} = k_t u_t, \\
&w = \bar{w} \quad \text{or} \quad V_n = \bar{V}_n \quad \text{or} \quad V_n = k_v w, \\
&\beta_n = \bar{\beta}_n \quad \text{or} \quad M_n = \bar{M}_n \quad \text{or} \quad M_n = k_m \beta_n,
\end{aligned} \tag{5.35}
$$

in which the overbar denotes the prescribed value, and k_n, k_t, k_v, and k_m are the given spring constants. T_{nt} and V_n are the well-known *Kirchhoff's effective shear force* defined by

$$
T_{nt} = N_{nt} + \frac{M_{nt}}{R_t}, \quad V_n = Q_n + \frac{\partial M_{nt}}{A_t \partial \alpha_t}, \tag{5.36}
$$

where n and t denote, respectively, the normal and tangential directions on a given edge.

5.3 SHELLS OF REVOLUTION

In the previous section, we have shown the basic equations which describe the static behavior of thin elastic shells. These equations form the basis of most of the shell analysis which have been carried out in the past. However, due to their generality and complexity, before embarking upon the solution of these equations we like to simplify them to special and commonly used shells – *shells of revolution*. With the notations and relations given in (5.17) and the facts that r_φ and r are independent of θ for the shells of revolution, the Gauss–Codazzi relations (5.14) and the basic equations shown in (5.22)–(5.36) can then be written as follows:

Gauss–Codazzi relations

$$\frac{\partial r}{\partial \varphi} = r_\varphi \cos\varphi. \tag{5.37}$$

Displacement fields

$$\begin{aligned} u_\varphi(\varphi,\theta,\zeta) &= u_\varphi^0(\varphi,\theta) + \zeta\beta_\varphi(\varphi,\theta), \\ u_\theta(\varphi,\theta,\zeta) &= u_\theta^0(\varphi,\theta) + \zeta\beta_\theta(\varphi,\theta), \\ u_3(\varphi,\theta,\zeta) &= w(\varphi,\theta), \end{aligned} \tag{5.38a}$$

where

$$\beta_\varphi = \frac{1}{r_\varphi}\left(u_\varphi^0 - \frac{\partial w}{\partial \varphi}\right), \quad \beta_\theta = \frac{1}{r}\left(u_\theta^0 \sin\varphi - \frac{\partial w}{\partial \theta}\right). \tag{5.38b}$$

Strain–displacement relations

$$\varepsilon_\varphi = \varepsilon_\varphi^0 + \zeta\kappa_\varphi, \quad \varepsilon_\theta = \varepsilon_\theta^0 + \zeta\kappa_\theta, \quad \gamma_{\varphi\theta} = \gamma_{\varphi\theta}^0 + \zeta\kappa_{\varphi\theta}, \tag{5.39a}$$

where

$$\varepsilon_\varphi^0 = \frac{1}{r_\varphi}\left(\frac{\partial u_\varphi^0}{\partial \varphi} + w\right), \quad \varepsilon_\theta^0 = \frac{1}{r}\left(\frac{\partial u_\theta^0}{\partial \theta} + u_\varphi^0 \cos\varphi + w \sin\varphi\right),$$
$$\gamma_{\varphi\theta}^0 = \frac{1}{r}\frac{\partial u_\varphi^0}{\partial \theta} + \frac{r}{r_\varphi}\frac{\partial}{\partial \varphi}\left(\frac{u_\theta^0}{r}\right),$$
$$\kappa_\varphi = \frac{1}{r_\varphi}\frac{\partial \beta_\varphi}{\partial \varphi}, \quad \kappa_\theta = \frac{1}{r}\left(\frac{\partial \beta_\theta}{\partial \theta} + \beta_\varphi \cos\varphi\right), \quad \kappa_{\varphi\theta} = \frac{1}{r}\frac{\partial \beta_\varphi}{\partial \theta} + \frac{r}{r_\varphi}\frac{\partial}{\partial \varphi}\left(\frac{\beta_\theta}{r}\right). \tag{5.39b}$$

Constitutive laws

$$\begin{Bmatrix} N_\varphi \\ N_\theta \\ N_{\varphi\theta} \\ M_\varphi \\ M_\theta \\ M_{\varphi\theta} \end{Bmatrix} = \begin{bmatrix} A_{11} & A_{12} & A_{16} & B_{11} & B_{12} & B_{16} \\ A_{12} & A_{22} & A_{26} & B_{12} & B_{22} & B_{26} \\ A_{16} & A_{26} & A_{66} & B_{16} & B_{26} & B_{66} \\ B_{11} & B_{12} & B_{16} & D_{11} & D_{12} & D_{16} \\ B_{12} & B_{22} & B_{26} & D_{12} & D_{22} & D_{26} \\ B_{16} & B_{26} & B_{66} & D_{16} & D_{26} & D_{66} \end{bmatrix} \begin{Bmatrix} \varepsilon_\varphi^0 \\ \varepsilon_\theta^0 \\ \gamma_{\varphi\theta}^0 \\ k_\varphi \\ k_\theta \\ k_{\varphi\theta} \end{Bmatrix}. \tag{5.40}$$

Equilibrium equations (if the external moments m_φ and m_θ are neglected)

$$\frac{\partial(rN_\varphi)}{\partial \varphi} + r_\varphi \frac{\partial N_{\varphi\theta}}{\partial \theta} - N_\theta r_\varphi \cos\varphi + rQ_\varphi + rr_\varphi q_\varphi = 0,$$
$$\frac{\partial(rN_{\varphi\theta})}{\partial \varphi} + r_\varphi \frac{\partial N_\theta}{\partial \theta} + N_{\varphi\theta} r_\varphi \cos\varphi + r_\varphi Q_\theta \sin\varphi + rr_\varphi q_\theta = 0,$$

$$\frac{\partial(rQ_\varphi)}{\partial \varphi} + r_\varphi \frac{\partial Q_\theta}{\partial \theta} - rN_\varphi - r_\varphi N_\theta \sin\varphi - rr_\varphi q = 0,$$
$$\frac{\partial(rM_\varphi)}{\partial \varphi} + r_\varphi \frac{\partial M_{\varphi\theta}}{\partial \theta} - M_\theta r_\varphi \cos\varphi - rr_\varphi Q_\varphi = 0,$$
$$\frac{\partial(rM_{\varphi\theta})}{\partial \varphi} + r_\varphi \frac{\partial M_\theta}{\partial \theta} + M_{\varphi\theta} r_\varphi \cos\varphi - rr_\varphi Q_\theta = 0. \tag{5.41}$$

Boundary conditions

$$
\begin{aligned}
&u_\varphi = \bar{u}_\varphi && \text{or} && N_\varphi = \bar{N}_\varphi && \text{or} && N_\varphi = k_n u_\varphi, \\
&u_\theta = \bar{u}_\theta && \text{or} && T_{\varphi\theta} = \bar{T}_{\varphi\theta} && \text{or} && T_{\varphi\theta} = k_t u_\theta, \\
&w = \bar{w} && \text{or} && V_\varphi = \bar{V}_\varphi && \text{or} && V_\varphi = k_v w, \\
&\beta_\varphi = \bar{\beta}_\varphi && \text{or} && M_\varphi = \bar{M}_\varphi && \text{or} && M_\varphi = k_m \beta_\varphi,
\end{aligned} \tag{5.42a}
$$

where

$$
T_{\varphi\theta} = N_{\varphi\theta} + \frac{M_{\varphi\theta}\sin\varphi}{r}, \quad V_\varphi = Q_\varphi + \frac{\partial M_{\varphi\theta}}{r\partial\theta}. \tag{5.42b}
$$

Unlike (5.35) which are the boundary conditions for the general curvilinear boundaries, equations (5.42) are the boundary conditions valid only on the edges of constant φ. Since for the shells of revolution we are concerned predominantly with shells that are closed with respect to the coordinate θ, most of the boundary conditions consider only the edges of constant φ.

Equations (5.37)–(5.42) are the basic equations which describe the static behavior of thin elastic shells of revolution. In practical applications, some specific types of shells of revolution are commonly used such as conical shells, circular cylindrical shells, spherical shells, shallow spherical shells, etc. The simplification of these specific shells are provided in the following sections.

5.3.1 Conical Shells

From the description stated before (5.18), we know that the basic equations for the conical shells can be reduced from (5.37)–(5.42) by letting $r_\varphi \to \infty$, $\varphi = \varphi_c$ = constant, $dx = r_\varphi d\varphi$ and $r = x\cos\varphi_c$. The results are as follows:

Displacement fields

$$
\begin{aligned}
u_x(x,\theta,\zeta) &= u_x^0(x,\theta) + \zeta\beta_x(x,\theta), \\
u_\theta(x,\theta,\zeta) &= u_\theta^0(x,\theta) + \zeta\beta_\theta(x,\theta), \\
u_3(x,\theta,\zeta) &= w(x,\theta),
\end{aligned} \tag{5.43a}
$$

where

$$\beta_x = -\frac{\partial w}{\partial x}, \quad \beta_\theta = \frac{1}{x}\left(u_\theta^0 \tan\varphi_c - \frac{\partial w}{\partial \theta}\sec\varphi_c\right). \tag{5.43b}$$

Strain–displacement relations

$$\varepsilon_x = \varepsilon_x^0 + \zeta\kappa_x, \quad \varepsilon_\theta = \varepsilon_\theta^0 + \zeta\kappa_\theta, \quad \gamma_{x\theta} = \gamma_{x\theta}^0 + \zeta\kappa_{x\theta}, \tag{5.44a}$$

where

$$\begin{aligned}
&\varepsilon_x^0 = \frac{\partial u_x^0}{\partial x}, \quad \varepsilon_\theta^0 = \frac{1}{x}\left(\frac{\partial u_\theta^0}{\partial \theta}\sec\varphi_c + u_x^0 + w\tan\varphi_c\right),\\
&\gamma_{x\theta}^0 = \frac{1}{x}\frac{\partial u_x^0}{\partial \theta}\sec\varphi_c + x\frac{\partial}{\partial x}\left(\frac{u_\theta^0}{x}\right),\\
&\kappa_x = \frac{\partial \beta_x}{\partial x}, \kappa_\theta = \frac{1}{x}\left(\frac{\partial \beta_\theta}{\partial \theta}\sec\varphi_c + \beta_x\right), \kappa_{x\theta} = \frac{1}{x}\frac{\partial \beta_x}{\partial \theta}\sec\varphi_c + x\frac{\partial}{\partial x}\left(\frac{\beta_\theta}{x}\right).
\end{aligned} \tag{5.44b}$$

Constitutive laws

$$\begin{Bmatrix} N_x \\ N_\theta \\ N_{x\theta} \\ M_x \\ M_\theta \\ M_{x\theta} \end{Bmatrix} = \begin{bmatrix} A_{11} & A_{12} & A_{16} & B_{11} & B_{12} & B_{16} \\ A_{12} & A_{22} & A_{26} & B_{12} & B_{22} & B_{26} \\ A_{16} & A_{26} & A_{66} & B_{16} & B_{26} & B_{66} \\ B_{11} & B_{12} & B_{16} & D_{11} & D_{12} & D_{16} \\ B_{12} & B_{22} & B_{26} & D_{12} & D_{22} & D_{26} \\ B_{16} & B_{26} & B_{66} & D_{16} & D_{26} & D_{66} \end{bmatrix} \begin{Bmatrix} \varepsilon_x^0 \\ \varepsilon_\theta^0 \\ \gamma_{x\theta}^0 \\ k_x \\ k_\theta \\ k_{x\theta} \end{Bmatrix}. \tag{5.45}$$

Equilibrium equations

$$\begin{aligned}
&\frac{\partial(xN_x)}{\partial x} + \frac{\partial N_{x\theta}}{\partial \theta}\sec\varphi_c - N_\theta + xq_x = 0,\\
&\frac{\partial(xN_{x\theta})}{\partial x} + \frac{\partial N_\theta}{\partial \theta}\sec\varphi_c + N_{x\theta} + Q_\theta\tan\varphi_c + xq_\theta = 0,
\end{aligned} \tag{5.46a}$$

$$\frac{\partial(xQ_x)}{\partial x}+\frac{\partial Q_\theta}{\partial\theta}\sec\varphi_c-N_\theta\tan\varphi_c-xq=0,$$
$$\frac{\partial(xM_x)}{\partial x}+\frac{\partial M_{x\theta}}{\partial\theta}\sec\varphi_c-M_\theta-xQ_x=0,$$
$$\frac{\partial(xM_{x\theta})}{\partial x}+\frac{\partial M_\theta}{\partial\theta}\sec\varphi_c+M_{x\theta}-xQ_\theta=0. \tag{5.46b}$$

Boundary conditions

$$\begin{aligned}
u_x&=\bar{u}_x &&\text{or}&& N_x=\bar{N}_x &&\text{or}&& N_x=k_n u_x,\\
u_\theta&=\bar{u}_\theta &&\text{or}&& T_{x\theta}=\bar{T}_{x\theta} &&\text{or}&& T_{x\theta}=k_t u_\theta,\\
w&=\bar{w} &&\text{or}&& V_x=\bar{V}_x &&\text{or}&& V_x=k_v w,\\
\beta_x&=\bar{\beta}_x &&\text{or}&& M_x=\bar{M}_x &&\text{or}&& M_x=k_m\beta_x,
\end{aligned} \tag{5.47a}$$

where

$$T_{x\theta}=N_{x\theta}+\frac{M_{x\theta}}{x}\tan\varphi_c,\quad V_x=Q_x+\frac{\partial M_{x\theta}}{x\partial\theta}\sec\varphi_c. \tag{5.47b}$$

5.3.2 Circular Cylindrical Shells

As described in the sentences ahead of (5.19), a circular cylindrical shell is a special case of conical shells with $\varphi=\pi/2$, $r=R=$ constant. Employing these relations into (5.43)–(5.47), the basic equations for the circular cylindrical shells can be simplified as follows:

Displacement fields: same as (5.43a) where

$$\beta_x=-\frac{\partial w}{\partial x},\quad \beta_\theta=\frac{1}{R}\left(u_\theta^0-\frac{\partial w}{\partial\theta}\right). \tag{5.48}$$

Strain–displacement relations: same as (5.44a) where

$$\varepsilon_x^0=\frac{\partial u_x^0}{\partial x},\ \varepsilon_\theta^0=\frac{1}{R}\left(\frac{\partial u_\theta^0}{\partial\theta}+w\right),\ \gamma_{x\theta}^0=\frac{\partial u_x^0}{R\partial\theta}+\frac{\partial u_\theta^0}{\partial x},$$
$$\kappa_x=\frac{\partial\beta_x}{\partial x},\ \kappa_\theta=\frac{\partial\beta_\theta}{R\partial\theta},\ \kappa_{x\theta}=\frac{\partial\beta_x}{R\partial\theta}+\frac{\partial\beta_\theta}{\partial x}. \tag{5.49}$$

Constitutive laws: same as (5.45).

Equilibrium equations:

$$\begin{aligned}
&\frac{\partial N_x}{\partial x}+\frac{\partial N_{x\theta}}{R\partial\theta}+q_x=0,\quad \frac{\partial N_{x\theta}}{\partial x}+\frac{\partial N_\theta}{R\partial\theta}+\frac{Q_\theta}{R}+q_\theta=0,\\
&\frac{\partial Q_x}{\partial x}+\frac{\partial Q_\theta}{R\partial\theta}-\frac{N_\theta}{R}-q=0,\\
&\frac{\partial M_x}{\partial x}+\frac{\partial M_{x\theta}}{R\partial\theta}-Q_x=0,\quad \frac{\partial M_{x\theta}}{\partial x}+\frac{\partial M_\theta}{R\partial\theta}-Q_\theta=0.
\end{aligned} \tag{5.50}$$

By means of the fourth and fifth equilibrium equations of (5.50), the transverse shear forces Q_x and Q_θ can be expressed in terms of the bending moments M_x, M_θ and $M_{x\theta}$. With this result, the third equilibrium equation can be written as

$$\frac{\partial^2 M_x}{\partial x^2}+\frac{2}{R}\frac{\partial^2 M_{x\theta}}{\partial x\partial\theta}+\frac{1}{R^2}\frac{\partial^2 M_\theta}{\partial\theta^2}-\frac{N_\theta}{R}-q=0. \tag{5.51}$$

Boundary conditions: same as (5.47).

5.3.3 Spherical Shells

As described in the sentence ahead of (5.20), a spherical shell is a special case of shells of revolution with $r_\varphi = R$ and $r = R\sin\varphi$. Employing these relations into (5.38)–(5.42), the basic equations for the spherical shells can be simplified as follows:

Displacement fields: same as (5.38a) where

$$\beta_\varphi=\frac{1}{R}\left(u_\varphi^0-\frac{\partial w}{\partial\varphi}\right),\quad \beta_\theta=\frac{1}{R}\left(u_\theta^0-\csc\varphi\frac{\partial w}{\partial\theta}\right). \tag{5.52}$$

Strain–displacement relations: same as (5.39a) where

$$\begin{aligned}
&\varepsilon_\varphi^0=\frac{1}{R}\left(\frac{\partial u_\varphi^0}{\partial\varphi}+w\right),\quad \varepsilon_\theta^0=\frac{1}{R}\left(\frac{\partial u_\theta^0}{\partial\theta}\csc\varphi+u_\varphi^0\cot\varphi+w\right),\\
&\kappa_\varphi=\frac{1}{R}\frac{\partial\beta_\varphi}{\partial\varphi},\quad \kappa_\theta=\frac{1}{R}\left(\frac{\partial\beta_\theta}{\partial\theta}\csc\varphi+\beta_\varphi\cot\varphi\right),
\end{aligned}$$

$$\kappa_\varphi = \frac{1}{R}\frac{\partial \beta_\varphi}{\partial \varphi}, \quad \kappa_\theta = \frac{1}{R}\left(\frac{\partial \beta_\theta}{\partial \theta}\csc\varphi + \beta_\varphi \cot\varphi\right),$$

$$\kappa_{\varphi\theta} = \frac{1}{R}\left(\frac{\partial \beta_\varphi}{\partial \theta}\csc\varphi + \frac{\partial \beta_\theta}{\partial \varphi} - \beta_\theta \cot\varphi\right). \tag{5.53}$$

Constitutive laws: same as (5.40).

Equilibrium equations:

$$\frac{\partial(N_\varphi \sin\varphi)}{\partial \varphi} + \frac{\partial N_{\varphi\theta}}{\partial \theta} - N_\theta \cos\varphi + Q_\varphi \sin\varphi + q_\varphi R \sin\varphi = 0,$$

$$\frac{\partial(N_{\varphi\theta} \sin\varphi)}{\partial \varphi} + \frac{\partial N_\theta}{\partial \theta} + N_{\varphi\theta} \cos\varphi + Q_\theta \sin\varphi + q_\theta R \sin\varphi = 0, \tag{5.54a}$$

$$\frac{\partial(Q_\varphi \sin\varphi)}{\partial \varphi} + \frac{\partial Q_\theta}{\partial \theta} - (N_\varphi + N_\theta)\sin\varphi - qR\sin\varphi = 0,$$

$$\frac{\partial(M_\varphi \sin\varphi)}{\partial \varphi} + \frac{\partial M_{\varphi\theta}}{\partial \theta} - M_\theta \cos\varphi - Q_\varphi R \sin\varphi = 0,$$

$$\frac{\partial(M_{\varphi\theta} \sin\varphi)}{\partial \varphi} + \frac{\partial M_\theta}{\partial \theta} + M_{\varphi\theta} \cos\varphi - Q_\theta R \sin\varphi = 0. \tag{5.54b}$$

Boundary conditions: same as (5.42).

5.3.4 Shallow Spherical Shells

Generally, a shell is considered to be shallow if its height h is less than one-eighth of its base diameter $2a$ (Figure 5.5). For a spherical shell, the equation of a meridian can be written as

$$z = \sqrt{R^2 - r^2} - (R - h). \tag{5.55}$$

On the base plane $z = 0$, we have $r = a$, and hence,

$$a^2 = 2hR - h^2. \tag{5.56}$$

For a shallow spherical shell, $h^2 << 2hR$ and (5.56) reduces to $a^2 \cong 2hR$. With this relation, the assumption $h/2a < 1/8$ can be restated as $a/R < 1/2$

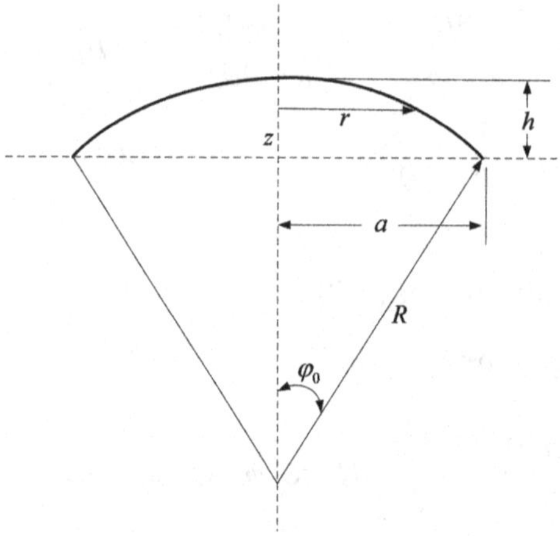

FIGURE 5.5 Geometry of a shallow spherical shell.

or equivalently as $\varphi_0 < 30^0$. Therefore, the consideration of shallow spherical shells is usually constrained in the range of $-30^0 < \varphi < 30^0$. Within this range, the projection of the radius on the z-axis, $\sqrt{3}/2R < R\cos\varphi < R$, can be approximated by R. Thus, the differential of $r = R\sin\varphi$, $dr = R\cos\varphi d\varphi$, can be approximated by $dr \cong Rd\varphi$. By using this approximation, it is convenient to replace the variable φ by r in the formulation of shallow spherical shells. With this replacement and $r = R\sin\varphi$ and $dr \cong Rd\varphi$, equations (5.52)–(5.54) can be further simplified as follows:

Displacement fields

$$\begin{aligned} u_r(r,\theta,\zeta) &= u_r^0(r,\theta) + \zeta\beta_r(r,\theta), \\ u_\theta(r,\theta,\zeta) &= u_\theta^0(r,\theta) + \zeta\beta_\theta(r,\theta), \\ u_3(r,\theta,\zeta) &= w(r,\theta), \end{aligned} \tag{5.57a}$$

where

$$\beta_r = \frac{u_r^0}{R} - \frac{\partial w}{\partial r}, \quad \beta_\theta = \frac{u_\theta^0}{R} - \frac{\partial w}{r\partial\theta}. \tag{5.57b}$$

Strain–displacement relations

$$\varepsilon_r = \varepsilon_r^0 + \zeta\kappa_r, \quad \varepsilon_\theta = \varepsilon_\theta^0 + \zeta\kappa_\theta, \quad \gamma_{r\theta} = \gamma_{r\theta}^0 + \zeta\kappa_{r\theta}, \tag{5.58a}$$

where

$$\varepsilon_r^0 = \frac{\partial u_r^0}{\partial r} + \frac{w}{R},\ \varepsilon_\theta^0 = \frac{1}{r}\left(\frac{\partial u_\theta^0}{\partial \theta} + u_r^0\right) + \frac{w}{R},\ \gamma_{r\theta}^0 = \frac{\partial u_r^0}{r\partial \theta} + \frac{\partial u_\theta^0}{\partial r} - \frac{u_\theta^0}{r},$$

$$\kappa_r = \frac{\partial \beta_r}{\partial r},\ \kappa_\theta = \frac{1}{r}\left(\frac{\partial \beta_\theta}{\partial \theta} + \beta_r\right),\ \kappa_{r\theta} = \frac{\partial \beta_r}{r\partial \theta} + \frac{\partial \beta_\theta}{\partial r} - \frac{\beta_\theta}{r}. \tag{5.58b}$$

Constitutive laws

$$\begin{Bmatrix} N_r \\ N_\theta \\ N_{r\theta} \\ M_r \\ M_\theta \\ M_{r\theta} \end{Bmatrix} = \begin{bmatrix} A_{11} & A_{12} & A_{16} & B_{11} & B_{12} & B_{16} \\ A_{12} & A_{22} & A_{26} & B_{12} & B_{22} & B_{26} \\ A_{16} & A_{26} & A_{66} & B_{16} & B_{26} & B_{66} \\ B_{11} & B_{12} & B_{16} & D_{11} & D_{12} & D_{16} \\ B_{12} & B_{22} & B_{26} & D_{12} & D_{22} & D_{26} \\ B_{16} & B_{26} & B_{66} & D_{16} & D_{26} & D_{66} \end{bmatrix} \begin{Bmatrix} \varepsilon_r^0 \\ \varepsilon_\theta^0 \\ \gamma_{r\theta}^0 \\ k_r \\ k_\theta \\ k_{r\theta} \end{Bmatrix}. \tag{5.59}$$

Equilibrium equations

$$\begin{aligned} &\frac{\partial (rN_r)}{\partial r} + \frac{\partial N_{r\theta}}{\partial \theta} - N_\theta + \frac{r}{R}Q_r + rq_r = 0, \\ &\frac{\partial (rN_{r\theta})}{\partial r} + \frac{\partial N_\theta}{\partial \theta} + N_{r\theta} + \frac{r}{R}Q_\theta + rq_\theta = 0, \\ &\frac{\partial (rQ_r)}{\partial r} + \frac{\partial Q_\theta}{\partial \theta} - \frac{r}{R}(N_r + N_\theta) - rq = 0, \\ &\frac{\partial (rM_r)}{\partial r} + \frac{\partial M_{r\theta}}{\partial \theta} - M_\theta - rQ_r = 0, \\ &\frac{\partial (rM_{r\theta})}{\partial r} + \frac{\partial M_\theta}{\partial \theta} + M_{r\theta} - rQ_\theta = 0. \end{aligned} \tag{5.60}$$

Boundary conditions

$$\begin{aligned} u_r &= \bar{u}_r \quad \text{or} \quad N_r = \bar{N}_r \quad \text{or} \quad N_r = k_n u_r, \\ u_\theta &= \bar{u}_\theta \quad \text{or} \quad T_{r\theta} = \bar{T}_{r\theta} \quad \text{or} \quad T_{r\theta} = k_t u_\theta, \\ w &= \bar{w} \quad \text{or} \quad V_r = \bar{V}_r \quad \text{or} \quad V_r = k_v w, \\ \beta_r &= \bar{\beta}_r \quad \text{or} \quad M_r = \bar{M}_r \quad \text{or} \quad M_r = k_m \beta_r, \end{aligned} \tag{5.61a}$$

where

$$T_{r\theta} = N_{r\theta} + \frac{M_{r\theta}}{R}, \quad V_r = Q_r + \frac{\partial M_{r\theta}}{r\partial\theta}. \tag{5.61b}$$

5.3.5 Solution Procedure and Examples

The basic equations for the shells of revolution with bending resistance under arbitrary load have been shown in (5.38)–(5.41) for the general case and shown in (5.43)–(5.61) for some special cases. These equations consist of 25 partial differential equations with 25 unknown functions $u_\varphi, u_\theta, u_3, u_\varphi^0, u_\theta^0, w, \beta_\varphi, \beta_\theta, \varepsilon_\varphi, \varepsilon_\theta, \gamma_{\varphi\theta}, \varepsilon_\varphi^0, \varepsilon_\theta^0, \gamma_{\varphi\theta}^0, \kappa_\varphi, \kappa_\theta, \kappa_{\varphi\theta}, N_\varphi, N_\theta, N_{\varphi\theta}, M_\varphi, M_\theta, M_{\varphi\theta}, Q_\varphi, Q_\theta$. To solve these equations, in general there are *two basic procedures* (Kraus, 1972). In the first, the basic equations are reduced to three differential equations based upon the mid-surface displacements u_φ^0, u_θ^0 and w. In the second, analogous to the Airy stress function for the two-dimensional problems, a stress function is introduced and the governing equations are reduced to two differential equations based upon the stress function and the transverse displacement w. Once either of the reductions has been accomplished, the unknown functions and all the related shell variables are expanded in Fourier series with respect to the polar coordinate θ. By means of these expansions, the dependence of the shell variables upon the polar coordinate is removed from consideration. The problem is thus reduced to a set of ordinary differential equations. However, due to the complexity of shell formulation, only specific cases have been solved completely. One may refer to (Kraus, 1972) for examples of isotropic shells and refer to Ambartsumyan (1961) for laminated shells.

To have a clear picture of the solution procedure, we now choose a simplest case for the shells of revolution with bending resistance. *Consider a circular cylindrical laminated shell under loads that are axially symmetric.* Under this condition, all the geometric quantities and external loads are independent of the polar angle θ, and hence the related equations can be simplified by letting $\partial(\ldots)/\partial\theta = 0$. If no surface shear stresses are applied on the shell surface, $q_x = q_\theta = 0$. Moreover, for simplicity we only consider the laminates such that the coupling terms vanish, that is, $B_{ij} = 0$ and $A_{16} = A_{26} = D_{16} = D_{26} = 0$. Substituting all these conditions into the basic equations shown in (5.48)–(5.50) for the *circular cylindrical shells*, we obtain the following simplified results:

Displacement fields

$$u_x(x,\zeta)=u_x^0(x)+\zeta\beta_x(x), u_\theta(x,\zeta)=u_\theta^0(x)+\zeta\beta_\theta(x), u_3(x,\zeta)=w(x), \tag{5.62a}$$

where

$$\beta_x=-\frac{\partial w}{\partial x},\quad \beta_\theta=\frac{u_\theta^0}{R}. \tag{5.62b}$$

Strain–displacement relations

$$\varepsilon_x^0=\frac{\partial u_x^0}{\partial x},\quad \varepsilon_\theta^0=\frac{w}{R},\quad \gamma_{x\theta}^0=\frac{\partial u_\theta^0}{\partial x},\quad \kappa_x=-\frac{\partial^2 w}{\partial x^2},\quad \kappa_\theta=0,\quad \kappa_{x\theta}=\frac{\partial u_\theta^0}{R\partial x}. \tag{5.63}$$

Constitutive laws

$$\begin{aligned} N_x&=A_{11}\frac{\partial u_x^0}{\partial x}+A_{12}\frac{w}{R},\quad N_\theta=A_{12}\frac{\partial u_x^0}{\partial x}+A_{22}\frac{w}{R},\quad N_{x\theta}=A_{66}\frac{\partial u_\theta^0}{\partial x},\\ M_x&=-D_{11}\frac{\partial^2 w}{\partial x^2},\quad M_\theta=-D_{12}\frac{\partial^2 w}{\partial x^2},\quad M_{x\theta}=D_{66}\frac{\partial u_\theta^0}{R\partial x}. \end{aligned} \tag{5.64}$$

Equilibrium equations

$$\begin{aligned} &\frac{\partial N_x}{\partial x}=0,\quad \frac{\partial N_{x\theta}}{\partial x}+\frac{Q_\theta}{R}=0,\\ &\frac{\partial Q_x}{\partial x}-\frac{N_\theta}{R}-q=0,\quad \frac{\partial M_x}{\partial x}-Q_x=0,\quad \frac{\partial M_{x\theta}}{\partial x}-Q_\theta=0. \end{aligned} \tag{5.65}$$

Boundary conditions

$$\begin{aligned} &u_x=\bar{u}_x \quad \text{or} \quad N_x=\bar{N}_x \quad \text{or} \quad N_x=k_n u_x,\\ &w=\bar{w} \quad \text{or} \quad Q_x=\bar{Q}_x \quad \text{or} \quad Q_x=k_v w,\\ &\beta_x=\bar{\beta}_x \quad \text{or} \quad M_x=\bar{M}_x \quad \text{or} \quad M_x=k_m\beta_x. \end{aligned} \tag{5.66}$$

From the first, fourth, and fifth equations of (5.65), we have

$$N_x=\text{constant},\quad Q_x=-D_{11}\frac{d^3 w}{dx^3},\quad Q_\theta=\frac{D_{66}}{R}\frac{d^2 u_\theta^0}{dx^2}. \tag{5.67}$$

Substituting (5.64) into the first three equations of (5.65), and using the results of (5.67), we can write down the governing differential equations in terms of the mid-surface displacements as

$$\frac{d^2u_x^0}{dx^2}+\frac{A_{12}}{RA_{11}}\frac{dw}{dx}=0,$$
$$\frac{d^4w}{dx^4}+4\lambda^4w=\frac{-1}{D_{11}}\left(q+\frac{A_{12}}{RA_{11}}N_x\right),\quad \frac{d^2u_\theta^0}{dx^2}=0, \tag{5.68a}$$

where

$$\lambda^4=\frac{A_{11}A_{22}-A_{12}^2}{4R^2A_{11}D_{11}}. \tag{5.68b}$$

The form of $(5.68a)_2$ is desirable since it is uncoupled from $(5.68a)_1$ and N_x is a constant to be determined from the boundary conditions. It is also observed that $(5.68a)_2$ has the same form as that of a beam on an elastic foundation. Thus, the physical intuition of the solutions for beams on an elastic foundation can be utilized.

By standard methods of ordinary differential equations, the general solution to $(5.68a)_2$ can be written as

$$w(x)=c_1e^{-\lambda x}\cos\lambda x+c_2e^{-\lambda x}\sin\lambda x+c_3e^{\lambda x}\cos\lambda x+c_4e^{\lambda x}\sin\lambda x+w_p(x), \tag{5.69}$$

where c_1, c_2, c_3, and c_4 are constants to be determined from the boundary conditions, and $w_p(x)$ is the particular solution associated with q and N_x. The in-plane displacement u_x^0 can then be obtained by integrating $(5.68a)_1$ as follows:

$$u_x^0(x)=\frac{N_x}{A_{11}}x-\frac{A_{12}}{RA_{11}}\int w\,dx+c_5, \tag{5.70}$$

where c_5 is a constant of integration. Now, employing the boundary conditions (5.66) for each end of the cylindrical shell, we will get six equations for the six unknown constants c_1, c_2, c_3, c_4, c_5, and N_x.

In solving $(5.68a)_2$, another form called *bending boundary layer solution* may be utilized to replace (5.69), which is written as (Vinson and Sierakowski, 1986).

$$w(x)=\frac{M_0}{2\lambda^2 D_{11}}e^{-\lambda x}\left(\sin\lambda x-\cos\lambda x\right)-\frac{Q_0}{2\lambda^3 D_{11}}e^{-\lambda x}\cos\lambda x$$
$$+\frac{M_\ell}{2\lambda^2 D_{11}}e^{-\lambda(l-x)}\left[\sin\lambda(\ell-x)-\cos\lambda(\ell-x)\right]$$
$$-\frac{Q_\ell}{2\lambda^3 D_{11}}e^{-\lambda(\ell-x)}\cos\lambda(\ell-x)+w_p(x) \tag{5.71}$$

where M_0, Q_0, M_ℓ, and Q_ℓ are constants to be determined. With the expression (5.71), the bending moment M_x and transverse shear force Q_x can be obtained from $(5.64)_4$ and $(5.65)_4$. When $d^2q(x)/dx^2 = d^3q(x)/dx^3 = 0$, it can be shown that $M_0 = M_x(0)$, $M_\ell = M_x(\ell)$, $Q_0 = Q_x(0)$, and $Q_\ell = Q_x(\ell)$. From (5.71) we see that each term of the homogeneous solution contains trigonometric terms, which oscillate between ±1, multiplied by an exponential term with a negative exponent. Hence, an exponential decay starts from two ends of the cylindrical shell. If we say any term is negligible when $e^{-\lambda x}$ or $e^{-\lambda(\ell-x)}$ is less than a very small number say ε, then in the region $|\ln\varepsilon|/\lambda \leq x \leq \ell - |\ln\varepsilon|/\lambda$ the terms involving M_0, Q_0, M_ℓ and Q_ℓ may be ignored. The regions $0 \leq x \leq |\ln\varepsilon|/\lambda$ and $\ell - |\ln\varepsilon|/\lambda \leq x \leq \ell$ are then called *bending boundary layers.*

5.4 MEMBRANE SHELLS

The general formulations presented in the previous section show that an arbitrary external loading may induce internal stress resultants and bending moments in a laminated shell. Under appropriate loading conditions, however, there are cases that the resulting bending moments are either zero or so small that they may be neglected. Such a state of stress is referred to as the membrane state of stress and the associated shells are called *membrane shells* because of the analogy to membranes which cannot support bending moments. Due to the neglect of the bending effects, the formulations shown previously can be greatly simplified. Moreover, it has also been shown that when a shell becomes thinner, regardless of its shape, the bending effects become more localized and the solutions obtained by the membrane theory become applicable over an increasingly greater region of the shell away from the sources of bending (Kraus, 1972). Thus, it is reasonable to start our analysis of shells from the membrane shells and make the following basic assumptions:

$$M_1 = M_2 = M_{12} = M_{21} = Q_1 = Q_2 = 0. \tag{5.72}$$

As a result, the basic equations of membrane shells can be reduced from (5.22)–(5.36) to the following set:

Displacement fields

$$u_1(\alpha_1,\alpha_2,\zeta)=u_1^0(\alpha_1,\alpha_2),$$
$$u_2(\alpha_1,\alpha_2,\zeta)=u_2^0(\alpha_1,\alpha_2),$$
$$u_3(\alpha_1,\alpha_2,\zeta)=w(\alpha_1,\alpha_2). \tag{5.73}$$

Strain–displacement relations

$$\varepsilon_{11}=\varepsilon_{11}^0,\quad \varepsilon_{22}=\varepsilon_{22}^0,\quad \gamma_{12}=\gamma_{12}^0, \tag{5.74a}$$

where

$$\varepsilon_{11}^0=\frac{1}{A_1}\frac{\partial u_1^0}{\partial\alpha_1}+\frac{u_2^0}{A_1A_2}\frac{\partial A_1}{\partial\alpha_2}+\frac{w}{R_1},$$
$$\varepsilon_{22}^0=\frac{1}{A_2}\frac{\partial u_2^0}{\partial\alpha_2}+\frac{u_1^0}{A_1A_2}\frac{\partial A_2}{\partial\alpha_1}+\frac{w}{R_2},$$
$$\gamma_{12}^0=\frac{A_1}{A_2}\frac{\partial}{\partial\alpha_2}\left(\frac{u_1^0}{A_1}\right)+\frac{A_2}{A_1}\frac{\partial}{\partial\alpha_1}\left(\frac{u_2^0}{A_2}\right). \tag{5.74b}$$

Constitutive laws

$$N_1=A_{11}\varepsilon_{11}^0+A_{12}\varepsilon_{22}^0+A_{16}\gamma_{12}^0,$$
$$N_2=A_{12}\varepsilon_{11}^0+A_{22}\varepsilon_{22}^0+A_{26}\gamma_{12}^0,$$
$$N_{12}=N_{21}=A_{16}\varepsilon_{11}^0+A_{26}\varepsilon_{12}^0+A_{66}\gamma_{12}^0. \tag{5.75}$$

Equilibrium equations

$$\frac{\partial(N_1A_2)}{\partial\alpha_1}+\frac{\partial(N_{21}A_1)}{\partial\alpha_2}+N_{12}\frac{\partial A_1}{\partial\alpha_2}-N_2\frac{\partial A_2}{\partial\alpha_1}+A_1A_2q_1=0,$$
$$\frac{\partial(N_{12}A_2)}{\partial\alpha_1}+\frac{\partial(N_2A_1)}{\partial\alpha_2}+N_{21}\frac{\partial A_2}{\partial\alpha_1}-N_1\frac{\partial A_1}{\partial\alpha_2}+A_1A_2q_2=0,$$
$$\frac{N_1}{R_1}+\frac{N_2}{R_2}+q=0. \tag{5.76}$$

Boundary conditions

$$
\begin{aligned}
&u_n = \bar{u}_n \quad \text{or} \quad N_n = \bar{N}_n \quad \text{or} \quad N_n = k_n u_n, \\
&u_t = \bar{u}_t \quad \text{or} \quad N_{nt} = \bar{N}_{nt} \quad \text{or} \quad N_{nt} = k_t u_t.
\end{aligned} \tag{5.77}
$$

5.4.1 Membrane Shells of Revolution

For membrane shells of revolution, by using the notations and relations given in (5.17), the system of basic equations (5.73)–(5.77) can be further simplified. Their associated mathematical expressions can be written directly from Section 5.3 by neglecting all the bending terms like β_φ, β_θ, κ_φ, κ_θ, $\kappa_{\varphi\theta}$, B_{ij}, D_{ij}, Q_φ, Q_θ, M_φ, M_θ, $M_{\varphi\theta}$, etc. By this way the basic equations simplified from (5.39)–(5.41) for the membrane shells of revolution can now be written as follows:

$$
\begin{aligned}
&\frac{\partial (rN_\varphi)}{\partial \varphi} + r_\varphi \frac{\partial N_{\varphi\theta}}{\partial \theta} - N_\theta r_\varphi \cos\varphi + r r_\varphi q_\varphi = 0, \\
&\frac{\partial (rN_{\varphi\theta})}{\partial \varphi} + r_\varphi \frac{\partial N_\theta}{\partial \theta} + N_{\varphi\theta} r_\varphi \cos\varphi + r r_\varphi q_\theta = 0, \\
&\frac{N_\varphi}{r_\varphi} + \frac{N_\theta}{r_\theta} + q = 0, \quad r_\theta = \frac{r}{\sin\varphi},
\end{aligned} \tag{5.78a}
$$

where N_φ, N_θ, and $N_{\varphi\theta}$ are related to the mid-surface strains $\varepsilon_\varphi^0, \varepsilon_\theta^0$ and $\gamma_{\varphi\theta}^0$ by

$$
\begin{aligned}
N_\varphi &= A_{11}\varepsilon_\varphi^0 + A_{12}\varepsilon_\theta^0 + A_{16}\gamma_{\varphi\theta}^0, \\
N_\theta &= A_{12}\varepsilon_\varphi^0 + A_{22}\varepsilon_\theta^0 + A_{26}\gamma_{\varphi\theta}^0, \\
N_{\varphi\theta} &= A_{16}\varepsilon_\varphi^0 + A_{26}\varepsilon_\theta^0 + A_{66}\gamma_{\varphi\theta}^0,
\end{aligned} \tag{5.78b}
$$

and the mid-surface strains are related to the mid-surface displacement by

$$
\begin{aligned}
&\varepsilon_\varphi^0 = \frac{1}{r_\varphi}\left(\frac{\partial u_\varphi^0}{\partial \varphi} + w\right), \quad \varepsilon_\theta^0 = \frac{1}{r}\left(\frac{\partial u_\theta^0}{\partial \theta} + u_\varphi^0 \cos\varphi + w\sin\varphi\right), \\
&\gamma_{\varphi\theta}^0 = \frac{1}{r}\frac{\partial u_\varphi^0}{\partial \theta} + \frac{r}{r_\varphi}\frac{\partial}{\partial \varphi}\left(\frac{u_\theta^0}{r}\right) = \frac{1}{r}\frac{\partial u_\varphi^0}{\partial \theta} + \frac{1}{r_\varphi}\frac{\partial u_\theta^0}{\partial \varphi} - \frac{u_\theta^0 \cos\varphi}{r}.
\end{aligned} \tag{5.78c}
$$

From the third equation of (5.78a), we get

$$N_\theta = -r_\theta q - \frac{r_\theta N_\varphi}{r_\varphi}. \tag{5.79}$$

The expression of (5.79) is used to eliminate N_θ from the first two equations of (5.78a), which leads to two equations for N_φ and $N_{\varphi\theta}$, that is,

$$\begin{aligned} &\frac{\partial(rN_\varphi)}{\partial\varphi} + r_\varphi \frac{\partial N_{\varphi\theta}}{\partial\theta} + N_\varphi r_\theta \cos\varphi + r_\theta r_\varphi (q_\varphi \sin\varphi + q\cos\varphi) = 0, \\ &\frac{\partial(rN_{\varphi\theta})}{\partial\varphi} + N_{\varphi\theta} r_\varphi \cos\varphi - r_\theta \frac{\partial N_\varphi}{\partial\theta} + r_\theta r_\varphi \left(q_\theta \sin\varphi - \frac{\partial q}{\partial\theta} \right) = 0. \end{aligned} \tag{5.80}$$

In general, to solve N_φ and $N_{\varphi\theta}$ from the above two partial differential equations, we may assume that the external loads be expanded in Fourier series of the type

$$\begin{aligned} q_\varphi &= \sum_{n=0}^{\infty} q_{\varphi n}(\varphi)\cos n\theta + \sum_{n=0}^{\infty} q_{\varphi n}^*(\varphi)\sin n\theta, \\ q_\theta &= \sum_{n=0}^{\infty} q_{\theta n}(\varphi)\cos n\theta + \sum_{n=0}^{\infty} q_{\theta n}^*(\varphi)\sin n\theta, \\ q &= \sum_{n=0}^{\infty} q_n(\varphi)\cos n\theta + \sum_{n=0}^{\infty} q_n^*(\varphi)\sin n\theta, \end{aligned} \tag{5.81}$$

and also the stress resultants be expanded in the form

$$\begin{aligned} N_\varphi &= \sum_{n=0}^{\infty} N_{\varphi n}(\varphi)\cos n\theta + \sum_{n=0}^{\infty} N_{\varphi n}^*(\varphi)\sin n\theta, \\ N_{\varphi\theta} &= \sum_{n=0}^{\infty} N_{n\varphi\theta}(\varphi)\cos n\theta + \sum_{n=0}^{\infty} N_{n\varphi\theta}^*(\varphi)\sin n\theta. \end{aligned} \tag{5.82}$$

We now substitute (5.81) and (5.82) into (5.80), and re-arrange the results into series of cos $n\theta$ and sin $n\theta$. Because the results are valid for any θ, all the coefficients associated with cos $n\theta$ and sin $n\theta$ should be identical to zero, which will lead two ordinary differential equations (differentiated with respect to φ) for $N_{\varphi n}$ and $N_{\varphi\theta n}$. The solutions of $N_{\varphi n}$ and $N_{\varphi\theta n}$ involving two unknown constants can be found by solving these two ordinary differential equations. Once the expressions of $N_{\varphi n}$ and $N_{\varphi\theta n}$ have been determined, the stress resultants can be obtained by (5.82), and the strains and displacements can be determined through (5.78b) and (5.78c). The unknown constants can then be determined through the satisfaction of boundary conditions. Examples following this procedure for membrane shells of revolution under arbitrary loads can be found in Kraus (1972).

5.4.2 Axisymmetrical Load

For membrane shells of revolution, if the external loads are independent of the polar angle θ, the terms containing $\partial(\ldots)/\partial\theta$ vanish and (5.80) reduce to

$$\frac{d\left(rN_{\varphi}\right)}{d\varphi}+N_{\varphi}r_{\theta}\cos\varphi+r_{\theta}r_{\varphi}\left(q_{\varphi}\sin\varphi+q\cos\varphi\right)=0,$$
$$\frac{d\left(rN_{\varphi\theta}\right)}{d\varphi}+N_{\varphi\theta}r_{\varphi}\cos\varphi+r_{\theta}r_{\varphi}q_{\theta}\sin\varphi=0. \tag{5.83}$$

If we consider the case that no shear forces exist, $N_{\varphi\theta}=q_{\theta}=0$. Multiplying $(5.83)_1$ by sin φ and then integrating the result with respect to φ, we find that

$$N_{\varphi}=\frac{-1}{r\sin\varphi}\left[c_1+\int_{\varphi_0}^{\varphi}rr_{\varphi}\left(q_{\varphi}\sin\varphi+q\cos\varphi\right)d\varphi\right], \tag{5.84}$$

in which the integration constant c_1 can be obtained by setting $\varphi = \varphi_0$, that is,

$$c_1=-\left[r\sin\varphi N_{\varphi}\right]_{\varphi=\varphi_0},\quad\text{or}\quad 2\pi c_1=-\left(2\pi r\left(\varphi_0\right)\right)\left(\sin\varphi_0 N_{\varphi}\left(\varphi_0\right)\right). \tag{5.85}$$

By $(5.85)_2$, we see that $2\pi c_1$ is the total force transmitted at $\varphi = \varphi_0$ in the axial direction.

Once N_{φ} is obtained from (5.84), N_{θ} can be determined from (5.79). If we consider the shell whose $A_{16} = A_{26} = 0$, the mid-surface strains obtained from (5.78b) are

$$\varepsilon_{\varphi}^{0}=\frac{A_{22}N_{\varphi}-A_{12}N_{\theta}}{A_{11}A_{22}-A_{12}^{2}},\quad \varepsilon_{\theta}^{0}=\frac{A_{11}N_{\theta}-A_{12}N_{\varphi}}{A_{11}A_{22}-A_{12}^{2}},\quad \gamma_{\varphi\theta}^{0}=0. \tag{5.86}$$

With the results of (5.86), the displacements can then be determined by (5.78c), which can be rearranged as

$$\frac{du_\varphi^0}{d\varphi} - u_\varphi^0 \cot\varphi = r_\varphi \varepsilon_\varphi^0 - r_\theta \varepsilon_\theta^0 = f(\varphi),$$
$$w = r_\theta \varepsilon_\theta^0 - u_\varphi^0 \cot\varphi, \tag{5.87a}$$

where

$$f(\varphi) = \frac{1}{A_{11}A_{22} - A_{12}^2}\left\{\left(r_\varphi A_{22} + r_\theta A_{12}\right)N_\varphi - \left(r_\varphi A_{12} + r_\theta A_{11}\right)N_\theta\right\}. \tag{5.87b}$$

Multiplying both sides of $(5.87\text{a})_1$ by $1/\sin\varphi$, the results can be arranged as

$$\frac{d\left(u_\varphi^0 / \sin\varphi\right)}{d\varphi} = \frac{f(\varphi)}{\sin\varphi}. \tag{5.88}$$

Integrating (5.88) with respect to φ, we find that

$$u_\varphi^0 = \sin\varphi \left\{ c_2 + \int_{\varphi_0}^{\varphi} \frac{f(\varphi)}{\sin\varphi} d\varphi \right\}, \tag{5.89}$$

where c_2 is an integration constant which represents the axial displacement of a reference point on the shell. Once u_φ^0 has been found, we can obtain w from $(5.87\text{a})_2$.

5.4.3 Conical Membrane Shells

By letting $r_\varphi \to \infty$, $\varphi = \varphi_c$ = constant, $dx = r_\varphi d\varphi$ and $r = x\cos\varphi_c$, the equilibrium equations (5.78a) become

$$\frac{\partial(xN_x)}{\partial x} + \frac{1}{\cos\varphi_c}\frac{\partial N_{x\theta}}{\partial\theta} - N_\theta + xq_x = 0,$$
$$\frac{\partial(xN_{x\theta})}{\partial x} + \frac{1}{\cos\varphi_c}\frac{\partial N_\theta}{\partial\theta} + N_{x\theta} + xq_\theta = 0,$$
$$N_\theta = -qx\cot\varphi_c. \tag{5.90}$$

For axisymmetrical loaded conical membrane shells, the results of N_x can be reduced from (5.84) as

$$N_x = \frac{-1}{x}\left[c_1 + \int_{x_0}^{x} x\left(q_x + q\cot\varphi_c\right)dx\right]. \tag{5.91}$$

With the results of (5.91), the mid-surface strains and displacements can then be obtained by following the same procedure as stated above for general shells of revolution.

5.4.4 Cylindrical Membrane Shells

The basic equations for cylindrical membrane shells can be obtained from (5.78) by letting $r_\varphi \to \infty$, $\varphi = \pi/2$, $dx = r_\varphi d\varphi$ and $r = R$, or directly reduced from (5.48)–(5.50) by neglecting all the bending terms. The results are

$$\varepsilon_x = \frac{\partial u_x}{\partial x}, \quad \varepsilon_\theta = \frac{1}{R}\left(\frac{\partial u_\theta}{\partial \theta} + w\right), \quad \gamma_{x\theta} = \frac{\partial u_x}{R\partial\theta} + \frac{\partial u_\theta}{\partial x}, \tag{5.92a}$$

$$\begin{aligned} N_x &= A_{11}\varepsilon_x + A_{12}\varepsilon_\theta + A_{16}\gamma_{x\theta}, \\ N_\theta &= A_{12}\varepsilon_x + A_{22}\varepsilon_\theta + A_{26}\gamma_{x\theta}, \\ N_{x\theta} &= A_{16}\varepsilon_x + A_{26}\varepsilon_\theta + A_{66}\gamma_{x\theta}, \end{aligned} \tag{5.92b}$$

$$\frac{\partial N_x}{\partial x} + \frac{\partial N_{x\theta}}{R\partial\theta} + q_x = 0, \quad \frac{\partial N_{x\theta}}{\partial x} + \frac{\partial N_\theta}{R\partial\theta} + q_\theta = 0, \quad N_\theta + qR = 0. \tag{5.92c}$$

The associated boundary conditions are

$$\begin{aligned} u_x &= \bar{u}_x \quad \text{or} \quad N_x = \bar{N}_x \quad \text{or} \quad N_x = k_n u_x, \\ u_\theta &= \bar{u}_\theta \quad \text{or} \quad N_{x\theta} = \bar{N}_{x\theta} \quad \text{or} \quad N_{x\theta} = k_t u_\theta. \end{aligned} \tag{5.93}$$

From the equilibrium equations shown in (5.92c) for the cylindrical membrane shells we see that the stress resultants can be solved directly without calling upon the displacements. Once the stress resultants are known, the strains can be obtained from the constitutive laws (5.92b). Substitution of these strain expressions into the strain–displacement relations (5.92a) followed by suitable integration of the results then leads to the displacements. In this way, the general solutions to (5.92) can be written formally as

$$N_\theta = -qR,$$
$$N_{x\theta} = -\int\left(q_\theta + \frac{\partial N_\theta}{R\partial\theta}\right)dx + f_1(\theta),$$
$$N_x = -\int\left(q_x + \frac{\partial N_{x\theta}}{R\partial\theta}\right)dx + f_2(\theta), \tag{5.94a}$$

and

$$u_x = \int\left(A_{11}^* N_x + A_{12}^* N_\theta + A_{16}^* N_{x\theta}\right)dx + f_3(\theta),$$
$$u_\theta = \int\left(A_{16}^* N_x + A_{26}^* N_\theta + A_{66}^* N_{x\theta} - \frac{\partial u_x}{R\partial\theta}\right)dx + f_4(\theta),$$
$$w = R\left(A_{12}^* N_x + A_{22}^* N_\theta + A_{26}^* N_{x\theta} - \frac{\partial u_\theta}{R\partial\theta}\right), \tag{5.94b}$$

where A_{ij}^* are the components of the matrix $\mathbf{A}^{-1}$ and four arbitrary functions $f_1(\theta)$, $f_2(\theta)$, $f_3(\theta)$, and $f_4(\theta)$ must be determined through the satisfaction of four boundary conditions set on edges of constant x.

5.5 VIBRATION OF SHELLS

To begin the discussion of vibration analysis, it is appropriate to set down the equations of motion of a thin elastic shell by adding the mass-acceleration terms into the equilibrium equations (5.33) as

$$\frac{\partial(N_1A_2)}{\partial\alpha_1} + \frac{\partial(N_{21}A_1)}{\partial\alpha_2} + N_{12}\frac{\partial A_1}{\partial\alpha_2} - N_2\frac{\partial A_2}{\partial\alpha_1} + A_1A_2\left(\frac{Q_1}{R_1} + q_1\right) = A_1A_2\rho h\frac{\partial^2 u_1^0}{\partial t^2},$$
$$\frac{\partial(N_{12}A_2)}{\partial\alpha_1} + \frac{\partial(N_2A_1)}{\partial\alpha_2} + N_{21}\frac{\partial A_2}{\partial\alpha_1} - N_1\frac{\partial A_1}{\partial\alpha_2} + A_1A_2\left(\frac{Q_2}{R_2} + q_2\right) = A_1A_2\rho h\frac{\partial^2 u_2^0}{\partial t^2},$$
$$\frac{\partial(Q_1A_2)}{\partial\alpha_1} + \frac{\partial(Q_2A_1)}{\partial\alpha_2} - A_1A_2\left(\frac{N_1}{R_1} + \frac{N_2}{R_2} + q\right) = A_1A_2\rho h\frac{\partial^2 w}{\partial t^2},$$
$$\frac{\partial(M_1A_2)}{\partial\alpha_1} + \frac{\partial(M_{21}A_1)}{\partial\alpha_2} + M_{12}\frac{\partial A_1}{\partial\alpha_2} - M_2\frac{\partial A_2}{\partial\alpha_1} + A_1A_2(m_1 - Q_1) = 0,$$
$$\frac{\partial(M_{12}A_2)}{\partial\alpha_1} + \frac{\partial(M_2A_1)}{\partial\alpha_2} + M_{21}\frac{\partial A_2}{\partial\alpha_1} - M_1\frac{\partial A_1}{\partial\alpha_2} + A_1A_2(m_2 - Q_2) = 0. \tag{5.95}$$

Three equations in terms of stress resultants N_1, N_2, N_{12} and stress couples M_1, M_2, M_{12} can be obtained by eliminating the shearing forces Q_1 and Q_2 from the first three equations of (5.95) with the use of the last two of (5.95). Then the use of constitutive laws (5.32) and the strain–displacement relations (5.25) will lead these three equations to be written in terms of the mid-surface displacements u_1^0, u_2^0, and w. Owing to their complexity, the explicit forms of these three equations will not be given here. Their symbolic form may be written as follows (Kraus, 1972):

$$L_1\left\{u_1^0,u_2^0,w\right\} = \rho h\frac{\partial^2 u_1^0}{\partial t^2},\ L_2\left\{u_1^0,u_2^0,w\right\} = \rho h\frac{\partial^2 u_2^0}{\partial t^2},\ L_3\left\{u_1^0,u_2^0,w\right\} = \rho h\frac{\partial^2 w}{\partial t^2}, \tag{5.96}$$

where L_i, $i = 1, 2, 3$, are the differential operators which can be obtained in the way stated earlier. The associated boundary conditions (5.35) for the edges of the shells can also be written symbolically as

$$B_k\left\{u_1^0,u_2^0,w\right\} = 0, \quad k = 1,2,\ldots\ldots N, \tag{5.97}$$

where B_k, $k = 1, 2, \ldots N$ are the differential operators which can be obtained from (5.35) and the necessary relations like (5.32) and (5.25) to express the forces in terms of the mid-surface displacements. From (5.35) we see that each edge of the shell will provide four boundary conditions. If there are four different edges bounding the shells, the number of the boundary conditions N may then be $4 \times 4 = 16$.

To consider the free vibration problems, all the external loads q_1, q_2, q, m_1, and m_2 in (5.95) should be discarded. To determine the normal modes and natural frequencies, a simple harmonic motion is assumed for each normal mode, and the spatial and temporal variations of the mid-surface displacements are assumed to be separable. Thus,

$$\begin{aligned} u_1^0(\alpha_1,\alpha_2,t) &= U_1(\alpha_1,\alpha_2)\cos\omega t, \\ u_2^0(\alpha_1,\alpha_2,t) &= U_2(\alpha_1,\alpha_2)\cos\omega t, \\ w(\alpha_1,\alpha_2,t) &= W(\alpha_1,\alpha_2)\cos\omega t, \end{aligned} \tag{5.98}$$

where U_1, U_2, W, and ω are the mode shapes and natural frequency to be determined. Substituting (5.98) into (5.96) and (5.97), the equations of motion and boundary conditions can be written as

$$\begin{aligned}&L_1\{U_1,U_2,W\}+\rho h\omega^2 U_1=0,\\&L_2\{U_1,U_2,W\}+\rho h\omega^2 U_2=0,\\&L_3\{U_1,U_2,W\}+\rho h\omega^2 U_3=0,\end{aligned}\tag{5.99a}$$

and

$$B_k\{U_1,U_2,W\}=0,\quad k=1,2,\ldots\ldots N.\tag{5.99b}$$

The system of equations (5.99a) may yield N independent solutions for U_1, U_2, and W which are functions of α_1, α_2, and ω. Superimposing these N independent solutions each multiplied by an arbitrary coefficient c_k, and substituting these solutions into the boundary conditions (5.99b), we obtain a system of N homogeneous linear equations in the N unknown constants c_k which can be written symbolically as

$$\sum_{k=1}^{N}K_{ik}(\omega)c_k=0,\quad k=1,2,\ldots\ldots,N.\tag{5.100}$$

To have nontrivial solutions for the constants c_k, the determinant of the system should be zero, that is, $\|K_{ik}(\omega)\|=0$. This equation known as the *frequency equation* is generally transcendental and has an infinite number of roots ω, each corresponding to a natural frequency of free vibration.

Due to the complexity, no detailed formulations for the specific shells will be given in this section. One may refer to Kraus (1972) for the cases of free vibration of isotropic cylindrical shells, shallow spherical shells, and non-shallow spherical shells. As to the problem of forced vibration of laminated shells, one may employ the *classical method of spectral representation* (or called *modal analysis*) wherein the dependent variables of the shells are expanded in infinite series of the normal modes of free vibration. In these expansions, each normal mode is multiplied by an unknown time-dependent coefficient known as the *generalized coordinate* whose determination represents the central aim of the forced vibration analysis. In other words, the spectral representation of shell variables may be expressed as

$$\begin{Bmatrix}u_1^0(\alpha_1,\alpha_2,t)\\u_2^0(\alpha_1,\alpha_2,t)\\w(\alpha_1,\alpha_2,t)\end{Bmatrix}=\sum_{k=1}^{\infty}g_k(t)\begin{Bmatrix}U_{1k}(\alpha_1,\alpha_2)\\U_{2k}(\alpha_1,\alpha_2)\\W_k(\alpha_1,\alpha_2)\end{Bmatrix}.\tag{5.101}$$

The employment of this kind of expansion is based on the premise that the normal modes of free vibration are orthogonal. With the assistance of the orthogonality condition of the modes of free vibration of the laminated shells, the system of equations for the forced vibration problems may then be reduced to a set of uncoupled second-order ordinary differential equations for the unknown generalized coordinates $g_k(t)$. To solve the differential equations, we must also account for the initial conditions.

REFERENCES

Ambartsumyan, S.A., 1961, *Theory of Anisotropic Shells*, National Aeronautics and Space Administration, Washington, DC.

Kraus, H., 1972, *Thin Elastic Shells*, John Wiley & Son, Inc., New York.

Love, A.E.H., 1888, "On the Small Free Vibrations and Deformations of Thin Elastic Shells," *Phil. Trans. Roy. Soc. (London)*, Vol. 17A, pp. 491–546.

Love, A.E.H., 1944, *A Treatise on the Mathematical Theory of Elasticity*, 4th ed., Dover Pub, New York.

Reissner, E., 1941, "A New Derivation of the Equations for the Deformation of Elastic Shells," *American Journal of Mathematics*, Vol. 14, pp. 177–184.

Sokolnikoff, I.S., 1956, *Mathematical Theory of Elasticity*, McGraw-Hill, New York.

Vinson, J.R. and Sierakowski, R.L., 1986, *The Behavior of Structure Composed of Composite Materials*, Martinus Nijhoff Pub., Dordrecht.

CHAPTER 6

Composite Sandwich Construction

SANDWICH CONSTRUCTION HAS MANY advantages over conventional structural constructions, for example, high bending stiffness, good weight savings, good surface finish, good fatigue properties, good thermal and acoustical insulation, and so on. Today, there is a renewed interest in using sandwich structures due to the introduction of new materials such as laminated composites for the faces of sandwich panels, which offer a long-awaited capability with both high stiffness and low specific weight. Similarly, new materials for the core are now available, such as nonmetallic honeycombs and plastic foams. To extend our previous discussions on the basic composite structural elements to their related sandwich constructions, in this chapter, we present the general formulation for composite sandwich plates, beams, and cylindrical shells separately in Sections 6.1.1, 6.2.1, and 6.3.1. After the introduction of general formulation for each composite sandwich structural element, in the first three sections, we also present some analytical solutions and numerical methods for their related buckling analysis and free vibration. Additional topics such as forced vibration and vibration suppression are presented in Sections 6.2.4 and 6.2.5 for the composite sandwich beams. Special topics on the delaminated composite sandwich beams are then presented in the last section of this chapter.

DOI: 10.1201/9781003470465-6

6.1 COMPOSITE SANDWICH PLATES

Due to the importance of composite sandwich plates, several works can be found in the literature such as Hoff (1966), Plantema (1966), Allen (1969), and Noor et al. (1996) for the textbooks and review articles. In this section we will focus on the general formulation of composite sandwich plates and their associated solutions for buckling analysis and free vibration. Some of the content has been adapted from our previously published articles, specifically Moh and Hwu (1997) for buckling analysis and Hwu et al. (2017) for free vibration.

6.1.1 General Formulation

Consider a sandwich plate with anisotropic laminated composite faces and an ideally orthotropic honeycomb core (Figure 6.1). To simplify the analysis, the following assumptions are usually made for sandwiches (Heath, 1960). (1) The faces are relatively thinner than the depth of the core.

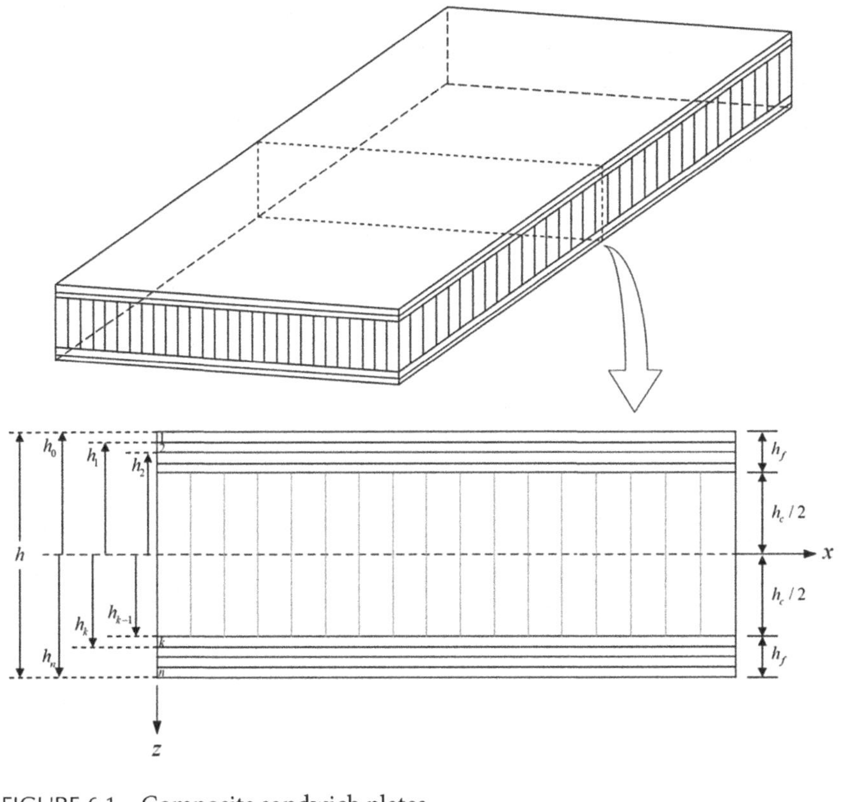

FIGURE 6.1 Composite sandwich plates.

(2) The transverse shear deformation is considered in the faces. However, the contribution of the transverse shear forces by the faces is neglected as compared to those contributed by the core. Hence, we may say that the faces carry loads in their own planes only. (3) The core has direct stiffness normal to the faces and shear stiffness in planes normal to the faces, but no other stiffness. (4) The face-to-core bond ensures that any displacement in the core adjacent to the faces is reproduced exactly in faces, and vice versa. According to these assumptions, the faces take almost all the in-plane loadings and bending moments, while the core undergoes only transverse shear and normal forces. Thus, the stress–strain relations for the orthotropic core are

$$\sigma_x = \sigma_y = \tau_{xy} = 0,\ \sigma_z = E_z \varepsilon_z,\ \tau_{xz} = G_{xz}\gamma_{xz},\ \tau_{yz} = G_{yz}\gamma_{yz}, \tag{6.1}$$

where E_z, G_{xz}, G_{yz} are, respectively, the Young's modulus in z-direction and the transverse shear moduli in x-z and y-z planes.

If the honeycomb cellular core is considered, G_{xz}, G_{yz}, E_z should be interpreted into the effective moduli which are obtained by treating honeycomb as a homogeneous orthotropic continuum and can be related to the geometry and actual material properties of the core. Penzien and Didriksson (1964) suggested a formula to predict the effective shear moduli of the honeycomb, which is

$$\frac{G_e}{G} = \frac{t\sin\theta\left(b + a\cos\theta\right)}{a\left(a+b\right)\sin^2\theta\cos^2\varphi + \left(b + a\cos\theta\right)^2\sin^2\varphi}, \tag{6.2}$$

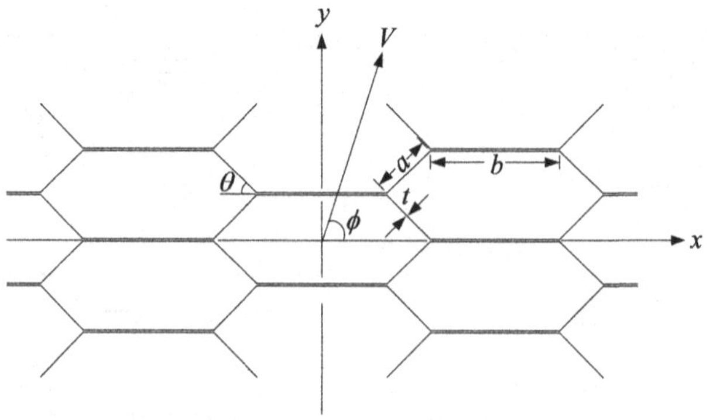

FIGURE 6.2 Geometry of honeycomb cells.

where a, b, t, θ are geometric properties of the honeycomb (Figure 6.2), and φ is the angle between the shear resultant force and the longitudinal x-axis. The effective transverse shear moduli G_{xz} and G_{yz} can therefore be obtained from (6.2) by letting $\varphi = 0°$ and $\varphi = 90°$. Note that in this formula the effect of prevention of warpage has been ignored.

Lacking three of the stress components, the strain–displacement relation and the equilibrium equation can be expressed as

$$\varepsilon_z = \frac{\partial w}{\partial z}, \quad \gamma_{yz} = \frac{\partial w}{\partial y} + \frac{\partial v}{\partial z}, \quad \gamma_{xz} = \frac{\partial w}{\partial x} + \frac{\partial u}{\partial z}, \tag{6.3}$$

$$\frac{\partial \tau_{xz}}{\partial z} = 0, \quad \frac{\partial \tau_{yz}}{\partial z} = 0, \quad \frac{\partial \tau_{xz}}{\partial x} + \frac{\partial \tau_{yz}}{\partial y} + \frac{\partial \sigma_z}{\partial z} = 0. \tag{6.4}$$

By equations $(6.4)_{1,2,3}$ it follows that τ_{xz} and τ_{yz} are functions of x and y only, and that

$$\sigma_z = -z\left(\frac{\partial \tau_{xz}}{\partial x} + \frac{\partial \tau_{yz}}{\partial y}\right) + \sigma_{z_0}, \tag{6.5}$$

where σ_{zo} is the value of σ_z at $z = 0$, and is thus a function of x and y only. By substituting the stresses for the strains in (6.1) and integrating (6.3), the following relations for the displacements are obtained:

$$\begin{aligned}
u &= \frac{z^3}{6E_z}\frac{\partial}{\partial x}\left(G_{xz}\frac{\partial \gamma_{xz}}{\partial x} + G_{yz}\frac{\partial \gamma_{yz}}{\partial y}\right) - \frac{z^2}{2E_z}\frac{\partial \sigma_{z0}}{\partial x} + z\left(\gamma_{xz} - \frac{\partial w_0}{\partial x}\right) + u_0, \\
v &= \frac{z^3}{6E_z}\frac{\partial}{\partial y}\left(G_{xz}\frac{\partial \gamma_{xz}}{\partial x} + G_{yz}\frac{\partial \gamma_{yz}}{\partial y}\right) - \frac{z^2}{2E_z}\frac{\partial \sigma_{z0}}{\partial y} + z\left(\gamma_{yz} - \frac{\partial w_0}{\partial y}\right) + v_0, \\
w &= \frac{-z^2}{2E_z}\left(G_{xz}\frac{\partial \gamma_{xz}}{\partial x} + G_{yz}\frac{\partial \gamma_{yz}}{\partial y}\right) + z\frac{\sigma_{z0}}{E_z} + w_0,
\end{aligned} \tag{6.6}$$

where u_0, v_0, and w_0 are the displacements of the plane $z = 0$, and σ_{zo}, γ_{xz}, γ_{yz} are functions of x and y only. The transverse normal stiffness E_z of the core is assumed to be infinitely large due to the fact that the transverse normal strains are usually negligible for sandwiches having honeycomb cores (Plantema, 1966). Based upon this assumption and equation (6.6), the core displacements can be expressed as

$$u = u_0 + z\beta_x, \quad v = v_0 + z\beta_y, \quad w = w_0, \tag{6.7a}$$

where β_x and β_y are the rotations with respect to the x and y axes, and are related to the transverse shear strain by

$$\beta_x = \gamma_{xz} - \frac{\partial w}{\partial x}, \quad \beta_y = \gamma_{yz} - \frac{\partial w}{\partial y}. \tag{6.7b}$$

The displacement field (6.7) is a model usually given by the *shear deformation plate theory* (Reissner, 1945; Mindlin, 1951), as stated in Section 3.4.

Since the faces are firmly bonded to the core, the displacements in the core adjacent to the faces are reproduced exactly in faces. That is, the displacements of the faces adjacent to the core can be obtained by substituting $z = \pm h_c/2$ into equation (6.7), where + and − are for lower and upper faces, respectively. Therefore, it is natural for us to assume that the displacements of the faces have the same form as those of the core described in equation (6.7). By this face displacement representation, the requirement of displacement continuity across the interface of face and core (Librescu, 1975) will then be satisfied automatically. It should be noted that the plane of $z = 0$ for the face displacements is still the mid-plane of the core, not the mid-plane of the face. Moreover, the transverse shear deformations are also included in this expression even though the transverse shear forces may be neglected as compared with those contributed by the core since the thicknesses of the faces are considered to be relatively thinner than the depth of the core. In other words, equation (6.7) is the expression of displacement for both of the core and the faces.

If small deformations are considered, the strains can be written in terms of the mid-surface displacements as follows:

$$\varepsilon_x = \frac{\partial u}{\partial x} = \varepsilon_x^0 + zk_x, \quad \varepsilon_y = \frac{\partial v}{\partial y} = \varepsilon_y^0 + zk_y, \quad \gamma_{xy} = \frac{\partial u}{\partial y} + \frac{\partial v}{\partial x} = \gamma_{xy}^0 + zk_{xy}, \tag{6.8a}$$

where

$$\varepsilon_x^0 = \frac{\partial u_0}{\partial x}, \quad \varepsilon_y^0 = \frac{\partial v_0}{\partial y}, \quad \gamma_{xy}^0 = \frac{\partial u_0}{\partial y} + \frac{\partial v_0}{\partial x},$$
$$k_x = \frac{\partial \beta_x}{\partial x}, \quad k_y = \frac{\partial \beta_y}{\partial y}, \quad k_{xy} = \frac{\partial \beta_x}{\partial y} + \frac{\partial \beta_y}{\partial x}. \tag{6.8b}$$

Because the deformation of the entire sandwich plates can be described by (6.7) and (6.8), in a way similar to the classical lamination theory presented in Section 2.4.1, we may now write down the constitutive laws of the sandwich plates in terms of resultant forces/moments and the mid-plane strains/curvatures as follows (Hwu and Hu, 1992):

$$\begin{Bmatrix} N_x \\ N_y \\ N_{xy} \\ M_x \\ M_y \\ M_{xy} \end{Bmatrix} = \begin{bmatrix} A_{11} & A_{12} & A_{16} & B_{11} & B_{12} & B_{16} \\ A_{12} & A_{22} & A_{26} & B_{12} & B_{22} & B_{26} \\ A_{16} & A_{26} & A_{66} & B_{16} & B_{26} & B_{66} \\ B_{11} & B_{12} & B_{16} & D_{11} & D_{12} & D_{16} \\ B_{12} & B_{22} & B_{26} & D_{12} & D_{22} & D_{26} \\ B_{16} & B_{26} & B_{66} & D_{16} & D_{26} & D_{66} \end{bmatrix} \begin{Bmatrix} \varepsilon_x^0 \\ \varepsilon_y^0 \\ \gamma_{xy}^0 \\ k_x \\ k_y \\ k_{xy} \end{Bmatrix} \tag{6.9}$$

where A_{ij}, B_{ij}, and D_{ij} are the extensional, coupling, and bending stiffness matrices calculated by

$$A_{ij} = \sum_{k=1}^{n} \left(Q_{ij}^*\right)_k \left(h_k - h_{k-1}\right),$$

$$B_{ij} = \frac{1}{2}\sum_{k=1}^{n} \left(Q_{ij}^*\right)_k \left(h_k^2 - h_{k-1}^2\right),$$

$$D_{ij} = \frac{1}{3}\sum_{k=1}^{n} \left(Q_{ij}^*\right)_k \left(h_k^3 - h_{k-1}^3\right). \tag{6.10}$$

Note that the plane $z = 0$ is the mid-plane of the sandwiches, not the mid-plane of the laminated faces. Hence, $B_{ij} = 0$ is for symmetric sandwiches not for symmetric laminated faces. Moreover, due to the assumption that the faces take almost all the in-plane loadings and bending moments, we may neglect the contribution of core in calculating the extensional, bending, and coupling matrices through the summation of (6.10). The number of layers n in (6.10) is therefore the number of laminae in both faces not including the core.

According to the assumption, the transverse shear forces contributed by the faces are negligible as compared to those contributed by the core. Thus, the total transverse shear force of the composite sandwich plates can be approximated by that of the core as

$$\begin{Bmatrix} Q_x \\ Q_y \end{Bmatrix} = h_c \begin{Bmatrix} G_{xz}\gamma_{xz} \\ G_{yz}\gamma_{yz} \end{Bmatrix}. \tag{6.11}$$

Note that the formula (6.11) is based upon the assumption that the transverse shear stress distribution is uniform across the thickness of the core, which may violate the traction-boundary condition along the core surfaces. For example, the delamination traction-free condition cannot be satisfied under the assumption of the uniform transverse shear stress distribution. To satisfy the boundary conditions of the core, we need to modify the assumption for the transverse shear stress distribution. Tracing back to (6.4), we find that the assumption of the uniform transverse shear stress is due to the neglect of the in-plane resistance of the core. To include the core in-plane capacity, we need to modify the entire mathematical formulation. With reference to the work of Cowper (1966), we may re-derive the governing equation of the composite sandwich plates by integration of the equations of three-dimensional elasticity theory. Admit that the core in-plane capacity is small as compared with that of the faces and accept the concept of mean transverse shear strain, the non-uniform distribution of the transverse shear stress may be accounted for by the introduction of the *shear coefficient* α. The relation between the transverse shear forces Q_x, Q_y and the mean transverse shear strains γ_{xz}, γ_{yz} can then be written as

$$\begin{Bmatrix} Q_x \\ Q_y \end{Bmatrix} = \begin{Bmatrix} A_{55}\gamma_{xz} \\ A_{44}\gamma_{yz} \end{Bmatrix}, \tag{6.12}$$

where A_{44} and A_{55} represent, respectively, the *transverse shear stiffness* in y-z and x-z planes, and are defined by

$$A_{44} = \alpha h_c G_{yz}, \quad A_{55} = \alpha h_c G_{xz}. \tag{6.13}$$

With reference to Cowper (1966), the shear coefficient α = 5/6 for rectangular cross-section. One may refer to Roy and Verchery (1993) for the prediction of the shear stiffness.

Similar to the equilibrium equations for the laminated plates shown in (3.9), the balance of forces and moments of the composite sandwich plates can be expressed as

$$\frac{\partial N_x}{\partial x} + \frac{\partial N_{yx}}{\partial y} = 0, \quad \frac{\partial N_{xy}}{\partial x} + \frac{\partial N_y}{\partial y} = 0,$$
$$\frac{\partial Q_x}{\partial x} + \frac{\partial Q_y}{\partial y} + q = 0, \quad \frac{\partial M_x}{\partial x} + \frac{\partial M_{xy}}{\partial y} - Q_x = 0, \quad \frac{\partial M_{xy}}{\partial x} + \frac{\partial M_y}{\partial y} - Q_y = 0. \tag{6.14}$$

By using (6.8b), (6.9), and (6.12), the five equilibrium equations for the composite sandwich plates given in (6.14) can be written in terms of five unknowns u_0, v_0, w_0, γ_{xz}, and γ_{yz}. Theoretically, the solutions can be found by solving these five coupled partial differential equations with the proper boundary conditions. For the general cases of composite sandwich plates, every boundary of the plates should be described by five prescribed values. Two of them correspond to the in-plane problems and the other three correspond to the bending problems. In applications, several different boundary conditions may occur. Generally, they can be classified as geometrical, static, and mixed boundary conditions. No matter which kind of boundary conditions, they all can be expressed as

$$k_{ui}\left(u_i-\hat{u}_i\right)=k_{Ni}\left(N_i-\hat{N}_i\right),\quad i=1,2,3,4,5, \tag{6.15a}$$

where u_i, i = 1, 2, 3, 4, 5 include the midplane displacements u_n, u_t, w_0 and the rotation angles β_n, β_t; N_i are the corresponding force terms N_n, N_{nt}, Q_n, M_n, M_{nt}. The subscripts n and t represent, respectively, the directions of the normal and tangent of the boundary. The overhat means the prescribed values at the boundaries. Except the category of elastic support and restraint, generally either k_{ui} or k_{Ni}, i = 1,2,3,4,5, are equal to zero. For the geometrical boundary condition all $k_{Ni}=0$, while for the static boundary condition all $k_{ui}=0$. As to the mixed boundary condition, only parts of k_{ui} and k_{Ni} are equal to zero. Besides (6.15a), it is also common to write the boundary conditions as

$$\begin{aligned}
u_n &= \hat{u}_n \quad &&\text{or}\quad N_n=\hat{N}_n \quad &&\text{or}\quad N_n=k_n u_n,\\
u_t &= \hat{u}_t \quad &&\text{or}\quad N_{nt}=\hat{N}_{nt} \quad &&\text{or}\quad N_{nt}=k_t u_t,\\
\beta_n &= \hat{\beta}_n \quad &&\text{or}\quad M_n=\hat{M}_n \quad &&\text{or}\quad M_n=k_m\beta_n,\\
\beta_t &= \hat{\beta}_t \quad &&\text{or}\quad M_{nt}=\hat{M}_{nt} \quad &&\text{or}\quad M_{nt}=k_{mt}\beta_t,\\
w &= \hat{w} \quad &&\text{or}\quad Q_n=\hat{Q}_n \quad &&\text{or}\quad Q_n=k_v w,
\end{aligned} \tag{6.15b}$$

where k_n, k_t, k_m, k_{mt}, and k_v are given spring constants.

Note that the general formulations we describe in (6.7)–(6.15) are basically the same as those presented in (3.96)–(3.99) for the first-order shear deformation theory of the thick laminated plates. Therefore, the matrix form expressions of governing equation shown in (3.101) can also be applied to the cases of composite sandwich plates and are also applicable to the sandwich plate with variable thickness.

6.1.2 Buckling Analysis

As discussed in the buckling of laminated composite plates, the force equilibrium in the lateral direction (z-direction) should be modified as

$$\frac{\partial Q_x}{\partial x}+\frac{\partial Q_y}{\partial y}+N_x\frac{\partial^2 w}{\partial x^2}+2N_{xy}\frac{\partial^2 w}{\partial x\partial y}+N_y\frac{\partial^2 w}{\partial y^2}=0. \tag{6.16}$$

Same as the bending analysis, by using (6.8b), (6.9), and (6.12), the five equilibrium equations for the composite sandwich plates given in (6.14) with the third equation replaced by (6.16) can be written in terms of five unknowns u_0, v_0, w_0, γ_{xz}, and γ_{yz}. Theoretically, the solutions can be found by solving these five coupled partial differential equations with the proper boundary conditions. However, due to mathematical infeasibility, only very few special cases can be solved analytically. Most of the problems should be solved numerically. In the following, we present analytical solutions for the composite sandwich plates with cross-ply symmetric laminate faces. After that, a general numerical procedure is established to solve the general buckling problems of composite sandwich plates. The accuracy of the numerical procedure is then verified by the analytical solution.

Sandwiches with Symmetric Cross-Ply Laminated Faces

In this subsection, a composite sandwich plate with symmetric cross-ply laminate faces will be treated analytically (Moh and Hwu, 1997). For this kind of composite sandwich plates, some of the stiffnesses will vanish such as $A_{16}=A_{26}=D_{16}=D_{26}=0$, $B_{ij}=0$, $i,j=1,2,6$, by (6.10). Under this special consideration, the in-plane and plate bending problems are uncoupled and can be treated individually. For plate bending problems, only (6.16) and the last two equations of (6.14) should be considered. By substituting (6.8b), (6.9), and (6.12) into $(6.14)_{4,5}$ and (6.16) with the above consideration, we obtain the governing equations as follows:

$$\left(N_x\frac{\partial^2}{\partial x^2}+N_y\frac{\partial^2}{\partial y^2}+2N_{xy}\frac{\partial^2}{\partial x\partial y}\right)w+A_{55}\frac{\partial\gamma_{xz}}{\partial x}+A_{44}\frac{\partial\gamma_{yz}}{\partial y}=0,$$

$$\frac{\partial}{\partial x}\left[D_{11}\frac{\partial^2}{\partial x^2}+\left(D_{12}+2D_{66}\right)\frac{\partial^2}{\partial y^2}\right]w-\left[D_{11}\frac{\partial^2}{\partial x^2}+D_{66}\frac{\partial^2}{\partial y^2}-A_{55}\right]\gamma_{xz}$$
$$-\left(D_{12}+D_{66}\right)\frac{\partial^2\gamma_{yz}}{\partial x\partial y}=0,$$

$$\frac{\partial}{\partial y}\left[D_{22}\frac{\partial^2}{\partial y^2}+(D_{12}+2D_{66})\frac{\partial^2}{\partial x^2}\right]w-\left[D_{22}\frac{\partial^2}{\partial y^2}+D_{66}\frac{\partial^2}{\partial x^2}-A_{44}\right]\gamma_{yz}$$
$$-(D_{12}+D_{66})\frac{\partial^2\gamma_{xz}}{\partial x\partial y}=0, \tag{6.17}$$

in which the in-plane loadings N_x, N_y, and N_{xy} are treated as known quantities to the bending problem and are assumed to be unchanged during bending.

Consider a rectangular plate of sides a and b simply supported on all edges and subjected to biaxial compressive loads $N_x = -P$ and $N_y = -kP$. The boundary conditions for simply supported edges can be described as

$$\begin{aligned} &w=0, \quad M_y=0, \quad \gamma_{xz}=0 \quad \text{at } y=0,b,\\ &w=0, \quad M_x=0, \quad \gamma_{yz}=0 \quad \text{at } x=0,a. \end{aligned} \tag{6.18}$$

Here an edge stiffener is considered to prevent shear strain. If no forces paralleled to the edges of the plate are applied to prevent shear strain, $\gamma_{xz} = 0$ and $\gamma_{yz} = 0$ should be replaced by $M_{xy} = 0$. By the fourth and fifth equations in both of (6.9) and (6.8b) with $D_{16} = D_{26} = 0$, $B_{ij} = 0$, $i, j = 1, 2, 6$, the boundary conditions (6.18) can now be rewritten as

$$\begin{aligned} &w=0, \quad \frac{\partial}{\partial y}\left(\gamma_{yz}-\frac{\partial w}{\partial y}\right)=0, \quad \gamma_{xz}=0, \text{ at } y=0,b,\\ &w=0, \quad \frac{\partial}{\partial x}\left(\gamma_{xz}-\frac{\partial w}{\partial x}\right)=0, \quad \gamma_{yz}=0, \text{ at } x=0,a, \end{aligned} \tag{6.19}$$

which will be satisfied if one assumes

$$\begin{aligned} &w=W\sin\frac{m\pi x}{a}\sin\frac{n\pi y}{b},\\ &\gamma_{xz}=\Gamma_x\cos\frac{m\pi x}{a}\sin\frac{n\pi y}{b},\\ &\gamma_{yz}=\Gamma_y\sin\frac{m\pi x}{a}\cos\frac{n\pi y}{b}. \end{aligned} \tag{6.20}$$

Substituting (6.20) into (6.17) with $N_x = -P$, $N_y = -kP$ and $N_{xy} = 0$, we obtain a system of three simultaneous linear equations in three unknowns W, Γ_x, and Γ_y. Nontrivial solutions exist only when the determinant of the coefficients vanishes, which leads to a solution for the buckling load of *P*. The final simplified result is

$$P_{cr} = \frac{1+\eta}{\left[1+k\left(\dfrac{an}{bm}\right)^2\right]\left(1+\eta_x+\eta_y+\eta_o\right)}P_o, \tag{6.21a}$$

where

$$\eta = \pi^2 D^*\left[\frac{1}{A_{44}}\left(\frac{m}{a}\right)^2+\frac{1}{A_{55}}\left(\frac{n}{b}\right)^2\right],$$

$$\eta_x = \frac{\pi^2}{A_{55}}\left[D_{11}\left(\frac{m}{a}\right)^2+D_{66}\left(\frac{n}{b}\right)^2\right],$$

$$\eta_y = \frac{\pi^2}{A_{44}}\left[D_{66}\left(\frac{m}{a}\right)^2+D_{22}\left(\frac{n}{b}\right)^2\right],$$

$$\eta_o = \frac{D^* P_o m^2 n^2}{a^2 A_{44} A_{55}}, \tag{6.21b}$$

and

$$P_o = \frac{\pi^2 a^2}{m^2}\left[D_{11}\left(\frac{m}{a}\right)^4+2\left(D_{12}+2D_{66}\right)\left(\frac{mn}{ab}\right)^2+D_{22}\left(\frac{n}{b}\right)^4\right],$$

$$D^* P_o = \frac{\pi^2 a^2}{m^2}\left[D_{11}D_{66}\left(\frac{m}{a}\right)^4+\left[D_{11}D_{12}-D_{12}\left(D_{12}+2D_{66}\right)\right]\left(\frac{mn}{ab}\right)^2 + D_{22}D_{66}\left(\frac{n}{b}\right)^4\right], \tag{6.21c}$$

in which only the lowest value of P_{cr} is of any importance usually. However, it is not clear which values of *m* and *n* result in the lowest critical buckling load. Values of *m* and *n* should be determined computationally.

Note that because the formulation listed above is for composite sandwich plates of which the core thickness is usually greater than the face thickness, the result obtained in (6.21) is not suitable for the case when h_c is relatively smaller than the face thickness. However, by letting the terms concerning the core properties vanish (i.e., $\eta = \eta_x = \eta_y = \eta_o = 0$), not by letting A_{44} and A_{55} be zero, equation (6.21) will be reduced to $P_{cr} = [1 + k(an/bm)^2]^{-1}P_o$. Under uniaxial loading $k = 0$, the result may be further reduced to $P_{cr} = P_o$ which is identical to that presented in (3.44) for the cases of laminated plates.

Rayleigh–Ritz Method

As we said previously, the equilibrium equation (6.14) together with the constitutive relations (6.9) and the kinematic relations (6.8) provide five equations in terms of five unknowns u_0, v_0, w, γ_{xz}, and γ_{yz}. Therefore, if one can find a set of functions u_0, v_0, w, γ_{xz}, and γ_{yz}, satisfying these five equations as well as the boundary and loading conditions, the solution to the buckling load may be found. Although the solution procedure stated above for the buckling load is straightforward, it is usually mathematically infeasible since they are coupled partial differential equations. Only the special cases like that shown above can be solved analytically. In engineering practice any kind of sandwich lay-up and boundary and loading conditions may occur. Especially for optimum structural design, we need a general scheme to deal with all the possible combinations. Thus, it is necessary for us to find an alternative approach. In this subsection, we choose the Rayleigh–Ritz method to find the buckling load of composite sandwich plate with arbitrary boundary conditions, which is a convenient procedure for determining solutions by the principle of minimum potential energy.

Consider a rectangular composite sandwich plate of sides a and b. The total potential energy Π of this sandwich plate consists of the strain energy U due to bending, stretching, and shearing, and the potential energy $-W$ of the external forces, that is,

$$\Pi = U - W, \tag{6.22a}$$

where

$$\begin{aligned} U &= \frac{1}{2}\int_0^b\int_0^a (M_x\kappa_x + M_y\kappa_y + M_{xy}\kappa_{xy} + N_x\varepsilon_x + N_y\varepsilon_y + N_{xy}\gamma_{xy} \\ &\quad + Q_x\gamma_{xz} + Q_y\gamma_{yz})dxdy, \\ W &= \frac{1}{2}\int_0^b\int_0^a \left[\hat{N}_x\left(\frac{\partial w}{\partial x}\right)^2 + \hat{N}_y\left(\frac{\partial w}{\partial y}\right)^2 + 2\hat{N}_{xy}\left(\frac{\partial w}{\partial x}\frac{\partial w}{\partial y}\right)\right]dxdy. \end{aligned} \tag{6.22b}$$

$\hat{N}_x, \hat{N}_y$, and $\hat{N}_{xy}$ denote external in-plane forces which should be distinguished with the internal in-plane forces N_x, N_y, and N_{xy}. Unlike the special cases discussed above where N_x, N_y, and N_{xy} are treated as known quantities and are assumed to be unchanged during bending, in general, N_x, N_y, and N_{xy} like all the other force components cannot be solved separately.

Substituting (6.8), (6.9), and (6.12) into (6.22), the total potential energy can be expressed as follows:

$$\Pi = \frac{1}{2}\int_0^b\int_0^a \left[f(x,y) - \lambda g(x,y)\right] dxdy, \tag{6.23}$$

where $f(x, y)$ is the strain energy density of which the explicit expression can be written in terms of u_0, v_0, w, γ_{xz}, and γ_{yz} (Moh and Hwu, 1997); $g(x, y)$ is a function related to the work done by the external force $\hat{N}_x = \lambda$, $\hat{N}_y = \lambda k_1$, and $\hat{N}_{xy} = \lambda k_2$. That is,

$$g(x,y) = \left(\frac{\partial w}{\partial x}\right)^2 + k_1\left(\frac{\partial w}{\partial y}\right)^2 + 2k_2\left(\frac{\partial w}{\partial x}\frac{\partial w}{\partial y}\right). \tag{6.24}$$

We now choose a solution for the deformation u_0, v_0, w, γ_{xz}, and γ_{yz} in the form of series containing undetermined parameters U_{ij}, V_{ij}, W_{ij}, Φ_{ij}, and Ψ_{ij}, $i = 1, 2, \ldots, m$, $j = 1, 2, \ldots, n$, for example,

$$u_0 = \sum_{i=1}^{m}\sum_{j=1}^{n} U_{ij}u_{xi}(x)u_{yj}(y), \quad v_0 = \sum_{i=1}^{m}\sum_{j=1}^{n} V_{ij}v_{xi}(x)v_{yj}(y),$$
$$w = \sum_{i=1}^{m}\sum_{j=1}^{n} W_{ij}w_{xi}(x)w_{yj}(y), \quad \gamma_{xz} = \sum_{i=1}^{m}\sum_{j=1}^{n} \Phi_{ij}\varphi_{xi}(x)\varphi_{yj}(y),$$
$$\gamma_{yz} = \sum_{i=1}^{m}\sum_{j=1}^{n} \Psi_{ij}\psi_{xi}(x)\psi_{yj}(y). \tag{6.25}$$

The deformation so selected must satisfy the geometric boundary conditions. The static boundary conditions need not be fulfilled. After employing the selected solution, the potential energy Π may now be expressed in terms of U_{ij}, V_{ij}, W_{ij}, Φ_{ij}, and Ψ_{ij}. This demonstrates that U_{ij},

V_{ij}, W_{ij}, Φ_{ij}, and Ψ_{ij} govern the variation of the potential energy. In order for the potential energy to be a minimum at equilibrium,

$$\frac{\partial \Pi}{\partial U_{ij}} = 0, \ \frac{\partial \Pi}{\partial V_{ij}} = 0, \ \frac{\partial \Pi}{\partial \Phi_{ij}} = 0, \ \frac{\partial \Pi}{\partial \Psi_{ij}} = 0, \ \frac{\partial \Pi}{\partial W_{ij}} = 0,$$
$$i = 1,2,\ldots,m, \ j = 1,2,\ldots,n, \qquad (6.26)$$

which yield a system of 5 mn algebraic equations. By proper mathematical manipulation, (6.26) can be organized into a standard eigenvalue problem which may yield numerous eigenvalues λ_{mn}. Usually, only the lowest value of λ_{mn} is important for the consideration of buckling of composite sandwich plates. One may refer to Moh and Hwu (1997) for detailed derivation and numerical illustration.

6.1.3 Free Vibration

The equations of motion can be obtained by adding the inertia forces to the right-hand side of the equilibrium equations (6.14), that is,

$$\frac{\partial N_x}{\partial x} + \frac{\partial N_{yx}}{\partial y} = \rho h \frac{\partial^2 u_0}{\partial t^2}, \quad \frac{\partial N_{xy}}{\partial x} + \frac{\partial N_y}{\partial y} = \rho h \frac{\partial^2 v_0}{\partial t^2},$$
$$\frac{\partial Q_x}{\partial x} + \frac{\partial Q_y}{\partial y} + q = \rho h \frac{\partial^2 w}{\partial t^2},$$
$$\frac{\partial M_x}{\partial x} + \frac{\partial M_{xy}}{\partial y} - Q_x = I \frac{\partial^2 \beta_x}{\partial t^2}, \quad \frac{\partial M_{xy}}{\partial x} + \frac{\partial M_y}{\partial y} - Q_y = I \frac{\partial^2 \beta_y}{\partial t^2}, \qquad (6.27)$$

where I, ρ, and h are, respectively, the moment of inertia (with respect to the midplane), mass density, and thickness of the composite sandwich plates. The mass density ρ for the sandwich may be approximated by the general mixture rule

$$\rho = \left(\rho_c h_c + 2\rho_f h_f\right) / \left(h_c + 2h_f\right), \qquad (6.28)$$

where ρ_c and ρ_f are, respectively, the mass density of core and faces, and h_c and h_f are, respectively, the thickness of core and faces. The moment of inertia I for rectangular cross-section with unit width is $I = \rho h^3/12$. Substituting (6.7)–(6.12) into (6.27), five governing equations can be obtained in terms of five basic functions u_0, v_0, w, β_x, β_y. Following the way suggested by Hwu and Gai (2003), these five governing equations can be rewritten in matrix form as

$$\mathbf{L}(\boldsymbol{\delta}) - \mathbf{K}_0\boldsymbol{\delta} + \mathbf{p} = \mathbf{I}_0\ddot{\boldsymbol{\delta}}, \tag{6.29a}$$

where

$$\begin{aligned}\mathbf{L}(\boldsymbol{\delta}) = \mathbf{K}_{xx}\boldsymbol{\delta}_{,xx} + (\mathbf{K}_{xy} + \mathbf{K}_{xy}^T)\boldsymbol{\delta}_{,xy} + \mathbf{K}_{yy}\boldsymbol{\delta}_{,yy} \\ + (\mathbf{K}_x - \mathbf{K}_x^T)\boldsymbol{\delta}_{,x} + (\mathbf{K}_y - \mathbf{K}_y^T)\boldsymbol{\delta}_{,y}\end{aligned} \tag{6.29b}$$

and

$$\mathbf{K}_{xx} = \begin{bmatrix} A_{11} & A_{16} & 0 & B_{11} & B_{16} \\ A_{16} & A_{66} & 0 & B_{16} & B_{66} \\ 0 & 0 & A_{55} & 0 & 0 \\ B_{11} & B_{16} & 0 & D_{11} & D_{16} \\ B_{16} & B_{66} & 0 & D_{16} & D_{66} \end{bmatrix},$$

$$\mathbf{K}_{yy} = \begin{bmatrix} A_{66} & A_{26} & 0 & B_{66} & B_{26} \\ A_{26} & A_{22} & 0 & B_{26} & B_{22} \\ 0 & 0 & A_{44} & 0 & 0 \\ B_{66} & B_{26} & 0 & D_{66} & D_{26} \\ B_{26} & B_{22} & 0 & D_{26} & D_{22} \end{bmatrix}, \boldsymbol{\delta} = \begin{Bmatrix} u_0 \\ v_0 \\ w \\ \beta_x \\ \beta_y \end{Bmatrix},$$

$$\mathbf{K}_{xy} = \begin{bmatrix} A_{16} & A_{12} & 0 & B_{16} & B_{12} \\ A_{66} & A_{26} & 0 & B_{66} & B_{26} \\ 0 & 0 & 0 & 0 & 0 \\ B_{16} & B_{12} & 0 & D_{16} & D_{12} \\ B_{66} & B_{26} & 0 & D_{66} & D_{26} \end{bmatrix},$$

$$\mathbf{I}_0 = \begin{bmatrix} \rho h & 0 & 0 & 0 & 0 \\ 0 & \rho h & 0 & 0 & 0 \\ 0 & 0 & \rho h & 0 & 0 \\ 0 & 0 & 0 & I & 0 \\ 0 & 0 & 0 & 0 & I \end{bmatrix}, \quad \mathbf{p} = \begin{Bmatrix} 0 \\ 0 \\ q \\ 0 \\ 0 \end{Bmatrix},$$

$$\mathbf{K}_x = A_{55}\mathbf{I}_{34}, \quad \mathbf{K}_y = A_{44}\mathbf{I}_{35}, \quad \mathbf{K}_0 = A_{44}\mathbf{I}_{55} + A_{55}\mathbf{I}_{44}. \tag{6.29c}$$

The subscript comma stands for differentiation with respect to the coordinates x and/or y; the overdot indicates differentiation with respect to time t; the superscript T denotes transpose of a matrix; $\mathbf{I}_{mn}$ is a 5×5 matrix with a unit value at component mn and all the other components are zero.

Theoretically, the solutions can be found by solving the governing equation (which contains five coupled partial differential equations) with proper boundary and initial conditions. For the general cases of composite sandwich plates, every boundary of the plates should be described by five prescribed values. Two of them correspond to the in-plane problems and the other three correspond to the bending problems. In applications, several different boundary conditions may occur. Generally, they can be expressed as (6.15b).

To know the natural frequency and its associated mode of vibration of the composite sandwich plates, we consider the case that the external load $q = 0$, that is, the problems of free vibration. To find the natural modes of vibration, the usual way is the method of separation of variables. By this method the displacement vector $\boldsymbol{\delta}(x, y, t)$ is written as a product of a spatial function vector $\boldsymbol{\Delta}(x, y)$ and a time function $f(t)$. Because of free vibration the time function $f(t)$ is assumed to be harmonic with frequency ω. Thus,

$$\boldsymbol{\delta}(x, y, t) = \boldsymbol{\Delta}(x, y)e^{i\omega t}, \tag{6.30a}$$

where

$$\boldsymbol{\Delta}(x, y) = \begin{Bmatrix} U(x, y) \\ V(x, y) \\ W(x, y) \\ \mathrm{B}(x, y) \\ \mathrm{B}^*(x, y) \end{Bmatrix}. \tag{6.30b}$$

Substituting (6.30a) into (6.29a) with $\mathbf{p} = \mathbf{0}$, we get

$$\mathbf{L}(\boldsymbol{\Delta}) - \left(\mathbf{K}_0 - \omega^2 \mathbf{I}_0\right)\boldsymbol{\Delta} = \mathbf{0}. \tag{6.31}$$

Equation (6.31) is a set of five partial differential equations whose solutions depend on the boundary conditions of the composite sandwich plates.

In general, even for an isotropic thin plate the exact solutions satisfying the governing equation as well as the associated boundary conditions are found only for a few simple cases. Although the exact analytical solutions have been found for certain simple cases of isotropic thin plates, very few of their corresponding solutions for composite sandwich plates have been presented. Following are some simple problems solved in Hwu et al. (2017) for the free vibration of composite sandwich plates by the classical methods of thin plate analysis.

Navier's Solution

In classical plate theory, Navier's solution is only applicable to the cases of rectangular isotropic plates with all edges simply supported. Here, the composite sandwich plate is restricted to be the one with symmetric cross-ply laminated faces whose $A_{16} = A_{26} = D_{16} = D_{26} = 0$, $B_{ij} = 0$, $i, j = 1, 2, 6$, and the simply supported boundary conditions are described as

$$\begin{aligned}
&N_x = 0, v_0 = 0, w = 0, M_x = 0, \beta_y = 0, \quad \text{at } x = 0, a, \\
&u_0 = 0, N_y = 0, w = 0, M_y = 0, \beta_x = 0, \quad \text{at } y = 0, b,
\end{aligned} \tag{6.32}$$

where a and b are the side lengths of the rectangular sandwich plates. All the boundary conditions will be satisfied automatically if the following double Fourier series are assumed for the displacement vector $\mathbf{\Delta}(x, y)$

$$\begin{aligned}
U(x,y) &= \sum_{m=1}^{\infty}\sum_{n=1}^{\infty} U_{mn} \cos\frac{m\pi x}{a} \sin\frac{n\pi y}{b}, \\
V(x,y) &= \sum_{m=1}^{\infty}\sum_{n=1}^{\infty} V_{mn} \sin\frac{m\pi x}{a} \cos\frac{n\pi y}{b}, \\
W(x,y) &= \sum_{m=1}^{\infty}\sum_{n=1}^{\infty} W_{mn} \sin\frac{m\pi x}{a} \sin\frac{n\pi y}{b}, \\
\mathrm{B}(x,y) &= \sum_{m=1}^{\infty}\sum_{n=1}^{\infty} \mathrm{B}_{mn} \cos\frac{m\pi x}{a} \sin\frac{n\pi y}{b}, \\
\mathrm{B}^*(x,y) &= \sum_{m=1}^{\infty}\sum_{n=1}^{\infty} \mathrm{B}^*_{mn} \sin\frac{m\pi x}{a} \cos\frac{n\pi y}{b}.
\end{aligned} \tag{6.33}$$

With the displacement vector assumed in (6.33), the governing equation (6.31) reduces to

$$\left(\mathbf{K}-\omega^2\mathbf{I}_0\right)\mathbf{d}=\mathbf{0}, \tag{6.34a}$$

where

$$\mathbf{K}=\begin{bmatrix} k_{11} & k_{12} & 0 & 0 & 0 \\ k_{21} & k_{22} & 0 & 0 & 0 \\ 0 & 0 & k_{33} & k_{34} & k_{35} \\ 0 & 0 & k_{43} & k_{44} & k_{45} \\ 0 & 0 & k_{53} & k_{54} & k_{55} \end{bmatrix}, \quad \mathbf{d}=\begin{Bmatrix} U_{mn} \\ V_{mn} \\ W_{mn} \\ \mathrm{B}_{mn} \\ \mathrm{B}^*_{mn} \end{Bmatrix}, \tag{6.34b}$$

and

$$\begin{aligned}
&k_{11}=A_{11}\lambda_m^2+A_{66}\lambda_n^2, \quad k_{22}=A_{66}\lambda_m^2+A_{22}\lambda_n^2,\\
&k_{12}=k_{21}=\left(A_{12}+A_{66}\right)\lambda_m\lambda_n,\\
&k_{33}=A_{55}\lambda_m^2+A_{44}\lambda_n^2, \quad k_{34}=k_{43}=A_{55}\lambda_m, \quad k_{35}=k_{53}=A_{44}\lambda_n,\\
&k_{44}=D_{11}\lambda_m^2+D_{66}\lambda_n^2+A_{55}, \quad k_{45}=k_{54}=\left(D_{12}+D_{66}\right)\lambda_m\lambda_n,\\
&k_{55}=D_{66}\lambda_m^2+D_{22}\lambda_n^2+A_{44}, \qquad \lambda_m=m\pi/a, \;\; \lambda_n=n\pi/b.
\end{aligned} \tag{6.34c}$$

Non-vanishing solutions **d** exist only when the determinant of the coefficient matrix becomes zero, that is, $|\mathbf{K}-\omega^2\mathbf{I}_0|=0$, by which we obtain the natural frequencies ω as

$$\omega_{mn}=\begin{cases} \left[\dfrac{\left(k_{11}+k_{22}\right)\pm\sqrt{\left(k_{11}+k_{22}\right)^2-4\left(k_{11}k_{22}-k_{12}^2\right)}}{2\rho h}\right]^{1/2}, & \text{in-plane mode,}\\[2ex] \sqrt{\dfrac{k_{33}k_{44}k_{55}+2k_{34}k_{35}k_{45}-k_{33}k_{45}^2-k_{44}k_{35}^2-k_{55}k_{34}^2}{\rho h\left(k_{44}k_{55}-k_{45}^2\right)}}, & \text{bending mode.} \end{cases} \tag{6.35}$$

The natural vibration mode shapes of the composite sandwich plates can then be obtained by substituting **d** of (6.34a) into (6.33). Note that the formula $(6.35)_2$ for the bending mode was derived by ignoring the inertia term $I=\rho h^3/12$ if its value is much smaller than ρh for thin sandwich plates. Equation (6.34a) is equivalent to the standard eigenvalue problem, $\left(\mathbf{I}_0^{-1}\mathbf{K}\right)\mathbf{d}=\omega^2\mathbf{d}$, whose eigenvalues and eigenvectors can be calculated directly from commercial software such as MATLAB.

If the faces of the sandwiches are made by nonsymmetric or antisymmetric laminates, the coupling between bending and extension may occur since $B_{ij} \neq 0$, and the formula (6.35) should be modified. A closed-form solution for the sandwiches with faces of antisymmetric cross-ply laminates has been obtained in Lin (2013).

Levy's Solution

It's known that Levy's type of solution can be applied to the rectangular plates which are simply supported at two opposite edges. Assuming that the simply supported edges are at $y = 0$ and $y = b$, the boundary conditions for the composite sandwich plates can be described as

$$u_0 = 0,\ N_y = 0,\ w = 0,\ M_y = 0,\ \beta_x = 0, \quad \text{at } y = 0, b. \tag{6.36}$$

If the composite sandwich plate is the one with symmetric cross-ply laminated faces, when the shape function of the displacement vector $\mathbf{\Delta}(x, y)$ takes the form of a single trigonometric series as

$$\begin{aligned} U(x,y) &= \sum_{n=1}^{\infty} U_m(x)\sin\frac{n\pi y}{b},\quad V(x,y) = \sum_{n=1}^{\infty} V_m(x)\cos\frac{n\pi y}{b}, \\ W(x,y) &= \sum_{n=1}^{\infty} W_m(x)\sin\frac{n\pi y}{b}, \\ \mathrm{B}(x,y) &= \sum_{n=1}^{\infty} \mathrm{B}_m(x)\sin\frac{n\pi y}{b},\quad \mathrm{B}^*(x,y) = \sum_{n=1}^{\infty} \mathrm{B}_m^*(x)\cos\frac{n\pi y}{b}, \end{aligned} \tag{6.37}$$

The simply supported boundary conditions at two opposite edges will be satisfied automatically. Substituting (6.37) into the governing equation (6.31), a set of five ordinary differential equations is obtained in terms of five unknown functions $U_m(x), V_m(x), W_m(x), \mathrm{B}_m(x), \mathrm{B}_m^*(x)$, which can be solved by letting

$$\mathbf{\Delta}_m(x) = \begin{Bmatrix} U_m(x) \\ V_m(x) \\ W_m(x) \\ \mathrm{B}_m(x) \\ \mathrm{B}_m^*(x) \end{Bmatrix} = e^{\lambda x} \begin{Bmatrix} U_0 \\ V_0 \\ W_0 \\ \mathrm{B}_0 \\ \mathrm{B}_0^* \end{Bmatrix} = e^{\lambda x}\mathbf{d}. \tag{6.38}$$

With (6.37) and (6.38), the governing equation (6.31) can also be reduced to (6.34a) in which the matrix **K** is a function of λ, that is, $\mathbf{K} = \mathbf{K}(\lambda)$ and keeps the form of $(6.34b)_1$ whose matrix components are

$$
\begin{aligned}
k_{11} &= -A_{11}\lambda^2 + A_{66}\lambda_n^2, \quad k_{22} = -A_{66}\lambda^2 + A_{22}\lambda_n^2, \\
k_{12} &= -k_{21} = \left(A_{12} + A_{66}\right)\lambda\lambda_n, \\
k_{33} &= -A_{55}\lambda^2 + A_{44}\lambda_n^2, \quad -k_{34} = k_{43} = A_{55}\lambda, \quad k_{35} = k_{53} = A_{44}\lambda_n, \\
k_{44} &= -D_{11}\lambda^2 + D_{66}\lambda_n^2 + A_{55}, \quad k_{45} = -k_{54} = \left(D_{12} + D_{66}\right)\lambda\lambda_n, \\
k_{55} &= -D_{66}\lambda^2 + D_{22}\lambda_n^2 + A_{44}, \qquad \lambda_n = n\pi / b.
\end{aligned} \tag{6.39}
$$

With the results of (6.39), we see that $|\mathbf{K}(\lambda) - \omega^2\mathbf{I}_0| = 0$ will lead to a 10th-order polynomial equation of λ which has 10 roots $\lambda_i(\omega)$, $i = 1, 2, \ldots, 10$. Each of the roots has an associated eigenvector $\mathbf{d}_i(\omega)$ determined from (6.34a). Linear superposition of these 10 homogeneous solutions now gives us

$$
\Delta_m\left(x\right) = \sum_{i=1}^{10} k_i \mathbf{d}_i e^{\lambda_i x}. \tag{6.40}
$$

Substituting (6.40) and (6.37) into the boundary conditions (6.15) will then set a system of 10 simultaneous linear algebraic equations with 10 unknown coefficients k_i. Because both of the eigenvalues λ_i and the eigenvectors $\mathbf{d}_i$ are functions of the natural frequency ω, the coefficient matrix of k_i is a function of the natural frequency ω. Again, non-vanishing solutions exist only when the determinant of the coefficient matrix becomes zero, by which we can then obtain the natural frequencies of the composite sandwich plates. With the determined natural frequency ω, the coefficients k_i can be calculated as the eigenvector, and hence the natural vibration mode shapes of the composite sandwich plates are obtained from (6.40) and (6.37).

Ritz Method

Although Navier's solution and Levy's solution have been successfully extended to the cases of composite sandwich plates, like the conventional plate problems these types of solutions are only applicable to the cases with all edges simply supported or two opposite edges simply supported and the faces of sandwiches should be the ones with symmetric cross-ply laminates. To deal with more complicated situations, an energy method – *Ritz method* is stated in this section to find approximate solutions. According to Hamilton's principle, we know that if the vibration system is conservative (no energy is added or lost), the line integral over the Lagrangian function (the difference between the potential and kinetic energies) between two arbitrary instants of time t_1 and t_2 must be a stationary value, that is,

$$\delta \int_{t_0}^{t_1} \left(\Pi - T \right) dt = 0, \tag{6.41}$$

where Π and T are, respectively, the potential energy and the kinetic energy of the composite sandwich plates. Since the harmonic motion is considered in free vibration, equation (6.41) can be further reduced to

$$\delta \left(\Pi_{\max} - T_{\max} \right) = 0, \tag{6.42}$$

in which the subscript max denotes the maximum value in the time period.

With the kinematic relations and constitutive laws given in (6.7)–(6.9), $\Pi_{\max}$ and $T_{\max}$ can be derived as

$$\begin{aligned} \Pi_{\max} &= \frac{1}{2} \iint_A (\Pi_a + \Pi_b + \Pi_d) dA, \\ T_{\max} &= \frac{\omega^2}{2} \iint_A [\rho h \, (U^2 + V^2 + W^2) + I(\mathrm{B}^2 + \mathrm{B}^{*2})] dA, \end{aligned} \tag{6.43a}$$

where

$$\begin{aligned} \Pi_a = {} & A_{11} \left(\frac{\partial U}{\partial x} \right)^2 + A_{22} \left(\frac{\partial V}{\partial y} \right)^2 + A_{66} \left(\frac{\partial U}{\partial y} + \frac{\partial V}{\partial x} \right)^2 \\ & + 2 \left[A_{12} \left(\frac{\partial U}{\partial x} \right) \left(\frac{\partial V}{\partial y} \right) + A_{16} \left(\frac{\partial U}{\partial x} \right) \left(\frac{\partial U}{\partial y} + \frac{\partial V}{\partial x} \right) + A_{26} \left(\frac{\partial V}{\partial y} \right) \left(\frac{\partial U}{\partial y} + \frac{\partial V}{\partial x} \right) \right] \end{aligned} \tag{6.43b}$$

$$\Pi_b/2 = B_{11}\left(\frac{\partial U}{\partial x}\right)\left(\frac{\partial \mathrm{B}}{\partial x}\right) + B_{22}\left(\frac{\partial V}{\partial y}\right)\left(\frac{\partial \mathrm{B}^*}{\partial y}\right) + B_{66}\left(\frac{\partial U}{\partial y} + \frac{\partial V}{\partial x}\right)\left(\frac{\partial \mathrm{B}}{\partial y} + \frac{\partial \mathrm{B}^*}{\partial x}\right)$$
$$+ B_{12}\left[\left(\frac{\partial U}{\partial x}\right)\left(\frac{\partial \mathrm{B}^*}{\partial y}\right) + \left(\frac{\partial V}{\partial y}\right)\left(\frac{\partial \mathrm{B}}{\partial x}\right)\right]$$
$$+ B_{16}\left[\left(\frac{\partial V}{\partial x} + \frac{\partial U}{\partial y}\right)\left(\frac{\partial \mathrm{B}}{\partial x}\right) + \left(\frac{\partial U}{\partial x}\right)\left(\frac{\partial \mathrm{B}}{\partial y} + \frac{\partial \mathrm{B}^*}{\partial x}\right)\right]$$
$$+ B_{26}\left[\left(\frac{\partial V}{\partial x} + \frac{\partial U}{\partial y}\right)\left(\frac{\partial \mathrm{B}^*}{\partial y}\right) + \left(\frac{\partial V}{\partial y}\right)\left(\frac{\partial \mathrm{B}}{\partial y} + \frac{\partial \mathrm{B}^*}{\partial x}\right)\right] \quad (6.43c)$$

$$\Pi_d = D_{11}\left(\frac{\partial \mathrm{B}}{\partial x}\right)^2 + D_{22}\left(\frac{\partial \mathrm{B}^*}{\partial y}\right)^2 + D_{66}\left(\frac{\partial \mathrm{B}}{\partial y} + \frac{\partial \mathrm{B}^*}{\partial x}\right)^2$$
$$+ 2\left[D_{12}\left(\frac{\partial \mathrm{B}}{\partial x}\right)\left(\frac{\partial \mathrm{B}^*}{\partial y}\right) + D_{16}\left(\frac{\partial \mathrm{B}}{\partial x}\right)\left(\frac{\partial \mathrm{B}^*}{\partial x} + \frac{\partial \mathrm{B}}{\partial y}\right)\right.$$
$$\left. + D_{26}\left(\frac{\partial \mathrm{B}^*}{\partial y}\right)\left(\frac{\partial \mathrm{B}^*}{\partial x} + \frac{\partial \mathrm{B}}{\partial y}\right)\right]$$
$$+ A_{55}\left(\mathrm{B} + \frac{\partial W}{\partial x}\right)^2 + A_{44}\left(\mathrm{B}^* + \frac{\partial W}{\partial y}\right)^2 \quad (6.43d)$$

Choose a solution for the deformation U, V, W, B, B* in the form of series containing the undetermined parameters $U_{ij}, V_{ij}, W_{ij}, \mathrm{B}_{ij}, \mathrm{B}^*_{ij}$, $i = 1, 2, \ldots, m$, $j = 1, 2, \ldots, n$, for example,

$$U(x,y) = \sum_{i=1}^{m}\sum_{j=1}^{n} U_{ij} u_{xi}(x) u_{yj}(y), \quad V(x,y) = \sum_{i=1}^{m}\sum_{j=1}^{n} V_{ij} v_{xi}(x) v_{yj}(y),$$
$$W(x,y) = \sum_{i=1}^{m}\sum_{j=1}^{n} W_{ij} w_{xi}(x) w_{yj}(y), \quad \mathrm{B}(x,y) = \sum_{i=1}^{m}\sum_{j=1}^{n} \mathrm{B}_{ij} \varphi_{xi}(x) \varphi_{yj}(y),$$
$$\mathrm{B}^*(x,y) = \sum_{i=1}^{m}\sum_{j=1}^{n} \mathrm{B}^*_{ij} \psi_{xi}(x) \psi_{yj}(y). \quad (6.44)$$

The deformation so selected must satisfy the geometric boundary conditions. The static boundary conditions need not be fulfilled. After employing the selected solution, the Lagrangian function $\Pi_{max} - T_{max}$ may now be expressed in terms of U_{ij}, V_{ij}, W_{ij}, B_{ij}, and B^*_{ij}. This demonstrates that U_{ij}, V_{ij}, W_{ij}, B_{ij}, and B^*_{ij} govern the variation of the Lagrangian function. In order for the Lagrangian function to be a stationary value,

$$\frac{\partial(\Pi_{max}-T_{max})}{\partial U_{ij}}=0,\quad \frac{\partial(\Pi_{max}-T_{max})}{\partial V_{ij}}=0,\quad \frac{\partial(\Pi_{max}-T_{max})}{\partial W_{ij}}=0,$$
$$\frac{\partial(\Pi_{max}-T_{max})}{\partial B_{ij}}=0,\quad \frac{\partial(\Pi_{max}-T_{max})}{\partial B^*_{ij}}=0,\quad i=1,2,\ldots,m,\ j=1,2,\ldots,n, \tag{6.45}$$

which yield a system of 5 mn algebraic equations. By proper mathematical manipulation, (6.45) can be organized into a standard eigenvalue problem such as

$$\left(\mathbf{P}-\omega^2\mathbf{Q}\right)\mathbf{d}=\mathbf{0}, \tag{6.46a}$$

where

$$\mathbf{d}=(U_{11},V_{11},W_{11},B_{11},B^*_{11},U_{12},V_{12},\cdots\cdots,U_{mn},V_{mn},W_{mn},B_{mn},B^*_{mn})^T, \tag{6.46b}$$

and **P** and **Q** are 5 mn × 5 mn matrices. The natural frequencies can then be obtained by solving (6.46).

Orthogonality Condition

In the cases that the motions in plane direction and thickness directions are uncoupled and the effects of rotary inertia and shear deformation are neglected, the mode shapes of deflections will be orthogonal to each other, that is, $\int W_i W_j dA = \delta_{ij}$ where δ_{ij} is the Kronecker delta. Similar to the way shown later in Section 6.2.3 for composite sandwich beams, we can prove that the natural vibration mode shapes of the composite sandwich plates obey the following orthogonality condition:

$$\int_A \left[\rho h\left(U_iU_j+V_iV_j+W_iW_j\right)+I\left(B_iB_j+B^*_iB^*_j\right)\right]dA=\delta_{ij}, \tag{6.47a}$$

where

$$U_i = U_i(x, y),\ V_i = V_i(x, y),\ W_i = W_i(x, y),\ \mathrm{B}_i = \mathrm{B}_i(x, y),\ \mathrm{B}_i^* = \mathrm{B}_i^*(x, y),$$
$$U_j = U_j(x, y),\ V_j = V_j(x, y),\ W_j = W_j(x, y),\ \mathrm{B}_j = \mathrm{B}_j(x, y),\ \mathrm{B}_j^* = \mathrm{B}_j^*(x, y),$$
(6.47b)

are two sets of mode shapes associated with the natural frequencies ω_i and ω_j.

In the above, classical methods for the analysis of isotropic plates such as Navier's solution, Levy's solution, and Ritz method are successfully extended to the free vibration analysis of composite sandwich plates. To check their correctness, several numerical examples have been solved in Hwu et al. (2017) and compared with each other or with the solutions obtained by the commercial finite element software ANSYS.

6.2 COMPOSITE SANDWICH BEAMS

This section will focus on the general formulation of composite sandwich beams and their related solutions for buckling analysis, free vibration, forced vibration, and vibration suppression. Some of the content has been adapted from our previously published articles such as Hwu et al. (2004).

6.2.1 General Formulation

Because the beams are one-dimensional counterparts of the plates, it seems that all the mathematical formulations used in composite sandwich plates may be employed in the composite sandwich beams just by neglecting all the terms related to y-coordinate. In this way, displacement fields assumed in (6.7), strain–displacement relations for small deformation (6.8), constitutive laws for laminates (6.9) and (6.12), equilibrium equations (6.14), and boundary conditions (6.15), can all be reduced to those for the composite sandwich beams as follows:

Displacement fields

$$u = u_0 + z\beta_x, \quad w = w_0, \tag{6.48a}$$

where

$$\beta_x = \gamma_{xz} - \frac{\partial w}{\partial x}. \tag{6.48b}$$

Strain–displacement relations

$$\varepsilon_x = \frac{\partial u}{\partial x} = \varepsilon_x^0 + zk_x, \tag{6.49a}$$

where

$$\varepsilon_x^0 = \frac{\partial u_0}{\partial x}, \quad k_x = \frac{\partial \beta_x}{\partial x}. \tag{6.49b}$$

Constitutive laws

$$\begin{Bmatrix} N_x \\ M_x \end{Bmatrix} = \begin{bmatrix} A_{11} & B_{11} \\ B_{11} & D_{11} \end{bmatrix} \begin{Bmatrix} \varepsilon_x^0 \\ k_x \end{Bmatrix}, \tag{6.50a}$$

$$Q_x = A_{55}\gamma_{xz}. \tag{6.50b}$$

Equilibrium equations

$$\frac{\partial N_x}{\partial x} = 0, \quad \frac{\partial Q_x}{\partial x} + q = 0, \quad \frac{\partial M_x}{\partial x} - Q_x = 0. \tag{6.51}$$

Boundary conditions

$$\begin{aligned} u_0 &= \hat{u}_0 \quad \text{or} \quad N_x = \hat{N}_x \quad \text{or} \quad N_x = k_x u_0, \\ \beta_x &= \hat{\beta}_x \quad \text{or} \quad M_x = \hat{M}_x \quad \text{or} \quad M_x = k_m \beta_x, \\ w &= \hat{w} \quad \text{or} \quad Q_x = \hat{Q}_x \quad \text{or} \quad Q_x = k_v w. \end{aligned} \tag{6.52}$$

Like the discussion given in Section 4.1 for the laminated composite beams, N_x, M_x, and Q_x are the quantities for the beam theory instead of the quantities for the plate theory, whose definitions and the associated ones for A_{11}, B_{11}, D_{11} are given in (4.8). Moreover, like those shown in (4.13) and (4.24), the constitutive laws (6.50) should also be modified for the consideration of narrow or wide beams. Thus, after we get the final solutions, the following replacements should be made to get the proper solutions:

$$\left.\begin{array}{l}A_{11} \to \widehat{D}_{11}/\Delta,\ D_{11} \to \widehat{A}_{11}/\Delta,\ B_{11} \to -\widehat{B}_{11}/\Delta \\ \Delta = \widehat{A}_{11}\widehat{D}_{11} - \widehat{B}_{11}^{2}\end{array}\right\}, \text{ for narrow beams,}$$

$$A_{11} \to \frac{1}{A_{11}^{v}},\ D_{11} \to D_{11}^{v} + \frac{B_{11}^{v2}}{A_{11}^{v}},\ B_{11} \to -\frac{B_{11}^{v}}{A_{11}^{v}}, \text{ for wide beams,} \tag{6.53}$$

where the representation of the symbols $\widehat{\bullet}$ and v can be found in (2.37) and (4.21).

Substituting (6.48b) and (6.49b) into (6.50), the resultant forces and moments, N_x, Q_x, M_x, can be expressed in terms of the mid-plane displacements u_0, w and transverse shear strain γ_{xz}. With this result, the three equilibrium equations (6.51) can then be written in terms of three unknowns u_0, w and γ_{xz} as

$$\begin{aligned}&A_{11}\frac{d^2u_o}{dx^2} + B_{11}\frac{d^2}{dx^2}\left(\gamma_{xz} - \frac{dw}{dx}\right) = 0,\\ &A_{55}\frac{d\gamma_{xz}}{dx} + q = 0,\\ &B_{11}\frac{d^2u_o}{dx^2} + D_{11}\frac{d^2}{dx^2}\left(\gamma_{xz} - \frac{dw}{dx}\right) - A_{55}\gamma_{xz} = 0.\end{aligned} \tag{6.54}$$

From $(6.54)_2$, we can obtain a simple equation for γ_{xz}. Multiplying $(6.54)_1$ by B_{11} and $(6.54)_3$ by $-A_{11}$, adding them together and using the results for γ_{xz}, we may get an ordinary differential equation for w. With these results, either $(6.54)_1$ or $(6.54)_3$ can provide the equation for u_0. The final simplified results are

$$\begin{aligned}&\frac{d\gamma_{xz}}{dx} = -\frac{q}{S},\quad \frac{d^4w}{dx^4} = -\frac{1}{S}\frac{d^2q}{dx^2} + \frac{q}{D},\\ &\frac{du_0}{dx} = B\frac{d^2w}{dx^2} + \frac{Bq}{S} + AN_x,\end{aligned} \tag{6.55a}$$

where

$$S = A_{55},\ A = \frac{1}{A_{11}},\ B = \frac{B_{11}}{A_{11}},\ D = D_{11} - \frac{B_{11}^2}{A_{11}}. \tag{6.55b}$$

After solving the governing equations (6.55) with six proper boundary conditions (6.52), three for each end of the beam, we may find the solutions for u_0, w, and γ_{xz}. With these solutions, the axial force N_x, bending moment M_x, and transverse shear force Q_x can then be obtained from (6.48b), (6.49b), and (6.50). Actually, it is much easier to get N_x, M_x, and Q_x directly from the equilibrium equation (6.51) if their boundary values are known. Equation $(6.51)_1$ reveals that N_x is a constant through the beam, which will be equal to the axial load applied at the ends of the beam. By treating N_x as a known value, the constitutive laws (6.50a) can be rewritten as

$$\varepsilon_x^0 = AN_x - B\kappa_x, \quad M_x = BN_x + D\kappa_x. \tag{6.56}$$

Substituting $(6.51)_3$ into $(6.51)_2$, and using (6.56) and (6.49b) we have

$$\frac{d^2 M_x}{dx^2} + q = 0, \quad \text{or,} \quad D\frac{d^2 \kappa_x}{dx^2} + q = 0, \quad \text{or,} \quad D\frac{d^3 \beta_x}{dx^3} + q = 0, \tag{6.57}$$

which are useful for solving the problem.

6.2.2 Buckling Analysis

Similar to (6.16) for the composite sandwich plates, in order to consider the buckling of the composite sandwich beams, the force equilibrium in the lateral direction, that is, equation $(6.51)_2$, should be modified as

$$\frac{\partial Q_x}{\partial x} + N_x \frac{\partial^2 w}{\partial x^2} = 0. \tag{6.58}$$

If the compressive load applied at the ends of the beam is P, from $(6.51)_1$ we know that N_x is a constant throughout the beams and

$$N_x = -P. \tag{6.59}$$

Through the comparison between (6.58) and $(6.51)_2$, we know that the governing equations for the buckling analysis can be obtained from (6.55a) with the replacement of q by $-P\partial^2 w/\partial x^2$, that is,

$$\begin{aligned} &\frac{d\gamma_{xz}}{dx} = \frac{P}{S}\frac{d^2 w}{dx^2}, \quad \frac{d^4 w}{dx^4} + \lambda^2 \frac{d^2 w}{dx^2} = 0, \\ &\frac{du_0}{dx} = B\left(1 - \frac{P}{S}\right)\frac{d^2 w}{dx^2} - AP, \end{aligned} \tag{6.60a}$$

where

$$\lambda^2 = \frac{P}{D\left(1-\dfrac{P}{S}\right)}. \tag{6.60b}$$

Integrating both sides of (6.60a), we get

$$\begin{aligned} \gamma_{xz} &= \frac{P}{S}\frac{dw}{dx} + d_1, \\ \frac{d^2w}{dx^2} &+ \lambda^2 w = c_1 x + c_2, \\ u_0 &= B\left(1-\frac{P}{S}\right)\frac{dw}{dx} - APx + c_3, \end{aligned} \tag{6.61}$$

where c_1, c_2, c_3, and d_1 are the integration constants to be determined. Substituting $(6.61)_1$ into (6.48b) and $(6.49b)_2$, we have

$$\beta_x = -\left(1-\frac{P}{S}\right)\frac{dw}{dx} + d_1, \quad \kappa_x = -\left(1-\frac{P}{S}\right)\frac{d^2w}{dx^2}. \tag{6.62}$$

The transverse shear force Q_x and bending moment M_x can then be obtained from the constitutive laws (6.50b) and $(6.50a)_2$ or $(6.56)_2$, which lead to

$$Q_x = P\frac{dw}{dx} + Sd_1, \quad M_x = -BP - D\left(1-\frac{P}{S}\right)\frac{d^2w}{dx^2}. \tag{6.63}$$

On the other hand, substituting the result of the transverse shear force obtained in $(6.63)_1$ into the equilibrium equation $(6.51)_3$ and integrating both sides of the equations with respect to x, we get

$$M_x = Pw + Sd_1 x + d_2. \tag{6.64}$$

Equating $(6.63)_2$ to (6.64), we obtain a second-order ordinary differential equation of w, which should be equivalent to $(6.61)_2$. From comparison, we get the following relations for the integration constants:

$$c_1 = -\frac{S\lambda^2}{P}d_1, \quad c_2 = -\lambda^2\left(\frac{d_2}{P} + B\right). \tag{6.65}$$

From the derivation given above, we know that the deflection w of the composite sandwich beam can be found by solving the following governing equation:

$$\frac{d^2w}{dx^2} + \lambda^2 w = c_1 x + c_2, \tag{6.66}$$

through the satisfaction of the boundary conditions. After finding the deflection, the transverse shear strain γ_{xz}, the rotation angle $\beta_x (= \gamma_{xz} - w_{,x})$, the curvature k_x, the axial displacement u_0, the transverse shear force Q_x, and the bending moment M_x can then be determined by the following relations:

$$\gamma_{xz} = \left(\frac{dw}{dx} - \frac{c_1}{\lambda^2}\right)\frac{P}{S}, \quad \beta_x = -\left(1 - \frac{P}{S}\right)\frac{dw}{dx} - \frac{Pc_1}{S\lambda^2}, \quad \kappa_x = -\left(1 - \frac{P}{S}\right)\frac{d^2w}{dx^2},$$
$$u_0 = \left(1 - \frac{P}{S}\right)B\frac{dw}{dx} - APx + c_3,$$
$$Q_x = \left(\frac{dw}{dx} - \frac{c_1}{\lambda^2}\right)P, \ M_x = \left(w - \frac{c_1 x + c_2}{\lambda^2} - B\right)P = -BP - \left(1 - \frac{P}{S}\right)D\frac{d^2w}{dx^2}. \tag{6.67}$$

Because the governing equation (6.66) is a second-order ordinary differential equation, its general solution can easily be written as

$$w = k_1 \cos \lambda x + k_2 \sin \lambda x + \frac{c_1}{\lambda^2}x + \frac{c_2}{\lambda^2}, \tag{6.68}$$

where k_1, k_2, c_1, and c_2 are four integration constants to be determined from the boundary conditions. Substituting (6.68) into (6.67), we have

$$\gamma_{xz} = \frac{P}{S}\left(-k_1\lambda \sin \lambda x + k_2\lambda \cos \lambda x\right),$$
$$\beta_x = -\left(1 - \frac{P}{S}\right)\left(-k_1\lambda \sin \lambda x + k_2\lambda \cos \lambda x\right) - \frac{c_1}{\lambda^2},$$
$$\kappa_x = \frac{P}{D}\left(k_1 \cos \lambda x + k_2 \sin \lambda x\right),$$
$$u_0 = B\left(1 - \frac{P}{S}\right)\left(-k_1\lambda \sin \lambda x + k_2\lambda \cos \lambda x + \frac{c_1}{\lambda^2}\right) - APx + c_3, \tag{6.69a}$$

$$Q_x = P\left(-k_1\lambda\sin\lambda x + k_2\lambda\cos\lambda x\right),$$
$$M_x = P\left(k_1\cos\lambda x + k_2\sin\lambda x - B\right). \qquad (6.69b)$$

EXAMPLE 1: BUCKLING OF A COMPOSITE SANDWICH BEAM SIMPLY SUPPORTED AT BOTH ENDS

Consider a composite sandwich beam of length ℓ simply supported at both ends, that is,

$$w = M_x = 0, \quad \text{at}\, x = 0 \text{ and}\, x = \ell. \qquad (6.70)$$

Substituting (6.68) and $(6.69\text{b})_2$ into (6.70), we have

$$\begin{aligned} &k_1 + \frac{c_2}{\lambda^2} = 0, \quad P\left(-B + k_1\right) = 0, \\ &k_1\cos\lambda\ell + k_2\sin\lambda\ell + \frac{c_1\ell}{\lambda^2} + \frac{c_2}{\lambda^2} = 0, \\ &P\left(-B + k_1\cos\lambda\ell + k_2\sin\lambda\ell\right) = 0, \end{aligned} \qquad (6.71)$$

which leads to

$$\frac{c_2}{\lambda^2} = -k_1 = -B, \quad k_1 = B, \quad c_1 = 0, \quad k_2 = \frac{B\left(1-\cos\lambda\ell\right)}{\sin\lambda\ell}. \qquad (6.72)$$

With the results given in (6.72), the deflection w of (6.68) now becomes

$$w = -B\left(\cos\lambda x + \frac{1-\cos\lambda\ell}{\sin\lambda\ell}\sin\lambda x - 1\right). \qquad (6.73)$$

Equation (6.73) shows that buckling will occur when $\sin\lambda\ell = 0$, that is, $\lambda\ell = n\pi$. Use of (6.60b) now gives us the buckling loads

$$P_{cr} = \frac{n^2\pi^2 D / \ell^2}{1 + n^2\pi^2 D / S\ell^2}. \qquad (6.74)$$

Obviously, the lowest value of (6.74) corresponds to $n = 1$.

EXAMPLE 2: BUCKLING OF A COMPOSITE SANDWICH BEAM CLAMPED AT BOTH ENDS

From (6.52), the boundary conditions of the clamped ends can be written as

$$w = \beta_x \left(= \gamma_{xz} - \frac{dw}{dx} \right) = 0, \quad \text{at } x = 0 \text{ and } x = \ell. \tag{6.75}$$

With the general solution obtained in (6.68) and (6.69), the boundary conditions (6.75) can be rewritten into the following simultaneous linear algebraic equations

$$\begin{aligned} &k_1 + \frac{c_2}{\lambda^2} = 0, \quad -k_2\lambda\left(1 - \frac{P}{S}\right) - \frac{c_1}{\lambda^2} = 0, \\ &k_1 \cos\lambda\ell + k_2 \sin\lambda\ell + \frac{c_1\ell}{\lambda^2} + \frac{c_2}{\lambda^2} = 0, \\ &-\left(1 - \frac{P}{S}\right)\left(-k_1\lambda\sin\lambda\ell + k_2\lambda\cos\lambda\ell\right) - \frac{c_1}{\lambda^2} = 0. \end{aligned} \tag{6.76}$$

Solving (6.76), we obtain

$$\frac{c_2}{\lambda^2} = -k_1, \quad \frac{c_1}{\lambda^2} = -\lambda\left(1 - \frac{P}{S}\right)k_2, \tag{6.77a}$$

and

$$\begin{bmatrix} -1 + \cos\lambda\ell & \sin\lambda\ell - \lambda\ell\left(1 - \dfrac{P}{S}\right) \\ \sin\lambda\ell & 1 - \cos\lambda\ell \end{bmatrix} \begin{Bmatrix} k_1 \\ k_2 \end{Bmatrix} = \begin{Bmatrix} 0 \\ 0 \end{Bmatrix}. \tag{6.77b}$$

A nontrivial solution to (6.77b) exists only when the determinant of the coefficients vanishes, which leads to

$$\tan\frac{\lambda\ell}{2} = 0, \quad \text{or} \quad \tan\frac{\lambda\ell}{2} = \frac{\lambda\ell}{2}\left(1 - \frac{P}{S}\right). \tag{6.78}$$

From (6.78), we have

$$\frac{\lambda\ell}{2} = n\pi, \quad \text{or} \quad \lambda = \lambda^*, \tag{6.79a}$$

where λ^* is the root of $(6.78)_2$, that is,

$$\tan\frac{\lambda^*\ell}{2}=\frac{\lambda^*\ell}{2}\left(1-\frac{P}{S}\right). \tag{6.79b}$$

Substituting (6.60b) into (6.79a), the buckling load of a composite sandwich beam with clamped ends can be obtained as

$$P_{cr}=\frac{4n^2\pi^2D/\ell^2}{1+4n^2\pi^2D/S\ell^2},\quad \text{or}\quad P_{cr}=\frac{\lambda^{*2}D}{1+\lambda^{*2}D/S}. \tag{6.80}$$

Their associated buckling modes can then be found from (6.68) with the coefficients obtained by substituting (6.79) into (6.77). The final simplified results are

$$\begin{aligned}
&w=k_1\left(\cos\frac{2n\pi}{\ell}x-1\right),\quad \text{for } \lambda=2n\pi/\ell,\\
&w=-k_2\tan\left(\lambda^*\ell/2\right)\left(\cos\lambda^*x-\frac{\sin\lambda^*x}{\tan\left(\lambda^*\ell/2\right)}+\frac{2x}{\ell}-1\right),\quad \text{for } \lambda=\lambda^*.
\end{aligned} \tag{6.81}$$

If we shift the origin of the coordinate to the beam center, that is, let $x = x' + (\ell/2)$, the buckling modes (6.81) can also be written as

$$\begin{aligned}
&w=k_1\left((-1)^n\cos\frac{2n\pi}{\ell}x'-1\right),\quad \text{for } \lambda=2n\pi/\ell,\\
&w=\frac{k_2}{\cos\left(\lambda^*\ell/2\right)}\left(\sin\left(\lambda^*x'\right)-\frac{\sin\left(\lambda^*\ell/2\right)}{\ell/2}x'\right),\quad \text{for } \lambda=\lambda^*.
\end{aligned} \tag{6.82}$$

From (6.82), we see that $\lambda = 2n\pi/\ell$ yields the symmetrical buckling modes, whereas $\lambda = \lambda^*$ yields the anti-symmetrical buckling modes. Moreover, it is easy to show that the nth anti-symmetrical buckling load is larger than the nth symmetrical buckling load, but smaller than the $(n + 1)$ th symmetrical buckling load.

6.2.3 Free Vibration

In order to consider the lateral vibration of composite sandwich beams, both the lateral and rotary inertia forces should be added to the equilibrium equation (6.51) to form the equations of motion, that is,

$$\frac{\partial N_x}{\partial x}=0,\quad \frac{\partial Q_x}{\partial x}+q=\rho h\frac{\partial^2 w}{\partial t^2},\quad \frac{\partial M_x}{\partial x}-Q_x=I\frac{\partial^2 \beta_x}{\partial t^2},\tag{6.83}$$

where I, ρ, and h are, respectively, the moment of inertia (with respect to the mid-plane), mass density, and thickness, as explained in Section 6.1.3. From the first equation of (6.83), we see that the axial force N_x is constant along the beam. If the composite sandwich beam is considered to be free of external axial load, $N_x = 0$ throughout the entire beam. The relations between transverse shear force/bending moment and transverse shear strain/rotation angle provided in (6.50b) and $(6.56)_2$ can then be written as

$$Q_x=S\gamma_{xz},\quad M_x=D\frac{\partial \beta_x}{\partial x},\tag{6.84}$$

in which the symbols defined in (6.55b) have been used.

Substituting (6.84) into (6.83) and expressing the equations in terms of the deflection w and the slope β_x, we get

$$S\left(\frac{\partial \beta_x}{\partial x}+\frac{\partial^2 w}{\partial x^2}\right)+q=\rho h\frac{\partial^2 w}{\partial t^2},\quad D\frac{\partial^2 \beta_x}{\partial x^2}=S\left(\beta_x+\frac{\partial w}{\partial x}\right)+I\frac{\partial^2 \beta_x}{\partial t^2}.\tag{6.85}$$

From the first equation of (6.85), we can further express $\partial\beta_x/\partial x$ in terms of w. Substituting this expression into the partial differential with respect to x of the second equation of (6.85), we obtain the equation of motion in terms of the transverse deflection only, which is

$$D\frac{\partial^4 w}{\partial x^4}-\left(I+\frac{\rho h D}{S}\right)\frac{\partial^4 w}{\partial x^2\partial t^2}+\frac{\rho h I}{S}\frac{\partial^4 w}{\partial t^4}+\rho h\frac{\partial^2 w}{\partial t^2}=q+\frac{I}{S}\frac{\partial^2 q}{\partial t^2}-\frac{D}{S}\frac{\partial^2 q}{\partial x^2}.\tag{6.86}$$

To know the natural frequency and its associated vibration mode of the composite sandwich beams, we consider the case that the external load $q(x, t) = 0$, that is, the free vibration problems. To find the natural modes of vibration, the usual way is the method of separation of variables. By this method we write the deflection $w(x, t)$ as a product of a function $W(x)$ of the spatial variables only and a function $f(t)$ depending on time only. Because of free vibration, the time function $f(t)$ is assumed to be harmonic and with frequency ω. Thus,

$$w(x,t)=W(x)e^{i\omega t}.\tag{6.87}$$

Through the use of (6.87), the equation of motion (6.86) can easily be reduced to an ordinary differential equation and the general solutions for $W(x)$ can be obtained as

$$W(x) = c_1 \cosh \lambda x + c_2 \sinh \lambda x + c_3 \cos \mu x + c_4 \sin \mu x, \qquad (6.88a)$$

where

$$\lambda^2 = \frac{\omega^2}{2D}\left(-\hat{I} + \sqrt{\hat{I}^2 + 4\hat{m}D}\right), \quad \mu^2 = \frac{\omega^2}{2D}\left(\hat{I} + \sqrt{\hat{I}^2 + 4\hat{m}D}\right), \qquad (6.88b)$$

and

$$\hat{I} = I + \frac{\rho h D}{S}, \quad \hat{m} = \frac{\rho h}{\omega^2} - \frac{\rho h I}{S}. \qquad (6.88c)$$

The above solution (6.88) is valid when $\hat{m} > 0$, or say $\omega^2 < S/I$, which is usually true for the beam whose transverse shear deformation is neglected since in that case the shear stiffness S is assumed to be infinite. For the case of $\hat{m} < 0$, similar results can be obtained and will not be discussed in this section. To find the natural frequency and the mode shape, we need to know the boundary conditions of the problems. The usual boundary conditions encountered in the vibration of the composite sandwich beams are

i. simply supported ends: $w = M_x = 0$ at $x = 0, \ell$; (6.89a)

ii. clamped-clamped ends: $w = \beta_x = 0$ at $x = 0, \ell$; (6.89b)

iii. clamped-free ends: $w = \beta_x = 0$ at $x = 0$ and $Q_x = M_x = 0$ at $x = \ell$. (6.89c)

Substituting the general solution (6.88a) into the boundary conditions (6.89), we obtain the natural frequency and mode shape of the composite sandwich beams as

i. simply supported ends:

$$\mu_j \ell = j\pi, \qquad (6.90a)$$

$$W_j(x) = \sin(\mu_j x), \quad j = 1, 2, \ldots \qquad (6.90b)$$

ii. clamped-clamped ends:

$$2\left(1-\cosh\lambda_j\ell\cos\mu_j\ell\right)+\left(\gamma_j-\frac{1}{\gamma_j}\right)\sinh\lambda_j\ell\sin\mu_j\ell=0, \qquad (6.91a)$$

$$W_j\left(x\right)=\cosh\lambda_jx-\cos\mu_jx-\alpha_j\left(\sinh\lambda_jx-\gamma_j\sin\mu_jx\right), \quad j=1,2,\ldots \qquad (6.91b)$$

iii. clamped-free ends:

$$2+\left(\frac{\lambda_j}{\mu_j}-\frac{\mu_j}{\lambda_j}\right)\sin\mu_j\ell\sinh\lambda_j\ell+\left(\frac{\gamma_j\lambda_j}{\mu_j}+\frac{\mu_j}{\gamma_j\lambda_j}\right)\cos\mu_j\ell\cosh\lambda_j\ell=0, \qquad (6.92a)$$

$$W_j\left(x\right)=\cosh\lambda_jx-\cos\mu_jx-\beta_j\left(\sinh\lambda_jx-\gamma_j\sin\mu_jx\right), \quad j=1,2,\ldots \qquad (6.92b)$$

In the above,

$$\alpha_j=\frac{\cosh\lambda_j\ell-\cos\mu_j\ell}{\sinh\lambda_j\ell-\gamma_j\sin\mu_j\ell}, \beta_j=\frac{\mu_j\sinh\lambda_j\ell-\lambda_j\sin\mu_j\ell}{\mu_j\cosh\lambda_j\ell+\gamma_j\lambda_j\cos\mu_j\ell}, \gamma_j=\frac{\lambda_j+\dfrac{\rho h\omega_j^2}{\lambda_jS}}{\mu_j-\dfrac{\rho h\omega_j^2}{\mu_jS}}. \qquad (6.93)$$

In the case of the laminated composite beams that can be considered as a sandwich without core, the thickness of the beam is usually small compared to its length. Therefore, it is reasonable to neglect the effects of rotary inertia and shear deformation for this special but common case. The natural frequency and mode shape of the laminated composite beams can then be obtained by specializing (6.90)–(6.93) with $I=0$ and $S\to\infty$.

Orthogonality Condition

If the family of natural vibration mode shapes $W_j(x)$ can constitute a complete set of orthonormal modes, most of the vibration problems can be solved by modal analysis through the use of the expansion theorem (Meirovitch, 1967). However, due to the complexity of the partial differential equation (6.86) which the deflection mode shape should satisfy, it is difficult to prove that they are orthogonal to each other. Even the direct use of the solutions shown in (6.90)–(6.93) cannot prove that the natural mode $W_j(x)$ only will constitute a complete orthonormal set. This difficulty leads us to think that maybe the orthogonal set contains not only the deflection but also the slope angle due to the inclusion of the effects of rotary inertia and shear deformation. With this consideration we now deal with (6.85) instead of (6.86). By assuming

$$w(x,t)=W(x)e^{i\omega t},\quad \beta_x(x,t)=B(x)e^{i\omega t}, \tag{6.94}$$

and introducing (6.94) into (6.85) with $q = 0$, we obtain

$$\begin{aligned}-\rho h\omega^2 W(x)&=S\left(B'(x)+W''(x)\right),\\ -I\omega^2 B(x)&=DB''(x)-SB(x)-SW'(x).\end{aligned} \tag{6.95}$$

Equation (6.95) contains two second-order ordinary differential equations, which must be supplemented by four boundary conditions, that is, two boundary conditions for each end. Usually, the boundary condition is either displacement-prescribed or forced-prescribed or mixed. For homogeneous boundary conditions, they may be expressed as (i) $w = 0$ or $Q_x = 0$ and (ii) $\beta_x = 0$ or $M_x = 0$. Through the use of the relations (6.84) and (6.94), we have

$$W(x_e)=0,\ \text{or}\ B(x_e)+W'(x_e)=0,\ x_e:\text{end point}; \tag{6.96a}$$

$$B(x_e)=0,\ \text{or}\ \ B'(x_e)=0,\ x_e:\text{end point}. \tag{6.96b}$$

Different combinations now provide us with the following four different types of end conditions:

i. Fixed end: $w = \beta_x = 0$ which leads to

$$W(x_e)=B(x_e)=0. \tag{6.97a}$$

ii. Free end: $Q_x = M_x = 0$ which leads to

$$B(x_e)+W'(x_e)=B'(x_e)=0. \tag{6.97b}$$

iii. Hinged end: $w = M_x = 0$ which leads to

$$W(x_e) = B'(x_e) = 0. \tag{6.97c}$$

iv. Moving end: $Q_x = \beta_x = 0$ which leads to

$$B(x_e) + W'(x_e) = B(x_e) = 0. \tag{6.97d}$$

By a simple mathematical manipulation, the natural frequencies ω_j and vibration mode shapes $W_j(x)$ for each different boundary condition can be obtained, which can be proved to be exactly the same as those shown in (6.90)–(6.93). With the results of $W_j(x)$, the natural mode shapes of rotation angles $B_j(x)$ can then be obtained by using $(6.95)_1$. Their solutions are now shown below:

i. Simply supported ends:

$$B_j(x) = \left(\frac{\rho h \omega_j^2}{\mu_j S} - \mu_j\right)\cos \mu_j x. \tag{6.98}$$

ii. Clamped-clamped ends:

$$B_j(x) = \left(-\frac{\rho h \omega_j^2}{\lambda_j S} - \lambda_j\right)\sinh \lambda_j x + \left(\frac{\rho h \omega_j^2}{\mu_j S} - \mu_j\right)\sin \mu_j x$$
$$+\left(\frac{\alpha_j \rho h \omega_j^2}{\lambda_j S} + \alpha_j \lambda_j\right)\cosh \lambda_j x + \left(\frac{\alpha_j \gamma_j \rho h \omega_j^2}{\mu_j S} - \alpha_j \gamma_j \mu_j\right)\cos \mu_j x. \tag{6.99}$$

iii. Clamped-free ends:

$$B_j(x) = \left(-\frac{\rho h \omega_j^2}{\lambda_j S} - \lambda_j\right)\sinh \lambda_j x + \left(\frac{\rho h \omega_j^2}{\mu_j S} - \mu_j\right)\sin \mu_j x$$
$$+\left(\frac{\beta_j \rho h \omega_j^2}{\lambda_j S} + \beta_j \lambda_j\right)\cosh \lambda_j x + \left(\frac{\beta_j \gamma_j \rho h \omega_j^2}{\mu_j S} - \beta_j \gamma_j \mu_j\right)\cos \mu_j x. \tag{6.100}$$

Let ω_i and ω_j be the two distinct natural frequencies and $W_i(x)$, $B_i(x)$ and $W_j(x)$, $B_j(x)$ be the corresponding natural modes of vibration resulting from the solution of the equations of motion (6.95) and its associated boundary conditions (6.96). Consider equations (6.95) corresponding to ω_i, $W_i(x)$, and $B_i(x)$. If we multiply $(6.95)_1$ by $W_j(x)$ and $(6.95)_2$ by $B_j(x)$, add them together and integrate both sides of the equation over the beam length ℓ, we obtain

$$\begin{aligned}
&-\omega_i^2 \int_0^{\ell} \left(\rho h W_i(x) W_j(x) + I B_i(x) B_j(x)\right) dx \\
&= \int_0^{\ell} \left[S\left(B_i'(x) + W_i''(x)\right) W_j(x) + \left(D B_i'' - S B_i(x) - S W_i'(x)\right) B_j(x) \right] dx.
\end{aligned} \tag{6.101}$$

Similarly, another equation can be obtained from (6.101) by interchanging the subscripts i and j. Subtracting these two equations, we get

$$\begin{aligned}
&\left(\omega_i^2 - \omega_j^2\right) \int_0^{\ell} \left(\rho h W_i(x) W_j(x) + I B_i(x) B_j(x)\right) dx \\
&= \int_0^{\ell} S\left[B_j'(x) W_i(x) + W_j''(x) W_i(x) - B_i'(x) W_j(x) - W_i''(x) W_j(x) \right] \\
&\quad + \left[\left(D B_j''(x) - S B_j(x) - S W_j'(x)\right) B_i(x) \right. \\
&\quad \left. - \left(D B_i'' - S B_i(x) - S W_i'(x)\right) B_j(x) \right] dx.
\end{aligned} \tag{6.102}$$

By means of integration by parts, for example,

$$\int_0^{\ell} B_j'(x) W_i(x) dx = W_i(x) B_j(x)\Big|_0^{\ell} - \int_0^{\ell} B_j(x) W_i'(x) dx, \tag{6.103}$$

the right-hand side of equation (6.102) can be rewritten as

$$\begin{aligned}
&\Big\{ S\left[W_i(x)\left(B_j(x) + W_j'(x)\right) - W_j(x)\left(B_i(x) + W_i'(x)\right) \right] \\
&\quad + D\left[B_i(x) B_j'(x) - B_j(x) B_i'(x) \right] \Big\}_0^{\ell}.
\end{aligned} \tag{6.104}$$

The substitution of the end conditions (6.96) now makes (6.104) equivalent to zero. Thus, (6.102) reduces to

$$\left(\omega_i^2-\omega_j^2\right)\int_0^{\ell}\left(\rho h W_i(x)W_j(x)+IB_i(x)B_j(x)\right)dx=0, \quad (6.105a)$$

or,

$$\int_0^{\ell}\left(\rho h W_i(x)W_j(x)+IB_i(x)B_j(x)\right)dx=0, \quad \text{when } \omega_i\neq\omega_j,$$

$$\neq 0, \quad \text{when } \omega_i=\omega_j. \quad (6.105b)$$

Through the normalization, (6.105b) can be combined into

$$\int_0^{\ell}\left(\rho h W_i(x)W_j(x)+IB_i(x)B_j(x)\right)dx=\delta_{ij}, \quad (6.106)$$

where δ_{ij} is the Kronecker delta. Unlike the usual orthogonality conditions for the cases that the effects of rotary inertia and shear deformation are neglected, the orthogonality found in (6.106) shows that the complete set includes not only the mode shapes of the deflection but also the slope angle.

6.2.4 Forced Vibration

After finding the orthogonality relation (6.106), the expansion theorem may be used to obtain the system response by *modal analysis*. Using the expansion theorem we write the solution of (6.85) as a superposition of the *natural modes*, $W_j(x)$ and $B_j(x)$, multiplying corresponding time-dependent *generalized coordinates* $\eta_j(t)$. Hence,

$$w(x,t)=\sum_{j=1}^{\infty}W_j(x)\eta_j(t), \quad \beta_x(x,t)=\sum_{j=1}^{\infty}B_j(x)\eta_j(t) \quad (6.107)$$

and introducing (6.107) into (6.85), we obtain

$$\rho h\sum_{j=1}^{\infty} W_j(x)\ddot{\eta}_j(t)=\sum_{j=1}^{\infty} S(B_j'(x)+W_j''(x))\eta_j(t)+q(x,t),$$
$$I\sum_{j=1}^{\infty} B_j(x)\ddot{\eta}_j(t)=\sum_{j=1}^{\infty}(DB_j''(x)-SB_j(x)-SW_j'(x))\eta_j(t). \quad (6.108)$$

Employing the results of (6.95) in (6.108), we get

$$\rho h\sum_{j=1}^{\infty} W_j(x)\ddot{\eta}_j(t)=-\sum_{j=1}^{\infty} \rho h\omega_j^2 W_j(x)\eta_j(t)+q(x,t),$$
$$I\sum_{j=1}^{\infty} B_j(x)\ddot{\eta}_j(t)=-\sum_{j=1}^{\infty} I\omega_j^2 B_j(x)\eta_j(t). \quad (6.109)$$

Multiplying $(6.109)_1$ by $W_i(x)$ and $(6.109)_2$ by $B_i(x)$, adding them together and integrating both sides of the equation over the beam length ℓ, we obtain

$$\sum_{j=1}^{\infty}\left(\ddot{\eta}_j(t)+\omega_j^2\eta_j(t)\right)\int_0^{\ell}\left(\rho h W_j(x)W_i(x)+IB_j(x)B_i(x)\right)dx$$
$$=\int_0^{\ell} q(x,t)W_i(x)dx. \quad (6.110)$$

Through the use of the orthogonality condition found in (6.106), an infinite set of *uncoupled* second-order ordinary differential equation system is obtained as

$$\ddot{\eta}_j(t)+\omega_j^2\eta_j(t)=N_j(t) \qquad j=1,2,\ldots\ldots \quad (6.111)$$

where $N_j(t)$ denotes a *generalized force* associated with the generalized coordinate $\eta_j(t)$ and is related to the transverse distributed load q by

$$N_j(t)=\int_0^{\ell} q(x,t)W_j(x)dx. \quad (6.112)$$

6.2.5 Vibration Suppression

Piezoelectric Sensors and Actuators

To suppress the vibration of composite sandwich beams, we consider the popular way of bonding piezoelectric sensors and actuators on the surfaces of the beam (see Figure 6.3). It is well known that piezoelectric materials produce an electric field when deformed and undergo deformation when subjected to an electric field. Due to this intrinsic coupling phenomenon, piezoelectric materials are widely used as sensors and actuators in intelligent advanced structure design. The piezoelectric sensors can respond to structural vibration and generate output voltage due to the direct piezoelectric effect. On the other hand, the actuators can induce force and moment and control the system due to the converse piezoelectric effect. The piezoelectric properties can be described as a constitutive relation which characterizes the coupling effects between mechanical and electrical properties as follows:

$$D_i = e_{ik}\varepsilon_k + \omega_{ij}^{\varepsilon} E_j,\ \varepsilon_k = S_{il}^{E}\sigma_l + d_{jk}E_j,\ i,j=1,2,3,\ k,l=1,2,\ldots\ldots,6, \quad (6.113)$$

where σ_l and ε_k represent the stress and strain, respectively; D_i and E_j represent the electric displacement and electric field, respectively. e_{ik}, $\omega_{ij}^{\varepsilon}$, S_{il}^{E}, and d_{jk} represent the piezoelectric stress constant, electric permittivity, elastic compliance, and piezoelectric strain coefficient, respectively.

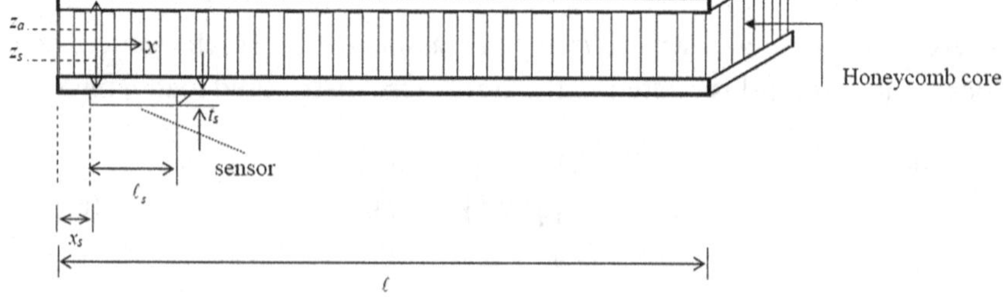

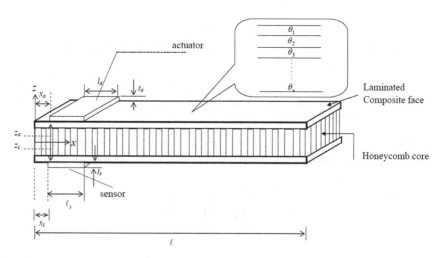

FIGURE 6.3 A composite sandwich beam with piezoelectric sensors and actuators (Hwu et al., 2004).

Note that in the following derivation of sensor and actuator equations, the assumption of the perfect bond between the piezoelectric sensors/actuators and the composite sandwich beams has been made. Therefore, the validity of our results will depend on this assumption. In real applications, this assumption may be influenced by the flexibility/rigidity of sensors and actuators. Hence, if a piezo-ceramics is selected to be an actuator, due to its significant large rigidity, the length of the actuator should be small enough to conform to the perfect bond assumption.

Sensor Equation

If the composite sandwich beams deform under a certain external load, the axial strain ε_1 on the surface of piezoelectric sensor which was attached on the beam will be $\varepsilon_1 = z_s \partial\beta_x/\partial x$ where z_s is the distance from the piezoelectric sensor to the mid-surface. By $(6.113)_1$, we know this strain will induce an electric displacement D_3 in the thickness direction as $D_3 = e_{31}\varepsilon_1$. The electric charge of the sensor region Ω_s corresponding to this electric displacement is $q_s = \int_{\Omega s} D_3 dx$. By using $(6.107)_2$ for the expansion of the rotation angle, the sensor charge can finally be expressed as a linear combination of the generalized coordinates η_j, which is

$$q_s(t) = \sum_{j=1}^{\infty} c_j \eta_j(t), \quad c_j = z_s e_{31} \int_{\Omega_s} B'_j(x) dx. \tag{6.114}$$

In the case of several sensors attached on the beam, the electric charge $q_s^{(i)}$ over each sensor region $\Omega_s^{(i)}$ can be expressed as

$$q_s^{(i)}(t) = \sum_{j=1}^{\infty} c_j^{(i)} \eta_j(t), \quad c_j^{(i)} = z_s e_{31} \int_{\Omega_s^{(i)}} B'_j(x) dx, \quad i = 1, \ldots, n_s, \tag{6.115}$$

where n_s is the number of the sensors.

Actuator Equation

To suppress the vibration, a control force is actuated through the piezoelectric actuators. By applying a voltage V_a in the thickness direction, an electric field E_3(= V_a/t_a, where t_a is the thickness of the piezoelectric actuators) is generated, which will induce an axial strain $\varepsilon_1^a = d_{31}E_3$ on the piezoelectric actuators. Due to the assumption of the perfect bonding, the axial strain of the composite sandwich beams caused by the deformation of the piezoelectric actuator may be written as $\varepsilon_1 = z\varepsilon_1^a / z_a$ where z_a is the distance from the piezoelectric actuator to the mid-surface and z is the coordinate in the thickness direction. Note that the linear variation of ε_1 with respect to z is the basic assumption of the formulation provided in (6.49). The bending moment M_x^a induced by the actuator can therefore be calculated by

$$M_x^a = \sum_{k=1}^{n} \int_{h_{k-1}}^{h_k} \sigma_1^{(k)} z dz = \sum_{k=1}^{n} \int_{h_{k-1}}^{h_k} \left(\bar{Q}_{11}^{(k)} \varepsilon_1\right) z dz = \sum_{k=1}^{n} \int_{h_{k-1}}^{h_k} \bar{Q}_{11}^{(k)} \left(\frac{z}{z_a} \varepsilon_1^a\right) z dz = k_a V_a, \tag{6.116a}$$

where

$$k_a = \frac{D_{11} d_{31}}{z_a \ t_a}. \tag{6.116b}$$

The equivalent transverse distributed load p_a can then be obtained by using the relation given in $(6.57)_1$, which says $p_a = -\ \partial^2 M_x / \partial x^2$. With this relation, the generalized force $N_j(t)$ of (6.112) induced by the applied voltage $V_a(t)$ can be found to be

$$N_j\left(t\right) = b_j V_a\left(t\right), \quad b_j = \int_{\Omega_a} k_a \frac{\partial^2 W_j\left(x\right)}{\partial x^2} dx, \tag{6.117}$$

and Ω_a is the actuator region that the voltage $V_a(t)$ applies. Note that in deriving (6.117) the technique of integration by part has been used and the conditions that voltage $V_a(t)$ along the actuator boundary and its derivative across the actuator boundary are assumed to be zero. In the case of several actuators attached on the beam, the generalized force $N_j(t)$ induced by all the applied voltage $V_a^{(i)}\left(t\right)$ over the region $\Omega_a^{(i)}$, $i = 1, 2, \ldots n_a$, is

$$N_j\left(t\right) = \sum_{i=1}^{n_a} b_j^{(i)} V_a^{(i)}\left(t\right), \quad b_j^{(i)} = \int_{\Omega_a^{(i)}} k_a \frac{\partial^2 W_j\left(x\right)}{\partial x^2} dx. \tag{6.118}$$

Dynamics of Feedback Control System

In control, it is customary to work with state equations instead of configuration equations (6.111). By applying the generalized force given in (6.118) and the sensor charge obtained in (6.115), and considering the process and measurement noise $\mathbf{v}(t)$ and $\mathbf{s}(t)$, the state space equation corresponding to equations (6.111), (6.115), and (6.118) for the composite sandwich beams can be written as

$$\begin{aligned} \dot{\mathbf{x}}(t) &= \mathbf{A}\mathbf{x}(t) + \mathbf{B}\mathbf{u}(t) + \mathbf{v}(t), \\ \mathbf{y}(t) &= \mathbf{C}\mathbf{x}(t) + \mathbf{s}(t), \end{aligned} \tag{6.119a}$$

where

$$\mathbf{x}(t) = \begin{Bmatrix} \mathbf{x}_1(t) \\ \mathbf{x}_2(t) \\ \mathbf{x}_3(t) \\ \vdots \end{Bmatrix}, \ \mathbf{u}(t) = \begin{Bmatrix} V_a^{(1)}(t) \\ V_a^{(2)}(t) \\ \cdot \\ \cdot \\ V_a^{(n_a)}(t) \end{Bmatrix}, \ \mathbf{y}(t) = \begin{Bmatrix} q_s^{(1)}(t) \\ q_s^{(2)}(t) \\ \cdot \\ \cdot \\ q_s^{(n_s)}(t) \end{Bmatrix}, \tag{6.119b}$$

$$\mathbf{A} = diag(\mathbf{A}_j), \ \mathbf{B} = \begin{Bmatrix} \mathbf{B}_1 \\ \mathbf{B}_2 \\ \mathbf{B}_3 \\ \vdots \end{Bmatrix}, \ \mathbf{C} = \begin{bmatrix} \mathbf{C}_1 & \mathbf{C}_2 & \mathbf{C}_3 & \cdots \end{bmatrix}, \tag{6.119c}$$

and

$$\mathbf{x}_j(t) = \begin{Bmatrix} \eta_j(t) \\ \dot{\eta}_j(t) \end{Bmatrix}, \mathbf{A}_j = \begin{bmatrix} 0 & 1 \\ -\omega_j^2 & 0 \end{bmatrix},$$

$$\mathbf{B}_j = \begin{bmatrix} 0 & 0 & \ldots\ldots & 0 \\ b_j^{(1)} & b_j^{(2)} & \ldots\ldots & b_j^{(n_a)} \end{bmatrix}, \mathbf{C}_j = \begin{bmatrix} c_j^{(1)} & 0 \\ c_j^{(2)} & 0 \\ \cdot & \cdot \\ \cdot & \cdot \\ c_j^{(n_s)} & 0 \end{bmatrix}. \tag{6.119d}$$

The problem now becomes how to apply the control voltage $\mathbf{u}(t)$ to suppress the vibration and how to estimate the full state vector $\mathbf{x}(t)$ by measuring the sensor output $\mathbf{y}(t)$. An effective approach to the control of structures is *feedback control.* In general, the nonlinear distributed control is not feasible, so that a linear control is usually considered. That is, the control force is linear proportional to the deflection and/or its velocity. Here, we consider

$$\mathbf{u}(t) = -\mathbf{G}\hat{\mathbf{x}}(t), \tag{6.120}$$

where $\hat{\mathbf{x}}(t)$ is an *estimated state* which is introduced to estimate the full state vector $\mathbf{x}(t)$ from the sensor output $\mathbf{y}(t)$; $\mathbf{G}$ is the *control gain* which should be determined such that the motion of the structure approaches zero asymptotically.

A state estimator also known as an *observer* for (6.119) is assumed to have the form

$$\dot{\hat{\mathbf{x}}}(t) = \mathbf{A}\hat{\mathbf{x}}(t) + \mathbf{B}\mathbf{u}(t) + \hat{\mathbf{K}}\left(\mathbf{y}(t) - \mathbf{C}\hat{\mathbf{x}}(t)\right), \tag{6.121}$$

where $\hat{\mathbf{K}}$ is the *Kalman filter gain matrix* and can be determined by minimizing the expected value, $E\left\{(\mathbf{x}-\hat{\mathbf{x}})^T(\mathbf{x}-\hat{\mathbf{x}})\right\}$. For steady-state case, the optimal observer gain matrix $\hat{\mathbf{K}}$ has been found to be (Meirovitch, 1990)

$$\hat{\mathbf{K}} = \mathbf{P}\mathbf{C}^T\mathbf{S}^{-1}, \tag{6.122a}$$

where the matrix $\mathbf{P}$ satisfying the *Riccati equation*

$$\mathbf{A}\mathbf{P} + \mathbf{P}\mathbf{A}^T + \mathbf{V} - \mathbf{P}\mathbf{C}^T\mathbf{S}^{-1}\mathbf{C}\mathbf{P} = \mathbf{0}. \tag{6.122b}$$

$\mathbf{V}(t)$ and $\mathbf{S}(t)$ are the intensities of assumed white noise $\mathbf{v}(t)$ and $\mathbf{s}(t)$ so that the correlation matrices have the forms $E\{\mathbf{v}(t_1)\mathbf{v}^T(t_2)\} = \mathbf{V}(t_1)\delta(t_2 - t_1)$ and $E\{\mathbf{s}(t_1)\mathbf{s}^T(t_2)\} = \mathbf{S}(t_1)\delta(t_2 - t_1)$, respectively.

The optimal control gain $\mathbf{G}$ is then determined by *minimizing the performance measure* $J = \int_0^{t_f}\left(\mathbf{x}^T\mathbf{Q}\mathbf{x} + \mathbf{u}^T\mathbf{R}\mathbf{u}\right)dt$ in which t_f is the final time, $\mathbf{Q}$ and $\mathbf{R}$ are, respectively, the *state weight matrix* and *control weight matrix*. For a more rapid vibration reduction, a larger value of $\mathbf{Q}$ can be selected, while for a smaller energy consumption a larger value of $\mathbf{R}$ can be selected. With an appropriate selection of the weight matrices, the optimal control gain can be found as (Meirovitch, 1990)

$$\mathbf{G} = \mathbf{R}^{-1}\mathbf{B}^T\mathbf{K}, \tag{6.123a}$$

where **K** satisfies the steady-state matrix Riccati equation, that is,

$$\mathbf{A}^T\mathbf{K}+\mathbf{K}\mathbf{A}+\mathbf{Q}-\mathbf{K}\mathbf{B}\mathbf{R}^{-1}\mathbf{B}^T\mathbf{K}=\mathbf{0}. \quad (6.123b)$$

Substituting (6.120) into (6.121) with the control gain **G** given in (6.123), the estimated state $\hat{\mathbf{x}}(t)$ can be computed by solving the ordinary differential equations (6.121). The control voltage $\mathbf{u}(t)$ calculated by (6.120) is then applied to suppress the vibration. The actual state $\mathbf{x}(t)$ can therefore be calculated from $(6.119a)_1$.

The characteristics of the control system can also be described in the following way. First, we obtain the observer error equation by subtracting (6.121) from $(6.119a)_1$. Then, with $\mathbf{u}(t)$ given in (6.120) we rewrite $(6.119a)_1$ in terms of the actual state $\mathbf{x}(t)$ and the error $\mathbf{e}(t)=\mathbf{x}(t)-\hat{\mathbf{x}}(t)$. By combining these two equations, the dynamics of the observed-state feedback control system can now be described as

$$\begin{Bmatrix}\dot{\mathbf{x}}(t)\\ \dot{\mathbf{e}}(t)\end{Bmatrix}=\begin{bmatrix}\mathbf{A}-\mathbf{B}\mathbf{G} & \mathbf{B}\mathbf{G}\\ \mathbf{0} & \mathbf{A}-\hat{\mathbf{K}}\mathbf{C}\end{bmatrix}\begin{Bmatrix}\mathbf{x}(t)\\ \mathbf{e}(t)\end{Bmatrix}+\begin{Bmatrix}\mathbf{v}(t)\\ \mathbf{v}(t)-\hat{\mathbf{K}}\mathbf{s}(t)\end{Bmatrix}. \quad (6.124)$$

By solving the ordinary differential equation system (6.124), the dynamic response $\mathbf{x}(t)$ of the control system can easily be calculated. The associated transverse deflection w and rotation angle β_x can then be obtained from (6.107).

The control procedure discussed in this section is called an *LQG/LTR (linear quadratic Gaussian with loop transfer recovery) control method* which uses a Kalman filter as an observer and a controller that minimizes an objective function of quadratic form (Meirovitch, 1990). To show the performance of the LQG/LTR controller, several examples have been done in Hwu et al. (2004). All these results show that the simulation obtained from the analytical model developed in this section can successfully suppress the vibration of composite sandwich beams.

6.3 COMPOSITE SANDWICH CYLINDRICAL SHELLS

In this section we will focus on the general formulation of composite sandwich cylindrical shells and the related solutions of free vibration. Part of the content is adapted from our previously published articles (Hwu et al., 2017).

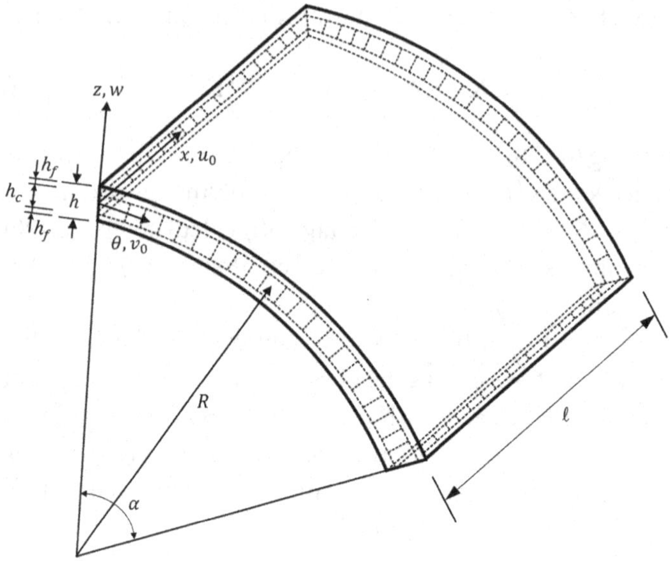

FIGURE 6.4 Geometry and nomenclature for a composite sandwich cylindrical panel.

6.3.1 General Formulation

Consider a sandwich cylindrical panel with laminated composite faces and an ideally orthotropic core (Figure 6.4). Applying the assumptions stated in Section 6.1.1 for the modeling of composite sandwich plates to the cylindrical panels, the kinematic relations, the constitutive laws, and the equations of motion for the vibration analysis of composite sandwich cylindrical panels can be expressed as follows:

Kinematic Relations

Let x, θ, and z be the axial, circumferential, and outward normal directions of the cylindrical panel (see Figure 6.4). The displacements u, v, and w in the directions of x, θ, and z, at time t can be expressed as

$$\begin{aligned} u(x,\theta,z,t) &= u_0(x,\theta,t) + z\beta_x(x,\theta,t), \\ v(x,\theta,z,t) &= v_0(x,\theta,t) + z\beta_\theta(x,\theta,t), \\ w(x,\theta,z,t) &= w_0(x,\theta,t), \end{aligned} \tag{6.125}$$

where u_0, v_0, and w_0 are the mid-plane displacements in the directions of x, θ, and z; β_x and β_θ are the rotation angles with respect to the x and θ directions, and are related to the transverse shear strains γ_{xz}, $\gamma_{\theta z}$ by

$$\beta_x = \gamma_{xz} - \frac{\partial w}{\partial x}, \quad \beta_\theta = \gamma_{\theta z} - \frac{\partial w}{R\partial \theta} + \frac{v_0}{R}, \tag{6.126}$$

in which R is the radius of cylindrical panel. If small deformations are considered, the strains can be written in terms of the mid-surface displacements as follows:

$$\varepsilon_x = \frac{\partial u}{\partial x} = \varepsilon_x^0 + zk_x, \quad \varepsilon_\theta = \frac{\partial v}{R\partial \theta} = \varepsilon_\theta^0 + zk_\theta, \quad \gamma_{x\theta} = \frac{\partial u}{R\partial \theta} + \frac{\partial v}{\partial x} = \gamma_{x\theta}^0 + zk_{x\theta}, \tag{6.127a}$$

where

$$\begin{aligned} \varepsilon_x^0 &= \frac{\partial u_0}{\partial x}, \quad \varepsilon_\theta^0 = \frac{\partial v_0}{R\partial \theta} + \frac{w}{R}, \quad \gamma_{x\theta}^0 = \frac{\partial u_0}{R\partial \theta} + \frac{\partial v_0}{\partial x}, \\ k_x &= \frac{\partial \beta_x}{\partial x}, \quad k_\theta = \frac{\partial \beta_\theta}{R\partial \theta}, \quad k_{x\theta} = \frac{\partial \beta_x}{R\partial \theta} + \frac{\partial \beta_\theta}{\partial x}. \end{aligned} \tag{6.127b}$$

Constitutive Laws

As shown in Figure 6.5, the stress resultants (N_x, N_θ, $N_{x\theta}$), bending moments (M_x, M_θ, $M_{x\theta}$), and transverse shear forces (Q_x, Q_θ) are related to the mid-plane strains $\left(\varepsilon_x^0, \varepsilon_\theta^0, \gamma_{x\theta}^0\right)$, curvatures ($\kappa_x$, κ_θ, $\kappa_{x\theta}$), and transverse shear strains (γ_{xz}, $\gamma_{\theta z}$) by

$$\begin{Bmatrix} N_x \\ N_\theta \\ N_{x\theta} \\ M_x \\ M_\theta \\ M_{x\theta} \end{Bmatrix} = \begin{bmatrix} A_{11} & A_{12} & A_{16} & B_{11} & B_{12} & B_{16} \\ A_{12} & A_{22} & A_{26} & B_{12} & B_{22} & B_{26} \\ A_{16} & A_{26} & A_{66} & B_{16} & B_{26} & B_{66} \\ B_{11} & B_{12} & B_{16} & D_{11} & D_{12} & D_{16} \\ B_{12} & B_{22} & B_{26} & D_{12} & D_{22} & D_{26} \\ B_{16} & B_{26} & B_{66} & D_{16} & D_{26} & D_{66} \end{bmatrix} \begin{Bmatrix} \varepsilon_x^0 \\ \varepsilon_\theta^0 \\ \gamma_{x\theta}^0 \\ k_x \\ k_\theta \\ k_{x\theta} \end{Bmatrix}, \quad \begin{Bmatrix} Q_x \\ Q_\theta \end{Bmatrix} = \begin{Bmatrix} A_{55}\gamma_{xz} \\ A_{44}\gamma_{\theta z} \end{Bmatrix}, \tag{6.128}$$

where A_{ij}, B_{ij}, D_{ij}, $i, j = 1, 2, 6$, and (A_{44}, A_{55}) are the extensional, coupling, bending stiffnesses, and transverse shear stiffnesses defined in (6.10) and (6.13).

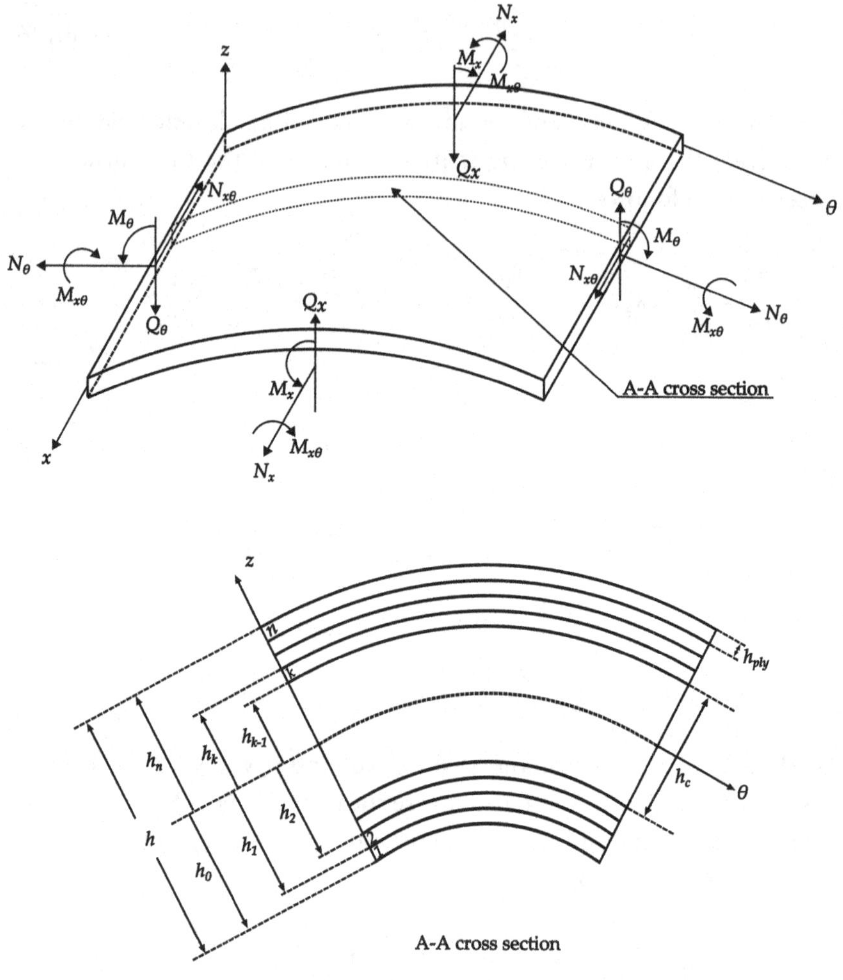

FIGURE 6.5 Stress resultants and bending moments on a panel element.

Equations of Motion

$$
\begin{aligned}
&\frac{\partial N_x}{\partial x}+\frac{\partial N_{\theta x}}{R\partial \theta}=\rho h\frac{\partial^2 u_0}{\partial t^2}, \quad \frac{\partial N_{x\theta}}{\partial x}+\frac{\partial N_{\theta}}{R\partial \theta}+\frac{Q_\theta}{R}=\rho h\frac{\partial^2 v_0}{\partial t^2},\\
&\frac{\partial Q_x}{\partial x}+\frac{\partial Q_{\theta}}{R\partial \theta}-\frac{N_\theta}{R}+q=\rho h\frac{\partial^2 w}{\partial t^2},\\
&\frac{\partial M_x}{\partial x}+\frac{\partial M_{x\theta}}{R\partial \theta}-Q_x=I\frac{\partial^2 \beta_x}{\partial t^2}, \quad \frac{\partial M_{x\theta}}{\partial x}+\frac{\partial M_{\theta}}{R\partial \theta}-Q_\theta=I\frac{\partial^2 \beta_\theta}{\partial t^2},
\end{aligned}
\tag{6.129}
$$

where q represents the transverse distributed load; I, ρ, and h are, respectively, the moment of inertia (with respect to the mid-surface), mass density, and thickness of the composite sandwich panel.

Substituting (6.126)–(6.128) into (6.129), five governing equations can be obtained in terms of five basic functions u_0, v_0, w, β_x, β_θ. Theoretically, the solutions can be found by solving the governing equations (which contain five coupled partial differential equations) with proper boundary conditions. For the general cases of composite sandwich cylindrical panels, every boundary of the panel should be described by five prescribed values. Two of them correspond to the in-plane problems and the other three correspond to the bending problems. In applications, several different boundary conditions may occur. Generally, they can be expressed as that shown in (6.15b).

Composite Sandwich Circular Cylindrical Shells (R = constant and $\alpha = 2\pi$)

The simplest of all shell geometries is that of the circular cylindrical shell, which is a closed cylindrical panel with R = constant and $\alpha = 2\pi$. If a composite sandwich circular cylindrical shell is considered, the governing equations for the vibration of composite sandwich circular cylindrical shells can be derived by substituting the constitutive laws (6.128) and kinematic relations (6.127) into the equations of motion (6.129). The final expressions can be written in terms of the mid-surface displacements u_0, v_0, w and the transverse shear strains γ_{xz}, $\gamma_{\theta z}$ such that

$$\begin{aligned}&L_1\left\{u_0,v_0,w,\gamma_{xz},\gamma_{\theta z}\right\}=\rho h\ddot{u}_0,\ L_2\left\{u_0,v_0,w,\gamma_{xz},\gamma_{\theta z}\right\}=\rho h\ddot{v}_0,\\&L_3\left\{u_0,v_0,w,\gamma_{xz},\gamma_{\theta z}\right\}=\rho h\ddot{w},\ L_4\left\{u_0,v_0,w,\gamma_{xz},\gamma_{\theta z}\right\}=I\ddot{\gamma}_{xz},\\&L_5\left\{u_0,v_0,w,\gamma_{xz},\gamma_{\theta z}\right\}=I\ddot{\gamma}_{\theta z},\end{aligned}\tag{6.130}$$

where L_i, $i = 1, 2, 3, 4, 5$ are the differential operators which can be obtained in the way stated above. Owing to their complexity, the explicit forms of these five operators are not given here.

To have a direct feeling about the governing equations, we consider the special case of a composite sandwich circular cylindrical shell whose faces are symmetric cross-ply laminates, that is, $A_{16} = A_{26} = D_{16} = D_{26} = 0$, $B_{ij} = 0$, $i, j = 1, 2, 6$. Under this special condition, by the way stated before (6.130) the governing equations can now be obtained as

$$A_{11}\frac{\partial^2 u_0}{\partial x^2}+\left(A_{12}+A_{66}\right)\frac{\partial^2 v_0}{R\partial\theta\partial x}+A_{66}\frac{\partial^2 u_0}{R^2\partial\theta^2}+\frac{A_{12}}{R}\frac{\partial w}{\partial x}=\rho h\frac{\partial^2 u_0}{\partial t^2},\tag{6.131a}$$

$$A_{66}\frac{\partial^2 v_0}{\partial x^2}+\left(A_{12}+A_{66}\right)\frac{\partial^2 u_0}{R\partial\theta\partial x}+A_{22}\frac{\partial^2 v_0}{R^2\partial\theta^2}+\left(A_{22}+A_{44}\right)\frac{\partial w}{R^2\partial\theta}$$
$$+\frac{A_{44}}{R}\left(\beta_\theta-\frac{v_0}{R}\right)=\rho h\frac{\partial^2 v_0}{\partial t^2}, \tag{6.131b}$$

$$A_{55}\frac{\partial^2 w}{\partial x^2}+A_{44}\frac{\partial^2 w}{R^2\partial\theta^2}+A_{55}\frac{\partial\beta_x}{\partial x}+A_{44}\frac{\partial\beta_\theta}{R\partial\theta}-\left(A_{22}+A_{44}\right)\frac{\partial v_0}{R^2\partial\theta}$$
$$-\frac{A_{12}}{R}\frac{\partial u_0}{\partial x}-\frac{A_{22}}{R^2}w+q=\rho h\frac{\partial^2 w}{\partial t^2}, \tag{6.131c}$$

$$D_{11}\frac{\partial^2\beta_x}{\partial x^2}+\left(D_{12}+D_{66}\right)\frac{\partial^2\beta_\theta}{R\partial\theta\partial x}+D_{66}\frac{\partial^2\beta_x}{R^2\partial\theta^2}-A_{55}\left(\beta_x+\frac{\partial w}{\partial x}\right)=I\frac{\partial^2\beta_x}{\partial t^2}, \tag{6.131d}$$

$$D_{66}\frac{\partial^2\beta_\theta}{\partial x^2}+\left(D_{12}+D_{66}\right)\frac{\partial^2\beta_x}{R\partial\theta\partial x}+D_{22}\frac{\partial^2\beta_\theta}{R^2\partial\theta^2}$$
$$-A_{44}\left(\beta_\theta+\frac{\partial w}{R\partial\theta}-\frac{v_0}{R}\right)=I\frac{\partial^2\beta_\theta}{\partial t^2}. \tag{6.131e}$$

If the sandwich shell is under axially symmetric load and its face lamination stacking sequence is such that there is no stretching–bending coupling and the material properties are independent of the variable θ, the terms $\partial(\ )/\partial\theta$ appeared in (6.126)–(6.129) will vanish. With this consideration, further reduction can be made on the governing equation, in which all the quantities depend on x and t only.

6.3.2 Free Vibration

Consider a composite sandwich circular cylindrical shell whose faces are symmetric cross-ply laminates, that is, $A_{16} = A_{26} = D_{16} = D_{26} = 0$, $B_{ij} = 0$, $i, j = 1,2,6$. If the ends of the cylindrical shell are *simply supported* by end plates,

$$N_x=0,\ u_\theta^0=0,\ w=0,\ M_x=0,\ \beta_\theta=0,\ \text{at } x=0 \text{ and } x=\ell. \tag{6.132}$$

To satisfy the boundary condition (6.132), the double Fourier series are assumed for the middle surface displacements as

$$
\begin{aligned}
u_x^0(x,\theta,t) &= \sum_{m=1}^{\infty}\sum_{n=1}^{\infty} U_{mn} \cos\frac{m\pi x}{\ell}\cos(n\theta)e^{i\omega_{mn}t},\\
u_\theta^0(x,\theta,t) &= \sum_{m=1}^{\infty}\sum_{n=1}^{\infty} V_{mn} \sin\frac{m\pi x}{\ell}\sin(n\theta)e^{i\omega_{mn}t},\\
w(x,\theta,t) &= \sum_{m=1}^{\infty}\sum_{n=1}^{\infty} W_{mn} \sin\frac{m\pi x}{\ell}\cos(n\theta)e^{i\omega_{mn}t},\\
\beta_x(x,\theta,t) &= \sum_{m=1}^{\infty}\sum_{n=1}^{\infty} \mathrm{B}_{mn} \cos\frac{m\pi x}{\ell}\cos(n\theta)e^{i\omega_{mn}t},\\
\beta_\theta(x,\theta,t) &= \sum_{m=1}^{\infty}\sum_{n=1}^{\infty} \mathrm{B}^*_{mn} \sin\frac{m\pi x}{\ell}\sin(n\theta)e^{i\omega_{mn}t}.
\end{aligned}
\tag{6.133}
$$

With the displacements assumed in (6.133), the governing equation (6.131) for the present problem can also be written in the form of (6.34a) in which the nonzero components of **K** are

$$
\begin{aligned}
&k_{11} = A_{11}\lambda_m^2 + \frac{A_{66}}{R^2}\lambda_n^2,\ k_{12} = k_{21} = -\left(\frac{A_{12}}{R} + \frac{A_{66}}{R}\right)\lambda_m\lambda_n,\ k_{13} = k_{31} = -\frac{A_{12}}{R}\lambda_m,\\
&k_{22} = \frac{A_{22}}{R^2}\lambda_n^2 + A_{66}\lambda_m^2 + \frac{A_{44}}{R^2},\ k_{23} = k_{32} = \left(\frac{A_{22}}{R^2} + \frac{A_{44}}{R^2}\right)\lambda_n,\ k_{25} = k_{52} = -\frac{A_{44}}{R},\\
&k_{33} = \frac{A_{22}}{R^2} + \frac{A_{44}}{R^2}\lambda_n^2 + A_{55}\lambda_m^2,\ k_{34} = k_{43} = A_{55}\lambda_m,\ k_{35} = k_{53} = -\frac{A_{44}}{R}\lambda_n,\\
&k_{44} = D_{11}\lambda_m^2 + \frac{D_{66}}{R^2}\lambda_n^2 + A_{55},\ k_{45} = k_{54} = -\left(\frac{D_{12}}{R} + \frac{D_{66}}{R}\right)\lambda_m\lambda_n,\\
&k_{55} = \frac{D_{22}}{R^2}\lambda_n^2 + D_{66}\lambda_m^2 + A_{44},\ \lambda_m = \frac{m\pi}{L},\ \lambda_n = n.
\end{aligned}
\tag{6.134}
$$

Like the composite sandwich plates discussed following (6.34), with the nonzero k_{ij} given in (6.134) the natural frequencies and mode shapes of composite sandwich cylindrical shells can be calculated by solving the standard eigenvalue problem of $\left(\mathbf{I}_0^{-1}\mathbf{K}\right)\mathbf{d} = \omega^2\mathbf{d}$.

To consider the most general composite sandwich cylindrical shells whose A_{ij}, B_{ij}, D_{ij}, $i, j = 1, 2, 6$ may all be nonzero, we may apply the Ritz method stated in (6.43)–(6.46), in which the following extra terms Π'_a, Π'_b, Π'_d should be added to Π_a, Π_b, Π_d of (6.43b,c,d)

$$\Pi'_a = \frac{A_{22}}{R^2}W^2 + \frac{2A_{22}}{R^2}\left(\frac{\partial V}{\partial \theta}\right)W + \frac{2A_{12}}{R}\left(\frac{\partial U}{\partial x}\right)W + \frac{2A_{26}}{R^2}\left(\frac{\partial U}{\partial \theta}\right)W + \frac{A_{26}}{R}\left(\frac{\partial V}{\partial x}\right)W,$$

$$\Pi'_b/2 = \frac{B_{22}}{R^2}W\left(\frac{\partial B^*}{\partial \theta}\right) + \frac{B_{12}}{R}W\left(\frac{\partial B}{\partial x}\right) + \frac{B_{26}}{R^2}W\left(\frac{\partial B}{\partial \theta}\right) + \frac{B_{26}}{R}W\left(\frac{\partial B^*}{\partial x}\right),$$

$$\Pi'_d = -\frac{2A_{44}}{R}VB^* + \frac{A_{44}}{R^2}V^2 - \frac{2A_{44}}{R^2}V\left(\frac{\partial W}{\partial \theta}\right). \tag{6.135}$$

EXAMPLE 3: A COMPOSITE SANDWICH CIRCULAR CYLINDRICAL SHELL [(0/θ)$_N$/CORE/(θ/0)$_N$] WITH TWO ENDS SIMPLY SUPPORTED

To show the applicability to the general cases whose A_{ij}, B_{ij}, D_{ij}, $i, j = 1, 2, 6$ may be nonzero, a circular cylindrical shell made by a composite sandwich [(0/θ)$_N$/core/(θ/0)$_N$] is solved by using the Ritz method for any possible fiber orientation θ. When $\theta = 0^\circ$ or $\theta = 90^\circ$, the sandwich becomes a symmetric cross-ply laminated sandwich, and the Navier's solution (6.133) will also be applied. The results presented in Figure 5 of Hwu et al. (2017) show that the natural frequencies obtained by the Ritz method and Navier's solution agree with each other for the special cases of $\theta = 0^\circ$ and $\theta = 90^\circ$. The maximum fundamental natural frequency occurs at $\theta = 60^\circ$ when $N = 2$ or 3, and $\theta = 65^\circ$ when $N = 1$. Also as expected, the thicker the sandwich face is laid up, the higher the natural frequency is obtained.

6.4 DELAMINATED COMPOSITE SANDWICH BEAMS

One of the most frequently encountered problems in composite laminates is interface cracking, sometimes known as *delamination*. For composite sandwiches, there is an extra interface between face and core, which may be weaker than those in layered composite faces. Delaminations may occur due to a variety of reasons such as low energy impact, manufacturing defects, or high stress concentrations at geometric or material discontinuities (e.g., the well-known free edge effects). The presence of delaminations is of major concern, especially in compressively loaded components where delaminations may grow under fatigue loading by out-of-plane distortion.

To study the delamination effects on the composite sandwich beams, the mathematical model developed in Section 6.2 can be applied. Through this model, several analytical closed-form solutions have been obtained, such as the solutions for the buckling loads (Hwu and Hu, 1992) and natural frequency (Hu and Hwu, 1995), etc. In these solutions, all the terms containing the transverse shear stiffness have been neglected in the regions adjacent to the delamination, and the transverse shear stress distribution is assumed to be uniform across the core thickness. By carefully studying their corresponding physical meaning, we found that the shear resistance of the core has been overestimated, which also leads to the overestimation of the buckling load and natural frequency. In order to be more close to the real situation, one may consider the transverse shear stiffness for the entire sandwich beams and a parabolic type distribution for the transverse shear stress.

6.4.1 Buckling Analysis

Consider a composite sandwich beam containing a delamination lying between the upper face and core. The beam has a constant width between two lateral edges and is subjected to compressive axial load P at the clamped ends $x = \pm \ell$. As shown in Figure 6.6, the delamination extends over an interval $-a \leq x \leq a$ and runs across the whole width of the beam. The delaminated sandwich beam starts to buckle when the axial load reaches a critical value. To analyze the delaminated composite sandwich beam, the entire beam is separated into three regions as shown in Figure 6.6, and each region is considered as a composite sandwich beam. Provided that the delamination remains completely open during the compression, the transverse shear stress will vanish along the delamination surface.

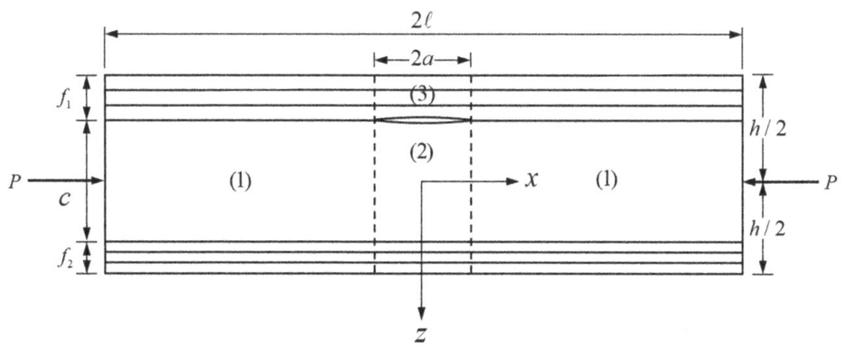

FIGURE 6.6 Delaminated composite sandwich beams subjected to compressive load.

If both ends of the sandwich beam are clamped, consideration of the symmetry with respect to the midpoint and the continuity of deflection and rotation angle at the crack tip will now lead to the following boundary conditions for the delaminated composite sandwich beams:

$$\begin{aligned} &\text{End condition:} && w_1 = \beta_{x1} = 0, && \text{at } x = \ell; \\ &\text{Symmetry condition:} && w_2' = w_3' = \gamma_{xz_2} = \gamma_{xz_3} = 0, && \text{at } x = 0; \\ &\text{Continuity condition:} && w_1 = w_2 = w_3,\ \beta_{x1} = \beta_{x2} = \beta_{x3}, && \text{at } x = a. \end{aligned} \tag{6.136}$$

The subscripts 1, 2, and 3 denote the region number. If the clamped ends are further restrained to prevent shear strain by edge stiffeners, the end conditions can be modified as

$$w_1 = w_1' = \gamma_{xz1} = 0, \quad \text{at } x = \ell. \tag{6.137}$$

In order to determine the buckling load for the problem described in (6.136) and (6.137), the governing equation (6.66) for the buckled composite sandwich beams is now employed for regions 1, 2, and 3 of Figure 6.6. Substituting the results obtained in (6.68) and (6.69) into the boundary conditions $(6.136)_{2,3}$ and (6.137), we obtain

$$\begin{aligned} w_1 &= \Gamma\left[1-\cos\lambda_1(\ell - x)\right], \quad a \le x \le \ell, \\ w_2 &= \Gamma\left[\frac{\lambda_1(1-P_1/S_1)\sin\lambda_1(\ell-a)}{\lambda_2(1-P_2/S_2)\sin\lambda_2 a}(\cos\lambda_2 x - \cos\lambda_2 a) + 1 - \cos\lambda_1(\ell-a)\right], \quad 0 \le x \le a, \\ w_3 &= \Gamma\left[\frac{\lambda_1(1-P_1/S_1)\sin\lambda_1(\ell-a)}{\lambda_3(1-P_3/S_3)\sin\lambda_3 a}(\cos\lambda_3 x - \cos\lambda_3 a) + 1 - \cos\lambda_1(\ell-a)\right], \quad 0 \le x \le a, \end{aligned} \tag{6.138a}$$

where

$$\lambda_i^2 = \frac{P_i}{D_i(1-P_i/S_i)}, \quad i = 1,2,3. \tag{6.138b}$$

In the above, Γ represents the amplitude of deflection; P_i is the compressive load of region i and $P_1 = P$ is given if the applied axial load P is known; S_i is the transverse shear stiffness of region i, which has been defined in $(6.13)_2$ and $(6.55b)_1$. Note that in region 3, the core thickness c used for the calculation of S_3 should be replaced by the face thickness f_1.

To determine P_2, P_3, and Γ, we consider the equilibrium of forces and moments at the crack tip $x = a$, and the compatibility of regions 1, 2, and 3, which are

$$P_1 = P_2 + P_3, \quad M_1 = M_2 + M_3, \quad u_1 = u_2 = u_3, \qquad \text{at the crack tip } x = a. \tag{6.139}$$

From the assumption of the displacement field given in (6.48) for all regions and the continuity of rotation angle given in $(6.136)_3$, the compatibility of regions 1, 2, and 3 required at $(6.139)_3$ will be satisfied automatically if

$$u_0^{(1)}(a) = u_0^{(2)}(a) = u_0^{(3)}(a). \tag{6.140}$$

Substituting (6.138a) into $(6.67)_{4,6}$, then applying $(6.139)_2$ and (6.140), we obtain

$$\begin{aligned} -P_1B_1 + P_2B_2 + P_3B_3 &= \gamma\left[\frac{P_1}{\lambda_1(1-P_1/S_1)\tan\lambda_1(\ell-a)}\right. \\ &\quad \left. + \frac{P_2}{\lambda_2(1-P_2/S_2)\tan\lambda_2 a} + \frac{P_3}{\lambda_3(1-P_3/S_3)\tan\lambda_3 a}\right], \\ B_1\gamma + A_1P_1a - c_3 &= B_2\gamma + A_2P_2a = B_3\gamma + A_3P_3a, \end{aligned} \tag{6.141a}$$

where

$$\gamma = \Gamma\lambda_1(1-P_1/S_1)\sin\lambda_1(\ell-a). \tag{6.141b}$$

From the first and second equalities of $(6.141\text{a})_2$, P_2 and P_3 can be expressed in terms of P_1 as

$$P_2 = \frac{A_1}{A_2}P_1 + \frac{B_1 - B_2}{aA_2}\gamma - \frac{c_3}{aA_2}, \quad P_3 = \frac{A_1}{A_3}P_1 + \frac{B_1 - B_3}{aA_3}\gamma - \frac{c_3}{aA_3}. \tag{6.142}$$

With the results of (6.142) and the definitions of A_i and B_i given in (6.55b), it can be proved that the force balance required in the first equation of (6.139) will be satisfied if

$$c_3 = 0. \tag{6.143}$$

Substituting (6.142) and (6.143) into $(6.141a)_1$, an equation for the unknown γ can be obtained as

$$P_1\left(\frac{A_1}{A_2}B_2+\frac{A_1}{A_3}B_3-B_1\right)+\frac{\gamma}{a}\left(\frac{B_1-B_2}{A_2}B_2+\frac{B_1-B_3}{A_3}B_3\right)$$
$$=\gamma\left[\frac{P_1}{\lambda_1\left(1-P_1/S_1\right)\tan\lambda_1\left(\ell-a\right)}+\frac{\bar{P}_2}{\bar{\lambda}_2\left(1-\bar{P}_2/S_2\right)\tan\bar{\lambda}_2 a}\right.$$
$$\left.+\frac{\bar{P}_3}{\bar{\lambda}_3\left(1-\bar{P}_3/S_3\right)\tan\bar{\lambda}_3 a}\right] \tag{6.144a}$$

where

$$\lambda_1^2=\frac{P_1}{D_1\left(1-P_1/S_1\right)},\quad \bar{\lambda}_2^2=\frac{\bar{P}_2}{D_2\left(1-\bar{P}_2/S_2\right)},\quad \bar{\lambda}_3^2=\frac{\bar{P}_3}{D_3\left(1-\bar{P}_3/S_3\right)},$$
$$\bar{P}_2=\frac{A_1}{A_2}P_1,\quad \bar{P}_3=\frac{A_1}{A_3}P_1. \tag{6.144b}$$

In (6.144a) the second-order term γ^2 has been neglected due to the consideration of small deflection at the stage immediately after the buckling. From the definitions of A_i, B_i, and D_i given in (6.55b), one can prove that

$$\frac{A_1}{A_2}B_2+\frac{A_1}{A_3}B_3-B_1=0,\qquad \frac{B_1-B_2}{A_2}B_2+\frac{B_1-B_3}{A_3}B_3=-D_1+D_2+D_3. \tag{6.145}$$

Substituting (6.145) into (6.144a), we observe that only the first-order term γ remains. When the compressive load $P(=P_1)$ reaches the critical buckling load, the amplitude of deflection Γ (hence γ) could be arbitrary. Therefore, the coefficient of γ in (6.144a) should be zero, which yields the following characteristic equation for the critical buckling load P_{cr}:

$$P_{cr}=\left(D_2+D_3-D_1\right)\left[\frac{a}{\left(1-P_{cr}/S_1\right)\bar{\lambda}_1\tan\bar{\lambda}_1\left(\ell-a\right)}+\frac{ka}{\left(1-kP_{cr}/S_2\right)\bar{\lambda}_2\tan\bar{\lambda}_2 a}\right.$$
$$\left.+\frac{\left(1-k\right)a}{\left(1-\left(1-k\right)P_{cr}/S_3\right)\bar{\lambda}_3\tan\bar{\lambda}_3 a}\right]^{-1} \tag{6.146a}$$

where

$$k=\frac{A_1}{A_2},\ \bar{\lambda}_1^2=\frac{P_{cr}}{D_1\left(1-\frac{P_{cr}}{S_1}\right)},\ \bar{\lambda}_2^2=\frac{kP_{cr}}{D_2\left(1-\frac{kP_{cr}}{S_2}\right)},\ \bar{\lambda}_3^2=\frac{(1-k)P_{cr}}{D_3\left(1-\frac{(1-k)P_{cr}}{S_3}\right)}. \tag{6.146b}$$

During the calculation of the buckling load by using formula (6.146), we observe that the solutions are usually obtained from the conditions that $\tan\bar{\lambda}_1(\ell-a)=0$, $\tan\bar{\lambda}_2 a=0$, or $\tan\bar{\lambda}_3 a=0$, which correspond to the following solutions:

$$P_{cr}=\frac{D_1\pi^2/(\ell-a)^2}{1+D_1\pi^2/S_1(\ell-a)^2}, \tag{6.147a}$$

$$\text{or } P_{cr}=\frac{D_2\pi^2/a^2}{k\left(1+D_2\pi^2/S_2a^2\right)}, \tag{6.147b}$$

$$\text{or } P_{cr}=\frac{D_3\pi^2/a^2}{(1-k)\left(1+D_3\pi^2/S_3a^2\right)}. \tag{6.147c}$$

The buckling load is then obtained from the lowest value of P_{cr} calculated from (6.147). Actually, the formulae shown in (6.147) correspond to the local buckling of regions 1, 2, and 3.

In order to verify the solutions (6.146) and (6.147), a comparison with the finite element solutions and the experimental results has been made in Hwu and Hsieh (1998), and Hwu and Bi (2001). In addition to the verification, several examples have also been done to study the effects of core, face, and delamination length on the buckling load (Hwu and Hu, 1992). The results show that (1) the buckling load increases when the transverse shear stiffness S increases until it reaches a certain value and becomes a constant; (2) the buckling load increases when the face thickness increases or the fiber is oriented to the direction of load; (3) the upper bound solution of the buckling load is approximated by that of perfect composite sandwich beams, while the combined axial load of two completely detached beam represents the lower bound; (4) if the energy release rate associated with the load immediately after buckling is less than the critical energy release rate, the delamination will not grow and the ultimate axial load capacity P_{ult} of the delaminated composite sandwich beam will be larger than P_{cr}, otherwise, $P_{ult} = P_{cr}$. The postbuckling analysis and the associated calculation of energy release rate will be shown in the next subsection.

Perfect Composite Sandwich Beams (Without Delamination)

By letting the delamination length $2a$ approach to zero, the characteristic equation for buckling load shown in (6.146) can be reduced to

$$\lim_{a\to 0} \frac{a}{\bar{\lambda}_1 \tan \bar{\lambda}_1 (\ell - a)} = -\frac{D_1}{P_{cr}}. \tag{6.148}$$

The solution to (6.148) exists only when $\bar{\lambda}_1 = \pi / \ell$. From $(6.146b)_2$, we have

$$P_{cr} = \frac{D_1 \pi^2 / \ell^2}{1 + D_1 \pi^2 / S_1 \ell^2}, \tag{6.149}$$

which is exactly the same as that obtained in (6.80) for the perfect composite sandwich beams (note that the total length here is 2ℓ instead of ℓ). The buckling load for perfect composite sandwich beams may be considered as an upper bound for the axial load capacity of delaminated composite sandwich beams.

Delaminated Composite Beams (Without Core)

The problem of delamination buckling and growth in composite laminates has received a considerable amount of attention. Analytical investigation by one dimensional model has been done by several researchers, for example, Chai et al. (1981) and Yin et al. (1986). It has been verified that the present results for the delaminated composite sandwich beams are also applicable to the cases of delaminated composite beams (without core) just by letting the terms containing S_i vanish, that is, the transverse shear strains are neglected (Hwu and Hu, 1992).

Thin Film Delamination

Because the faces are relatively thinner than the depth of the core, the delamination laid on the interface can usually be treated as thin-film delamination. Because the delamination is relatively slender in comparison with the whole plate, the buckling may be initiated by local buckling of the thin delamination. Since the thin layer of delamination has elastically supported ends, the buckling load may be close to but less than that of a fixed-end beam of length $2a$, (6.147c). In other words,

$$P_{cr} \leq \frac{D_3\pi^2 / a^2}{(1-k)\left(1+D_3\pi^2 / S_3a^2\right)}. \tag{6.150}$$

6.4.2 Postbuckling Analysis

When the applied axial force exceeds the critical buckling load, the buckled delamination may grow. To study the postbuckling behavior of the delaminated composite sandwich beams, a consideration of large deformation should be included. Under this consideration, the strain–displacement relation shown in the first equation of (6.49b) should be modified to include the higher-order term such that

$$\varepsilon_x^0 = \frac{\partial u_0}{\partial x} + \frac{1}{2}\left(\frac{\partial w}{\partial x}\right)^2. \tag{6.151}$$

Except for the modification introduced in (6.151), all the other basic equations shown in (6.48)–(6.52) and (6.58) remain unchanged. By carefully reviewing the derivation procedure for the governing equation (6.66) and its associated relations obtained in (6.67), we find that all these equations also remain unchanged except the relation for the axial displacement u_0, which should be modified as

$$u_0 = B\left(1-\frac{P}{S}\right)\frac{dw}{dx} - APx - \frac{1}{2}\int\left(\frac{dw}{dx}\right)^2 dx + c_3. \tag{6.152}$$

With this modification, we see that the deflection of the postbuckled sandwich beams can also be expressed as those obtained in (6.138) for the buckling analysis because in the derivation of (6.138) no condition related to the axial displacements has been used. After getting the expressions for the deflection, the determination of the deflection amplitude Γ requires the use of crack tip equilibrium and compatibility as shown in (6.139). Because of the modification given in (6.152), the results obtained in the second equation of (6.141a) should be modified as

$$B_1\gamma + A_1P_1a + b_1\gamma^2 - c_3 = B_2\gamma + A_2P_2a + b_2\gamma^2 = B_3\gamma + A_3P_3a + b_3\gamma^2, \quad (6.153a)$$

where

$$b_1 = \frac{2\lambda_1 a + \sin 2\lambda_1(\ell - a)}{8\lambda_1(1 - P_1/S_1)^2 \sin^2 \lambda_1(\ell - a)},$$
$$b_2 = \frac{2\lambda_2 a - \sin 2\lambda_2 a}{8\lambda_2(1 - P_2/S_2)^2 \sin^2 \lambda_2 a}, \quad b_3 = \frac{2\lambda_3 a - \sin 2\lambda_3 a}{8\lambda_3(1 - P_3/S_3)^2 \sin^2 \lambda_3 a}. \quad (6.153b)$$

By following the same steps as those for obtaining (6.142), we now get

$$P_2 = \frac{A_1}{A_2}P_1 + \frac{B_1 - B_2}{aA_2}\gamma + \frac{b_1 - b_2}{aA_2}\gamma^2, \quad P_3 = \frac{A_1}{A_3}P_1 + \frac{B_1 - B_3}{aA_3}\gamma + \frac{b_1 - b_3}{aA_3}\gamma^2. \quad (6.154)$$

When the applied axial force $P(= P_1)$ that exceeds the critical buckling load is given, substitution of (6.154) into (6.141a)$_1$ will yield an algebraic equation for the unknown γ. After the determination of γ, by the use of (6.141b), (6.138), and (6.67), we can find all the physical responses in the postbuckling stage such as the amplitude Γ, the transverse deflection w, the transverse shear strain γ_{xz}, the rotation angle β_x, the curvature κ_x, the axial displacement u_0, the transverse shear force Q_x, and the bending moment M_x, etc.

If the growth of a buckled delamination is governed by a Griffith-type criterion of a critical energy release rate, the prediction of whether delamination will grow requires an evaluation of energy release rate G. By applying the previous postbuckling solution, evaluating the total potential energy and differentiating the result with respect to the delamination length $2a$, we can obtain an explicit expression for G. The function $G(a)$ can then be used to study the initiation and stability of delamination growth.

The total potential energy Π of the buckled delaminated sandwich beams consists of the contribution from strain energy U due to bending/stretching of the faces and shearing of the core as well as the potential energy $-W$ of the external forces, that is,

$$\Pi = U - W, \quad (6.155a)$$

where

$$U = 2\left\{\frac{1}{2}\int_0^\ell \left(M_x k_x + N_x \varepsilon_x^0 + Q_x \gamma_{xz}\right)dx\right\}, \quad W = -2Pu_0\left(\ell\right). \quad (6.155b)$$

Note that factor 2 in (6.155b) is due to the symmetry condition considered in our problem. From (6.59), (6.56), (6.50b), and (6.55b), we have

$$N_x = -P, \quad M_x = -BP + D\kappa_x, \quad \varepsilon_x^0 = -AP - B\kappa_x, \quad Q_x = S\gamma_{xz}. \quad (6.156)$$

Substituting (6.156) into (6.155), an alternative expression for the total potential energy can be written as

$$\Pi = \int_0^\ell \left(D\kappa_x^2 + S\gamma_{xz}^2 + AP^2\right)dx + 2Pu_0\left(\ell\right), \quad (6.157)$$

in which κ_x, γ_{xz}, and u_0 can be found from (6.138) and $(6.67)_{1,3,4}$. The results are

$$\kappa_x = -\frac{P_1\Gamma}{D_1}\cos\lambda_1\left(\ell - x\right), \gamma_{xz} = -\frac{P_1\Gamma}{S_1}\lambda_1 \sin\lambda_1\left(\ell - x\right), \text{when } a \le x \le \ell, \text{region 1},$$
$$\kappa_x = \frac{P_2\Gamma_2}{D_2}\cos\lambda_2 x, \ \gamma_{xz} = \frac{P_2\Gamma_2}{S_2}\lambda_2 \sin\lambda_2 x, \ \text{when } 0 \le x \le a, \text{region 2},$$
$$\kappa_x = \frac{P_3\Gamma_3}{D_3}\cos\lambda_3 x, \ \gamma_{xz} = \frac{P_3\Gamma_3}{S_3}\lambda_3 \sin\lambda_3 x, \ \text{when } 0 \le x \le a, \text{region 3},$$
$$(6.158a)$$

and

$$u_0\left(\ell\right) = -A_1 P_1 \ell - \frac{1}{4}\lambda_1^2 \Gamma^2 \ell, \quad (6.158b)$$

where

$$\Gamma_2 = \frac{\lambda_1\left(1 - P_1/S_1\right)\sin\lambda_1\left(\ell - a\right)}{\lambda_2\left(1 - P_2/S_2\right)\sin\lambda_2 a}\Gamma, \quad \Gamma_3 = \frac{\lambda_1\left(1 - P_1/S_1\right)\sin\lambda_1\left(\ell - a\right)}{\lambda_3\left(1 - P_3/S_3\right)\sin\lambda_3 a}\Gamma.$$
$$(6.158c)$$

Substituting (6.158) into (6.157) and performing the integration for each region and adding them together, we get an explicit expression for the total potential energy Π as

$$\Pi = 2\gamma^2\left(-b_1P_1 + b_2P_2 + b_3P_3\right) + \gamma\left(-B_1P_1 + B_2P_2 + B_3P_3\right)$$
$$+a\left(-A_1P_1^2 + A_2P_2^2 + A_3P_3^2\right) - A_1P_1^2\ell. \tag{6.159}$$

where b_1, b_2, and b_3 are defined in (6.153b). Note that the relation $(6.141a)_1$ has been used to get the simplified results (6.159).

After finding the explicit expression for the total potential energy, the energy release rate G can be calculated by

$$G = -\frac{\partial \Pi}{2\partial a}. \tag{6.160}$$

6.4.3 Free Vibration

To study the delamination effect on free vibration, same geometry of the delaminated composite sandwich beams as that shown in Figure 6.6 will be considered in this subsection. Because the shift in natural frequency due to the delamination may provide the basis for nondestructive testing via vibration technique, it is desired to know not only the effects of delamination length but also its position and the associated end conditions. Therefore, unlike the examples discussed in the previous subsection for buckling problems, in this subsection the delaminations are not restricted to lie on the center part of the beam. Moreover, several different end conditions will be considered.

Like the buckling analysis, to study the free vibration of the delaminated composite sandwich beams, the entire beam is separated into four regions as shown in Figure 6.7, and each region is considered as a composite sandwich beam. As suggested by Mujumdar and Suryanarayan (1988), even for the free vibration analysis the transverse distributed load q and the axial force P will not be zero in regions 2 and 3. They showed that, for the cases of delaminated isotropic beams, the free mode model, though mathematically admissible, is not physically feasible, since it gives vibration modes with overlaps of deformation which violate compatibility. In reality, however, the tendency of one of the delaminated layers to overlap on the other will be resisted by the development of a contact pressure distribution between the adjacent layers. Such a pressure distribution would constrain the transverse deformation of these adjacent layers to be identical and thus

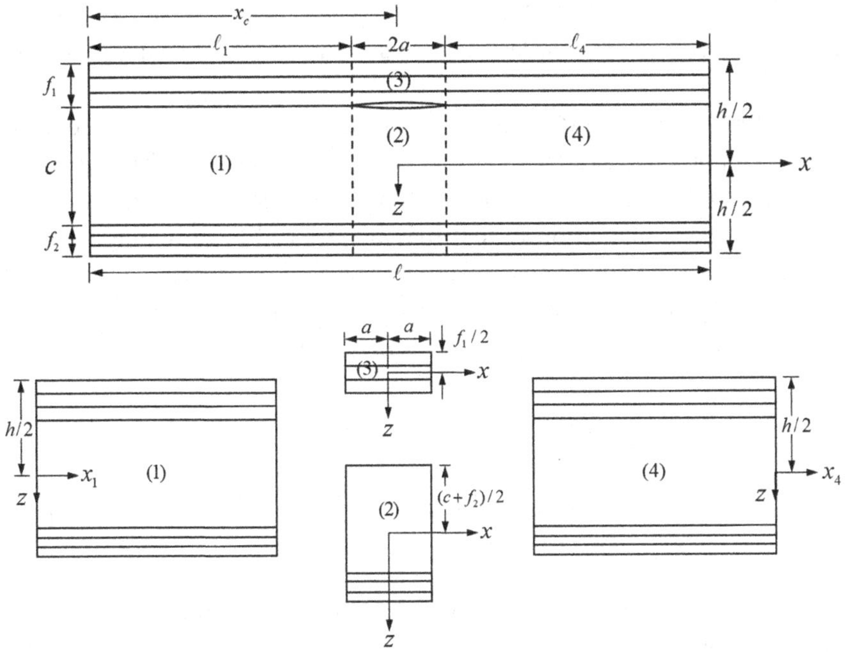

FIGURE 6.7 Geometry and notation of delaminated composite sandwich beam.

ensure compatibility. The two segments in the delamination region, though having identical transverse displacements, are assumed to be free to slide over each other in the axial direction except at their ends, which are connected to the integral segments. The contact is assumed to be frictionless and uniform. Thus, for regions 2 and 3, we assume

$$w_2 = w_3, \; P_2 = -P_3 = P = \text{constant}, \quad q_2 = -q_3 = q, \tag{6.161}$$

where the normal pressure q can be represented in harmonic form as

$$q = q_0 \sin \omega t, \quad q_0 = \text{constant}. \tag{6.162}$$

Under the consideration that both transverse distributed load and axial force will exist on some parts of the delaminated composite sandwich beams, the equation of motion (6.83) should be modified as

$$\begin{aligned} &\frac{\partial N_x}{\partial x} = 0, \quad \frac{\partial Q_x}{\partial x} + N_x \frac{\partial^2 w}{\partial x^2} + q = \rho h \frac{\partial^2 w}{\partial t^2}, \\ &\frac{\partial M_x}{\partial x} - Q_x = I \frac{\partial^2 \beta_x}{\partial t^2}, \end{aligned} \tag{6.163}$$

where the longitudinal inertia term $\rho \ddot{u}_0$ has been neglected since it is small for the lower flexural modes of beams (Mujumdar and Suryanarayan, 1988).

The first equation of (6.163) shows that N_x is a constant and can be set to be the compressive axial load P, that is, $N_x = -P$. Like the free vibration discussed in Section 6.2.3, if the axial load P is treated as a known value, three equations of motion (6.163) may be reduced to only one equation expressed by the transverse displacement w. Use of equation $(6.163)_2$ and $(6.84)_1$ may provide the relation between γ_{xz} and w. With this relation, the curvature κ_x can be expressed in terms of w. Thus,

$$\begin{aligned} \frac{\partial \gamma_{xz}}{\partial x} &= \frac{P}{S}\frac{\partial^2 w}{\partial x^2} + \frac{\rho h}{S}\frac{\partial^2 w}{\partial t^2} - \frac{q}{S}, \\ k_x &= \frac{\partial \beta_x}{\partial x} = \frac{\partial}{\partial x}\left(\gamma_{xz} - \frac{\partial w}{\partial x}\right) = -\left(1 - \frac{P}{S}\right)\frac{\partial^2 w}{\partial x^2} + \frac{\rho h}{S}\frac{\partial^2 w}{\partial t^2} - \frac{q}{S}. \end{aligned} \tag{6.164}$$

Substituting $(6.84)_1$, $(6.56)_2$, (6.48b) into $(6.163)_3$ and differentiating both sides of the equation with respect to x, the equation of motion can be written as

$$D\frac{\partial^2 k_x}{\partial x^2} = S\frac{\partial \gamma_{xz}}{\partial x} + I\frac{\partial^2 k_x}{\partial t^2}. \tag{6.165}$$

With the relations provided in (6.164), the equation of motion (6.165) for the composite sandwich beams can be expressed by only one parameter w as

$$\begin{aligned} &D\left(1 - \frac{P}{S}\right)\frac{\partial^4 w}{\partial x^4} - \left[I\left(1 - \frac{P}{S}\right) + \frac{\rho h D}{S}\right]\frac{\partial^4 w}{\partial x^2 \partial t^2} + \frac{\rho h I}{S}\frac{\partial^4 w}{\partial t^4} \\ &+ P\frac{\partial^2 w}{\partial x^2} + \rho h\frac{\partial^2 w}{\partial t^2} = q + \frac{I}{S}\frac{\partial^2 q}{\partial t^2} - \frac{D}{S}\frac{\partial^2 q}{\partial x^2}, \end{aligned} \tag{6.166}$$

which can be reduced to (6.86) when $P = 0$.

We now apply the equation of motion (6.166) to each region of the delaminated beams shown in Figure 6.7. Because no external load is applied for free vibration problems, the axial force P and transverse distributed load q are zero in regions 1 and 4, whereas in regions 2 and 3 they are assumed as (6.161). Moreover, by the fact that the rotary inertia I is higher order in h which is small even for the sandwich beam, and the in-plane forces are usually far smaller than the transverse shear forces for the flexural vibration problems, the terms $\omega^2 I/S$ and P/S might be far less than unity and may be neglected for the convenience of mathematical manipulation. For the sake of prudence, we will check this assumption after the natural frequencies and the mode shapes are obtained. With this understanding, the equations of motion for each region can be expressed as

$$
\begin{aligned}
&D_i \frac{\partial^4 w_i}{\partial x_i^4} - \left[I_i + \frac{\rho_i h_i D_i}{S_i} \right] \frac{\partial^4 w_i}{\partial x_i^2 \partial t^2} + \frac{\rho_i h_i I_i}{S_i} \frac{\partial^4 w_i}{\partial t^4} + \rho_i h_i \frac{\partial^2 w_i}{\partial t^2} = 0, \quad i = 1, 4, \\
&D_2 \frac{\partial^4 w_2}{\partial x^4} - \left(I_2 + \frac{\rho_2 h_2 D_2}{S_2} \right) \frac{\partial^4 w_2}{\partial x^2 \partial t^2} + \frac{\rho_2 h_2 I_2}{S_2} \frac{\partial^4 w_2}{\partial t^4} + P \frac{\partial^2 w_2}{\partial x^2} + \rho_2 h_2 \frac{\partial^2 w_2}{\partial t^2} = q, \\
&D_3 \frac{\partial^4 w_2}{\partial x^4} - \left(I_3 + \frac{\rho_3 h_3 D_3}{S_3} \right) \frac{\partial^4 w_2}{\partial x^2 \partial t^2} + \frac{\rho_3 h_3 I_3}{S_3} \frac{\partial^4 w_2}{\partial t^4} - P \frac{\partial^2 w_2}{\partial x^2} + \rho_3 h_3 \frac{\partial^2 w_2}{\partial t^2} = -q,
\end{aligned}
\tag{6.167}
$$

in which the local coordinates x_2 and x_3 are chosen to be the same as the global coordinate x (see Figure 6.7). By adding the last two equations of (6.167), the unknown normal contact pressure q and the axial load P can be eliminated. The result is

$$
\begin{aligned}
&(D_2 + D_3) \frac{\partial^4 w_2}{\partial x^4} - \left(I_2 + I_3 + \frac{\rho_2 h_2 D_2}{S_2} + \frac{\rho_3 h_3 D_3}{S_3} \right) \frac{\partial^4 w_2}{\partial x^2 \partial t^2} \\
&+ \left(\frac{\rho_2 h_2 I_2}{S_2} + \frac{\rho_3 h_3 I_3}{S_3} \right) \frac{\partial^4 w_2}{\partial t^4} + (\rho_2 h_2 + \rho_3 h_3) \frac{\partial^2 w_2}{\partial t^2} = 0.
\end{aligned}
\tag{6.168}
$$

For harmonic motion,

$$
w_i(x_i, t) = W_i(x_i) \sin \omega t, \quad i = 1, 2, 4, \tag{6.169}
$$

in which ω is the natural frequency and W_i is the mode shape for the ith region. Substituting (6.169) into $(6.167)_1$ and (6.168), the general solutions of the differential equations $(6.167)_1$ and (6.168) can be obtained as

$$
\begin{aligned}
&W_i(x_i) = \Gamma_{i1} \cosh \lambda_i x_i + \Gamma_{i2} \sinh \lambda_i x_i + \Gamma_{i3} \cos \mu_i x_i + \Gamma_{i4} \sin \mu_i x_i, \ i = 1, 4, \\
&W_2(x) = \Gamma_{21} \cosh \lambda_2 x + \Gamma_{22} \sinh \lambda_2 x + \Gamma_{23} \cos \mu_2 x + \Gamma_{24} \sin \mu_2 x,
\end{aligned}
\tag{6.170a}
$$

where

$$\lambda_i^2 = \frac{\omega^2}{2D_i}\left\{-\hat{I}_i + \sqrt{\hat{I}_i^2 + 4\hat{m}_i D_i}\right\},\quad \mu_i^2 = \frac{\omega^2}{2D_i}\left\{\hat{I}_i + \sqrt{\hat{I}_i^2 + 4\hat{m}_i D_i}\right\},\quad i = 1, 4,$$
$$\lambda_2^2 = \frac{\omega^2}{2(D_2 + D_3)}\left\{-(\hat{I}_2 + \hat{I}_3) + \sqrt{(\hat{I}_2 + \hat{I}_3)^2 + 4(\hat{m}_2 + \hat{m}_3)(D_2 + D_3)}\right\},$$
$$\mu_2^2 = \frac{\omega^2}{2(D_2 + D_3)}\left\{(\hat{I}_2 + \hat{I}_3) + \sqrt{(\hat{I}_2 + \hat{I}_3)^2 + 4(\hat{m}_2 + \hat{m}_3)(D_2 + D_3)}\right\},$$
$$(6.170b)$$

and

$$\hat{I}_i = I_i + \frac{\rho_i h_i D_i}{S_i},\qquad \hat{m}_i = \frac{\rho_i h_i}{\omega^2} - \frac{\rho_i h_i I_i}{S_i}. \tag{6.170c}$$

Γ_{ij} are the unknown coefficients to be determined by the continuity and boundary conditions.

Usually, the continuity conditions in the beam problems include the continuity of the transverse displacements, slopes, bending moments, and shear forces. Since the transverse shear effects are considered in all regions, the slopes of all regions are expressed in the form $\gamma_{xz} - \partial w/\partial x$. Because the face made up of composite laminate is much thinner compared with the core thickness, the slope of region 3 is dominated by the derivative of deflection $\partial w_2/\partial x$, and the transverse shear strain γ_{xz3} is of little effect. Moreover, region 2 is assumed to vibrate together with region 3. Therefore, the slope of regions 2 and 3 may now be approximated by $\partial w_2/\partial x$ for the convenience of mathematical manipulation, which should be verified after the natural frequency and mode shape are calculated.

With the above approximation, the continuity conditions at the delamination tips for the present problem can then be expressed as

$$w_1\big|_{x_1=\ell_1} = w_2\big|_{x=-a},\quad \left[\frac{\partial w_1}{\partial x_1} - (\gamma_{xz})_1\right]_{x_1=\ell_1} = \frac{\partial w_2}{\partial x}\bigg|_{x=-a},$$
$$w_2\big|_{x=a} = w_4\big|_{x_4=-\ell_4},\quad \frac{\partial w_2}{dx}\bigg|_{x=a} = \left[\frac{\partial w_4}{dx_4} - (\gamma_{xz})_4\right]_{x_4=-\ell_4}, \tag{6.171a}$$

$$\begin{aligned}(M_x)_1\big|_{x_1=\ell_1} &= (M_x)_2\big|_{x=-a} + (M_x)_3\big|_{x=-a} + Ph/2,\\ (M_x)_4\big|_{x_4=-\ell_4} &= (M_x)_2\big|_{x=a} + (M_x)_3\big|_{x=a} + Ph/2,\end{aligned} \tag{6.171b}$$

$$\begin{aligned}(Q_x)_1\big|_{x_1=\ell_1} &= (Q_x)_2\big|_{x=-a} + (Q_x)_3\big|_{x=-a},\\ (Q_x)_4\big|_{x_4=-\ell_4} &= (Q_x)_2\big|_{x=a} + (Q_x)_3\big|_{x=a}.\end{aligned} \tag{6.171c}$$

It should be noted that the transverse shear force Q_x in all regions is related to γ_{xz} by $(6.84)_1$. The bending moment M_x and the transverse shear strain γ_{xz} are related to the transverse displacement w by $(6.56)_2$, (6.48b), $(6.49b)_2$, and (6.164). With these relations, the continuity conditions (6.171) can all be expressed in terms of the transverse displacement w_i except that the bending and shear force continuity have the extra unknown axial load P and normal pressure q in which the axial load P may be found by the compatibility of axial displacements at the tip of delamination, that is,

$$\begin{aligned}\left[(u_0)_2 + \frac{1}{2}(c+f_2)\frac{\partial w_2}{\partial x}\right]_{x=-a} &= \left[(u_0)_3 - \frac{f_1}{2}\frac{\partial w_2}{\partial x}\right]_{x=-a},\\ \left[(u_0)_2 + \frac{1}{2}(c+f_2)\frac{\partial w_2}{\partial x}\right]_{x=a} &= \left[(u_0)_3 - \frac{f_1}{2}\frac{\partial w_2}{\partial x}\right]_{x=a}.\end{aligned} \tag{6.172}$$

The midplane axial displacement u_0 of regions 2 and 3 can also be expressed in terms of w as shown in (6.152). Substituting (6.152) into (6.172) and subtracting $(6.172)_2$ by $(6.172)_1$, we obtain

$$P = \frac{(B_3 - B_2) - h/2}{2a(A_2 + A_3)}\left(\frac{\partial w_2}{\partial x}\big|_{x=a} - \frac{\partial w_2}{\partial x}\big|_{x=-a}\right). \tag{6.173}$$

By using the relation given in (6.173), the other unknown q may be found by the compatibility of transverse shear strain at the crack tip of delamination. However, this approach will lead to a nonlinear equation of Γ_{ij}, which may cause trouble in mathematical manipulation. To avoid that, we ignore q_i/S_i, $i = 2,3$, since they may be far smaller than the other terms shown in $(6.164)_1$. This assumption will also be checked after the natural frequencies and mode shapes are obtained.

The continuity conditions (6.171) will now provide 8 linear homogeneous algebraic equations in 12 unknown coefficients Γ_{ij} (i = 1,2,4; j = 1,2,3,4). The remaining four equations come from the boundary conditions for both ends of the sandwich beams. Three different boundary conditions are considered as follows:

i. Simply supported ends:

$$w_1 = 0,\ (M_x)_1 = 0,\ \text{at}\ \ x_1 = 0,$$
$$w_4 = 0,\ (M_x)_4 = 0,\ \text{at}\ \ x_4 = 0. \tag{6.174a}$$

ii. Clamped-clamped ends:

$$w_1 = 0,\ (\gamma_{xz})_1 - \frac{\partial w_1}{\partial x_1} = 0,\ \text{at}\ x_1 = 0,$$
$$w_4 = 0,\ (\gamma_{xz})_4 - \frac{\partial w_4}{\partial x_4} = 0,\ \text{at}\ x_4 = 0. \tag{6.174b}$$

iii. Clamped-free ends:

$$w_1 = 0,\ (\gamma_{xz})_1 - \frac{\partial w_1}{\partial x_1} = 0,\ \text{at}\ x_1 = 0,$$
$$(M_x)_4 = 0,\ (Q_x)_4 = 0,\quad \text{at}\ x_4 = 0. \tag{6.174c}$$

In the above, all the boundary conditions provide four equations. Combining these 4 equations with the 8 continuity equations, we obtain 12 linear homogeneous algebraic equations in 12 unknown coefficients Γ_{ij}. The natural frequencies and mode shapes of the delaminated composite sandwich beams can then be obtained as the eigenvalues and eigenvectors of this equation set. After getting the solutions, we should check all the assumptions stated in this section, that is, (1) normal pressure $q(=q_0 \sin \omega t)$ is uniform along the contact region; (2) P_i/S_i and $\omega_0^2\, I_i/S_i$, $i = 2,3$ are far smaller than unity; (3) γ_{xz_2} and γ_{xz_3} are far smaller than $\partial w_2/\partial x$; and (4) $2aq_i/S_i$, $i = 2,3$ are far smaller than γ_{xz_2} and γ_{xz_3}. The numerical results show that these assumptions are valid under small flexural vibration, usually $w/h \le 0.1$. Under this situation, the solutions obtained by the present way are in good agreement with the experimental results (Hwu and Bi, 2001).

By varying the parameters related to delamination, core, and face, their effects on the free vibration can be studied. The results show that (Hu and

Hwu, 1995): (1) the existence of delamination will lower the natural frequencies; (2) the upper bound and lower bound of the natural frequencies can be represented by the perfect sandwich beam and the detached sandwich beam, respectively; (3) the natural frequency of a delaminated composite sandwich beam decreases gradually when the core thickness decreases and will approach that of a delaminated composite laminate as the core thickness is reduced to zero; (4) the mode shapes of delaminated composite sandwich beams differ from those of perfect composite sandwich beams especially in the region of delamination; (5) the natural frequency depends significantly on the transverse shear stiffness when the delamination length is short, while there is almost no influence on frequency for longer delamination; and (6) the natural frequency increases when the face thickness increases, or when the fiber is oriented to the axial direction.

REFERENCES

Allen, H.G., 1969, *Analysis and Design of Structural Sandwich Panels*, Pergamon Press, Oxford.

Chai, H., Babcock, C.D. and Knauss, W.G., 1981, "One Dimensional Modelling of Failure in Laminated Plates by Delamination Buckling," *International Journal of Solids Structures*, Vol. 17, No. 11, pp. 1069–1083.

Cowper, G.R., 1966, "The Shear Coefficient in Timoshenko's Beam Theory," *Journal of Applied Mechanics*, pp. 335–340.

Heath, W.G., 1960, "Sandwich Construction - Correlation and Extension of Existing Theory of Flat Panels Subjected to Lengthwise Compression, Part 1: The Strength of Flat Sandwich Panels," *Aircraft Engineering*, Vol. 32, pp. 186–191.

Hoff, N.J., 1966, *Sandwich Constructions*, John Wiley and Sons, Inc., New York.

Hu, J.S. and Hwu, C., 1995, "Free Vibration of Delaminated Sandwich Beams," *AIAA Journal*, Vol. 33, No. 10, pp. 1911–1918.

Hwu, C. and Hu, J.S., 1992, "Buckling and Postbuckling of Delaminated Composite Sandwich Beams," *AIAA Journal*, Vol. 30, No. 7, pp. 1901–1909.

Hwu, C. and Hsieh, C.H., 1998, "The Effect of Transverse Shear Stress on the Buckling of the Delaminated Composite Sandwich Beams," *Proceedings of the First Asian-Australasian Conference on Composite Materials*, Osaka, Vol. I, pp. 303.1–303.4.

Hwu, C. and Bi, T.Y., 2001, "Re-Examination of the Buckling and Vibration of the Delaminated Composite Sandwich Beams," *The Chinese Journal of Mechanics, Series B*, Vol. 17, No. 1, pp. 27–38. (in Chinese).

Hwu, C. and Gai, H.S., 2003, "Vibration Analysis of Composite Wing Structures by a Matrix Form Comprehensive Model," *AIAA Journal*, Vol. 41, No. 11, pp. 2261–2273.

Hwu, C., Chang, W.C. and Gai, H.S., 2004, "Vibration Suppression of Composite Sandwich Beams," *Journal of Sound and Vibration*, Vol. 272, pp. 1–20.

Hwu, C., Hsu, H.W. and Lin, Y.H., 2017, "Free Vibration of Composite Sandwich Plates and Cylindrical Shells," *Composite Structures*, Vol. 171, pp. 528–537.

Librescu, L., 1975, *Elastostatics and Kinetics of Anisotropic and Hetergeneores Shell-Type Structures*, Noordhoff, Leyden.

Lin, Y.H., 2013, Free vibration analysis of composite laminated sandwich shells, *M.S. Thesis*, Institute of Aeronautics and Astronautics, National Cheng Kung University.

Meirovitch, L., 1967, *Analytical Methods in Vibration*, Macmillan, New York.

Meirovitch, L., 1990, *Dynamics and Control of Structures*, John Wiley & Sons, New York.

Mindlin, R.D., 1951, "Influence of Rotatory Inertia and Shear on Flexural Motions of Isotropic, Elastic Plates," *ASME Journal of Applied Mechanics*, pp. 31–38.

Moh, J.S. and Hwu, C., 1997, "Optimization for Buckling of Composite Sandwich Plates", *AIAA Journal*, Vol. 35, No. 5, pp. 863–868.

Mujumdar, P.M. and Suryanarayan, S., 1988, "Flexural Vibrations of Beams with Delaminations," *Journal of Sound and Vibration*, Vol. 125, No. 3, pp. 441–461.

Noor, A.K., Burton W.S., and Bert, C.W., 1996, "Computational Models for Sandwich Panels and Shells," *Applied Mechanics Review*, Vol. 49, No. 3, pp. 155–199.

Penzien, J. and Didriksson, T., 1964, "Effective Shear Modulus of Honeycomb Cellular Structure," *AIAA Journal*, Vol. 2, No. 3, pp. 531–535.

Plantema, F.J., 1966, *Sandwich Construction – The Bending and Buckling of Sandwich Beams, Plates and Shells*, John Wiley & Sons, Inc., New York.

Reissner, E., 1945, "The Effect of Transverse Shear Deformation on the Bending of Elastic Plates," *ASME Journal of Applied Mechanics*, pp. A69–A77.

Roy, A.K. and Verchery, G., 1993, "Approximate Methods for Predicting Interlaminar Shear Stiffness of Laminated and Sandwich Beams," *SAMPE Quarterly*, Vol. 24, pp. 22–27.

Yin, W.L., Sallam, S.N. and Simitses, G.L., 1986, "Ultimate Axially Loaded Capacity of a Delaminated Beam-Plate," *AIAA Journal*, Vol. 24, No. 1, pp. 123–128.

CHAPTER 7

Composite Wing Structures

THE PRIMARY FUNCTION OF the wing structure is to provide lift for an aircraft, which is governed by aerodynamic consideration. In addition to the aerodynamic pressure, there are other forces resisted by the wing structures such as the weight of the structures, fuels, engines, undercarriage system, and/or possible carried weapons, and the thrust of engines, etc. To sustain these loads, the wing structures are usually designed as shown in Figure 7.1, which consists of axial members in stringers, bending members in spars, shear panels in the cover skin and spar webs, and planar members in ribs. If the cover skin of the wing is made of composite laminates, the entire wing structure may be simulated by a composite sandwich plate in which the wing skins and stringers (including the spar flanges) are simulated as the faces while the spar webs and ribs are simulated as the core of the sandwich plates. Because the wing cross-section must have a streamlined shape commonly referred to as an *airfoil* section, unlike the usual uniform thickness modeling it is more appropriate to simulate the wings as variable thickness sandwich plates where the thickness is a function of the airfoil. Moreover, as the usual assumptions for sandwich plates, the overall thickness of the wing structures will not be too small to neglect the transverse shear deformation. With this consideration, a mathematical model for the composite sandwich plates discussed in Chapter 6 will be applied in the wing structures. Based upon the model for composite sandwich plates, a comprehensive beam model will

DOI: 10.1201/9781003470465-7

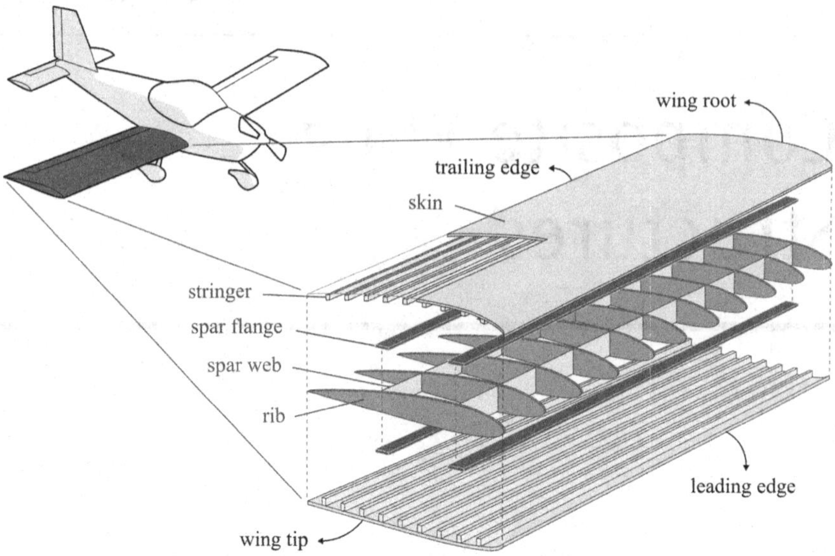

FIGURE 7.1 Composite wing structure arrangement.

be introduced in Section 7.1, which can be applied to the wings with uniform cross-section and discussed in Section 7.2. With this beam model, a comprehensive finite element method is developed and presented in Section 7.3 for tapered wings. To be more compliant with the real situation, a variable thickness plate model and its associated finite element method will then be presented in Section 7.4.

7.1 COMPREHENSIVE BEAM MODEL

As stated at the beginning of this chapter, by treating wing skins, stringers, and the spar flanges as the sandwich faces, and wing spar webs and ribs as the sandwich cores, the stiffened composite multicell wing structures may be modeled as a composite sandwich plate. The basic equations for the composite sandwich plates shown in Chapter 6 are: (6.7a, 6.8b) for the displacement fields; (6.8a, 6.8b) for the strain–displacement relations; (6.9)–(6.13) for the constitutive laws; (6.14) for the equilibrium equations; and (6.15a, 6.15b) for the boundary conditions. In this chapter, we will employ these basic equations together with the postulation for the wing structures to develop a mathematical model for the stiffened composite multicell wing structures, called *comprehensive beam model*. For the convenience of presentation this section will be divided into four subsections. The first two subsections are for static analysis and dynamic analysis of general cases. Section 7.1.3 introduces matrix form expressions to rewrite the component form expressions into simple matrix forms. Section 7.1.4 will then discuss some special cases reduced from the present general case. The contents presented in this section and later in Section 7.2 are adapted from Hwu and Tsai (2002) and Hwu and Gai (2003).

7.1.1 Static Analysis

Rigid Wing Chordwise Section

Because of the closely spaced stringers and the transverse stiffening members like wing spars and ribs, in aircraft analysis it is usually assumed that the wing chordwise section is rigid (Megson, 1990). Consistent with the chordwise-rigid postulation, the mid-plane displacement and rotation angles shown in (6.7) may be assumed to be

$$\begin{aligned} &u_0 = 0, \;\; v_0 = v_0(y), \;\; w_0 = w_f(y) - x\theta(y), \\ &\beta_x = \theta(y), \;\; \beta_y = \beta_f(y) + x\beta_r(y), \end{aligned} \tag{7.1}$$

where $w_f(y)$ denotes the deflection (positive upward) measured at the line of flexural center, which is now selected as the reference axis shown in Figure 7.2; $\theta(y)$ is the twist around the flexural axis (positive nose up). The relation between β_x and θ is due to the chordwise-rigid assumption which leads to $\gamma_{xz} = 0$. By use of $(6.7b)_1$ and $(7.1)_3$ we can get $(7.1)_4$. However, in the spanwise direction the transverse shear deformation cannot be neglected for the thick plates. Hence, two extra functions β_f and β_r are needed for the representation of β_y, where β_f denotes the rotation angle measured at the flexural axis and β_r stands for the rate of angle change in the x-direction.

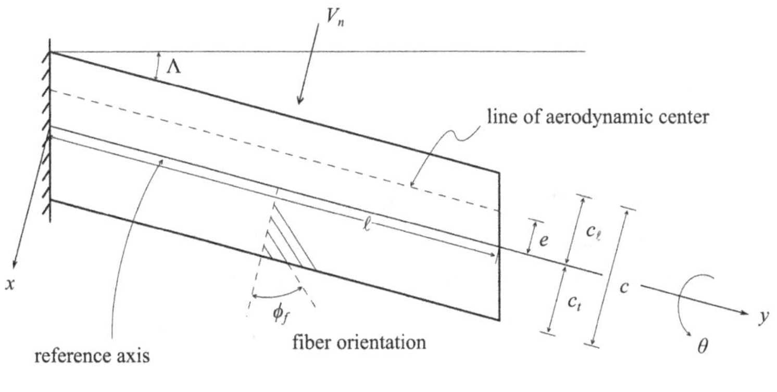

FIGURE 7.2 Geometry of the composite cantilever swept wings.

By the assumption given in (7.1), the mid-plane strains and curvatures defined in (6.8b) become

$$\varepsilon_x^0 = 0,\ \ \varepsilon_y^0 = v_0',\ \ \gamma_{xy}^0 = 0,\ \ \kappa_x = 0,\ \ \kappa_y = \beta_f' + x\beta_r',\ \ \kappa_{xy} = \theta' + \beta_r, \quad (7.2)$$

where the prime •′ means differentiation with respect to y.

With the expressions assumed in (7.1) and (7.2), the displacements and the strains of the entire composite wings can be expressed as

$$\begin{aligned} u(x,y,z,t) &= z\theta(y,t), \\ v(x,y,z,t) &= v_0(y,t) + z\{\beta_f(y,t) + x\beta_r(y,t)\}, \\ w(x,y,z,t) &= w_f(y,t) - x\theta(y,t), \end{aligned} \quad (7.3a)$$

and

$$\begin{aligned} &\varepsilon_x = 0,\ \ \varepsilon_y = v_0' + z\left(\beta_f' + x\beta_r'\right),\ \ \gamma_{xy} = z\left(\beta_r + \theta'\right), \\ &\gamma_{xz} = 0,\ \ \gamma_{yz} = \beta_f + w_f' + x\left(\beta_r - \theta'\right). \end{aligned} \quad (7.3b)$$

Modeling of Wing Skins, Stringers, and Spar Flanges

In the wing structure design, most of the bending and axial loads are resisted by the stringers and spar flanges, whereas the wing skins resist almost all of the in-plane shear forces. Hence, it is reasonable to model the wing skins, stringers, and spar flanges as the sandwich faces. If the stringers and spar flanges are considered to be the fibers of a pseudo-lamina, by the rule of mixture (Jones, 1974) the equivalent material properties of this pseudo-lamina may be written as

$$E_L = E_s\frac{A_s}{A_p} + E_f\frac{A_f}{A_p},\quad \nu_{LT} = \nu_s\frac{A_s}{A_p} + \nu_f\frac{A_f}{A_p},\quad E_T = 0,\quad G_{LT} = 0, \quad (7.4)$$

where E, ν, G, and A denote, respectively, the Young's modulus, Poisson's ratio, shear modulus, and cross-section area. The subscript L, T, s, and f denote the longitudinal direction, transverse direction, stringer, and spar flange, respectively. A_p stands for the cross-section area of pseudo-lamina. By adding this pseudo-lamina to the laminated composite skin, the face properties may be represented by A_{ij}, B_{ij}, and D_{ij} defined in (6.10). Note that due to the shape of the airfoil, h_k in (6.10) is a function of x. If the upper and lower surfaces of the airfoil are represented by $f_u(x)$ and $f_l(x)$, the stiffness matrices A_{ij}, B_{ij}, and D_{ij} of the wings are related to those of the corresponding laminated composite flat plate by (Tsai, 2000)

$$\begin{aligned} A_{ij}(x) &= A_{ij}^F, \\ B_{ij}(x) &= B_{ij}^F + f_u(x)A_{ij}^u + f_l(x)A_{ij}^l, \\ D_{ij}(x) &= D_{ij}^F + 2f_u(x)B_{ij}^u + 2f_l(x)B_{ij}^l + f_u^2(x)A_{ij}^u + f_l^2(x)A_{ij}^l, \quad i,j = 1,2,6. \end{aligned} \tag{7.5}$$

Here, the superscript F denotes the properties associated with the flat composite plates, while the superscripts u and l denote those of the upper and lower parts of the flat plate.

Because of the assumption given in (7.1), it is desirable to reduce the two-dimensional formulation given for the composite sandwich plates to an equivalent one-dimensional formulation. With this consideration, we like to integrate the stress resultants and bending moments of (6.9) with respect to x. Their relations with the mid-plane strains and curvatures may then be written as

$$\begin{Bmatrix} \tilde{N}_y \\ \tilde{M}_y \\ \tilde{M}_{xy} \end{Bmatrix} = \begin{bmatrix} \tilde{A}_{22} & \tilde{B}_{22} & \tilde{B}_{22}^* & \tilde{B}_{26} \\ \tilde{B}_{22} & \tilde{D}_{22} & \tilde{D}_{22}^* & \tilde{D}_{26} \\ \tilde{B}_{26} & \tilde{D}_{26} & \tilde{D}_{26}^* & \tilde{D}_{66} \end{bmatrix} \begin{Bmatrix} v_0' \\ \beta_f' \\ \beta_r' \\ \theta' + \beta_r \end{Bmatrix} \tag{7.6}$$

Here, the tilde ~ denotes integration with respect to x, the superscript * denotes multiplication by x, and the prime •′ means differentiation with respect to y. For example,

$$\begin{aligned} &\tilde{D}_{22} = \int_{-c_l}^{c_t} D_{22}dx, \quad \tilde{D}_{22}^* = \int_{-c_l}^{c_t} D_{22}xdx, \\ &\tilde{D}_{22}^{**} = \int_{-c_l}^{c_t} D_{22}x^2dx, \quad v_0' = \frac{dv_0}{dy}, \ldots\ldots, \text{etc.} \end{aligned} \tag{7.7}$$

The lower and upper limits $-c_l$ and c_t denote, respectively, the location of leading and trailing edges (Figure 7.2). The relations for the N_x, N_{xy}, and M_x are not shown because they play no roles in one-dimensional representation.

Modeling of Spar Webs and Ribs for Multicell Wings

A multicell wing is made by proper arrangement of spars and ribs (Figure 7.1). The main function of wing spar webs and ribs is to resist the transverse shear force. Moreover, the multicellular arrangement looks like a sandwich honeycomb core. Hence, it is suitable to model them as the sandwich core. To simplify the analysis and to get the overall effects, the multicellular arrangement of spar webs and ribs is modeled by using an equivalent shear modulus that should be related to the material properties and sizes of the wing spars and ribs. By assuming uniform transverse shear strain over the wing cross-section, the equivalent transverse shear modulus G_{yz} may be estimated by $G_{yz} = \tau_{yz}/\gamma_{yz}$ where τ_{yz} is the average transverse shear stress and may be calculated by dividing the total transverse shear force $\tilde{Q}_y$ over the wing cross-section A_w, that is, $\tau_{yz} = \tilde{Q}_y / A_w$. As to the transverse shear strains, it can be calculated by $\gamma_{yz} = \tau_k / G_k = \tau_k A_k / G_k A_k = \sum \tau_k A_k / \sum G_k A_k = \tilde{Q}_y / \sum G_k A_k$, $k = 1,2,\ldots, n_s$. Here, τ_k, G_k, and A_k denotes the shear stress, shear modulus, and section area of the kth spar web, and n_s is the number of the wing spars. By this calculation, the equivalent transverse shear modulus G_{yz} may be estimated as

$$G_{yz} = \sum_{k=1}^{n_s} G_k A_k / A_w. \tag{7.8}$$

As for the transverse shear modulus G_{xz} that will be contributed by the wing ribs, no estimation is needed due to the assumption that $\gamma_{xz} = 0$. This also means that G_{xz} is assumed to be infinite under the construction of wing ribs. By $(6.7\text{b})_2$, $(6.12)_2$, $(6.13)_1$, $(7.1)_{3,5}$, and (7.8), we now have

$$\tilde{Q}_y = \tilde{A}_{44}(\beta_f + w_f') + \tilde{A}_{44}^*(\beta_r - \theta'). \tag{7.9}$$

Equilibrium Equations and Boundary Conditions

According to the postulation given in (7.1), the basic functions describing the deformation of the composite wing structures become v_0, w_f, θ, β_f, and β_r. The equilibrium equations corresponding to these basic functions will then be obtained by integrating (6.14) with respect to x. The associated integration and the results are

$$
\begin{aligned}
&\int\left(\sum F_y=0\right)dx: && \frac{d\tilde{N}_y}{dy}=0,\\
&\int\left(\sum F_z=0\right)dx: && \frac{d\tilde{Q}_y}{dy}+\tilde{q}=0,\\
&\int\left[\left(\sum M_x=0\right)-x\left(\sum F_z=0\right)\right]dx: && \frac{d(\tilde{M}_{xy}-\tilde{Q}_y^*)}{dy}-\tilde{q}^*=0,\\
&\int\left(\sum M_y=0\right)dx: && \frac{d\tilde{M}_y}{dy}=\tilde{Q}_y,\\
&\int\left(\sum M_y=0\right)xdx: && -\tilde{M}_{xy}+\frac{d\tilde{M}_y^*}{dy}=\tilde{Q}_y^*.
\end{aligned} \tag{7.10}
$$

During the integration, the boundary values of N_{xy}, Q_x, M_x, and M_{xy} have been assumed to be zero for the present one-dimensional model (see (7.7) for the definition of tilde ~ and superscript *). $\tilde{N}_y, \tilde{Q}_y, \tilde{M}_y$, and $\tilde{M}_{xy}$ are related to the basic functions v_0, w_f, θ, β_f, and β_r by (7.6) and (7.9). With (6.9), (6.12), and (7.2), $\tilde{Q}_y^*$ and $\tilde{M}_y^*$ can also be expressed in terms of the basic functions as

$$
\begin{aligned}
\tilde{Q}_y^* &= \tilde{A}_{44}^*(\beta_f+w_f')+\tilde{A}_{44}^{**}(\beta_r-\theta'),\\
\tilde{M}_y^* &= \tilde{B}_{22}^*v_0'+\tilde{D}_{22}^*\beta_f'+\tilde{D}_{22}^{**}\beta_r'+\tilde{D}_{26}^*(\theta'+\beta_r).
\end{aligned} \tag{7.11}
$$

The boundary conditions along $y = constant$ can then be expressed as

$$
\begin{aligned}
&\tilde{N}_y=\hat{\tilde{N}}_y, \quad \text{or} \quad v_0=\hat{v}_0,\\
&\tilde{Q}_y=\hat{\tilde{Q}}_y, \quad \text{or} \quad w_f=\hat{w}_f,\\
&\tilde{M}_{xy}-\tilde{Q}_y^*=\hat{\tilde{M}}_{xy}-\hat{\tilde{Q}}_y^*, \quad \text{or} \quad \theta=\hat{\theta},\\
&\tilde{M}_y=\hat{\tilde{M}}_y, \quad \text{or} \quad \beta_f=\hat{\beta}_f,\\
&\tilde{M}_y^*=\hat{\tilde{M}}_y^*, \quad \text{or} \quad \beta_r=\hat{\beta}_r.
\end{aligned} \tag{7.12}
$$

By using the relations given in (7.6), (7.9), and (7.11), the equilibrium equations derived in (7.10) can be expressed in terms of five basic unknown functions v_0, w_f, θ, β_f, and β_r that constitute a 10th-order system of five ordinary differential equations. This system of equations can be completely solved with ten boundary conditions described in (7.12) which consist of five boundary conditions per each edge.

Warping Restraint Effects

The importance of the warping restraint effects has been discussed vastly in Librescu's series work (Librescu and Simovich, 1988; Librescu and Khdeir, 1988; Librescu and Thangjitham, 1991; Karpouzian and Librescu, 1994). This effect has been considered when we include β_r' in the expression of ε_y, $(7.3\text{b})_2$. In other words, the free warping assumption can easily be done by inserting a tracer λ in the coefficient of $x\beta_r'$ of ε_y to identify the warping effect. That is,

$$\varepsilon_y = v_0' + z\left(\beta_f' + \lambda x \beta_r'\right), \tag{7.13}$$

where λ takes the values 0 or 1, according to whether the free warping or warping restraint are implied, respectively (Karpouzian and Librescu, 1994). All the related equations following (7.13) should include λ as a tracer of the warping restraint effect.

7.1.2 Dynamic Analysis

The main difference between the static and dynamic analyses is the consideration of the time variable which will then make the right hand side of the equilibrium equations (7.10) proportional to the acceleration instead of zero. In order to establish the equations of motion corresponding to the basic functions v_0, w_f, θ, β_f, and β_r, like the static analysis we first consider the equations of motions for composite sandwich plates, which can be expressed in terms of the stress resultants (N_x, N_y, N_{xy}, Q_x, Q_y) and bending moments (M_x, M_y, M_{xy}) as

$$\begin{aligned} &\frac{\partial N_x}{\partial x} + \frac{\partial N_{xy}}{\partial y} + q_x = \int_{z_l}^{z_u} \rho \ddot{u} dz, \\ &\frac{\partial N_{xy}}{\partial x} + \frac{\partial N_y}{\partial y} + q_y = \int_{z_l}^{z_u} \rho \ddot{v} dz, \end{aligned} \tag{7.14a}$$

$$\begin{aligned} &\frac{\partial Q_x}{\partial x} + \frac{\partial Q_y}{\partial y} + q = \int_{z_l}^{z_u} \rho \ddot{w} dz, \\ &\frac{\partial M_x}{\partial x} + \frac{\partial M_{xy}}{\partial y} - Q_x + m_x = \int_{z_l}^{z_u} \rho \ddot{u} z dz, \\ &\frac{\partial M_{xy}}{\partial x} + \frac{\partial M_y}{\partial y} - Q_y + m_y = \int_{z_l}^{z_u} \rho \ddot{v} z dz, \end{aligned} \tag{7.14b}$$

where q_x, q_y, q and m_x, m_y are the total distributed loads and moments applied on the upper and lower surfaces of the sandwich plates; ρ is the mass density and the overdot $\dot{}$ denotes the time derivative; z_l and z_u are the locations of the upper and lower surfaces of the plates.

Substituting (7.3a) into (7.14a, 7.14b) and performing the integration with respect to x in the following way: $\int(7.14a)_2 dx$, $\int(7.14b)_1 dx$, $\int[(7.14b)_2 - x(7.14b)_1]dx$, $\int(7.14b)_3 dx$ and $\int(7.14b)_3 x dx$, we get

$$\frac{\partial \tilde{N}_y}{\partial y} + \tilde{q}_y = m\ddot{v}_0 + mz_c\ddot{\beta}_f + I_{xz}\ddot{\beta}_r,$$
$$\frac{\partial \tilde{Q}_y}{\partial y} + \tilde{q} = m\ddot{w}_f - mx_c\ddot{\theta}, \tag{7.15a}$$

$$\frac{\partial(\tilde{M}_{xy} - \tilde{Q}_y^*)}{\partial y} + \tilde{m}_x - \tilde{q}^* = I_y\ddot{\theta} - mx_c\ddot{w}_f,$$
$$\frac{\partial \tilde{M}_y}{\partial y} - \tilde{Q}_y + \tilde{m}_y = mz_c\ddot{v}_0 + I_x\ddot{\beta}_f + I_{xz^2}\ddot{\beta}_r,$$
$$\lambda\frac{\partial \tilde{M}_y^*}{\partial y} - \tilde{M}_{xy} - \tilde{Q}_y^* + \tilde{m}_y^* = I_{xz}\ddot{v}_0 + I_{xz^2}\ddot{\beta}_f + I_{x^2z^2}\ddot{\beta}_r, \tag{7.15b}$$

where m is the mass per unit spanwise length; (x_c, z_c) is the coordinate of the center of gravity; and I stands for the mass moment of inertia. They are defined by

$$m = \int_A \rho dA,\quad mx_c = \int_A \rho x dA,\quad mz_c = \int_A \rho z dA,$$
$$I_x = \int_A \rho z^2 dA,\quad I_y = \int_A \rho\left(x^2 + z^2\right)dA,$$
$$I_{xz} = \int_A \rho xz dA,\quad I_{xz^2} = \int_A \rho xz^2 dA,\quad I_{x^2z^2} = \int_A \rho x^2 z^2 dA, \tag{7.16}$$

where A is the area of chordwise cross-section.

From the boundary conditions shown in (7.12), we see that the corresponding force terms to the basic variables v_0, w_f, θ, β_f, and β_r are $\tilde{N}_y, \tilde{Q}_y, \tilde{M}_{xy} - \tilde{Q}_y^*, \tilde{M}_y$, and $\tilde{M}_y^*$. The constitutive relations between the basic variables and their corresponding forces can be found in (7.6) and (7.11), that is,

$$\tilde{N}_y = \tilde{A}_{22}v_0' + \tilde{B}_{22}\beta_f' + \tilde{B}_{22}^*\lambda\beta_r' + \tilde{B}_{26}(\beta_r + \theta'),$$
$$\tilde{M}_y = \tilde{B}_{22}v_0' + \tilde{D}_{22}\beta_f' + \tilde{D}_{22}^*\lambda\beta_r' + \tilde{D}_{26}(\beta_r + \theta'),$$
$$\tilde{M}_{xy} = \tilde{B}_{26}v_0' + \tilde{D}_{26}\beta_f' + \tilde{D}_{26}^*\lambda\beta_r' + \tilde{D}_{66}(\beta_r + \theta'),$$
$$\tilde{Q}_y^* = \tilde{A}_{44}^*(\beta_f + w_f') + \tilde{A}_{44}^{**}(\beta_r - \theta'),$$
$$\tilde{Q}_y = \tilde{A}_{44}(\beta_f + w_f') + \tilde{A}_{44}^*(\beta_r - \theta'),$$
$$\tilde{M}_y^* = \tilde{B}_{22}^*v_0' + \tilde{D}_{22}^*\beta_f' + \tilde{D}_{22}^{**}\lambda\beta_r' + \tilde{D}_{26}^*(\beta_r + \theta'). \quad (7.17)$$

7.1.3 Matrix Form

By using the constitutive relations given in (7.17), the equations of motion (7.15) can be expressed in terms of five basic unknown functions v_0, w_f, θ, β_f, and β_r that constitute a 10th-order system of five partial differential equations. This system of equations can be completely solved with 10 boundary conditions (7.12) together with 10 initial conditions, which consist of 5 conditions per each edge and 2 conditions at $t = 0$ for each basic function. Due to the complexity of the equations, most of the studies choose not to express the final governing equations in detail or not to solve its associated problems in general. In the following, the detailed expressions will be given in a compact matrix form and then its associated problems will be solved by analogy with the scalar form expressions. With this understanding, we now rewrite the equations of motion (7.15), constitutive relations (7.17), and boundary conditions (7.12) as follows:

$$\begin{aligned}&\mathbf{F}' - \mathbf{F}_0 + \mathbf{q} = \mathbf{I}_0\ddot{\boldsymbol{\delta}},\\ &\mathbf{F} = \mathbf{K}_1\boldsymbol{\delta} + \mathbf{K}_2\boldsymbol{\delta}', \quad \mathbf{F}_0 = \mathbf{K}_0\boldsymbol{\delta} + \mathbf{K}_1^T\boldsymbol{\delta}',\\ &\mathbf{F} = \hat{\mathbf{F}}, \quad \text{or} \quad \boldsymbol{\delta} = \hat{\boldsymbol{\delta}}, \quad \text{along } y = \text{constant}\end{aligned} \quad (7.18a)$$

where

$$\mathbf{F} = \begin{Bmatrix} \tilde{N}_y \\ \tilde{Q}_y \\ \tilde{M}_{xy} - \tilde{Q}_y^* \\ \tilde{M}_y \\ \lambda\tilde{M}_y^* \end{Bmatrix}, \quad \mathbf{F}_0 = \begin{Bmatrix} 0 \\ 0 \\ 0 \\ \tilde{Q}_y \\ \tilde{M}_{xy} + \tilde{Q}_y^* \end{Bmatrix}, \quad \mathbf{q} = \begin{Bmatrix} \tilde{q}_y \\ \tilde{q} \\ \tilde{m}_x - \tilde{q}^* \\ \tilde{m}_y \\ \tilde{m}_y^* \end{Bmatrix}, \quad \boldsymbol{\delta} = \begin{Bmatrix} v_0 \\ w_f \\ \theta \\ \beta_f \\ \beta_r \end{Bmatrix}, \quad (7.18b)$$

and

$$\mathbf{K}_2 = \begin{bmatrix} \tilde{A}_{22} & 0 & \tilde{B}_{26} & \tilde{B}_{22} & \lambda\tilde{B}^*_{22} \\ 0 & \tilde{A}^{**}_{44} & -\tilde{A}^*_{44} & 0 & 0 \\ \tilde{B}_{26} & -\tilde{A}^*_{44} & \tilde{D}_{66}+\tilde{A}^{**}_{44} & \tilde{D}_{26} & \lambda\tilde{B}^*_{26} \\ \tilde{B}_{22} & 0 & \tilde{D}_{26} & \tilde{D}_{22} & \lambda\tilde{D}^*_{22} \\ \lambda\tilde{B}^*_{22} & 0 & \lambda\tilde{D}^*_{26} & \lambda\tilde{D}^*_{22} & \lambda\tilde{D}^{**}_{22} \end{bmatrix}, \mathbf{K}_1 = \begin{bmatrix} 0 & 0 & 0 & 0 & \tilde{B}_{26} \\ 0 & 0 & 0 & \tilde{A}_{44} & \tilde{A}^*_{44} \\ 0 & 0 & 0 & -\tilde{A}^*_{44} & \tilde{D}_{66}+\tilde{A}^{**}_{44} \\ 0 & 0 & 0 & 0 & \lambda\tilde{D}_{26} \\ 0 & 0 & 0 & 0 & \lambda\tilde{D}^*_{26} \end{bmatrix}$$

$$\mathbf{K}_0 = \begin{bmatrix} 0 & 0 & 0 & 0 & 0 \\ 0 & 0 & 0 & 0 & 0 \\ 0 & 0 & 0 & 0 & 0 \\ 0 & 0 & 0 & \tilde{A}_{44} & \tilde{A}^*_{44} \\ 0 & 0 & 0 & -\tilde{A}^*_{44} & \tilde{D}_{66}+\tilde{A}^{**}_{44} \end{bmatrix}, \mathbf{I}_0 = \begin{bmatrix} m & 0 & 0 & mz_c & I_{xz} \\ 0 & m & -mz & 0 & 0 \\ 0 & -mz & I_y & 0 & 0 \\ mz_c & 0 & 0 & I_x & I_{xz^2} \\ I_{xz} & 0 & 0 & I_{xz^2} & I_{x^2z^2} \end{bmatrix} \tag{7.18c}$$

Substituting the second and third equations of (7.18a) into the first equation of (7.18a), we can then write down the governing equations for the stiffened composite wing structures in terms of the basic function vector $\boldsymbol{\delta}$ as

$$\mathbf{K}_2\boldsymbol{\delta}''(y,t)+(\mathbf{K}_1-\mathbf{K}_1^T)\boldsymbol{\delta}'(y,t)-\mathbf{K}_0\boldsymbol{\delta}(y,t)+\mathbf{q}(y,t)=\mathbf{I}_0\ddot{\boldsymbol{\delta}}(y,t). \tag{7.19}$$

If the wing is fixed at the root ($y = 0$) and free at the tip ($y = L$), by the last two equations of (7.18a) the boundary conditions of the wing structures can now be written as

$$\boldsymbol{\delta}(0)=\mathbf{0}, \text{ and } \mathbf{K}_1\boldsymbol{\delta}(L)+\mathbf{K}_2\boldsymbol{\delta}'(L)=\mathbf{0}. \tag{7.20}$$

Absence of In-Plane Spanwise Loads

If the influence of the in-plane spanwise surface loads $\tilde{q}_y$ as well as of in-plane and rotary inertia terms $m\ddot{v}_0, mz_c\ddot{\beta}_f, I_{xz}\ddot{\beta}_r$ can be disregarded, the first equation of motion shown in (7.15a) will lead to $\tilde{N}_y = \text{constant}$. If no in-plane spanwise loads are applied at the wingtip, this constant value will then be identical to zero. That is $\tilde{N}_y = 0$ along the entire wing. When this result is substituted into the first equation of the constitutive relations (7.17), v_0' may be expressed in terms of β_f', β_r' and $\beta_r + \theta'$. Thus, the 10th-order system of equations can be reduced to an equivalent 8th-order system of equations, which should bear exactly the same form as (7.19) except that the dimensions of $\boldsymbol{\delta}$ and $\mathbf{q}$ will be reduced to 4×1 and those of $\mathbf{K}_2$, $\mathbf{K}_1$, and $\mathbf{K}_0$ to 4×4 as,

$$\boldsymbol{\delta}(y,t) = \begin{Bmatrix} w_f(y,t) \\ \theta(y,t) \\ \beta_f(y,t) \\ \beta_r(y,t) \end{Bmatrix}, \quad \mathbf{q} = \begin{Bmatrix} \tilde{q} \\ \tilde{m}_x - \tilde{q}^* \\ \tilde{m}_y \\ \tilde{m}_y^* \end{Bmatrix}, \tag{7.21a}$$

and

$$\mathbf{K}_2 = \begin{bmatrix} \tilde{A}_{44} & -\tilde{A}_{44}^* & 0 & 0 \\ -\tilde{A}_{44}^* & \breve{D}_{66} + \tilde{A}_{44}^{**} & \breve{D}_{26} & \lambda\breve{D}_{26}^* \\ 0 & \breve{D}_{26} & \breve{D}_{22} & \lambda\breve{D}_{22}^* \\ 0 & \lambda\breve{D}_{26}^* & \lambda\breve{D}_{22}^* & \lambda\breve{D}_{22}^{**} \end{bmatrix}, \mathbf{K}_1 = \begin{bmatrix} 0 & 0 & -\tilde{A}_{44} & \tilde{A}_{44}^* \\ 0 & 0 & -\tilde{A}_{44}^* & \breve{D}_{66} - \tilde{A}_{44}^{**} \\ 0 & 0 & 0 & \breve{D}_{26} \\ 0 & 0 & 0 & \lambda\breve{D}_{26}^* \end{bmatrix}$$

$$\mathbf{K}_0 = \begin{bmatrix} 0 & 0 & 0 & 0 \\ 0 & 0 & 0 & 0 \\ 0 & 0 & \tilde{A}_{44} & \tilde{A}_{44}^* \\ 0 & 0 & \tilde{A}_{44}^* & \breve{D}_{66} + \tilde{A}_{44}^{**} \end{bmatrix}, \mathbf{I}_0 = \begin{bmatrix} m & -mx_c & 0 & 0 \\ -mx_c & I_y & 0 & 0 \\ 0 & 0 & 0 & 0 \\ 0 & 0 & 0 & 0 \end{bmatrix}. \tag{7.21b}$$

In the above, $\breve{D}_{ij}$, $\breve{D}_{ij}^*$, and $\breve{D}_{ij}^{**}$ are defined by

$$\breve{D}_{ij} = \tilde{D}_{ij} - \frac{\tilde{B}_{i2}\tilde{B}_{2j}}{\tilde{A}_{22}}, \quad \breve{D}_{ij}^* = \tilde{D}_{ij}^* - \frac{\tilde{B}_{i2}^*\tilde{B}_{2j}}{\tilde{A}_{22}}, \quad \breve{D}_{ij}^{**} = \tilde{D}_{ij}^{**} - \frac{\tilde{B}_{i2}^*\tilde{B}_{2j}^*}{\tilde{A}_{22}}, \quad i,j = 2,6. \tag{7.22}$$

7.1.4 Some Reductions

Because the displacement fields assumed in (7.3a) are quite general, several special conditions considered in the literature can be covered by the present formulations. For example, (i) neglect of the transverse shear deformation can be formulated by letting $\gamma_{yz} = 0$ which will lead to $\beta_f = -w_f'$, $\beta_r = \theta'$; (ii) neglect of the warping restraint effect can be formulated by letting $\varepsilon_y = v_0' + z\beta_f'$ which will lead to $\lambda = 0$; (iii) reduction to the conventional composite sandwich beams can be formulated by letting $\theta = \beta_r = 0$; and (iv) reduction to the conventional laminated beams can be formulated by letting $\theta = \beta_r = 0$, $\beta_f = -w_f'$, and $I_x = 0$. Following are the descriptions for these reductions.

Neglect of Transverse Shear Deformation ($\gamma_{yz} = 0 \Rightarrow \beta_f = -w_f', \beta_r = \theta'$)
Usually when the plate thickness is thin enough or the transverse shear modulus is large enough, the transverse shear deformation may be neglected, that is, $\gamma_{yz} = 0$, and the problem formulation may be further simplified. From (7.3b) we see that $\gamma_{yz} = 0$ will lead to $\beta_f = -w_f'$ and $\beta_r = \theta'$. With these two relations for the cases of absence of in-plane spanwise loads, the number of the independent functions will only remain two, that is, w_f and θ. The equations of motion corresponding to the deflection w_f

and the slope θ can then be obtained by substituting $(7.15b)_2$ into $(7.15a)_2$ and $(7.15b)_3$ into $(7.15b)_1$ with $\beta_f = -w'_f$ and $\beta_r = \theta'$. The results are, if the distributed moments $\tilde{m}_x, \tilde{m}_y$, and $\tilde{m}^*_y$ are neglected,

$$\begin{aligned} &\frac{\partial^2 \tilde{M}_y}{\partial y^2} + \tilde{q} = m\ddot{w}_f - mx_c\ddot{\theta}, \\ &2\frac{\partial \tilde{M}_{xy}}{\partial y} - \lambda \frac{\partial^2 \tilde{M}^*_y}{\partial y^2} = \tilde{q}^* - mx_c\ddot{w}_f + I_y\ddot{\theta}. \end{aligned} \tag{7.23}$$

Their associated boundary conditions become

$$\begin{aligned} &\tilde{M}'_y = \hat{\tilde{M}}'_y, \quad \text{or} \quad w_f = \hat{w}_f, \\ &2\tilde{M}_{xy} - \lambda \tilde{M}^{*\prime}_y = 2\hat{\tilde{M}}_{xy} - \lambda \hat{\tilde{M}}^{*\prime}_y, \quad \text{or} \quad \theta = \hat{\theta}, \\ &\tilde{M}_y = \hat{\tilde{M}}_y, \quad \text{or} \quad w'_f = \hat{w}'_f, \\ &\lambda \tilde{M}^*_y = \lambda \hat{\tilde{M}}^*_y, \quad \text{or} \quad \lambda\theta' = \lambda\hat{\theta}'. \end{aligned} \tag{7.24}$$

Employment of the constitutive relations $(7.18a)_2$ with $(7.21b)_{1,2}$ into (7.23) and (7.24) now gives us the governing equations and the boundary conditions as

$$\begin{aligned} &2\breve{D}_{26}\theta''' - [\breve{D}_{22}w''_f - \lambda \breve{D}^*_{22}\theta'']'' - m\ddot{w}_f + mx_c\ddot{\theta} + \tilde{q} = 0, \\ &4\breve{D}_{66}\theta'' - 2\breve{D}_{26}w'''_f + \lambda[\breve{D}^*_{22}w''_f - \lambda \breve{D}^{**}_{22}\theta'']'' + mx_c\ddot{w}_f - I_y\ddot{\theta} - \tilde{q}^* = 0, \end{aligned} \tag{7.25}$$

and

$$\begin{aligned} &2\breve{D}_{26}\theta'' - \breve{D}_{22}w'''_f + \lambda \breve{D}^*_{22}\theta''' = \hat{\tilde{M}}'_y, \quad \text{or} \quad w_f = \hat{w}_f, \\ &4\breve{D}_{66}\theta' - 2\breve{D}_{26}w''_f + \breve{D}^*_{22}\lambda w'''_f - \breve{D}^{**}_{22}\lambda\theta''' = 2\hat{\tilde{M}}_{xy} - \lambda \hat{\tilde{M}}^{*\prime}_y, \quad \text{or} \quad \theta = \hat{\theta}, \\ &2\breve{D}_{26}\theta' - \breve{D}_{22}w''_f + \breve{D}^*_{22}\lambda\theta'' = \hat{\tilde{M}}_y, \quad \text{or} \quad w'_f = \hat{w}'_f, \\ &\lambda[2\breve{D}^*_{26}\theta' - \breve{D}^*_{22}w''_f + \breve{D}^{**}_{22}\theta''] = \lambda \hat{\tilde{M}}^*_y, \quad \text{or} \quad \lambda\theta' = \lambda\hat{\theta}'. \end{aligned} \tag{7.26}$$

Neglect of Warping Restraint Effects ($\varepsilon_y = v'_0 + z\beta'_f \Rightarrow \lambda = 0$)

The importance of the warping restraint effects has been discussed in the series works of Librescu such as Librescu and Khdeir (1988). As mentioned earlier in Section 7.1.1, the inclusion of the warping restraint effect can be achieved by letting $\lambda = 1$. On the other hand, $\lambda = 0$ denotes the free warping condition. When $\lambda = 0$ is substituted into the governing equations (7.19) with absence of in-plane spanwise loads, it can be observed that β_r is no more an independent function but a dependent function of w_f, θ, and β_f. Thus, the system of governing equations can be further reduced from eighth order to sixth order.

Reduction to Conventional Composite Sandwich Beams ($\theta = \beta_r = 0$)

The difference of the present model with the conventional composite sandwich beams discussed in Chapter 6 is the consideration of the torsional angle θ and the rate of angle change β_r. If we neglect these two variations, the formulations should be exactly the same as those of conventional composite sandwich beams. Substituting $\theta = \beta_r = 0$ into the equations of motion $(7.15a)_2$ and $(7.15b)_2$ corresponding to the deflection w_f and rotation β_f, we get

$$\frac{\partial \tilde{Q}_y}{\partial y} + \tilde{q} = m\ddot{w}_f, \quad \frac{\partial \tilde{M}_y}{\partial y} = \tilde{Q}_y + I_x\ddot{\beta}_f. \tag{7.27}$$

Their associated boundary conditions are

$$\begin{aligned} &\tilde{Q}_y = \hat{\tilde{Q}}_y, \quad \text{or} \quad w_f = \hat{w}_f, \\ &\tilde{M}_y = \hat{\tilde{M}}_y, \quad \text{or} \quad \beta_f = \hat{\beta}_f. \end{aligned} \tag{7.28}$$

Use of the constitutive relations will then provide us the governing equations

$$\begin{aligned} &\tilde{A}_{44}\beta_f' + \tilde{A}_{44}w_f'' - m\ddot{w}_f + \tilde{q} = 0, \\ &\tilde{A}_{44}\beta_f + \tilde{A}_{44}w_f' - \breve{D}_{22}\beta_f'' + I_x\ddot{\beta}_f = 0, \end{aligned} \tag{7.29}$$

and the boundary conditions

$$\begin{aligned} &\tilde{A}_{44}(\beta_f + w_f') = \hat{\tilde{Q}}_y, \quad \text{or} \quad w_f = \hat{w}_f, \\ &\breve{D}_{22}\beta_f' = \hat{\tilde{M}}_y, \quad \text{or} \quad \beta_f = \hat{\beta}_f. \end{aligned} \tag{7.30}$$

Reduction to Laminated Composite Beams ($\theta = \beta_r = 0$, $\beta_f = -w_f'$, $I_x = 0$)

When the thickness of the laminated composite beam is considered to be much smaller relative to its length, the transverse shear deformation and the rotary inertia are usually ignored, that is, $\gamma_{yz} = 0$ and $I_x = 0$, in which the former leads to $\beta_f = -w_f'$. Substituting $\beta_f = -w_f'$ and $I_x = 0$ into

(7.27)–(7.30) now provides us the equation of motion and its associated boundary conditions as

$$\frac{\partial^2 \tilde{M}_y}{\partial y^2} + \tilde{q} = m\ddot{w}_f \Rightarrow [\breve{D}_{22} w_f'']'' + m\ddot{w}_f - \tilde{q} = 0, \tag{7.31}$$

and

$$\begin{aligned} \tilde{M}_y' &= -\breve{D}_{22} w_f''' = \hat{\tilde{M}}_y', \quad \text{or} \quad w_f = \hat{w}_f, \\ \tilde{M}_y &= -\breve{D}_{22} w_f'' = \hat{\tilde{M}}_y, \quad \text{or} \quad w_f' = \hat{w}_f'. \end{aligned} \tag{7.32}$$

7.2 WINGS WITH UNIFORM CROSS-SECTION

7.2.1 Free Vibration

To know the natural frequency and its associated vibration mode of the stiffened composite wing structures, we consider the case that the external forces $\tilde{q}$ and $\tilde{q}_y$, torsional moment $\tilde{q}^*$ as well as the distributed moments $\tilde{m}_x, \tilde{m}_y, \tilde{m}_y^*$ are all zero, that is, the force vector $\mathbf{q} = \mathbf{0}$ in (7.18b)$_3$. To find the natural modes of vibration, the usual way is the method of separation of variables. By this method we write the displacement vector $\boldsymbol{\delta}(y, t)$ as a product of a function $\boldsymbol{\Delta}(y)$ of the spatial variables only and a function $f(t)$ depending on time only. Furthermore, because of free vibration the function $f(t)$ is harmonic and of frequency ω. Thus,

$$\boldsymbol{\delta}(y,t) = \boldsymbol{\Delta}(y) e^{i\omega t}, \tag{7.33a}$$

where

$$\boldsymbol{\Delta}(y) = \begin{Bmatrix} V_0(y) \\ W_f(y) \\ \Theta(y) \\ B_f(y) \\ B_r(y) \end{Bmatrix}. \tag{7.33b}$$

Through the use of (7.33), the equation of motion (7.19) can be reduced to a system of ordinary differential equations

$$\mathbf{K}_2 \boldsymbol{\Delta}''(y) + (\mathbf{K}_1 - \mathbf{K}_1^T)\boldsymbol{\Delta}'(y) - \mathbf{K}_0 \boldsymbol{\Delta}(y) + \omega^2 \mathbf{I}_0 \boldsymbol{\Delta}(y) = \mathbf{0}, \tag{7.34}$$

which can be solved by letting

$$\Delta(y) = \mathbf{d}e^{ry}. \tag{7.35}$$

Substituting (7.35) into (7.34), we get

$$\left\{\mathbf{K}_2 r^2 + \left(\mathbf{K}_1 - \mathbf{K}_1^T\right) r - \mathbf{K}_0 + \omega^2 \mathbf{I}_0\right\} \mathbf{d} = \mathbf{0}, \tag{7.36}$$

whose non-vanishing solutions exist only when the determinant of the coefficient matrix of **d** becomes zero, that is,

$$\left\| \mathbf{K}_2 r^2 + \left(\mathbf{K}_1 - \mathbf{K}_1^T\right) r - \mathbf{K}_0 + \omega^2 \mathbf{I}_0 \right\| = 0. \tag{7.37}$$

As shown in (7.18c), the coefficient matrix of **d** is a 5×5 matrix for the general cases. Thus, (7.37) is a 10th-order polynomial equation which will have 10 roots $r_i(\omega)$, $i = 1, 2, \ldots, 10$. Each of the roots has an associated eigenvector $\mathbf{d}_i(\omega)$ determined from (7.36). Linear superposition of these 10 homogeneous solutions now gives us

$$\Delta(y) = \sum_{i=1}^{10} k_i \mathbf{d}_i e^{r_i y}. \tag{7.38}$$

With (7.38) and (7.33), the boundary conditions (7.20) will then set a system of 10 simultaneous linear algebraic equations with 10 unknown coefficients k_i as

$$\begin{bmatrix} \mathbf{K}_1 \mathbf{D}\left\langle e^{r_i L}\right\rangle + \mathbf{K}_2 \mathbf{D}\left\langle r_i e^{r_i L}\right\rangle \\ \mathbf{D} \end{bmatrix} \mathbf{k} = \mathbf{0}, \tag{7.39a}$$

where

$$\mathbf{D} = \left[\mathbf{d}_1\ \mathbf{d}_2 \ldots\ldots \mathbf{d}_{10}\right], \quad \mathbf{k} = \begin{Bmatrix} k_1 \\ k_2 \\ \vdots \\ k_{10} \end{Bmatrix}, \tag{7.39b}$$

and the angular bracket <> stands for a diagonal matrix in which each component is varied according to its subscript i, for example, $<r_i e^{r_i L}> = \text{diag}\,[r_1 e^{r_1 L}, r_2 e^{r_2 L}, \ldots\ldots, r_{10} e^{r_{10} L}]$.

Because both of the eigenvalues r_i and the eigenvectors $\mathbf{d}_i$ are functions of the natural frequency ω, the coefficient matrix of $\mathbf{k}$ in (7.39a) is a function of the natural frequency ω. Again, non-vanishing solutions exist only when the determinant of the coefficient matrix of $\mathbf{k}$ becomes zero, by which we can then obtain the natural frequencies of the stiffened composite wing structures. With the determined natural frequency ω, the coefficients k_i can be calculated from (7.39a) as the eigenvector, and hence the natural vibration mode shapes of the composite wing structures are obtained from (7.38).

Orthogonality Condition

If the family of natural vibration mode shapes $\mathbf{\Delta}_j(y)$ can constitute a complete set of orthonormal modes, most of the vibration problems can be solved by modal analysis through the use of the expansion theorem (Meirovitch, 1990). Let ω_i and ω_j be the two distinct natural frequencies and $\mathbf{\Delta}_i(y)$ and $\mathbf{\Delta}_j(y)$ be the corresponding natural modes of vibration resulting from the solution of the equations of motion (7.34) and its associated boundary conditions (7.20). Consider equation (7.34) corresponding to ω_i and $\mathbf{\Delta}_i(y)$, and multiplied by $\mathbf{\Delta}_j^T(y)$, and another equation (7.34) corresponding to ω_j and $\mathbf{\Delta}_j(y)$, and multiplied by $\mathbf{\Delta}_i^T(y)$. Subtracting these two equations and integrating the results over the wing spanwise length L, we obtain

$$
\begin{aligned}
\left(\omega_i^2 - \omega_j^2\right)\int_0^L \mathbf{\Delta}_j^T \mathbf{I}_0 \mathbf{\Delta}_i dy = \int_0^L (&-\mathbf{\Delta}_j^T \mathbf{K}_0 \mathbf{\Delta}_i + \mathbf{\Delta}_i^T \mathbf{K}_0 \mathbf{\Delta}_j \\
&+ \mathbf{\Delta}_j^T\left(\mathbf{K}_1 - \mathbf{K}_1^T\right)\mathbf{\Delta}_i' - \mathbf{\Delta}_i^T\left(\mathbf{K}_1 - \mathbf{K}_1^T\right)\mathbf{\Delta}_j' \\
&+ \mathbf{\Delta}_j^T \mathbf{K}_2 \mathbf{\Delta}_i'' - \mathbf{\Delta}_i^T \mathbf{K}_2 \mathbf{\Delta}_j'')dy
\end{aligned}
\tag{7.40}
$$

From (7.18c), we see that $\mathbf{K}_2$, $\mathbf{K}_0$, and $\mathbf{I}_0$ are symmetric matrices. Knowing that the transpose of a symmetric matrix (including a scalar) is identical to the matrix itself, we have

$$
\mathbf{K}_2 = \mathbf{K}_2^T,\quad \mathbf{K}_0 = \mathbf{K}_0^T,\quad \mathbf{I}_0 = \mathbf{I}_0^T,\quad \mathbf{\Delta}_j^T \mathbf{K}_0 \mathbf{\Delta}_i = (\mathbf{\Delta}_j^T \mathbf{K}_0 \mathbf{\Delta}_i)^T, \ldots\ldots, \text{etc.,} \tag{7.41}
$$

With the relation (7.41), equation (7.40) can be further reduced to

$$
\begin{aligned}
&(\omega_i^2 - \omega_j^2)\int_0^L \Delta_j^T \mathbf{I}_0 \Delta_i dy \\
&= \int_0^L \left\{[\Delta_j^T(\mathbf{K}_1 - \mathbf{K}_1^T)\Delta_i]' + \Delta_j^T \mathbf{K}_2 \Delta_i'' - \Delta_i^T \mathbf{K}_2 \Delta_j''\right\} dy \\
&= \Delta_j^T(\mathbf{K}_1 - \mathbf{K}_1^T)\Delta_i \Big|_0^L + \int_0^L \left\{(\Delta_j^T \mathbf{K}_2 \Delta_i')' - \Delta_j'^T \mathbf{K}_2 \Delta_i' - (\Delta_i^T \mathbf{K}_2 \Delta_j')' + \Delta_i'^T \mathbf{K}_2 \Delta_j'\right\} dy \\
&= \left\{\Delta_j^T(\mathbf{K}_1 \Delta_i + \mathbf{K}_2 \Delta_i') - \Delta_i^T(\mathbf{K}_1 \Delta_j + \mathbf{K}_2 \Delta_j')\right\}\Big|_0^L \\
&= (\Delta_j^T \mathbf{F}_i - \Delta_i^T \mathbf{F}_j)\Big|_0^L,
\end{aligned} \tag{7.42}
$$

in which the last equality comes from $(7.18\text{a})_2$ and (7.33). Because the boundary conditions are either displacement-prescribed or force-prescribed, we have

$$
\begin{aligned}
\mathbf{F}(0) = \mathbf{0}, \quad &\text{or} \quad \Delta(0) = \mathbf{0}, \\
\mathbf{F}(L) = \mathbf{0}, \quad &\text{or} \quad \Delta(L) = \mathbf{0}.
\end{aligned} \tag{7.43}
$$

Substituting (7.43) into (7.42) for both of the ith and jth modes, we get

$$
\left(\omega_i^2 - \omega_j^2\right)\int_0^L \Delta_j^T \mathbf{I}_0 \Delta_i dy = 0, \tag{7.44a}
$$

or

$$
\begin{aligned}
\int_0^L \Delta_j^T \mathbf{I}_0 \Delta_i dy &= 0, \quad \text{when } i \neq j, \\
&\neq 0, \quad \text{when } i = j.
\end{aligned} \tag{7.44b}
$$

Through the normalization, (7.44b) can be combined into

$$
\int_0^L \Delta_j^T \mathbf{I}_0 \Delta_i dy = \delta_{ij}, \tag{7.45}
$$

where δ_{ij} is the Kronecker delta. Expansion of (7.45) leads to

$$\int_0^L [mV_{0_i}V_{0_j} + mW_{f_i}W_{f_j} + I_y\Theta_i\Theta_j + I_xB_{f_i}B_{f_j} + I_{x^2z^2}B_{r_i}B_{r_j}$$
$$-mx_c\left(\Theta_iW_{f_j} + W_{f_i}\Theta_j\right) + mz_c\left(B_{f_i}V_{0_j} + V_{0_i}B_{f_j}\right)$$
$$+I_{xz}\left(B_{r_i}V_{0_j} + V_{0_i}B_{r_j}\right) + I_{xz^2}\left(B_{r_i}B_{f_j} + B_{f_i}B_{r_j}\right)]dy = \delta_{ij}. \quad (7.46)$$

Unlike the usual orthogonality conditions that only the lateral deflection is considered, the orthogonality found in (7.46) shows that the complete set includes not only the mode shapes of the deflection but also the mode shapes of all the other basic functions.

EXAMPLE: A NACA 2412 COMPOSITE WING

As discussed in Section 7.1.4, the present wing model can be reduced to several special cases. Therefore, before performing the wing structural analysis, simple verification can be done for these specialized conditions (Hwu and Gai, 2003). After verifying the results by using the flat composite beam model, the analysis for the composite wing structure with NACA 2412 airfoil is now illustrated below. From the data given in Hunsaker (1949), the shape of airfoil can be approximated by a ninth-order polynomial as

$$f_u(\overline{x})/c = 0.0719 - 0.0588\overline{x} - 0.0769\overline{x}^2 - 0.767\overline{x}^3 - 2.599\overline{x}^4 + 15.382\overline{x}^5 + 18.536\overline{x}^6 - 92.174\overline{x}^7 - 45.673\overline{x}^8 + 186.078\overline{x}^9,$$
$$f_u(\overline{x})/c = -0.0329 + 0.0405\overline{x} - 0.0239\overline{x}^2 + 0.675\overline{x}^3 + 1.815\overline{x}^4 - 12.310\overline{x}^5 - 13.870\overline{x}^6 + 73.449\overline{x}^7 + 36.096\overline{x}^8 - 149.862\overline{x}^9, \qquad \overline{x} = x/c,$$

where $f_u(\overline{x})$ and $f_l(\overline{x})$ are, respectively, the approximate functions for the upper and lower surfaces of the airfoil in which the coordinate origin is located at the mid-point of the chord line. The wing chordwise length c = 0.1 m and spanwise length L = 0.4 m. The wing skin is made of graphite/epoxy fiber-reinforced composite whose mechanical properties are: E_L = 200 GPa, E_T = 5GPa, ν_{LT} = 0.25, G_{LT} = 2.5 GPa, ρ = 1.9 g/cm³ and ply thickness t = 0.025 mm. The laminate lay-up is [90/ − 45/45/0] for upper skin and [0/45/ − 45/90] for lower skin. Two wing spars made of isotropic materials with shear modulus G = 8 GPa and thickness 0.6 mm are located at ±0.25c from the mid-chord line. Eight stringers and two ribs, made of Aluminum with material properties E = 69 GPa, ν = 0.33, G = 26 GPa, ρ = 2.8 g/cm³, are equally spaced on the wing.

Through the numerical results presented in Hwu and Gai (2003), we know that the assumptions made for the case of "absence of in-plane spanwise loads" are reasonable and acceptable. The results also show a physically reasonable tendency that the larger the aspect ratio is, the smaller the natural frequency becomes. Based upon this equivalent reduced model, the mode shapes of NACA 2412 composite wing can be found through (7.36)–(7.39). Some influential factors have been studied and their numerical results show that (1) the larger the transverse shear modulus the higher the natural frequency, and after a certain value of G_{yz} the natural frequency will approach to a constant value which is exactly the one obtained by neglecting the transverse shear deformation; (2) the more slender the wing the more influential the warping restraint effects; (3) the approximation of the airfoil shape within a tolerant error is important for obtaining the accurate natural frequencies.

7.2.2 Forced Vibration

After finding the orthogonality relation (7.46), the expansion theorem can be used to obtain the system response by modal analysis. Using the expansion theorem we write the solution of (7.19) as a superposition of the natural modes $\mathbf{\Delta}(y)$ multiplying corresponding time-dependent generalized coordinates $\eta_j(t)$, and hence,

$$\boldsymbol{\delta}(y,t)=\sum_{j=1}^{\infty}\mathbf{\Delta}_j(y)\eta_j(t). \tag{7.47}$$

Substituting (7.47) into (7.19), we obtain

$$\sum_{j=1}^{\infty}\left\{[\mathbf{K}_2\mathbf{\Delta}_j''(y)+\left(\mathbf{K}_1-\mathbf{K}_1^T\right)\mathbf{\Delta}_j'(y)-\mathbf{K}_0\mathbf{\Delta}_j(y)]\eta_j(t)\right\}+\mathbf{q}(y,t)$$
$$=\mathbf{I}_0\sum_{j=1}^{\infty}\mathbf{\Delta}_j(y)\ddot{\eta}_j(t). \tag{7.48}$$

Employing the results of (7.34) in (7.48), we get

$$\sum_{j=1}^{\infty}\left\{-\omega_j^2\mathbf{I}_0\mathbf{\Delta}_j(y)\eta_j(t)\right\}+\mathbf{q}(y,t)=\mathbf{I}_0\sum_{j=1}^{\infty}\mathbf{\Delta}_j(y)\ddot{\eta}_j(t). \tag{7.49}$$

Multiplying both sides of (7.49) by $\mathbf{\Delta}_i^T(y)$, integrating over the spanwise length L, and utilizing the orthogonality relation (7.45), we obtain an infinite set of *uncoupled* second-order ordinary differential equation system as

$$\ddot{\eta}_j(t)+\omega_j^2\eta_j(t)=N_j(t),\quad j=1,2,\ldots\ldots, \tag{7.50a}$$

where $N_j(t)$ denotes a generalized force associated with the generalized coordinate $\eta_j(t)$ and is related to the load vector $\mathbf{q}$ by

$$N_j(t)=\int_0^L \mathbf{\Delta}_j^T(y)\mathbf{q}(y,t)dy. \tag{7.50b}$$

To solve the generalized coordinate $\eta_j(t)$ through the uncoupled second-order ordinary differential equation (7.50), we need to know the initial value of $\eta_j(t)$ and $\dot{\eta}_j(t)$. From the relation (7.47) and the orthogonality condition (7.45), we have

$$\eta_j(0)=\int_0^L \mathbf{\Delta}_j^T(y)\mathbf{I}_0\boldsymbol{\delta}(y,0)dy,\ \dot{\eta}_j(0)=\int_0^L \mathbf{\Delta}_j^T(y)\mathbf{I}_0\dot{\boldsymbol{\delta}}(y,0)dy \tag{7.51}$$

To illustrate the capability of the present wing model for the forced vibration analysis, we now focus our attention to the uncoupled second-order differential equation system (7.50). Because this equation system is exactly the same as that of the simple beam theory shown in (6.111) for composite sandwich beams, the same approach for the forced vibration analysis can be applied to the present model. A typical and useful example of forced vibration analysis is vibration suppression. In Section 6.2.5, an LQG/LTR controller is designed for the composite sandwich beams by bonding piezoelectric sensors and actuators on the beam surfaces. Because the governing equation system (7.50) for the generalized coordinate $\eta_j(t)$ is exactly the same as that of the composite sandwich beams, the sensor equation, actuator equation, and the dynamics of the observed-state feedback control system shown in Section 6.2.5 can all be applied here for the vibration suppression of composite wing structures. Detailed numerical simulation results can be found in Hwu and Gai (2003).

7.2.3 Aeroelastic Divergence

The fundamental work concerning the divergence instability of swept metallic wings was done about 70 years ago (Diederich and Budiansky, 1948). It was shown that bending deflections have a destabilizing effect on the swept-forward wings. Hence the swept-forward wing aircraft as a possible option was completely eliminated for a long time, until the aeroelastic tailoring concept for the composite wing structures was raised and studied by Krone (1975). Following his work, many different approaches and considerations have been studied, such as the works done by Lerner and Markowitz (1979), Weisshaar (1980), Oyibo (1984), Lottati (1985), and a series works done by Librescu and his co-workers (1988–1994). They first considered the warping restraint effects (Librescu and Simovich, 1988; Librescu and Khdeir, 1988; Librescu and Thangjitham, 1991), then considered the effects of transverse shear strains (Karpouzian and Librescu, 1994). All these results support that a composite swept-forward wing can be tailored to overcome this adverse instability phenomenon.

To study the aeroelastic divergence phenomena, the spanwise loads as well as the distributed moments are neglected, that is, $\tilde{q}_y = \hat{\tilde{N}}_y = 0$ and $\tilde{m}_x = \tilde{m}_y = \tilde{m}_y^* = 0$. The applied forces $\tilde{q}$ and $-\tilde{q}^*$ are considered to be the lift and the aerodynamic nose-up torsional moment (per unit length), respectively. Here, we use the aerodynamic strip theory (Bisplinghoff et al., 1955) to approximate the lift and moment. In this approximation, one employs the known results for two-dimensional flow (infinite span airfoil) to calculate the aerodynamic forces on a lifting surface of finite span. Although this approximation may not be good for low aspect ratio wings, it is generally accepted to have a preliminary study by using this approximation theory. By replacing the expressions for the lift and torsional moment with more accurate ones, and following the same steps described in this section, one may get updated results for the aeroelastic divergence problems. With this understanding, we now express $\tilde{q}(y)$ and $-\tilde{q}^*(y)$ for the static case considered here, as (Oyibo, 1984)

$$\begin{aligned} \tilde{q}(y) &= a q_n c \theta_{eff}(y), \\ -\tilde{q}^*(y) &= a q_n c e \theta_{eff}(y) + q_n c^2 C_{m,ac} \end{aligned} \tag{7.52a}$$

where

$$\theta_{eff}(y) = \theta_0 + \theta(y) - w_f'(y) \tan\Lambda \tag{7.52b}$$

a is the lift curve slope coefficient; q_n denotes the dynamic pressure component normal to the leading edge; c is the wing chord length and e is the distance between the lines of aerodynamic and flexural centers, both of which are considered to be constant; θ_0 denotes the angle of attack corresponding to the rigid wing assumption; Λ is the angle of sweep (positive for swept-back and negative for swept-forward); $C_{m,\,ac}$ is the pitching moment coefficient about the aerodynamic center, which will be constant with respect to the angle of attack. The related equations about q_n and a are

$$q_n = \frac{1}{2}\rho_0 V_n^2 = q_0 \cos^2\Lambda, \quad a = \frac{dC_L}{d\theta} = a_0 \frac{AR}{AR + 4\cos\Lambda}, \tag{7.52c}$$

where ρ_0 is the density of the airflow; V_n is the airflow velocity component normal to the leading edge; q_0 is the dynamic pressure; C_L is the lift coefficient; a_0 is the corresponding two-dimensional lift-curve slope; AR is the wing aspect ratio defined as $AR \equiv (2\ell)^2/S = 2\ell/c$ where 2ℓ is the wingspan measured as the distance between the two wingtips and $S(=2c\ell$ in the case that c is independent of $y)$ is the total area of the wing in the planform (x-y plane) view.

From (7.52b), a well-known fact can be seen that the bending deformation tends to reduce the effective angle of attack for swept-back wings ($\Lambda > 0$), while in the case of swept-forward wings ($\Lambda < 0$) the opposite effect is valid. In other words, the bending deflections have a destabilizing effect on the swept-forward wings. That is why most of the studies devote their efforts to the aeroelastic tailoring of composite aircraft swept-forward wings.

With the lift and moment given in (7.52), the governing equation and boundary conditions shown in (7.19) and (7.20) for the static condition can now be written as

$$\mathbf{K}_2\boldsymbol{\delta}''(y) + \left(\mathbf{K}_1 - \mathbf{K}_1^T\right)\boldsymbol{\delta}'(y) - \mathbf{K}_0\boldsymbol{\delta}(y) + \mathbf{q}(y) = \mathbf{0}, \tag{7.53a}$$

$$\boldsymbol{\delta}(0) = \mathbf{0}, \text{ and } \mathbf{K}_1\boldsymbol{\delta}(L) + \mathbf{K}_2\boldsymbol{\delta}'(L) = \mathbf{0}. \tag{7.53b}$$

where

$$\mathbf{q}(y) = q_n c \begin{Bmatrix} a\{\theta_0 + \theta(y) - w_f'(y)\tan\Lambda\} \\ ae\{\theta_0 + \theta(y) - w_f'(y)\tan\Lambda\} + cC_{m,ac} \\ 0 \\ 0 \end{Bmatrix}. \tag{7.53c}$$

With the boundary condition and the load vector given in (7.53b)$_1$ and (7.53c), the Laplace transform of (7.53a) leads to

$$\left\{s^2\mathbf{K}_2 + s\left(\mathbf{K}_1 - \mathbf{K}_1^T\right) - \mathbf{K}_0 - \boldsymbol{\Gamma}\right\}\breve{\boldsymbol{\delta}}(s) = \mathbf{K}_2\boldsymbol{\delta}'(0) - \frac{1}{s}\mathbf{q}_0, \tag{7.54a}$$

where

$$\mathbf{q}_0 = q_n c\begin{Bmatrix} a\theta_0 \\ ae\theta_0 + cC_{m,ac} \\ 0 \\ 0 \end{Bmatrix}, \quad \boldsymbol{\Gamma} = aq_n c\begin{bmatrix} s\tan\Lambda & -1 & 0 & 0 \\ se\tan\Lambda & -e & 0 & 0 \\ 0 & 0 & 0 & 0 \\ 0 & 0 & 0 & 0 \end{bmatrix},$$

$$\breve{\boldsymbol{\delta}}(s) = \int_0^\infty \boldsymbol{\delta}(y)e^{-ey}dy. \tag{7.54b}$$

Let

$$\breve{\mathbf{K}}_s(s) = \left\{s^2\mathbf{K}_2 + s\left(\mathbf{K}_1 - \mathbf{K}_1^T\right) - \mathbf{K}_0 - \boldsymbol{\Gamma}\right\}^{-1}, \tag{7.55}$$

equation (7.54a) now gives us

$$\breve{\boldsymbol{\delta}}(s) = \breve{\mathbf{K}}_s(s)\left[\mathbf{K}_2\boldsymbol{\delta}'(0) - \frac{1}{s}\mathbf{q}_0\right]. \tag{7.56}$$

In the above, $\breve{\mathbf{K}}_s(s)$ is a four-by-four matrix, and its inversion can be manipulated with the assistance of symbolic computational software like Mathematica or MATLAB. Implementing the inverse Laplace transform for (7.56), we get

$$\boldsymbol{\delta}(y) = \mathbf{K}_s(y)\mathbf{K}_2\boldsymbol{\delta}'(0) - \left[\int_0^y \mathbf{K}_s(y)dy\right]\mathbf{q}_0, \tag{7.57}$$

where $\mathbf{K}_s(y)$ is the inverse Laplace transform of $\breve{\mathbf{K}}_s(s)$. The unknown vector $\boldsymbol{\delta}'$ (0) can then be determined by substituting (7.57) into the boundary condition (7.53b)$_2$, which leads to

$$\left[\mathbf{K}_1\mathbf{K}_s(L)\mathbf{K}_2+\mathbf{K}_2\mathbf{K}'_s(L)\mathbf{K}_2\right]\boldsymbol{\delta}'(0)=\left[\mathbf{K}_1\left(\int_0^L\mathbf{K}_s(y)dy\right)+\mathbf{K}_2\mathbf{K}_s(L)\right]\mathbf{q}_0. \tag{7.58}$$

Equation (7.58) can address the problems of the static subcritical aeroelastic response, and also that of the divergence instability. For the former problem, in the subcritical flight speed range, $\boldsymbol{\delta}'(0)$ can be uniquely determined from the non-homogeneous equation (7.58). The four basic functions $\boldsymbol{\delta}(y) = (w_f, \theta, \beta_f, \beta_r)$ are then determined by (7.57). The resultant forces and bending moments $\mathbf{F}(y)=\left(\tilde{Q}_y,\tilde{M}_{xy}-\tilde{Q}_y^*,\tilde{M}_y,\tilde{M}_y^*\right)$ can also be found by the second equation of (7.18a). On the other hand, when the flight speed reaches a certain value the determinant of the coefficient matrix of $\boldsymbol{\delta}'(0)$ in (7.58) may become zero, that is,

$$\left\|\mathbf{K}_1\mathbf{K}_s(L)\mathbf{K}_2+\mathbf{K}_2\mathbf{K}'_s(L)\mathbf{K}_2\right\|=0, \tag{7.59}$$

which will lead to the results that $\boldsymbol{\delta}'(0)$ is infinite. In other words, under a certain condition, the deflection of the wing structures may become infinite, which is the situation of aeroelastic divergence. The lowest value of q_0 for which the determinant vanishes corresponds to the critical (divergence) dynamic pressure q_d, and its associated velocity V_d is called *divergence speed*. The determination of divergence speed from the vanishing of the determinant shown in (7.59) can also be explained via the use of the eigenvalue concept. That is, for a homogeneous equation (7.58) (its right hand side is taken to be zero), nontrivial solutions $\boldsymbol{\delta}'(0)$ exist only when (7.59) is fulfilled. Under this condition, non-unique values of deflections exist, which is the condition of instability. The variation of the determinant with respect to the flow speed can be seen from the following relations: V_n is related to q_n through $(7.52c)_1$; q_n will influence the determinant through $\boldsymbol{\Gamma}$, (7.54b), which in turn is related to $\hat{\mathbf{K}}_s(s)$ through (7.55).

From equation (7.59) and the relation (7.55) we see that the divergence speed will be influenced by $\mathbf{K}_0$, $\mathbf{K}_1$, $\mathbf{K}_2$, and $\boldsymbol{\Gamma}$. By (7.21b), it is observed that $\mathbf{K}_0$, $\mathbf{K}_1$, and $\mathbf{K}_2$ contain the information of bending stiffness $\breve{D}_{ij}$ and shearing stiffness $\tilde{A}_{44}$. From the definitions given in (6.10), (7.4), (7.5), (7.7), and (7.22), we know that $\breve{D}_{ij}$ will be influenced by the properties of the composite wing skin, stringers, spar flanges, and the shape of airfoil. From (6.13) and (7.8), we see that $\tilde{A}_{44}$ will be influenced by the properties of the spar webs and the shape of airfoil. As to the matrix $\mathbf{\Gamma}$ defined in (7.54b), it contains the information related to the aerodynamic characteristics such as c, e, a, Λ, and q_n defined in the paragraph between equations (7.52) and (7.53). All these factors can be studied numerically through (7.59) to see their influence on the divergence speed. Detailed numerical results and discussions can be found in Hwu and Tsai (2002), in which a composite wing with NACA 2412 airfoil is illustrated and the effects of spars, skins (including the ply orientation and stacking sequence), stringers, swept angles, aspect ratio, shape of airfoil, and the warping restraints on the divergence dynamic pressures and the lift loads redistribution are all studied. Their results show that all these effects preserve the same trend as those of the flat composite wing structure assumptions except that their values will be influenced by the shape of airfoil. Moreover, the adding of stringers and spars will increase the divergence dynamic pressure and lower the effective angle of attack by an amount lower than the stiffness increment.

7.3 TAPERED WINGS

Although (7.19) is a result of a comprehensive wing model which considers several effects such as bending-torsion coupling, warping restraint, transverse shear deformation, shape of airfoil, rotary inertia, etc., it cannot be applied to the tapered wings because in this model the wing cross-section is considered to be independent of y (spanwise direction). In other words, in (7.19) the stiffness and inertia matrices, $\mathbf{K}_0$, $\mathbf{K}_1$, $\mathbf{K}_2$, and $\mathbf{I}_0$ are all constant matrices, which is not true for tapered wings. In order to extend the comprehensive wing model to the tapered wings or more general wing cases, a concept like finite element model is adopted in our previous study (Hwu and Yu, 2010) and will be presented in this section. That is, the tapered wing is cut into several elements and each element is approximated by a uniform wing section. Since the finite element formulation can be established by minimizing the *Lagrangian function* (Reddy, 1993), to have a smooth transition from the comprehensive wing model to comprehensive finite element wing model, we now re-derive (7.19) through the use of *Hamilton's principle*. It states that the motion of a continuum acted on by conservative forces between two arbitrary instants of time t_1 and t_2 is such that the line integral over the Lagrangian function is an extremum for the path motion. The Lagrangian function is the difference between kinetic and total potential energies, and the total potential energy is the sum of the strain energy and potential energy of external forces. Thus, the Hamilton's principle can be expressed by

$$\delta\int_{t_1}^{t_2}(\Pi - T)dt = 0, \tag{7.60}$$

where δ is the variational operator, t_1 and t_2 are the integration limits of time. Π is the total potential energy and T is the kinetic energy, which can be written as

$$\Pi = \int_V (W - f_i u_i)dV - \int_{S_\sigma} \hat{t}_i u_i dS, \quad T = \frac{1}{2}\int_V \rho \dot{u}_i \dot{u}_i dV, \tag{7.61}$$

where W, f_i, u_i, and $\hat{t}_i$ are, respectively, the strain energy density, body forces, displacements, and prescribed surface tractions; V and S_σ are the regions for volume and surface integrals, respectively. If the integral is performed in the sequence of thickness direction (z), chordwise direction (x), and spanwise direction (y), each term of the potential and kinetic energy shown in (7.61) can be integrated as follows:

$$\begin{aligned}
\int_V W dV &= \frac{1}{2}\int_V \sigma_{ij}\varepsilon_{ij} dV \\
&= \frac{1}{2}\int_A \{N_x\varepsilon_{x_0} + N_y\varepsilon_{y_0} + N_{xy}\gamma_{xy_0} + M_x\kappa_x \\
&\quad + M_y\kappa_y + M_{xy}\kappa_{xy} + Q_x\gamma_{xz} + Q_y\gamma_{yz}\}dxdy \\
&= \frac{1}{2}\int_A \{N_y(v_0') + M_y(\beta_f' + x\beta_r') \\
&\quad + M_{xy}(\beta_r + \theta') + Q_y[\beta_f + w_f' + x(\beta_r - \theta')]\}dxdy \\
&= \frac{1}{2}\int_y \{\tilde{N}_y v_0' + \tilde{M}_y\beta_f' + \tilde{M}_y^*\beta_r' \\
&\quad + \tilde{M}_{xy}(\beta_r + \theta') + \tilde{Q}_y(\beta_f + w_f') + \tilde{Q}_y^*(\beta_r - \theta')\}dy \\
&= \frac{1}{2}\int_y \{\mathbf{F}^T\mathbf{\Delta}' + \mathbf{F}_0^T\mathbf{\Delta}\}dy
\end{aligned} \tag{7.62a}$$

$$\int_V f_i u_i dV = \int_A \{q_x u + q_y v + qw\} dxdy$$
$$= \int_A \{q_x (z\theta) + q_y [v_0 + z(\beta_f + x\beta_r)] + q(w_f - x\theta)\} dxdy$$
$$= \int_A \{m_x \theta + q_y v_0 + m_y (\beta_f + x\beta_r) + q(w_f - x\theta)\} dxdy$$
$$= \int_y \{\tilde{m}_x \theta + \tilde{q}_y v_0 + \tilde{m}_y \beta_f + \tilde{m}_y^* \beta_r + \tilde{q} w_f - \tilde{q}^* \theta\} dy$$
$$= \int_y \{\mathbf{q}^T \boldsymbol{\Delta}\} dy \tag{7.62b}$$

$$\int_{S_\sigma} \hat{t}_i u_i dS = \left[\int_x \{\hat{M}_{xy}\theta + \hat{N}_y v_0 + \hat{M}_y(\beta_f + x\beta_r) + \hat{Q}_y(w_f - x\theta)\} dx \right]_{y_r}^{y_t}$$
$$= \left[\hat{\tilde{M}}_{xy}\theta + \hat{\tilde{N}}_y v_0 + \hat{\tilde{M}}_y \beta_f + \hat{\tilde{M}}_y^* \beta_r + \hat{\tilde{Q}}_y w_f - \hat{\tilde{Q}}_y^* \theta \right]_{y_r}^{y_t}$$
$$= [\hat{\mathbf{F}}^T \boldsymbol{\Delta}]_{y_r}^{y_t} \tag{7.62c}$$

$$\frac{1}{2}\int_V \rho \dot{u}_i \dot{u}_i dV = \frac{1}{2}\int_V \rho \{\dot{u}^2 + \dot{v}^2 + \dot{w}^2\} dV$$
$$= \frac{1}{2}\int_V \rho \{(z\dot{\theta})^2 + (\dot{v}_0 + z\dot{\beta}_f + zx\dot{\beta}_r)^2 + (\dot{w}_f - x\dot{\theta})^2\} dzdxdy$$
$$= \frac{1}{2}\int_y \{I_x(\dot{\theta}^2 + \dot{\beta}_f^2) + I_z\dot{\theta}^2 + 2I_{xz}\dot{v}_0\dot{\beta}_r + 2I_{xz^2}\dot{\beta}_f\dot{\beta}_r + I_{x^2z^2}\dot{\beta}_r^2$$
$$+ m(\dot{v}_0^2 + \dot{w}_f^2) - 2mx_c\dot{w}_f\dot{\theta} + 2mz_c\dot{v}_0\dot{\beta}_f\} dy$$
$$= \frac{1}{2}\int_V \{\dot{\boldsymbol{\Delta}}^T \mathbf{I}_0 \dot{\boldsymbol{\Delta}}\} dV \tag{7.62d}$$

With the results of (7.62a–d), the difference between the total potential energy and kinetic energy can now be expressed in matrix form as

$$\Pi - T = \frac{1}{2}\int_y \{\mathbf{F}^T\boldsymbol{\Delta}' + \mathbf{F}_0^T\boldsymbol{\Delta} - 2\mathbf{q}^T\boldsymbol{\Delta} - \dot{\boldsymbol{\Delta}}^T\mathbf{I}_0\dot{\boldsymbol{\Delta}}\} dy - \left[\hat{\mathbf{F}}^T\boldsymbol{\Delta}\right]_{y_1}^{y_2}, \tag{7.63}$$

where $\mathbf{F}$ and $\mathbf{F}_0$ are the force vectors defined in (7.18b) with $\lambda = 1$; $\hat{\mathbf{F}}$ denotes the prescribed value of $\mathbf{F}$ on the boundary $y = y_1$ or $y = y_2$, which is the integration limits of y, and $\left[f\right]_{y_1}^{y_2} = f(y_2) - f(y_1)$. Note that to avoid confusion about the variational operator δ and the vector $\boldsymbol{\delta}$ defined in (7.18b), in this section $\boldsymbol{\Delta}$ is used to represent $\boldsymbol{\delta}$, that is,

$$\boldsymbol{\Delta} = \boldsymbol{\Delta}(y,t) = \begin{Bmatrix} v_0(y,t) \\ w_f(y,t) \\ \theta(y,t) \\ \beta_f(y,t) \\ \beta_r(y,t) \end{Bmatrix}. \tag{7.64}$$

Substituting $(7.18\text{a})_{2,3}$ into (7.63), we obtain

$$\Pi - T = \frac{1}{2}\int_y \left\{2\boldsymbol{\Delta}^T\mathbf{K}_1{}^T\boldsymbol{\Delta}' + \boldsymbol{\Delta}'^T\mathbf{K}_2\boldsymbol{\Delta}' + \boldsymbol{\Delta}^T\mathbf{K}_0\boldsymbol{\Delta} - \dot{\boldsymbol{\Delta}}^T\mathbf{I}_0\dot{\boldsymbol{\Delta}} - 2\mathbf{q}^T\boldsymbol{\Delta}\right\}dy - \left[\hat{\mathbf{F}}^T\boldsymbol{\Delta}\right]_{y_1}^{y_2}. \tag{7.65}$$

The variation of the first term of (7.65) can be derived as follows:

$$\delta\int_y \boldsymbol{\Delta}^T\mathbf{K}_1{}^T\boldsymbol{\Delta}'dy = \int_y \delta\left\{\boldsymbol{\Delta}^T\mathbf{K}_1{}^T\boldsymbol{\Delta}'\right\}dy = \int_y \left\{\left(\delta\boldsymbol{\Delta}^T\right)\mathbf{K}_1{}^T\boldsymbol{\Delta}' + \boldsymbol{\Delta}^T\mathbf{K}_1{}^T\delta\left(\boldsymbol{\Delta}'\right)\right\}dy, \tag{7.66}$$

in which the second term of the last equality can be rewritten as

$$\begin{aligned}\int_y \boldsymbol{\Delta}^T\mathbf{K}_1^T\delta\left(\boldsymbol{\Delta}'\right)dy &= \int_y \boldsymbol{\Delta}^T\mathbf{K}_1^T\left(\delta\boldsymbol{\Delta}\right)'dy = \int_y \left\{\left(\boldsymbol{\Delta}^T\mathbf{K}_1^T\delta\boldsymbol{\Delta}\right)' - \left(\boldsymbol{\Delta}'^T\mathbf{K}_1^T\delta\boldsymbol{\Delta}\right)\right\}dy \\ &= \left[\boldsymbol{\Delta}^T\mathbf{K}_1^T\delta\boldsymbol{\Delta}\right]_{y_r}^{y_t} - \int_y \boldsymbol{\Delta}'^T\mathbf{K}_1^T\delta\boldsymbol{\Delta}dy = \left[\boldsymbol{\Delta}^T\mathbf{K}_1^T\delta\boldsymbol{\Delta}\right]_{y_r}^{y_t} - \int_y \left(\delta\boldsymbol{\Delta}^T\right)\mathbf{K}_1\boldsymbol{\Delta}'dy.\end{aligned} \tag{7.67}$$

Therefore,

$$\delta\int_y \boldsymbol{\Delta}^T\mathbf{K}_1^T\boldsymbol{\Delta}'dy = \int_y \left(\delta\boldsymbol{\Delta}^T\right)\left(\mathbf{K}_1^T - \mathbf{K}_1\right)\boldsymbol{\Delta}'dy + \left[\left(\delta\boldsymbol{\Delta}\right)^T\mathbf{K}_1\boldsymbol{\Delta}\right]_{y_r}^{y_t}. \tag{7.68}$$

Similarly,

$$\delta\int_y \Delta'^T \mathbf{K}_2\Delta' dy = -2\int_y \left(\delta\Delta^T\right)\mathbf{K}_2\Delta'' dy + 2\left[\left(\delta\Delta^T\right)\mathbf{K}_2\Delta'\right]_{y_r}^{y_t},$$

$$\delta\int_y \Delta^T \mathbf{K}_0\Delta dy = 2\int_y \left(\delta\Delta^T\right)\mathbf{K}_0\Delta dy,$$

$$\delta\int_y \dot{\Delta}^T \mathbf{I}_0\dot{\Delta} dy = -2\int_y \left(\delta\Delta^T\right)\mathbf{I}_0\ddot{\Delta} dy + 2\frac{\partial}{\partial t}\int_y \left(\delta\Delta^T\right)\mathbf{I}_0\Delta dy. \tag{7.69}$$

With the results of (7.68) and (7.69), the first variation of the Lagrangian, $\Pi - T$, can be obtained as

$$\delta\left(\Pi - T\right) = \int_y \left(\delta\Delta^T\right)\left[\left(\mathbf{K}_1^T - \mathbf{K}_1\right)\Delta' - \mathbf{K}_2\Delta'' + \mathbf{K}_0\Delta + \mathbf{I}_0\ddot{\Delta} - \mathbf{q}\right]dy$$
$$+\left(\delta\Delta^T\right)\left[\mathbf{K}_1\Delta + \mathbf{K}_2\Delta' - \hat{\mathbf{F}}\right]_{y_1}^{y_2} - \frac{\partial}{\partial t}\int_y \left(\delta\Delta^T\right)\mathbf{I}_0\Delta dy. \tag{7.70}$$

Using (7.70), the Hamilton's principle (7.60) can then provide the governing equation and boundary conditions as follows:

$$\mathbf{K}_2\Delta'' + \left(\mathbf{K}_1 - \mathbf{K}_1^T\right)\Delta' - \mathbf{K}_0\Delta + \mathbf{q} = \mathbf{I}_0\ddot{\Delta}, \quad \text{for all } y,$$
$$\Delta = \hat{\Delta} \ \text{ or } \ \mathbf{K}_1\Delta + \mathbf{K}_2\Delta' = \hat{\mathbf{F}}, \quad \text{on } y = y_r \text{ and } y = y_t. \tag{7.71}$$

where $y = y_t$ or $y = y_r$, which are the locations of wing tip or wing root. Equation (7.71) is the same as those shown in (7.19) and (7.20) derived by different approach.

7.3.1 Comprehensive Finite Element Model

As stated previously the comprehensive wing model cannot be applied to the tapered wings directly. In order to extend its applicability to the case of tapered wings, a comprehensive dynamic finite element model is introduced in this subsection. By this new model, the tapered wings will be divided into a series of elements which are connected at a finite number of nodal points. Within each element, the cross-section is assumed to be uniform in spanwise direction (y-direction) as shown in Figure 7.3. Since the cross-section can be different for different elements, general wing shapes including tapered wings can be handled through this new model.

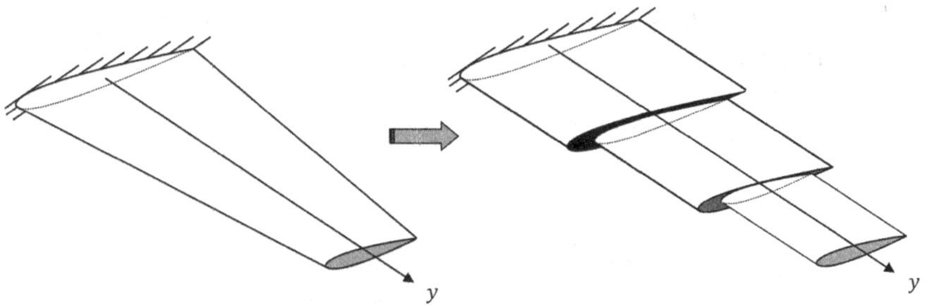

FIGURE 7.3 Discretization of the tapered wings.

In the finite element displacement method, the displacement is assumed to have unknown values only at the nodal points, so that the variation within any element is described in terms of the nodal values by means of interpolation function. With this understanding, within each element the generalized displacement vector $\mathbf{\Delta}$ composed of five basic functions v_0, w_f, θ, β_f, and β_r is now assumed as

$$\mathbf{\Delta}(y,t) = g(t)\mathbf{N}(y)\mathbf{u}_e, \tag{7.72}$$

where, $\mathbf{u}_e$ is a 15×1 vector of nodal displacements of the element, $\mathbf{N}(y)$ is a 5×15 matrix containing a set of shape functions, and $g(t)$ is a time function assumed to be the same for all degrees of freedom.

In this model, each element is assumed to have three nodal points as shown in Figure 7.4 and each basic function is assumed to be interpolated

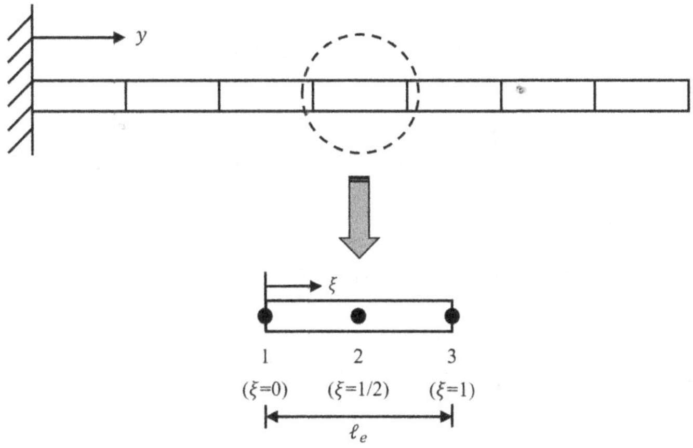

FIGURE 7.4 Finite element discretization and element local coordinate.

by the same shape functions within element. Thus, $\mathbf{u}_e$ and $\mathbf{N}(y)$ can be written in a partition form as

$$\mathbf{u}_e = \begin{Bmatrix} \mathbf{\Delta}_1 \\ \mathbf{\Delta}_2 \\ \mathbf{\Delta}_3 \end{Bmatrix}, \quad \mathbf{\Delta}_i = \begin{Bmatrix} v_0 \\ w_f \\ \theta \\ \beta_f \\ \beta_r \end{Bmatrix}_i,$$

$$\mathbf{N}(y) = \left[\mathbf{N}_1(\xi)\ \mathbf{N}_2(\xi)\ \mathbf{N}_3(\xi)\right], \quad \mathbf{N}_i(\xi) = N_i(\xi)\mathbf{I}, \quad i = 1,2,3, \tag{7.73}$$

where $\mathbf{\Delta}_i$, i = 1, 2, 3, are the sub-vectors of nodal displacements corresponding to the three points within the element, $\mathbf{I}$ is a 5 × 5 unit matrix and $N_i(\xi)$, i = 1, 2, 3, are quadratic interpolation functions defined by

$$N_1(\xi) = 2(\xi - 1)\left(\xi - \frac{1}{2}\right), \ N_2(\xi) = -4\xi(\xi - 1), \ N_3(\xi) = 2\xi\left(\xi - \frac{1}{2}\right),$$
$$0 \le \xi = \frac{y_e}{\ell_e} \le 1, \tag{7.74}$$

in which ℓ_e is the length of the element, y_e is the local coordinate starting from node 1 of the element denoted in Figure 7.4, and ξ is its associated local non-dimensional coordinate.

Substituting (7.72) into (7.65), we get the Lagrangian function in single element

$$\Pi - T = \frac{1}{2}\mathbf{u}_e^T\left[g^2(t)\mathbf{K}_e - \dot{g}^2(t)\mathbf{J}_e\right]\mathbf{u}_e - g(t)\mathbf{u}_e^T\mathbf{f}_e, \tag{7.75}$$

where $\mathbf{K}_e$, $\mathbf{J}_e$, and $\mathbf{f}_e$ are, respectively, the *element stiffness matrix*, *element inertia matrix*, and *element force vector*, and are defined as

$$\mathbf{K}_e = \int_0^1 \left\{\mathbf{N}^T(\xi)\mathbf{K}_0\mathbf{N}(\xi) + 2\mathbf{N}^T(\xi)\mathbf{K}_1^T\mathbf{N}'(\xi) + \mathbf{N}'^T(\xi)\mathbf{K}_2\mathbf{N}'(\xi)\right\}\ell_e d\xi,$$

$$\mathbf{J}_e = \int_0^1 \left\{\mathbf{N}^T(\xi)\mathbf{I}_0\mathbf{N}(\xi)\right\}\ell_e d\xi,$$

$$\mathbf{f}_e = \int_0^1 \left\{\mathbf{N}^T(\xi)\mathbf{q}(\xi)\right\}\ell_e d\xi + \left[\mathbf{N}^T(\xi)\hat{\mathbf{F}}(\xi)\right]_0^1. \tag{7.76}$$

By substituting $(7.73)_{3,4}$ and (7.74) into (7.76) and assuming uniform pressure with constant load vector $\mathbf{q}$ within each element, the explicit forms of the element stiffness matrix, inertia matrix, and element force vector can be obtained as

$$\mathbf{K}_e = \ell_e \begin{bmatrix} \frac{2}{15}\mathbf{K}_0 & \frac{1}{15}\mathbf{K}_0 & -\frac{1}{30}\mathbf{K}_0 \\ \frac{1}{15}\mathbf{K}_0 & \frac{8}{15}\mathbf{K}_0 & \frac{1}{15}\mathbf{K}_0 \\ -\frac{1}{30}\mathbf{K}_0 & \frac{1}{15}\mathbf{K}_0 & \frac{2}{15}\mathbf{K}_0 \end{bmatrix} + 2 \begin{bmatrix} -\frac{1}{2}\mathbf{K}_1^T & \frac{2}{3}\mathbf{K}_1^T & -\frac{1}{6}\mathbf{K}_1^T \\ -\frac{2}{3}\mathbf{K}_1^T & 0\mathbf{K}_1^T & \frac{2}{3}\mathbf{K}_1^T \\ \frac{1}{6}\mathbf{K}_1^T & -\frac{2}{3}\mathbf{K}_1^T & \frac{1}{2}\mathbf{K}_1^T \end{bmatrix}$$
$$+ \frac{1}{\ell_e} \begin{bmatrix} \frac{7}{3}\mathbf{K}_2 & -\frac{8}{3}\mathbf{K}_2 & \frac{1}{3}\mathbf{K}_2 \\ -\frac{8}{3}\mathbf{K}_2 & \frac{16}{3}\mathbf{K}_2 & -\frac{8}{3}\mathbf{K}_2 \\ \frac{1}{3}\mathbf{K}_2 & -\frac{8}{3}\mathbf{K}_2 & \frac{7}{3}\mathbf{K}_2 \end{bmatrix} \tag{7.77a}$$

$$\mathbf{J}_e = \ell_e \begin{bmatrix} \frac{2}{15}\mathbf{I}_0 & \frac{1}{15}\mathbf{I}_0 & -\frac{1}{30}\mathbf{I}_0 \\ \frac{1}{15}\mathbf{I}_0 & \frac{8}{15}\mathbf{I}_0 & \frac{1}{15}\mathbf{I}_0 \\ -\frac{1}{30}\mathbf{I}_0 & \frac{1}{15}\mathbf{I}_0 & \frac{2}{15}\mathbf{I}_0 \end{bmatrix}, \quad \mathbf{f}_e = \ell_e \begin{Bmatrix} \frac{1}{6}\mathbf{q} \\ \frac{2}{3}\mathbf{q} \\ \frac{1}{6}\mathbf{q} \end{Bmatrix} + \begin{Bmatrix} -\hat{\mathbf{F}}(0) \\ 0 \\ \hat{\mathbf{F}}(1) \end{Bmatrix}. \tag{7.77b}$$

Note that $\mathbf{K}_e$, $\mathbf{J}_e$, and $\mathbf{f}_e$ calculated from (7.77) will have different values for different elements if a tapered wing is considered, since $\mathbf{K}_0$, $\mathbf{K}_1$, $\mathbf{K}_2$, and $\mathbf{I}_0$ defined in (7.18c) are related to the chord length of element which is a function of spanwise location, that is, $c = c(y)$. The dimensions of $\mathbf{K}_e$, $\mathbf{J}_e$, and $\mathbf{f}_e$ are, respectively, 15 × 15, 15 × 15, and 15 × 1.

7.3.2 Static Analysis

The foregoing model is general enough to deal with both static and dynamic analysis in tapered composite wings. If we only consider the static model, the function $g(t)$ should be independent of time and can be set as $g(t) = 1$. With this substitution, (7.75) can be rewritten as

$$\Pi = \frac{1}{2}\mathbf{u}_e^T\mathbf{K}_e\mathbf{u}_e - \mathbf{u}_e^T\mathbf{f}_e. \tag{7.78}$$

By taking the first variation of the total potential energy Π to be zero, the equilibrium equation within a single element can be obtained as

$$\frac{1}{2}\left(\mathbf{K}_e+\mathbf{K}_e^T\right)\mathbf{u}_e=\mathbf{f}_e. \tag{7.79}$$

Once we divide the wing into n elements, there are $2n + 1$ nodes in the structure. Because there is a common node shared by two adjacent elements, the total number of the common nodes of the wing structure will be $n-1$. As every nodal displacement sub-vector $\boldsymbol{\Delta}_i$ contains five components (v_0, w_f, θ, β_f, β_r), the dimension of the global nodal displacement vector $\mathbf{u}_g$ becomes $(10n + 5) \times 1$. The global stiffness matrix $\mathbf{K}_g$ is a $(10n + 5) \times (10n + 5)$ matrix and the global force vector $\mathbf{f}_g$ is a $(10n + 5) \times 1$ vector. With the concept of finite element method, we can move all the element stiffness and force vectors into the corresponding position in global ones. The portions related to the common nodes between two elements will overlap. The global stiffness matrices $\mathbf{K}_g$ and force vector $\mathbf{f}_g$ are further fulfilled by superimposing all the values of the element stiffness matrices and force vectors, and finally the equilibrium equation for the entire wing structures can be written as follows

$$\frac{1}{2}\left(\mathbf{K}_g+\mathbf{K}_g^T\right)\mathbf{u}_g=\mathbf{f}_g. \tag{7.80}$$

Equation (7.80) is a standard form of the system of linear algebraic equations. With the known values of $\mathbf{K}_g$ and the prescribed values of $\mathbf{f}_g$ or $\mathbf{u}_g$ set for the problems, all the unknowns $\mathbf{u}_g$ or $\mathbf{f}_g$ can be solved in principle.

7.3.3 Free Vibration

In this subsection, free vibration of the tapered wing structure is considered. The pressure $\mathbf{q}$ and nodal force $\hat{\mathbf{F}}$ are both set to be zero and hence the element force vector $\mathbf{f}_e = \mathbf{0}$. A harmonic motion is assumed for free vibration analysis, that is, $g(t) = e^{i\omega t}$ where ω is the natural frequency. With these assumptions, the Lagrangian function (7.75) can be reduced to

$$\Pi - T=\frac{e^{2i\omega t}}{2}\mathbf{u}_e^T\left[\mathbf{K}_e+\omega^2\mathbf{J}_e\right]\mathbf{u}_e. \tag{7.81}$$

By taking the first variation of the Lagrangian function Π − T to be zero, the equation of motion for a single element can be derived as

$$\frac{1}{2}\left[\left(\mathbf{K}_e+\mathbf{K}_e^T\right)+\omega^2\left(\mathbf{J}_e+\mathbf{J}_e^T\right)\right]\mathbf{u}_e=\mathbf{0}. \tag{7.82}$$

Following the standard procedure for element assembly, global stiffness matrix $\mathbf{K}_g$, and global inertia matrix $\mathbf{J}_g$ can be obtained, and the equation of motion for the entire wing structures can be written as

$$\frac{1}{2}\left[\left(\mathbf{K}_g+\mathbf{K}_g^T\right)+\omega^2\left(\mathbf{J}_g+\mathbf{J}_g^T\right)\right]\mathbf{u}_g=\mathbf{0}. \tag{7.83}$$

Equation (7.83) is a typical form of eigenvalue problem. After embedding the boundary conditions, the system of equations (7.83) will be reduced. The natural frequency ω can then be solved by letting the determinant of the reduced coefficient matrix of $\mathbf{u}_g$ be zero. The mode shape $\mathbf{u}_g$ corresponding to the natural frequency ω is obtained via (7.83).

7.3.4 Aeroelastic Divergence

With the lift and the pitching moment given in (7.52), the external aerodynamic force vector $\mathbf{f}_e$ of (7.76) can be simplified further as

$$\mathbf{f}_e=q_n\left(\mathbf{q}_0+\mathbf{K}_a\mathbf{u}_e\right), \tag{7.84}$$

where $\mathbf{q}_0$ is a constant vector and $\mathbf{K}_a$ is the aerodynamic stiffness matrix resulting from the interaction between aerodynamic force $\left(\tilde{q},\tilde{q}^*\right)$ and deformation (θ, w_f). Because divergence is a static phenomenon in aeroelasticity, we can neglect the time effect by letting $g(t) = 1$ in (7.75), which leads to

$$\Pi-T=\frac{1}{2}\mathbf{u}_e^T\mathbf{K}_e\mathbf{u}_e-\mathbf{u}_e^T q_n\left(\mathbf{q}_0+\mathbf{K}_a\mathbf{u}_e\right). \tag{7.85}$$

By taking the first variation of the Lagrangian function Π − T to be zero, the equilibrium equation of divergence analysis for a single element can be derived as

$$\left[\frac{1}{2}\left(\mathbf{K}_e+\mathbf{K}_e^T\right)-q_n\left(\mathbf{K}_a+\mathbf{K}_a^T\right)\right]\mathbf{u}_e=q_n\mathbf{q}_0. \tag{7.86}$$

Similar to the problem of free vibration, after the assembly for getting the global matrices the divergence value of q_n can be obtained from

$$\left\| \frac{1}{2}\left(\mathbf{K}_e + \mathbf{K}_e^T\right) - q_n\left(\mathbf{K}_a + \mathbf{K}_a^T\right) \right\| = 0. \tag{7.87}$$

7.4 VARIABLE THICKNESS PLATE MODEL

7.4.1 General Formulation

As stated in Section 3.4.1, based upon the first-order shear deformation theory (FSDT) the basic equations of thick composite laminated plate: displacement fields, strain–displacement relations, constitutive laws, and equilibrium equations, can be expressed by (3.96)–(3.99), from which the governing equation can be written in matrix form as that shown in (3.101) and (3.102), that is,

$$\mathcal{L}^{*}\boldsymbol{\delta} + \mathbf{p} = \mathbf{0}, \tag{7.88a}$$

where

$$\mathcal{L}^{*} = \begin{bmatrix} L_{11}^{(A)} & L_{12}^{(A)} & L_{11}^{(B)} & L_{12}^{(B)} & 0 \\ L_{21}^{(A)} & L_{22}^{(A)} & L_{21}^{(B)} & L_{22}^{(B)} & 0 \\ L_{11}^{(B)} & L_{12}^{(B)} & L_{11}^{(D)} - A_{55} & L_{12}^{(D)} - A_{45} & -L_{1}^{(G)} \\ L_{21}^{(B)} & L_{22}^{(B)} & L_{21}^{(D)} - A_{45} & L_{22}^{(D)} - A_{44} & -L_{2}^{(G)} \\ 0 & 0 & L_{1*}^{(G)} & L_{2*}^{(G)} & L_{11}^{(G)} \end{bmatrix},$$

$$\boldsymbol{\delta} = \begin{Bmatrix} u_0 \\ v_0 \\ \beta_x \\ \beta_y \\ w \end{Bmatrix}, \quad \mathbf{p} = \begin{Bmatrix} q_x \\ q_y \\ m_x \\ m_y \\ q_z \end{Bmatrix}. \tag{7.88b}$$

and $L_{ij}^{(X)}, i, j = 1,2, X = A,B,D,\ L_{i}^{(G)}, L_{i*}^{(G)}, i = 1,2,$ and $L_{11}^{(G)}$ are defined in (3.102b) and (3.102c). As noted in (3.103), in this matrix form expression the thickness of the plate is not required to be uniform. Therefore, the general formulation stated in Section 3.4.1 can be applied to the cases of variable thickness plate model.

Reduction to Comprehensive Beam Model of Wing Structures

Since the beams can be considered to be one-dimensional reduction of the plates, the comprehensive beam model stated in the previous three sections may be reduced from the variable thickness plate model. To see whether the comprehensive beam model can be treated as a special case of the plate model, we reduce the governing equation (7.88) of the various thickness plates by considering the displacement field described in (7.3a) for the comprehensive beam model. That is, we consider the plates whose displacement fields assumed in (3.96a) are further specialized to

$$\begin{aligned} &u_0(x,y)=0,\ \ v_0(x,y)=v_0(y),\ \ w_0(x,y)=w_f(y)-x\theta(y),\\ &\beta_x(x,y)=\theta(y),\ \ \beta_y(x,y)=\beta_f(y)+x\beta_r(y). \end{aligned} \tag{7.89}$$

For the convenience of the following derivation, we now rewrite (7.89) in matrix form as

$$\boldsymbol{\delta}(x,y)=\mathbf{X}(x)\boldsymbol{\delta}_B(y), \tag{7.90a}$$

where

$$\boldsymbol{\delta}(x,y)=\begin{Bmatrix} u_0(x,y)\\ v_0(x,y)\\ \beta_x(x,y)\\ \beta_y(x,y)\\ w(x,y) \end{Bmatrix},\ \ \mathbf{X}(x)=\begin{bmatrix} 0&0&0&0&0\\ 1&0&0&0&0\\ 0&0&1&0&0\\ 0&0&0&1&x\\ 0&1&-x&0&0 \end{bmatrix},$$

$$\boldsymbol{\delta}_B(y)=\begin{Bmatrix} v_0(y)\\ w_f(y)\\ \theta(y)\\ \beta_f(y)\\ \beta_r(y) \end{Bmatrix}. \tag{7.90b}$$

In the beam model, the force vector $\mathbf{p}_B$ associated with the displacement vector $\boldsymbol{\delta}_B$ is defined by

$$\mathbf{p}_B(y)=\int_{x_0}^{x_1}\mathbf{X}^*(x)\mathbf{p}(x,y)\,dx, \tag{7.91a}$$

where

$$\mathbf{p}_B(y)=\begin{Bmatrix}\tilde{q}_y(y)\\ \tilde{q}_z(y)\\ \tilde{m}_x(y)-\tilde{q}_z^*(y)\\ \tilde{m}_y(y)\\ \tilde{m}_y^*(y)\end{Bmatrix},\ \mathbf{X}^*(x)=\begin{bmatrix}0&1&0&0&0\\0&0&0&0&1\\0&0&1&0&-x\\0&0&0&1&0\\0&0&0&x&0\end{bmatrix},$$

$$\mathbf{p}(x,y)=\begin{Bmatrix}q_x(x,y)\\ q_y(x,y)\\ m_x(x,y)\\ m_y(x,y)\\ q_z(x,y)\end{Bmatrix}. \tag{7.91b}$$

The lower and upper limits x_0 and x_1 denote, respectively, the location of leading and trailing edges of the wing.

With the relations of (7.90a) and (7.91a), the governing equation (7.88a) now becomes

$$\int_{x_0}^{x_1}\mathbf{X}^*\boldsymbol{\mathcal{L}}^*\mathbf{X}\boldsymbol{\delta}_B dx+\mathbf{p}_B=\mathbf{0}. \tag{7.92}$$

Substituting (3.102), $(7.90b)_2$, and $(7.91b)_2$ into (7.92), and performing carefully on the integration for the first term of (7.92), we get

$$\mathbf{K}_2\boldsymbol{\delta}_B''+\mathbf{K}_1\boldsymbol{\delta}_B'+\mathbf{K}_0\boldsymbol{\delta}_B+\mathbf{p}_B=\mathbf{0}, \tag{7.93a}$$

where

$$\mathbf{K}_2=\begin{bmatrix}\tilde{A}_{22}&0&\tilde{B}_{26}&\tilde{B}_{22}&\tilde{B}_{22}^*\\0&\tilde{A}_{44}&-\tilde{A}_{44}^*&0&0\\ \tilde{B}_{26}&-\tilde{A}_{44}^*&\tilde{D}_{66}+\tilde{A}_{44}^{**}&\tilde{D}_{26}&\tilde{D}_{26}^*\\ \tilde{B}_{22}&0&\tilde{D}_{26}&\tilde{D}_{22}&\tilde{D}_{22}^*\\ \tilde{B}_{22}^*&0&\tilde{D}_{26}^*&\tilde{D}_{22}^*&\tilde{D}_{22}^{**}\end{bmatrix}, \tag{7.93b}$$

$$\mathbf{K}_1 = \begin{bmatrix} 0 & 0 & 0 & 0 & \tilde{B}_{26} \\ 0 & 0 & 0 & \tilde{A}_{44} & \tilde{A}^*_{44} \\ 0 & 0 & 0 & -\tilde{A}^*_{44} & \tilde{D}_{66} - \tilde{A}^{**}_{44} \\ 0 & -\tilde{A}_{44} & \tilde{A}^*_{44} & 0 & \tilde{D}_{26} \\ -\tilde{B}_{26} & -\tilde{A}^*_{44} & -\tilde{D}_{66} + \tilde{A}^{**}_{44} & -\tilde{D}_{26} & 0 \end{bmatrix}, \quad (7.93c)$$

$$\mathbf{K}_0 = \begin{bmatrix} 0 & 0 & 0 & 0 & 0 \\ 0 & 0 & 0 & 0 & 0 \\ 0 & 0 & 0 & 0 & 0 \\ 0 & 0 & 0 & -\tilde{A}_{44} & -\tilde{A}^*_{44} \\ 0 & 0 & 0 & -\tilde{A}^*_{44} & -\tilde{D}_{66} - \tilde{A}^{**}_{44} \end{bmatrix}. \quad (7.93d)$$

In the above, the symbols: tilde ~, the superscript *, and the prime •′ are defined in (7.7). The governing equation obtained in (7.93) is exactly the same as that shown in (7.19) for the comprehensive beam model, which is derived by a different approach. Note that here the symbols $\mathbf{K}_1$ and $\mathbf{K}_0$ are different from those used in (7.19).

7.4.2 Finite Element Method

In general, to solve a problem by FSDT of variable thick plate, the governing differential equation (7.88) is required to be satisfied at every structural point and is generally called *strong form*. Since satisfaction at every point is not easy to be met, an integral expression shown by an average sense, which is generally called *weak form*, is usually adopted to be an alternative solution method. With this consideration, as presented in (3.106)–(3.113) the governing equation (7.88) has been re-derived by minimization of the total potential energy. Since the re-derivation is successful, we now take the total potential energy (which is an integral expression) as a basis to construct the system of equations via finite element formulation.

In finite element formulation, the variable thickness composite laminated plate is discretized into a series of elements which are connected at a finite number of nodal points. Consider an isoparametric eight-node quadratic quadrilateral finite element, in each element

$$\mathbf{x} = \sum_{k=1}^{8} \varpi_k \mathbf{x}_m^{(k)}, \boldsymbol{\Delta} = \sum_{k=1}^{8} \varpi_k \boldsymbol{\Delta}_m^{(k)}, \text{on the } m\text{th element} \quad (7.94a)$$

where

$$\boldsymbol{\Delta} = \left(u_0, v_0, w, \beta_x, \beta_y\right)^T, \tag{7.94b}$$

$$\begin{aligned}
\varpi_1 &= -\frac{1}{4}(1-\xi)(1-\eta)(1+\xi+\eta), \quad \varpi_2 = \frac{1}{2}\left(1-\xi^2\right)(1-\eta),\\
\varpi_3 &= -\frac{1}{4}(1+\xi)(1-\eta)(1-\xi+\eta), \quad \varpi_4 = \frac{1}{2}(1+\xi)\left(1-\eta^2\right),\\
\varpi_5 &= -\frac{1}{4}(1+\xi)(1+\eta)(1-\xi-\eta), \quad \varpi_6 = \frac{1}{2}\left(1-\xi^2\right)(1+\eta),\\
\varpi_7 &= -\frac{1}{4}(1-\xi)(1+\eta)(1+\xi-\eta), \quad \varpi_8 = \frac{1}{2}(1-\xi)\left(1-\eta^2\right).
\end{aligned} \tag{7.94c}$$

We may rewrite $(7.94\text{a})_2$ in matrix form as

$$\begin{aligned}
&\boldsymbol{\Delta} = \boldsymbol{\varpi}(\xi,\eta)\mathbf{u}_e, \quad \mathbf{u}_e = \begin{Bmatrix} \boldsymbol{\Delta}_1 \\ \boldsymbol{\Delta}_2 \\ \vdots \\ \boldsymbol{\Delta}_8 \end{Bmatrix},\\
&\boldsymbol{\varpi}(\xi,\eta) = \left[\boldsymbol{\varpi}_1(\xi,\eta), \boldsymbol{\varpi}_2(\xi,\eta), \cdots, \boldsymbol{\varpi}_8(\xi,\eta)\right],\\
&\boldsymbol{\varpi}_i(\xi,\eta) = \varpi_i(\xi,\eta)\mathbf{I}_{5\times5},
\end{aligned} \tag{7.95}$$

where $\boldsymbol{\Delta}$, $\boldsymbol{\varpi}$, and $\mathbf{u}_e$ are, respectively, 5 × 1, 5 × 40, and 40 × 1 matrices. With (7.95), the mid-surface strain $\boldsymbol{\varepsilon}$, curvature $\boldsymbol{\kappa}$, and the transverse shear $\boldsymbol{\gamma}$ of (3.110) in each element can be written as

$$\boldsymbol{\varepsilon} = \boldsymbol{\varpi}_\varepsilon \mathbf{u}_e, \quad \boldsymbol{\kappa} = \boldsymbol{\varpi}_\kappa \mathbf{u}_e, \quad \boldsymbol{\gamma} = \boldsymbol{\varpi}_\lambda \mathbf{u}_e, \tag{7.96a}$$

where

$$\begin{aligned}
\boldsymbol{\varpi}_\varepsilon &= \left[\boldsymbol{\varpi}_1^\varepsilon(\xi,\eta), \boldsymbol{\varpi}_1^\varepsilon(\xi,\eta), \cdots, \boldsymbol{\varpi}_1^\varepsilon(\xi,\eta)\right],\\
\boldsymbol{\varpi}_\kappa &= \left[\boldsymbol{\varpi}_1^\kappa(\xi,\eta), \boldsymbol{\varpi}_1^\kappa(\xi,\eta), \cdots, \boldsymbol{\varpi}_1^\kappa(\xi,\eta)\right],\\
\boldsymbol{\varpi}_\gamma &= \left[\boldsymbol{\varpi}_1^\gamma(\xi,\eta), \boldsymbol{\varpi}_1^\gamma(\xi,\eta), \cdots, \boldsymbol{\varpi}_1^\gamma(\xi,\eta)\right],
\end{aligned} \tag{7.96b}$$

and

$$\boldsymbol{\varpi}_i^{\varepsilon}(\xi,\eta)=\begin{bmatrix}\varpi_{i,x} & 0 & 0 & 0 & 0\\ 0 & \varpi_{i,y} & 0 & 0 & 0\\ \varpi_{i,y} & \varpi_{i,x} & 0 & 0 & 0\end{bmatrix},$$

$$\boldsymbol{\varpi}_i^{\kappa}(\xi,\eta)=\begin{bmatrix}0 & 0 & 0 & \varpi_{i,x} & 0\\ 0 & 0 & 0 & 0 & \varpi_{i,y}\\ 0 & 0 & 0 & \varpi_{i,y} & \varpi_{i,x}\end{bmatrix}, \tag{7.96c}$$

$$\boldsymbol{\varpi}_i^{\gamma}(\xi,\eta)=\begin{bmatrix}0 & 0 & \varpi_{i,x} & \varpi_i & 0\\ 0 & 0 & \varpi_{i,y} & 0 & \varpi_i\end{bmatrix}.$$

Substituting (7.95) and (7.96) into (3.109), the Hamilton's principle (3.106) now leads to

$$\delta\int_{t_1}^{t_2}(\Pi-T)dt=\delta\int_{t_1}^{t_2}\sum_{e=1}^{m}\left\{\frac{1}{2}\mathbf{u}_e^T\mathbf{K}_e\mathbf{u}_e-\mathbf{u}_e^T\mathbf{f}_e-\frac{1}{2}\dot{\mathbf{u}}_e^T\mathbf{J}_e\dot{\mathbf{u}}_e\right\}dt=0, \tag{7.97a}$$

where

$$\mathbf{K}_e=\int_{A_e}\left\{\boldsymbol{\varpi}_\varepsilon^T\boldsymbol{A}\boldsymbol{\varpi}_\varepsilon+\boldsymbol{\varpi}_\varepsilon^T\boldsymbol{B}\boldsymbol{\varpi}_\kappa+\boldsymbol{\varpi}_\kappa^T\boldsymbol{B}\boldsymbol{\varpi}_\varepsilon+\boldsymbol{\varpi}_\kappa^T\boldsymbol{D}\boldsymbol{\varpi}_\kappa+\boldsymbol{\varpi}_\gamma^T\boldsymbol{G}\boldsymbol{\varpi}_\gamma\right\}dA,$$

$$\mathbf{f}_e=\int_{A_e}\boldsymbol{\varpi}_\varepsilon^T\mathbf{f}dA+\int_{\Gamma_\sigma}\boldsymbol{\varpi}^T\hat{\mathbf{F}}d\Gamma,$$

$$\mathbf{J}_e=\int_{A_e}\boldsymbol{\varpi}^T\mathbf{I}_0\boldsymbol{\varpi}dA, \tag{7.97b}$$

$$\mathbf{f}=\begin{Bmatrix}q_x\\ q_y\\ q_z\\ m_x\\ m_y\end{Bmatrix},\quad \hat{\mathbf{F}}=\begin{Bmatrix}\hat{T}_x\\ \hat{T}_y\\ Q_n\\ \hat{H}_x\\ \hat{H}_y\end{Bmatrix},\quad \mathbf{I}_0=\begin{bmatrix}g_0 & 0 & 0 & g_1 & 0\\ 0 & g_0 & 0 & 0 & g_1\\ 0 & 0 & g_0 & 0 & 0\\ g_1 & 0 & 0 & g_2 & 0\\ 0 & g_1 & 0 & 0 & g_2\end{bmatrix}. \tag{7.97c}$$

Taking the variation with respect to $\mathbf{u}_e$, we get

$$\mathbf{K}_e\mathbf{u}_e + \mathbf{J}_e\ddot{\mathbf{u}}_e = \mathbf{f}_e. \tag{7.98}$$

Note that in (7.96c), the differentiation can be made by using chain rule with the assistance of inverse of Jacobian matrix, for example,

$$\begin{aligned} \varpi_{i,x} &= \frac{\partial \varpi_i}{\partial x} = \frac{\partial \varpi_i}{\partial \xi}\frac{\partial \xi}{\partial x} + \frac{\partial \varpi_i}{\partial \eta}\frac{\partial \eta}{\partial x}, \\ \varpi_{i,y} &= \frac{\partial \varpi_i}{\partial y} = \frac{\partial \varpi_i}{\partial \xi}\frac{\partial \xi}{\partial y} + \frac{\partial \varpi_i}{\partial \eta}\frac{\partial \eta}{\partial y}, \end{aligned} \tag{7.99a}$$

where

$$\begin{bmatrix} \dfrac{\partial \xi}{\partial x} & \dfrac{\partial \xi}{\partial y} \\ \dfrac{\partial \eta}{\partial x} & \dfrac{\partial \eta}{\partial y} \end{bmatrix} = \mathbf{J}^{-1} = \begin{bmatrix} \dfrac{\partial x}{\partial \xi} & \dfrac{\partial x}{\partial \eta} \\ \dfrac{\partial y}{\partial \xi} & \dfrac{\partial y}{\partial \eta} \end{bmatrix}^{-1} = \frac{1}{J_0}\begin{bmatrix} \dfrac{\partial y}{\partial \eta} & -\dfrac{\partial x}{\partial \eta} \\ -\dfrac{\partial y}{\partial \xi} & \dfrac{\partial x}{\partial \xi} \end{bmatrix},$$

$$J_0 = \frac{\partial x}{\partial \xi}\frac{\partial y}{\partial \eta} - \frac{\partial x}{\partial \eta}\frac{\partial y}{\partial \xi}, \tag{7.99b}$$

and $\partial x/\partial \xi$, $\partial x/\partial \eta$, $\partial y/\partial \xi$, $\partial y/\partial \eta$ can be obtained by the differentiation made for $(7.94\text{a})_1$. And, the integration of (7.97b) can be made by the standard Gaussian quadrature rule, that is,

$$\int_{A_e} f(x,y)\,dA = \int_{-1}^{1}\int_{-1}^{1} F(\xi,\eta)J_0\,d\xi\,d\eta \cong \sum_{i=1}^{n}\sum_{j=1}^{m} w_i w_j F(\xi_i,\eta_j)J_0(\xi_i,\eta_j). \tag{7.100}$$

REFERENCES

Bisplinghoff, R.L., Asheley, H., and Halfman, R.L., 1955, *Aeroelasticity*, Addison-Wesley Publishing Company, Cambridge.

Diederich, F.W. and Budiansky, B., 1948, "Divergence of Swept Wings," NASA TN 1680.

Hunsaker, J. C., 1949, *Theory of Wing Sections*, McGraw-Hill, New York.

Hwu, C. and Tsai, Z.S., 2002, "Aeroelastic Divergence of the Stiffened Multicell Composite Wing Structures," *Journal of Aircraft*, Vol. 39, No. 2, pp. 242–251.

Hwu, C. and Gai, H.S., 2003, "Vibration Analysis of Composite Wing Structures by a Matrix Form Comprehensive Model," *AIAA Journal*, Vol. 41, No. 11, pp. 2261–2273.

Hwu, C. and Yu, M.C., 2010, "A Comprehensive Finite Element Model for Tapered Composite Wing Structures," *Computer Modeling in Engineering & Sciences*, Vol. 67, No. 2, pp. 151–173.

Jones, R.M., 1974, *Mechanics of Composite Materials*, Scripta, Washington, D.C..

Karpouzian, G. and Librescu, L., 1994, "Comprehensive Model of Anisotropic Composite Aircraft Wings Suitable for Aeroelastic Analyses," *Journal of Aircraft*, Vol. 31, No. 3, pp. 703–712.

Krone, Jr. N. J., 1975, "Divergence Elimination with Advanced Composites," AIAA Paper 75-1009, Los Angeles.

Lerner, E. and Markowitz, J., 1979, "An Efficient Structural Resizing Procedure for Meeting Static Aeroelastic Design Objectives," *Journal of Aircraft*, Vol. 16, pp. 65–71.

Librescu, L. and Khdeir, A. A., 1988, "Aeroelastic Divergence of Swept-Forward Composite Wings Including Warping Restraint Effect," *AIAA Journal*, Vol. 26, No. 11, pp. 1373–1377.

Librescu, L. and Simovich, J., 1988, "General Formulation for the Aeroelastic Divergence of Composite Swept-Forward Wing Structures," *Journal of Aircraft*, Vol. 25, No. 4, pp. 364–371.

Librescu, L. and Thangjitham, S., 1991, "Analytical Studies on Static Aeroelastic Behavior of Forward-Swept Composite Wing Structures," *Journal of Aircraft*, Vol. 28, No. 2, pp. 151–157.

Lottati, I., 1985, "Flutter and Divergence Aeroelastic Characteristics for Composite Forward Swept Cantilevered Wing," *Journal of Aircraft*, Vol. 22, No. 11, pp. 1001–1007.

Megson, T. H. G., 1990, *Aircraft Structures - for Engineering Students*, 2nd Ed. Edward Arnold, London.

Meirovitch, L., 1990, *Dynamics and Control of Structures*, John Wiley & Sons, New York.

Oyibo, G. A., 1984, "Generic Approach to Determine Optimum Aeroelastic Characteristics for Composite Forward-Swept-Wing Aircraft," *AIAA Journal*, Vol. 22, No. 1, pp. 117–123.

Reddy, J. N., 1993, *Applied Functional Analysis and Variational Methods in Engineering*, McGraw-Hill, New York.

Tsai, Z. S., 2000, Aeroelastic Divergence of Composite Wing Structures, *M.S. Thesis*, Department of Aeronautics and Astronautics, National Cheng Kung University.

Weisshaar, T. A., 1980, "Divergence of Forward Swept Composite Wings," *Journal of Aircraft*, Vol. 17, No. 6, pp. 442–448.

CHAPTER 8

Anisotropic Elasticity

As stated at the beginning of this book, due to the nature of anisotropy composite materials are usually modeled as anisotropic elastic solids. To study the behavior of an elastic continuous medium, the theory of elasticity is a generally accepted model. A simple idealized linear stress–strain relationship gives a good description of the mechanical properties of many elastic materials around us. By this relation, we need 21 elastic constants to describe a linear anisotropic elastic material if the materials do not possess any symmetry properties. Consideration of the material symmetry may reduce the number of elastic constants. If the two-dimensional deformation is considered, the number of elastic constants used in the theory of elasticity can be further reduced. If the materials are under thermal environment, additional thermal properties are needed to express the temperature effects on the stress–strain relation. If the materials exhibit piezoelectric and/or piezomagnetic effects, the stress–strain relation should be further expanded to include the electric displacements, electric fields, and/or magnetic fields and magnetic flux. If not only the in-plane deformation but also the out-of-plane deflection are considered for the laminates made by laying up various unidirectional fiber-reinforced composites, the elastic constants will generally be reorganized into the extensional, coupling, and bending stiffnesses to suit for the classical lamination theory.

To deal with the problems of linear anisotropic elasticity, in the literature there are two different complex variable formulations. One is the *Lekhnitskii formalism* which starts with the equilibrated stress functions followed by compatibility equations, and the other is the *Stroh formalism* which starts with the compatible displacements followed by equilibrium equations.

DOI: 10.1201/9781003470465-8

One may refer to the textbooks Lekhnitskii (1963, 1968), Ting (1996), and Hwu (2010, 2021) for detailed description of these two formalisms. In this chapter only the finalized general solutions of Stroh formalism and their extensions are presented as those classified in Hwu and Becker (2022). Section 8.1 presents Stroh formalism for two-dimensional analysis, Section 8.2 presents Stroh-like formalism for coupled stretching–bending analysis, and Section 8.3 presents Radon–Stroh formalism for three-dimensional analysis. In the follow-up three sections, Sections 8.4, 8.5, and 8.6, several Green's functions are presented, respectively, for two-dimensional, coupled stretching–bending, and three-dimensional analysis.

8.1 TWO-DIMENSIONAL ANALYSIS

8.1.1 Stroh Formalism for Anisotropic Elastic Materials

In a fixed rectangular coordinate system (x_1, x_2, x_3), let u_i, ε_{ij}, σ_{ij}, $i, j = 1, 2, 3$, be, respectively, the displacements, strains, and stresses. The strain–displacement relations, the stress–strain laws, and the equilibrium equations for linear anisotropic elasticity can be expressed as

$$\varepsilon_{ij} = \frac{1}{2}\left(u_{i,j} + u_{j,i}\right),\quad \sigma_{ij} = C_{ijkl}\varepsilon_{kl},\ \sigma_{ij,j} = 0,\quad i,j,k,l = 1,2,3, \tag{8.1}$$

where repeated indices imply summation, a comma stands for differentiation, and C_{ijkl} are the elastic constants which are assumed to be fully symmetric and positive definite. These three equation sets, (8.1), constitute 15 partial differential equations with 15 unknown functions u_i, ε_{ij}, σ_{ij}, $i, j = 1, 2, 3$, in terms of three coordinate variables x_i, $i = 1, 2, 3$.

If only the two-dimensional deformation is considered, with the compatible displacements u_i as the basic functions the general solutions satisfying (8.1) can be obtained and expressed in matrix form as (Hwu, 2010)

$$\mathbf{u} = 2\,\mathrm{Re}\{\mathbf{A}\mathbf{f}(z)\},\quad \boldsymbol{\phi} = 2\,\mathrm{Re}\{\mathbf{B}\mathbf{f}(z)\}, \tag{8.2a}$$

where

$$\mathbf{A} = \left[\mathbf{a}_1\ \mathbf{a}_2\ \mathbf{a}_3\right],\qquad \mathbf{B} = \left[\mathbf{b}_1\ \mathbf{b}_2\ \mathbf{b}_3\right],$$
$$\mathbf{u} = \begin{Bmatrix} u_1 \\ u_2 \\ u_3 \end{Bmatrix},\ \boldsymbol{\phi} = \begin{Bmatrix} \phi_1 \\ \phi_2 \\ \phi_3 \end{Bmatrix},\ \mathbf{f}(z) = \begin{Bmatrix} f_1(z_1) \\ f_2(z_2) \\ f_3(z_3) \end{Bmatrix},$$
$$z_k = x_1 + \mu_k x_2,\ k = 1,2,3. \tag{8.2b}$$

In (8.2), Re denotes the real part of a complex number; $f_k(z_k)$, $k = 1, 2, 3$ are three holomorphic functions of complex variables z_k; ϕ_i, $i = 1, 2, 3$ are the *stress functions* related to the stresses σ_{ij} by

$$\sigma_{i1} = -\phi_{i,2}, \quad \sigma_{i2} = \phi_{i,1}. \tag{8.3}$$

μ_k and $(\mathbf{a}_k, \mathbf{b}_k)$, $k = 1, 2, 3$ are, respectively, the *material eigenvalues and material eigenvectors* (sometimes called *Stroh's eigenvalues and Stroh's eigenvectors*) of the following eigenrelation:

$$\mathbf{N}\boldsymbol{\xi} = \mu\boldsymbol{\xi}, \tag{8.4a}$$

where $\mathbf{N}$ is the *fundamental elasticity matrix* and $\boldsymbol{\xi}$ is its eigenvector defined by

$$\mathbf{N} = \begin{bmatrix} \mathbf{N}_1 & \mathbf{N}_2 \\ \mathbf{N}_3 & \mathbf{N}_1^T \end{bmatrix}, \quad \boldsymbol{\xi} = \begin{Bmatrix} \mathbf{a} \\ \mathbf{b} \end{Bmatrix}, \tag{8.4b}$$

and

$$\mathbf{N}_1 = -\mathbf{T}^{-1}\mathbf{R}^T, \quad \mathbf{N}_2 = \mathbf{T}^{-1} = \mathbf{N}_2^T, \quad \mathbf{N}_3 = \mathbf{R}\mathbf{T}^{-1}\mathbf{R}^T - \mathbf{Q}. \tag{8.4c}$$

The superscript T stands for the transpose and $\mathbf{Q}$, $\mathbf{R}$, $\mathbf{T}$ are the 3×3 real matrices defined by

$$Q_{ik} = C_{i1k1}, \quad R_{ik} = C_{i1k2}, \quad T_{ik} = C_{i2k2}. \tag{8.5}$$

By using the contracted notation for the fourth-order elastic tensor C_{ijks}, the three real matrices $\mathbf{Q}$, $\mathbf{R}$, $\mathbf{T}$ can also be expressed in matrix form as

$$\begin{aligned} \mathbf{Q} &= \begin{bmatrix} C_{11} & C_{16} & C_{15} \\ C_{16} & C_{66} & C_{56} \\ C_{15} & C_{56} & C_{55} \end{bmatrix}, \\ \mathbf{R} &= \begin{bmatrix} C_{16} & C_{12} & C_{14} \\ C_{66} & C_{26} & C_{46} \\ C_{56} & C_{25} & C_{45} \end{bmatrix}, \\ \mathbf{T} &= \begin{bmatrix} C_{66} & C_{26} & C_{46} \\ C_{26} & C_{22} & C_{24} \\ C_{46} & C_{24} & C_{44} \end{bmatrix}. \end{aligned} \tag{8.6}$$

From the definition (8.5) we see that **Q** and **T** are symmetric and positive definite if the strain energy is positive.

Note that although the fundamental elasticity matrix **N** is a real matrix, all its eigenvalues μ_k have been proved to be complex numbers. In the general solution (8.2), these eigenvalues have been assumed to be three pairs of distinct eigenvalues and the first three are set to be the one with positive imaginary part. If a material whose eigenvalues have repeated roots, to employ the general solution (8.2) one may use the technique of small perturbation to get distinct roots of μ_k.

8.1.2 Stroh Formalism in Laplace Domain for Viscoelastic Materials

Viscoelastic materials exhibit a time and rate dependence that is completely absent in elastic materials. With time-dependent *relaxation function* $C_{ijkl}(t)$, the stress–strain relation of viscoelastic materials can be expressed as

$$\sigma_{ij}(t) = C_{ijkl}(t)\varepsilon_{kl}(0) + \int_0^t C_{ijkl}(t-\tau)\frac{\partial \varepsilon_{kl}(\tau)}{\partial \tau}d\tau,\ i,j,k,l = 1,2,3. \quad (8.7)$$

If the condition that $\varepsilon_{kl}(t) = 0$ when $t < 0$ is considered, equation (8.7) can be re-written by using the notation $*$ for Stieltjes convolution, that is,

$$\sigma_{ij}(t) = C_{ijkl}(t) * d\varepsilon_{kl}(t) = \int_{-\infty}^{t} C_{ijkl}(t-\tau)d\varepsilon_{kl}(\tau). \quad (8.8)$$

Taking the Laplace transform on (8.8) leads to

$$\breve{\sigma}_{ij}(s) = s\breve{C}_{ijkl}(s)\breve{\varepsilon}_{kl}(s), \quad (8.9)$$

where s is the transform variable and the Laplace transform $\breve{f}(s)$ of $f(t)$ is defined as

$$\breve{f}(s) = \int_0^{\infty} f(t)e^{-st}dt. \quad (8.10)$$

With the replacement of the elastic stiffness tensor C_{ijkl} by $s\breve{C}_{ijkl}(s)$, it can be proved that all the basic equations in the Laplace domain required for anisotropic viscoelasticity are identical to the basic equations for linear anisotropic elasticity, which is the origin of the so-called *elastic-viscoelastic correspondence principle*. And hence, using the Stroh formalism shown in (8.2) for two-dimensional linear anisotropic elasticity, the general solutions for anisotropic viscoelasticity can be written as

$$\breve{\mathbf{u}}(\mathbf{x},s)=2\mathrm{Re}\{\mathbf{A}_s(s)\mathbf{f}_s(z,s)\},\quad \breve{\boldsymbol{\phi}}(\mathbf{x},s)=2\mathrm{Re}\{\mathbf{B}_s(s)\mathbf{f}_s(z,s)\},\tag{8.11a}$$

where

$$\breve{\mathbf{u}}=\begin{Bmatrix}\breve{u}_1\\ \breve{u}_2\\ \breve{u}_3\end{Bmatrix},\quad \breve{\boldsymbol{\phi}}=\begin{Bmatrix}\breve{\phi}_1\\ \breve{\phi}_2\\ \breve{\phi}_3\end{Bmatrix},\quad \mathbf{f}_s(z,s)=\begin{Bmatrix}f_1^s(z_1,s)\\ f_2^s(z_2,s)\\ f_3^s(z_3,s)\end{Bmatrix},$$
$$\mathbf{A}_s(s)=\left[\mathbf{a}_1^s(s)\ \mathbf{a}_2^s(s)\ \mathbf{a}_3^s(s)\right],\quad \mathbf{B}_s(s)=\left[\mathbf{b}_1^s(s)\ \mathbf{b}_2^s(s)\ \mathbf{b}_3^s(s)\right],$$
$$z_k=x_1+\mu_k^s x_2,\qquad k=1,2,3.\tag{8.11b}$$

$\breve{\mathbf{u}}$ and $\breve{\boldsymbol{\phi}}$ are the displacement and stress function vectors in the Laplace domain, and $\breve{\phi}_i, i=1,2,3$ are related to the stresses in the Laplace domain by

$$\breve{\sigma}_{i1}=-\breve{\phi}_{i,2},\quad \breve{\sigma}_{i2}=\breve{\phi}_{i,1}.\tag{8.12}$$

$\mathbf{f}_s(z,s)$ is a function vector composed of $f_k^s(z_k,s), k=1,2,3$, which are three holomorphic complex functions to be determined through the satisfaction of boundary conditions. μ_k^s and $(\mathbf{a}_k^s,\mathbf{b}_k^s), k=1,2,3$, are the material eigenvalues and their associated eigenvectors in the Laplace domain.

After the displacements $\breve{u}_i$, stress functions $\breve{\phi}_i$, and stresses $\breve{\sigma}_{ij}$ in the Laplace domain are solved, their associated solutions in real-time domain can be determined by numerical inversion of Laplace transform (Schapery 1962; Stehfest, 1970).

8.1.3 Expanded Stroh Formalism for Piezoelectric Materials

It is well known that piezoelectric materials produce an electric field when deformed and undergo deformation when subjected to an electric field. For electromechanical analysis of piezoelectric materials, the basic equations can be written as (Rogacheva, 1994)

$$\begin{cases}\sigma_{ij}=C_{ijkl}^{E}\varepsilon_{kl}-e_{kij}E_k,\\ D_j=e_{jkl}\varepsilon_{kl}+\omega_{jk}^{\varepsilon}E_k,\end{cases}\quad \varepsilon_{ij}=\frac{1}{2}\left(u_{i,j}+u_{j,i}\right),\quad \begin{cases}\sigma_{ij,j}=0,\\ D_{i,i}=0,\end{cases}\quad i,j,k,s=1,2,3, \tag{8.13}$$

where D_j and E_k denote, respectively, the electric displacement and electric field; C_{ijkl}^{E}, e_{kij}, and $\omega_{jk}^{\varepsilon}$ are, respectively, the elastic stiffness tensor at constant electric field, piezoelectric stress tensor, and dielectric permittivity tensor at constant strain. These tensors have symmetry properties, that is, $C_{ijkl}^{E}=C_{jikl}^{E}=C_{klij}^{E}$, $e_{kij}=e_{kji}$, and $\omega_{jk}^{\varepsilon}=\omega_{kj}^{\varepsilon}$. By letting

$$\begin{aligned}&D_j=\sigma_{4j},\ -E_j=u_{4,j}=2\varepsilon_{4j},\\ &C_{ijkl}=C_{ijkl}^{E},\ C_{ij4l}=e_{lij},\\ &C_{4jkl}=e_{jkl},\ C_{4j4l}=-\omega_{jl}^{\varepsilon},\quad i,j,k,l=1,2,3,\end{aligned} \tag{8.14}$$

(8.13) can be rewritten in an expanded tensor notation as

$$\sigma_{IJ}=C_{IJKL}\varepsilon_{KL},\quad \varepsilon_{IJ}=\left(u_{I,J}+u_{J,I}\right)/2,\quad \sigma_{IJ,J}=0,\qquad I,J,K,L=1,2,3,4, \tag{8.15}$$

in which the assumptions of $u_{J,4}=0$, $C_{IJKL}=C_{JIKL}=C_{IJLK}=C_{KLIJ}$, $\sigma_{IJ}=\sigma_{JI}$, and $\varepsilon_{IJ}=\varepsilon_{JI}$ have been made, and the components of σ_{44}, ε_{44}, C_{ij44}, and C_{44ij} are set to be zero.

Since the mathematical form of the basic equations (8.15) for piezoelectric materials is identical to (8.1) for elastic materials except for the expansion of index from 3 to 4, the general solutions for anisotropic piezoelectricity can therefore be written in the form of Stroh formalism and is called *expanded Stroh formalism*. With the solution given in (8.2), we now have the following solution for anisotropic piezoelectricity:

$$\mathbf{u}=2\,\mathrm{Re}\{\mathbf{Af}(z)\},\quad \boldsymbol{\phi}=2\,\mathrm{Re}\{\mathbf{Bf}(z)\}, \tag{8.16a}$$

where

$$\mathbf{u}=\begin{Bmatrix}u_1\\u_2\\u_3\\u_4\end{Bmatrix},\ \boldsymbol{\phi}=\begin{Bmatrix}\phi_1\\\phi_2\\\phi_3\\\phi_4\end{Bmatrix},\mathbf{f}(z)=\begin{Bmatrix}f_1(z_1)\\f_2(z_2)\\f_3(z_3)\\f_4(z_4)\end{Bmatrix},z_k=x_1+\mu_k x_2,k=1,2,3,4,$$

$$\mathbf{A}=\begin{bmatrix}\mathbf{a}_1 & \mathbf{a}_2 & \mathbf{a}_3 & \mathbf{a}_4\end{bmatrix},\ \mathbf{B}=\begin{bmatrix}\mathbf{b}_1 & \mathbf{b}_2 & \mathbf{b}_3 & \mathbf{b}_4\end{bmatrix}. \tag{8.16b}$$

The *generalized stress functions* ϕ_i are related to the stresses and electric displacements by

$$\begin{aligned}&\sigma_{i1}=-\phi_{i,2},\quad \sigma_{i2}=\phi_{i,1},\ i=1,2,3,\\&D_1=\sigma_{41}=-\phi_{4,2},\ D_2=\sigma_{42}=\phi_{4,1}.\end{aligned} \tag{8.17}$$

8.1.4 Expanded Stroh Formalism for MEE Materials

Like the piezoelectric materials discussed in the previous section, the piezomagnetic materials also have the ability of converting energy from one form (between magnetic and mechanical energies) to the other. If a multilayered composite is made up of different layers such as a fiber-reinforced composite layer and a composite layer consisting of piezoelectric materials and/or piezomagnetic materials, it may exhibit magnetoelectric effects that are more complicated than those of single-phase piezoelectric or piezomagnetic materials. Because of this intrinsic coupling phenomenon, piezoelectric, piezomagnetic, and magneto-electro-elastic (MEE) materials are widely used as sensors and actuators in intelligent advanced structure design. To study the magneto-electro-mechanical behaviors of MEE materials, suitable mathematical modeling becomes important. The basic equations for a linearly MEE solid can be written as (Soh and Liu, 2005; Xie, et al., 2016; Hwu, 2021)

$$\begin{cases}\sigma_{ij}=C_{ijkl}^{E,H}\varepsilon_{kl}-e_{kij}^{H}E_k-q_{kij}^{E}H_k\\D_j=e_{jkl}^{H}\varepsilon_{kl}+\omega_{jk}^{\varepsilon,H}E_k+m_{jk}^{\varepsilon}H_k\\B_j=q_{jkl}^{E}\varepsilon_{kl}+m_{jk}^{\varepsilon}E_k+\xi_{jk}^{\varepsilon,E}H_k\end{cases},\ \varepsilon_{ij}=\frac{1}{2}\left(u_{i,j}+u_{j,i}\right),\begin{cases}\sigma_{ij,j}=0\\D_{i,i}=0\\B_{j,j}=0\end{cases},i,j,k,l=1,2,3, \tag{8.18}$$

where the repeated indices imply summation through index ranges; σ_{ij}, ε_{ij}, D_j, E_j, B_j and H_j denote, respectively, the stress, strain, electric displacement, electric field, magnetic flux, and magnetic field; C_{ijkl} are the elastic stiffness; ω_{jk} and ξ_{jk} are the permittivity and permeability; e_{kij} are the piezoelectric coefficients; q_{kij} are the piezomagnetic coefficients; m_{jk} are the magneto-electric coefficients; the superscripts E, H, and ε mean that the associated quantities are measured at the condition of constant electric field, magnetic field, and elastic strain, respectively. Like the anisotropic elastic and piezoelectric materials, full symmetry is also assumed for all the above tensors. In addition to the symbol change introduced in (8.14) for piezoelectric materials, if we further let

$$B_j = \sigma_{5j},\ -H_j = u_{5,j} = 2\varepsilon_{5j},\ C_{ijkl} = C_{ijkl}^{E,H},$$
$$C_{ij5l} = C_{5lij} = q_{lij}^{E},\ C_{5j5l} = -\xi_{jl}^{\varepsilon,E},\ C_{4j5l} = C_{5j4l} = -m_{jl}^{\varepsilon},\ i,j,k,l = 1,2,3, \quad (8.19a)$$

(8.18) can also be written in an expanded tensor notation as

$$\sigma_{IJ} = C_{IJKL}\varepsilon_{KL},\ \varepsilon_{IJ} = \frac{1}{2}\left(u_{I,J} + u_{J,I}\right),\ \sigma_{IJ,J} = 0,\ I,J = 1,2,3,4,5, \quad (8.19b)$$

in which the assumptions of $u_{J,4} = u_{J,5} = 0$, $C_{IJKL} = C_{JIKL} = C_{IJLK} = C_{KLIJ}$, $\sigma_{IJ} = \sigma_{JI}$, and $\varepsilon_{IJ} = \varepsilon_{JI}$ have been made, and the extra components σ_{44}, ε_{44}, C_{ij44}, C_{44ij} as well as σ_{55}, ε_{55}, C_{ij55}, and C_{55ij} are all set to be zero. Thus, the expanded Stroh formalism for MEE materials can be written as

$$\mathbf{u} = 2\,\mathrm{Re}\{\mathbf{A}\mathbf{f}(z)\},\quad \boldsymbol{\phi} = 2\,\mathrm{Re}\{\mathbf{B}\mathbf{f}(z)\}, \quad (8.20a)$$

where

$$\mathbf{u} = \begin{Bmatrix} u_1 \\ u_2 \\ u_3 \\ u_4 \\ u_5 \end{Bmatrix},\ \boldsymbol{\phi} = \begin{Bmatrix} \phi_1 \\ \phi_2 \\ \phi_3 \\ \phi_4 \\ \phi_5 \end{Bmatrix}, \mathbf{f}(z) = \begin{Bmatrix} f_1(z_1) \\ f_2(z_2) \\ f_3(z_3) \\ f_4(z_4) \\ f_5(z_5) \end{Bmatrix}, z_k = x_1 + \mu_k x_2, k = 1,2,3,4,5,$$
$$\mathbf{A} = \begin{bmatrix} \mathbf{a}_1 & \mathbf{a}_2 & \mathbf{a}_3 & \mathbf{a}_4 & \mathbf{a}_5 \end{bmatrix},\ \mathbf{B} = \begin{bmatrix} \mathbf{b}_1 & \mathbf{b}_2 & \mathbf{b}_3 & \mathbf{b}_4 & \mathbf{b}_5 \end{bmatrix}. \quad (8.20b)$$

The generalized stress functions ϕ_i are related to the stresses, electric displacements, and magnetic induction by

$$\sigma_{i1} = -\phi_{i,2},\quad \sigma_{i2} = \phi_{i,1},\ i = 1,2,3,$$
$$D_1 = \sigma_{41} = -\phi_{4,2},\ D_2 = \sigma_{42} = \phi_{4,1},$$
$$B_1 = \sigma_{51} = -\phi_{5,2},\ B_2 = \sigma_{52} = \phi_{5,1}. \quad (8.21)$$

8.1.5 Extended Stroh Formalism for Thermoelastic Problems

For the uncoupled steady-state thermoelastic problems, to include the thermal effects in the anisotropic elastic solids the constitutive relation stated in $(8.1)_2$ should be modified as

$$\sigma_{ij} = C_{ijkl}\varepsilon_{kl} - \beta_{ij}T, \quad i,j,k,l = 1,2,3, \tag{8.22}$$

where β_{ij} are the thermal moduli which are related to the thermal expansion coefficients α_{ij} by

$$\beta_{ij} = C_{ijkl}\alpha_{kl}. \tag{8.23}$$

T denotes the change in temperature from the stress-free state, which is related to the heat flux h_i by

$$h_i = -k_{ij}T_{,j}, \tag{8.24}$$

where k_{ij} are the heat conduction coefficients.

For two-dimensional deformations in which u_k, $k = 1, 2, 3$ and T depend on x_1 and x_2 only, a general solution for the temperature change T, the heat flux h_i, and the displacements and stresses has been obtained as (Hwu, 2010)

$$\begin{aligned} &T = 2\,\mathrm{Re}\{g'(z_t)\}, \quad h_i = -2\,\mathrm{Re}\{(k_{i1} + \tau k_{i2})g''(z_t)\}, \quad z_t = x_1 + \tau x_2, \\ &\mathbf{u} = 2\,\mathrm{Re}\{\mathbf{A}\mathbf{f}(z) + \mathbf{c}g(z_t)\}, \quad \boldsymbol{\phi} = 2\,\mathrm{Re}\{\mathbf{B}\mathbf{f}(z) + \mathbf{d}g(z_t)\}, \end{aligned} \tag{8.25}$$

where $g(z_t)$ is an arbitrary function of complex variable z_t and the prime $\bullet'$ denotes differentiation with respect to its argument z_t. τ is the *thermal eigenvalue* calculated by

$$k_{22}\tau^2 + (k_{12} + k_{21})\tau + k_{11} = 0. \tag{8.26}$$

The fact that heat always flows from higher temperature to lower temperature tells us that the roots of (8.26) cannot be real (Ting, 1996). There is one pair of complex conjugates for τ, and we let τ in (8.25) be the one with positive imaginary part. Like the eigenrelation established in (8.4) for the determination of the material eigenvectors **a** and **b**, the *thermal eigenvectors* **c** and **d** can be determined from the following eigenrelation:

$$\mathbf{N}\boldsymbol{\eta} = \tau\boldsymbol{\eta} + \boldsymbol{\gamma}. \tag{8.27}$$

In this eigenrelation, **N** is the fundamental elasticity matrix defined in (8.4b,c), and $\boldsymbol{\eta}$ and $\boldsymbol{\gamma}$ are defined by

$$\boldsymbol{\eta}=\begin{Bmatrix}\mathbf{c}\\ \mathbf{d}\end{Bmatrix},\quad \boldsymbol{\gamma}=-\begin{bmatrix}\mathbf{0} & \mathbf{N}_2\\ \mathbf{I} & \mathbf{N}_1^T\end{bmatrix}\begin{Bmatrix}\boldsymbol{\beta}_1\\ \boldsymbol{\beta}_2\end{Bmatrix}, \tag{8.28a}$$

where

$$\boldsymbol{\beta}_1=\begin{Bmatrix}\beta_{11}\\ \beta_{21}\\ \beta_{31}\end{Bmatrix},\quad \boldsymbol{\beta}_2=\begin{Bmatrix}\beta_{12}\\ \beta_{22}\\ \beta_{32}\end{Bmatrix}. \tag{8.28b}$$

Note that the solutions shown in (8.25) to (8.28) can be used not only for the anisotropic elastic materials but also for the piezoelectric/MEE materials. The only difference is that all the related matrices or vectors should be expanded, such as

$$\boldsymbol{\beta}_1=\begin{Bmatrix}\beta_{11}\\ \beta_{21}\\ \beta_{31}\\ \beta_{41}\end{Bmatrix},\quad \boldsymbol{\beta}_2=\begin{Bmatrix}\beta_{12}\\ \beta_{22}\\ \beta_{32}\\ \beta_{42}\end{Bmatrix},\quad \text{for piezoelectric materials,}$$

$$\boldsymbol{\beta}_1=\begin{Bmatrix}\beta_{11}\\ \beta_{21}\\ \beta_{31}\\ \beta_{41}\\ \beta_{51}\end{Bmatrix},\quad \boldsymbol{\beta}_2=\begin{Bmatrix}\beta_{12}\\ \beta_{22}\\ \beta_{32}\\ \beta_{42}\\ \beta_{52}\end{Bmatrix},\quad \text{for MEE materials.} \tag{8.29}$$

8.2 COUPLED STRETCHING–BENDING ANALYSIS

In the previous section only the two-dimensional problems are considered, that is, all the physical responses depend only on two variables x_1 and x_2. Thus, the complex variable, $z_\alpha = x_1 + \mu_\alpha x_2$, which is a combination of two variables, is an appropriate variable to develop the formalisms for two-dimensional anisotropic elasticity. When the *anisotropic plates* or *laminates* are subjected to in-plane loads as well as lateral loads or bending moments, their deformations and stresses will depend on the third variable x_3. If the thickness of the plates or laminates is relatively small compared to their other dimensions, Kirchhoff assumptions are usually made in classical plate theory or lamination theory. In this section the most general laminates, *symmetric or unsymmetric*, are considered. When the laminates are symmetric, the in-plane and out-of-plane deformation will decouple. If the laminates are unsymmetric they will be stretched as well as bent even under pure in-plane forces or pure bending moments. Thus, in Section 8.2.1 we will present the Stroh-like formalism for the most general cases of laminated plates, which can be specialized to the cases of symmetric laminates for plate bending analysis stated in Section 8.2.2. The expanded and extended formalisms for electro-elastic laminated plates, MEE laminated plates, and thermal stresses will then be presented, respectively, in Sections 8.2.3, 8.2.4, and 8.2.5.

8.2.1 Stroh-Like Formalism for General Laminated Plates

For the coupled stretching–bending analysis of thin laminated plates, the strain–displacement relations, the constitutive laws, and the equilibrium equations can be written in terms of tensor notation as follows (Hwu, 2003, 2010):

$$
\begin{aligned}
&\varepsilon_{ij} = \frac{1}{2}\left(u_{i,j} + u_{j,i}\right), \ \kappa_{ij} = \frac{1}{2}\left(\beta_{i,j} + \beta_{j,i}\right), \ \beta_1 = -w_{,1}, \ \beta_2 = -w_{,2}, \\
&N_{ij} = A_{ijkl}\varepsilon_{kl} + B_{ijkl}\kappa_{kl}, \ M_{ij} = B_{ijkl}\varepsilon_{kl} + D_{ijkl}\kappa_{kl}, \\
&N_{ij,j} = 0, \ M_{ij,ij} + q = 0, \ Q_i = M_{ij,j}, \qquad i,j,k,l = 1,2,
\end{aligned} \tag{8.30}
$$

where u_1, u_2, and w are the midplane displacements; and β_1 and β_2 are the negative slopes in x_1 and x_2 directions; ε_{ij} and κ_{ij} are the midplane strains and curvatures; N_{ij}, M_{ij}, and Q_i denote the stress resultants, bending moments, and transverse shear forces; A_{ijkl}, B_{ijkl}, and D_{ijkl} are, respectively, the extensional, coupling, and bending stiffness tensors; q is the lateral distributed load applied on the laminates. A general solution satisfying all the basic equations shown in (8.30) has been obtained as (Hwu, 2010)

$$
\mathbf{u}_d = 2\,\mathrm{Re}\{\mathbf{Af}(z)\}, \quad \boldsymbol{\phi}_d = 2\,\mathrm{Re}\{\mathbf{Bf}(z)\}, \tag{8.31a}
$$

where

$$\mathbf{u}_d = \begin{Bmatrix} \mathbf{u} \\ \boldsymbol{\beta} \end{Bmatrix} = \begin{Bmatrix} u_1 \\ u_2 \\ \beta_1 \\ \beta_2 \end{Bmatrix}, \quad \boldsymbol{\phi}_d = \begin{Bmatrix} \boldsymbol{\phi} \\ \boldsymbol{\psi} \end{Bmatrix} = \begin{Bmatrix} \phi_1 \\ \phi_2 \\ \psi_1 \\ \psi_2 \end{Bmatrix}, \quad \mathbf{f}(z) = \begin{Bmatrix} f_1(z_1) \\ f_2(z_2) \\ f_3(z_3) \\ f_4(z_4) \end{Bmatrix},$$

$$\mathbf{A} = \begin{bmatrix} \mathbf{a}_1 & \mathbf{a}_2 & \mathbf{a}_3 & \mathbf{a}_4 \end{bmatrix}, \quad \mathbf{B} = \begin{bmatrix} \mathbf{b}_1 & \mathbf{b}_2 & \mathbf{b}_3 & \mathbf{b}_4 \end{bmatrix},$$

$$z_\alpha = x_1 + \mu_\alpha x_2, \quad \alpha = 1,2,3,4. \tag{8.31b}$$

In (8.31b), ϕ_i, $i = 1, 2$, are the stress functions related to in-plane forces N_{ij}, and ψ_i, $i = 1, 2$, are the stress functions related to bending moments M_{ij}, transverse shear forces Q_i and effective transverse shear forces V_i. Their relations are

$$\begin{aligned} &N_{11} = -\phi_{1,2}, \quad N_{22} = \phi_{2,1}, \quad N_{12} = \phi_{1,1} = -\phi_{2,2} = N_{21}, \\ &M_{11} = -\psi_{1,2}, \quad M_{22} = \psi_{2,1}, \quad M_{12} = (\psi_{1,1} - \psi_{2,2})/2 = M_{21}, \\ &Q_1 = -\eta_{,2}, \quad Q_2 = \eta_{,1}, \quad \eta = (\psi_{1,1} + \psi_{2,2})/2, \\ &V_1 = -\psi_{2,22}, \quad V_2 = \psi_{1,11}, \end{aligned} \tag{8.32a}$$

or

$$\begin{aligned} &N_{ss} = -\mathbf{s}^T\boldsymbol{\phi}_{,n}, \quad N_{nn} = \mathbf{n}^T\boldsymbol{\phi}_{,s}, \quad N_{sn} = -\mathbf{n}^T\boldsymbol{\phi}_{,n} = \mathbf{s}^T\boldsymbol{\phi}_{,s} = N_{ns}, \\ &M_{ss} = -\mathbf{s}^T\boldsymbol{\psi}_{,n}, \quad M_{nn} = \mathbf{n}^T\boldsymbol{\psi}_{,s}, \quad M_{sn} = \left(\mathbf{s}^T\boldsymbol{\psi}_{,s} - \mathbf{n}^T\boldsymbol{\psi}_{,n}\right)/2 = M_{ns}, \\ &Q_s = -\eta_{,n}, \quad Q_n = \eta_{,s}, \quad \eta = \left(\mathbf{s}^T\boldsymbol{\psi}_{,s} + \mathbf{n}^T\boldsymbol{\psi}_{,n}\right)/2, \\ &V_s = -\left(\mathbf{n}^T\boldsymbol{\psi}_{,n}\right)_{,n}, \quad V_n = \left(\mathbf{s}^T\boldsymbol{\psi}_{,s}\right)_{,s}, \end{aligned} \tag{8.32b}$$

where

$$\mathbf{s}^T = (\cos\theta, \sin\theta), \quad \mathbf{n}^T = (-\sin\theta, \cos\theta), \tag{8.32c}$$

are the unit vectors denoting the tangential and normal directions, respectively.

The material eigenvalues μ_α and their associated eigenvectors $\mathbf{a}_\alpha$ and $\mathbf{b}_\alpha$ can be determined from the following eigenrelation:

$$\mathbf{N}\boldsymbol{\xi} = \mu\boldsymbol{\xi}, \tag{8.33a}$$

where

$$\mathbf{N} = \begin{bmatrix} \mathbf{N}_1 & \mathbf{N}_2 \\ \mathbf{N}_3 & \mathbf{N}_1^T \end{bmatrix} = \mathbf{I}_t \mathbf{N}_m \mathbf{I}_t, \quad \boldsymbol{\xi} = \begin{Bmatrix} \mathbf{a} \\ \mathbf{b} \end{Bmatrix}, \tag{8.33b}$$

and

$$\mathbf{N}_m = \begin{bmatrix} (\mathbf{N}_m)_1 & (\mathbf{N}_m)_2 \\ (\mathbf{N}_m)_3 & (\mathbf{N}_m)_1^T \end{bmatrix}, \quad \mathbf{I}_t = \begin{bmatrix} \mathbf{I}_1 & \mathbf{I}_2 \\ \mathbf{I}_2 & \mathbf{I}_1 \end{bmatrix},$$
$$(\mathbf{N}_m)_1 = -\mathbf{T}_m^{-1}\mathbf{R}_m^T, \quad (\mathbf{N}_m)_2 = \mathbf{T}_m^{-1} = (\mathbf{N}_m)_2^T,$$
$$(\mathbf{N}_m)_3 = \mathbf{R}_m\mathbf{T}_m^{-1}\mathbf{R}_m^T - \mathbf{Q}_m = (\mathbf{N}_m)_3^T, \quad \mathbf{I}_1 = \begin{bmatrix} \mathbf{I} & \mathbf{0} \\ \mathbf{0} & \mathbf{0} \end{bmatrix}, \quad \mathbf{I}_2 = \begin{bmatrix} \mathbf{0} & \mathbf{0} \\ \mathbf{0} & \mathbf{I} \end{bmatrix}. \tag{8.33c}$$

I is the 2 × 2 identity matrix, and the three 4 × 4 real matrices $\mathbf{Q}_m$, $\mathbf{R}_m$, and $\mathbf{T}_m$ are defined by

$$\mathbf{Q}_m = \begin{bmatrix} \tilde{A}_{11} & \tilde{A}_{16} & \tilde{B}_{16}/2 & \tilde{B}_{12} \\ \tilde{A}_{16} & \tilde{A}_{66} & \tilde{B}_{66}/2 & \tilde{B}_{62} \\ \tilde{B}_{16}/2 & \tilde{B}_{66}/2 & -\tilde{D}_{66}/4 & -\tilde{D}_{26}/2 \\ \tilde{B}_{12} & \tilde{B}_{62} & -\tilde{D}_{26}/2 & -\tilde{D}_{22} \end{bmatrix}, \tag{8.34a}$$

$$\mathbf{R}_m = \begin{bmatrix} \tilde{A}_{16} & \tilde{A}_{12} & -\tilde{B}_{11} & -\tilde{B}_{16}/2 \\ \tilde{A}_{66} & \tilde{A}_{26} & -\tilde{B}_{61} & -\tilde{B}_{66}/2 \\ \tilde{B}_{66}/2 & \tilde{B}_{26}/2 & \tilde{D}_{16}/2 & \tilde{D}_{66}/4 \\ \tilde{B}_{62} & \tilde{B}_{22} & \tilde{D}_{12} & \tilde{D}_{26}/2 \end{bmatrix}, \tag{8.34b}$$

$$\mathbf{T}_m = \begin{bmatrix} \tilde{A}_{66} & \tilde{A}_{26} & -\tilde{B}_{61} & -\tilde{B}_{66}/2 \\ \tilde{A}_{26} & \tilde{A}_{22} & -\tilde{B}_{21} & -\tilde{B}_{26}/2 \\ -\tilde{B}_{61} & -\tilde{B}_{21} & -\tilde{D}_{11} & -\tilde{D}_{16}/2 \\ -\tilde{B}_{66}/2 & -\tilde{B}_{26}/2 & -\tilde{D}_{16}/2 & -\tilde{D}_{66}/4 \end{bmatrix}. \tag{8.34c}$$

$\tilde{A}_{ij}, \tilde{B}_{ij}$, and $\tilde{D}_{ij}$ are components of the matrices $\tilde{\boldsymbol{A}}$, $\tilde{\boldsymbol{B}}$, and $\tilde{\boldsymbol{D}}$ which are related to the extensional, coupling, and bending stiffness matrices $\boldsymbol{A}$, $\boldsymbol{B}$, and $\boldsymbol{D}$ by

$$\tilde{\boldsymbol{A}} = \boldsymbol{A} - \boldsymbol{B}\boldsymbol{D}^{-1}\boldsymbol{B},\ \tilde{\boldsymbol{B}} = \boldsymbol{B}\boldsymbol{D}^{-1},\ \tilde{\boldsymbol{D}} = \boldsymbol{D}^{-1}. \tag{8.35}$$

In the Stroh-like formalism presented above, once the complex function vector $\mathbf{f}(z)$ is determined, the displacements/slopes and stress functions can be provided by (8.31), and the stress resultants, bending moments, transverse shear forces can then be obtained by (8.32) through appropriate differentiation of $\mathbf{f}(z)$. On the other hand, the calculation of plate deflection w requires the integration of $\mathbf{f}(z)$, which comes from its relations with negative slopes, $(8.30)_{3,4}$. With this understanding, extracting the slope β_1 (or β_2) from the displacement vector $\mathbf{u}_d$ of (8.31a) and performing the integration would lead to

$$w = -2\mathbf{i}_3^T \operatorname{Re}\{\mathbf{A}\tilde{\mathbf{f}}(z)\} = -2\mathbf{i}_4^T \operatorname{Re}\{\mathbf{A} < \mu_\alpha^{-1} > \tilde{\mathbf{f}}(z)\},\quad \tilde{\mathbf{f}}(z) = \int \mathbf{f}(z)dz, \tag{8.36}$$

where the angular bracket $<\bullet_\alpha>$ stands for an 4 × 4 diagonal matrix in which each component is varied according to its subscript α, for example, $\langle \mu_\alpha^{-1} \rangle = diag\left[\mu_1^{-1}, \mu_2^{-1}, \mu_3^{-1}, \mu_4^{-1}\right]$; the vector $\mathbf{i}_k$, k = 1, 2, 3, 4 denotes a 4 × 1 column vector with unit value at the kth component and all the other components are zero. The second equality of $(8.36)_1$ comes from the compatibility condition of plate slopes implied in $(8.30)_{3,4}$.

8.2.2 Specialization to Plate Bending Analysis

If a symmetric laminate is considered, the coupling stiffness becomes zero and the in-plane problem and plate bending problem can be decoupled. Considering only the plate bending problem, the basic equations shown in (8.30) with $B_{ijkl} = 0$ can now be specialized to

$$M_{ij} = -D_{ijkl}w_{,kl},\ \ M_{ij,ij} + q = 0,\ \ Q_i = M_{ij,j},\ i,j,k,l = 1,2. \tag{8.37}$$

A general solution satisfying (8.37) can then be specialized from (8.31) to

$$\boldsymbol{\beta} = 2\operatorname{Re}\{\mathbf{Af}(z)\},\ \ \boldsymbol{\psi} = 2\operatorname{Re}\{\mathbf{Bf}(z)\}, \tag{8.38a}$$

where

$$\boldsymbol{\beta}=\begin{Bmatrix}\beta_1\\ \beta_2\end{Bmatrix}=\begin{Bmatrix}-w_{,1}\\ -w_{,2}\end{Bmatrix},\ \boldsymbol{\psi}=\begin{Bmatrix}\psi_1\\ \psi_2\end{Bmatrix},\ \mathbf{f}(z)=\begin{Bmatrix}f_1(z_1)\\ f_2(z_2)\end{Bmatrix},$$
$$\mathbf{A}=\begin{bmatrix}\mathbf{a}_1 & \mathbf{a}_2\end{bmatrix},\ \mathbf{B}=\begin{bmatrix}\mathbf{b}_1 & \mathbf{b}_2\end{bmatrix},\ z_\alpha=x_1+\mu_\alpha x_2,\ \alpha=1,2. \quad (8.38b)$$

The fundamental elasticity matrix **N** used to determine the material eigenvalues μ_α and eigenvectors $\mathbf{a}_\alpha$ and $\mathbf{b}_\alpha$, as shown in (8.33b), can now be specialized to

$$\mathbf{N}_1=\begin{bmatrix}0 & 1\\ -\dfrac{D_{12}}{D_{22}} & -\dfrac{2D_{26}}{D_{22}}\end{bmatrix},\quad \mathbf{N}_2=\begin{bmatrix}0 & 0\\ 0 & \dfrac{1}{D_{22}}\end{bmatrix},$$
$$\mathbf{N}_3=\begin{bmatrix}-D_{11}+\dfrac{D_{12}^2}{D_{22}} & -2D_{16}+\dfrac{2D_{12}D_{26}}{D_{22}}\\ -2D_{16}+\dfrac{2D_{12}D_{26}}{D_{22}} & -4D_{66}+\dfrac{4D_{26}^2}{D_{22}}\end{bmatrix}, \quad (8.39)$$

in which D_{ij} is the contracted notation of bending stiffness tensor D_{ijkl}.

8.2.3 Expanded Stroh-Like Formalism for Electro-Elastic Laminates

If a multilayered composite is made up of different layers such as fiber-reinforced composite layers and layers consisting of piezoelectric materials, it may exhibit electric effects that are more complicated than those of single-phase piezoelectric materials. Like the extension of Stroh formalism to piezoelectric materials discussed in Section 8.1.3, the Stroh-like formalism can also be expanded to the coupled mechanical-electrical analysis. Let

$$u_{4,j}+x_3\beta_{4,j}=-E_j^{(0)}-x_3E_j^{(1)}=-E_j,$$
$$N_{4j}=\int_{-h/2}^{h/2}\sigma_{4j}dx_3=\int_{-h/2}^{h/2}D_jdx_3=\tilde{D}_j,$$
$$M_{4j}=\int_{-h/2}^{h/2}\sigma_{4j}x_3dx_3=\int_{-h/2}^{h/2}D_jx_3dx_3=\tilde{D}_j^*,\quad j=1,2, \quad (8.40)$$

where E_j and D_j are, respectively, the electric field and electric displacement. The constitutive laws and the equilibrium equations for the coupled stretching–bending analysis can be written in terms of tensor notation as follows (Hwu, 2010):

$$N_{pq} = A_{pqrs}u_{r,s} + B_{pqrs}\beta_{r,s}, \quad M_{pq} = B_{pqrs}u_{r,s} + D_{pqrs}\beta_{r,s},$$
$$N_{pj,j} = 0, \quad M_{ij,ij} = 0, \quad M_{4j,j} = 0, \quad p,q,r,s = 1,2,4; i,j = 1,2, \tag{8.41}$$

in which the tensors A_{pqrs}, B_{pqrs}, and D_{pqrs} are related to their associated contracted notation by the following rules:

$$11 \leftrightarrow 1, \quad 22 \leftrightarrow 2, \quad 12 \text{ or } 21 \leftrightarrow 6, \quad 41 \leftrightarrow 7, \quad 42 \leftrightarrow 8. \tag{8.42}$$

The general solutions satisfying (8.41) can be expressed in matrix form as

$$\mathbf{u}_d = 2\,\mathrm{Re}\{\mathbf{A}\mathbf{f}(z)\}, \quad \boldsymbol{\phi}_d = 2\,\mathrm{Re}\{\mathbf{B}\mathbf{f}(z)\}, \tag{8.43a}$$

where

$$\mathbf{u}_d = \begin{Bmatrix} \mathbf{u} \\ \boldsymbol{\beta} \end{Bmatrix} = \begin{Bmatrix} u_1 \\ u_2 \\ u_4 \\ \beta_1 \\ \beta_2 \\ \beta_4 \end{Bmatrix}, \quad \boldsymbol{\phi}_d = \begin{Bmatrix} \boldsymbol{\phi} \\ \boldsymbol{\psi} \end{Bmatrix} = \begin{Bmatrix} \phi_1 \\ \phi_2 \\ \phi_4 \\ \psi_1 \\ \psi_2 \\ \psi_4 \end{Bmatrix}, \quad \mathbf{f}(z) = \begin{Bmatrix} f_1(z_1) \\ f_2(z_2) \\ f_3(z_3) \\ f_4(z_4) \\ f_5(z_5) \\ f_6(z_6) \end{Bmatrix},$$
$$\mathbf{A} = \begin{bmatrix} \mathbf{a}_1 & \mathbf{a}_2 & \mathbf{a}_3 & \mathbf{a}_4 & \mathbf{a}_5 & \mathbf{a}_6 \end{bmatrix}, \quad \mathbf{B} = \begin{bmatrix} \mathbf{b}_1 & \mathbf{b}_2 & \mathbf{b}_3 & \mathbf{b}_4 & \mathbf{b}_5 & \mathbf{b}_6 \end{bmatrix},$$
$$z_\alpha = x_1 + \mu_\alpha x_2, \quad \alpha = 1,2,3,4,5,6. \tag{8.43b}$$

The stress resultants, bending moments, transverse shear forces, and effective transverse shear forces are related to the stress functions by (8.32). The relations for the generalized stress resultants/bending moments are

$$N_{41} = -\phi_{4,2}, \quad N_{42} = \phi_{4,1}, \quad M_{41} = -\psi_{4,2}, \quad M_{42} = \psi_{4,1}. \tag{8.44}$$

The material eigenvalues μ_α and their associated eigenvectors $\mathbf{a}_\alpha$ and $\mathbf{b}_\alpha$ can also be determined from the eigenrelation shown in (8.33) with $\mathbf{Q}_m$, $\mathbf{R}_m$, and $\mathbf{T}_m$ be defined by

$$\mathbf{Q}_m = \begin{bmatrix} \tilde{A}_{11} & \tilde{A}_{16} & \tilde{A}_{17} & \tilde{B}_{16}/2 & \tilde{B}_{12} & \tilde{B}_{18}/2 \\ \tilde{A}_{16} & \tilde{A}_{66} & \tilde{A}_{67} & \tilde{B}_{66}/2 & \tilde{B}_{62} & \tilde{B}_{68}/2 \\ \tilde{A}_{17} & \tilde{A}_{67} & \tilde{A}_{77} & \tilde{B}_{76}/2 & \tilde{B}_{72} & \tilde{B}_{78}/2 \\ \tilde{B}_{16}/2 & \tilde{B}_{66}/2 & \tilde{B}_{76}/2 & -\tilde{D}_{66}/4 & -\tilde{D}_{26}/2 & -\tilde{D}_{68}/4 \\ \tilde{B}_{12} & \tilde{B}_{62} & \tilde{B}_{72} & -\tilde{D}_{26}/2 & -\tilde{D}_{22} & -\tilde{D}_{28}/2 \\ \tilde{B}_{18}/2 & \tilde{B}_{68}/2 & \tilde{B}_{78}/2 & -\tilde{D}_{68}/4 & -\tilde{D}_{28}/2 & -\tilde{D}_{88}/4 \end{bmatrix}, \tag{8.45a}$$

$$\mathbf{R}_m = \begin{bmatrix} \tilde{A}_{16} & \tilde{A}_{12} & \tilde{A}_{18} & -\tilde{B}_{11} & -\tilde{B}_{16}/2 & -\tilde{B}_{17}/2 \\ \tilde{A}_{66} & \tilde{A}_{26} & \tilde{A}_{68} & -\tilde{B}_{61} & -\tilde{B}_{66}/2 & -\tilde{B}_{67}/2 \\ \tilde{A}_{67} & \tilde{A}_{27} & \tilde{A}_{78} & -\tilde{B}_{71} & -\tilde{B}_{76}/2 & -\tilde{B}_{77}/2 \\ \tilde{B}_{66}/2 & \tilde{B}_{26}/2 & \tilde{B}_{86}/2 & \tilde{D}_{16}/2 & \tilde{D}_{66}/4 & \tilde{D}_{67}/4 \\ \tilde{B}_{62} & \tilde{B}_{22} & \tilde{B}_{82} & \tilde{D}_{12} & \tilde{D}_{26}/2 & \tilde{D}_{27}/2 \\ \tilde{B}_{68}/2 & \tilde{B}_{28}/2 & \tilde{B}_{88}/2 & \tilde{D}_{18}/2 & \tilde{D}_{68}/4 & \tilde{D}_{78}/4 \end{bmatrix}, \tag{8.45b}$$

$$\mathbf{T}_m = \begin{bmatrix} \tilde{A}_{66} & \tilde{A}_{26} & \tilde{A}_{68} & -\tilde{B}_{61} & -\tilde{B}_{66}/2 & -\tilde{B}_{67}/2 \\ \tilde{A}_{26} & \tilde{A}_{22} & \tilde{A}_{28} & -\tilde{B}_{21} & -\tilde{B}_{26}/2 & -\tilde{B}_{27}/2 \\ \tilde{A}_{68} & \tilde{A}_{28} & \tilde{A}_{88} & -\tilde{B}_{81} & -\tilde{B}_{86}/2 & -\tilde{B}_{87}/2 \\ -\tilde{B}_{61} & -\tilde{B}_{21} & -\tilde{B}_{81} & -\tilde{D}_{11} & -\tilde{D}_{16}/2 & -\tilde{D}_{17}/2 \\ -\tilde{B}_{66}/2 & -\tilde{B}_{26}/2 & -\tilde{B}_{86}/2 & -\tilde{D}_{16}/2 & -\tilde{D}_{66}/4 & -\tilde{D}_{67}/4 \\ -\tilde{B}_{67}/2 & -\tilde{B}_{27}/2 & -\tilde{B}_{87}/2 & -\tilde{D}_{17}/2 & -\tilde{D}_{67}/4 & -\tilde{D}_{77}/4 \end{bmatrix}. \tag{8.45c}$$

Note that for the electro-elastic laminates, the transformation matrix $\mathbf{I}_t$ of (8.33c) should be changed to

$$\mathbf{I}_t = \begin{bmatrix} \mathbf{I}_1 & \mathbf{I}_2 \\ \mathbf{I}_3 & \mathbf{I}_1 \end{bmatrix}, \tag{8.46a}$$

where

$$\mathbf{I}_1 = \langle 1,1,1,0,0,0 \rangle,\quad \mathbf{I}_2 = \langle 0,0,0,1,1,2 \rangle,\quad \mathbf{I}_3 = \langle 0,0,0,1,1,0.5 \rangle, \tag{8.46b}$$

and the symbol $< \bullet >$ denotes a diagonal matrix.

8.2.4 Expanded Stroh-Like Formalism for MEE Laminated Plates

The constitutive laws for three-dimensional linearly anisotropic MEE solids can generally be expressed in tensor notation as shown in $(8.18)_1$. When we model the thin MEE laminated plates by Kirchhoff's hypotheses, like the matrix form shown in (3.53a), the relation $(8.18)_1$ can be converted to the following tensor notation (Hsieh and Hwu, 2003):

$$N_{pq} = A_{pqrs}\varepsilon_{rs} + B_{pqrs}\kappa_{rs},\ M_{pq} = B_{pqrs}\varepsilon_{rs} + D_{pqrs}\kappa_{rs},\ p,q,r,s = 1,2,4,5, \quad (8.47)$$

where ε_{pq} and κ_{pq} denote the generalized mid-plane strains and plate curvatures; N_{pq} and M_{pq} represent, respectively, the generalized stress resultants and bending moments; A_{pqrs}, B_{pqrs} and D_{pqrs} are the generalized extensional, coupling, and bending stiffness tensors related to the material properties of each layer and stacking sequence.

In addition to the constitutive relation (8.47), in order to study the magneto-electro-mechanical behavior of the thin MEE laminated plates with coupled stretching–bending deformation we also need to know the kinematic relations and equilibrium equations as shown in (3.57), which can be expressed in tensor notation as

$$\begin{aligned}
&\varepsilon_{pq} = \left(u_{p,q} + u_{q,p}\right)/2,\ \kappa_{pq} = \left(\beta_{p,q} + \beta_{q,p}\right)/2,\ \beta_j = -w_{,j}.\\
&N_{pj,j} + q_p = 0,\ M_{pj,j} + m_p = Q_p,\ Q_{j,j} + q_3 = 0,\\
&\qquad p,q,r,s = 1,2,4,5, \qquad j = 1,2,
\end{aligned} \quad (8.48)$$

where the subscript comma stands for partial differentiation and $u_{j,4} = \beta_{j,4} = u_{j,5} = \beta_{j,5} = 0$ and $Q_4 = Q_5 = 0$ have been assumed; u_p and β_p are the generalized mid-plane displacements and plate slopes, and w is the mid-plane deflection; q_p, $p = 1, 2, 4, 5$, and q_3 are the generalized distributed forces applied on the laminate surfaces; m_p, $p = 1, 2, 4, 5$, are the generalized distributed moments applied on the laminate surfaces; Q_1 and Q_2 are the transverse shear forces.

In the above, the indices 1, 2, and 3 denote the mechanical responses in the direction of x_1, x_2, and x_3, whereas the indices 4 and 5 represent, respectively, the responses due to the electric and magnetic effects. Their relations to the electric field E_j, magnetic field H_j, electric displacement D_j, and magnetic flux B_j are

$$\begin{aligned}
&u_4 = -\int E_1^0 dx_1 = -\int E_2^0 dx_2,\ \beta_4 = -\int E_1^* dx_1 = -\int E_2^* dx_2,\\
&u_5 = -\int H_1^0 dx_1 = -\int H_2^0 dx_2,\ \beta_5 = -\int H_1^* dx_1 = -\int H_2^* dx_2,
\end{aligned} \quad (8.49a)$$

$$N_{4j}=\int_{-h/2}^{h/2}D_j dx_3,\ M_{4j}=\int_{-h/2}^{h/2}D_j x_3 dx_3,\ Q_4=\int_{-h/2}^{h/2}D_3 dx_3,$$
$$N_{5j}=\int_{-h/2}^{h/2}B_j dx_3,\ M_{5j}=\int_{-h/2}^{h/2}B_j x_3 dx_3,\ Q_5=\int_{-h/2}^{h/2}B_3 dx_3, \tag{8.49b}$$

where h is the laminate thickness, E_j^0, H_j^0 are the mid-plane electric and magnetic fields, and E_j^*, H_j^* are the rate changes of electric and magnetic fields along thickness direction. Note that based upon the Kirchhoff's hypotheses, like the displacement fields here the electric and magnetic fields are also assumed to vary linearly along the thickness direction and can be expressed as those shown in $(3.52\text{a})_{2,3}$.

By analogy of the basic equations for electro-elastic laminates as shown in (8.41), if all the generalized distributed forces and moments are ignored, that is, $q_p = q_3 = m_p = 0$, $p = 1, 2, 4, 5$, a general solution satisfying all the basic equations written in (8.47) and (8.48) can be expressed as

$$\mathbf{u}_d = 2\,\text{Re}\{\mathbf{A}\mathbf{f}(z)\},\ \boldsymbol{\phi}_d = 2\,\text{Re}\{\mathbf{B}\mathbf{f}(z)\}, \tag{8.50a}$$

where

$$\mathbf{u}_d=\begin{Bmatrix}\mathbf{u}\\ \boldsymbol{\beta}\end{Bmatrix}=\begin{Bmatrix}u_1\\u_2\\u_4\\u_5\\\beta_1\\\beta_2\\\beta_4\\\beta_5\end{Bmatrix},\ \boldsymbol{\phi}_d=\begin{Bmatrix}\boldsymbol{\phi}\\ \boldsymbol{\psi}\end{Bmatrix}=\begin{Bmatrix}\phi_1\\\phi_2\\\phi_4\\\phi_5\\\psi_1\\\psi_2\\\psi_4\\\psi_5\end{Bmatrix},\ \mathbf{f}(z)=\begin{Bmatrix}f_1(z_1)\\f_2(z_2)\\f_3(z_3)\\f_4(z_4)\\f_5(z_5)\\f_6(z_6)\\f_7(z_7)\\f_8(z_8)\end{Bmatrix},$$
$$\mathbf{A}=[\mathbf{a}_1\ \mathbf{a}_2\ \mathbf{a}_3\ \mathbf{a}_4\ \mathbf{a}_5\ \mathbf{a}_6\ \mathbf{a}_7\ \mathbf{a}_8],\quad \mathbf{B}=[\mathbf{b}_1\ \mathbf{b}_2\ \mathbf{b}_3\ \mathbf{b}_4\ \mathbf{b}_5\ \mathbf{b}_6\ \mathbf{b}_7\ \mathbf{b}_8],$$
$$z_\alpha = x_1+\mu_\alpha x_2,\quad \alpha=1,2,3,4,5,6,7,8. \tag{8.50b}$$

In addition to the relations of stress functions shown in (8.32) and (8.44), the relations for the magnetic stress resultants/bending moments are

$$N_{51}=-\phi_{5,2},\ N_{52}=\phi_{5,1},\ M_{51}=-\psi_{5,2},\ M_{52}=\psi_{5,1}. \tag{8.51}$$

The material eigenvalues μ_α and their associated eigenvectors $\mathbf{a}_\alpha$ and $\mathbf{b}_\alpha$ can also be determined from the eigenrelation shown in (8.33) with $\mathbf{Q}_m$, $\mathbf{R}_m$, and $\mathbf{T}_m$ be three 8×8 real matrices defined by

$$\mathbf{Q}_m = \begin{bmatrix} \mathbf{Q}_{\tilde{A}} & \mathbf{R}_{\tilde{B}} \\ \mathbf{R}_{\tilde{B}}^T & -\mathbf{T}_{\tilde{D}} \end{bmatrix}, \quad \mathbf{R}_m = \begin{bmatrix} \mathbf{R}_{\tilde{A}} & -\mathbf{Q}_{\tilde{B}} \\ \mathbf{T}_{\tilde{B}}^T & \mathbf{R}_{\tilde{D}}^T \end{bmatrix}, \quad \mathbf{T}_m = \begin{bmatrix} \mathbf{T}_{\tilde{A}} & -\tilde{\mathbf{R}}_{\tilde{B}} \\ -\tilde{\mathbf{R}}_{\tilde{B}}^T & -\mathbf{Q}_{\tilde{D}} \end{bmatrix}, \tag{8.52a}$$

where

$$\mathbf{Q}_{\tilde{A}} = \begin{bmatrix} \tilde{A}_{11} & \tilde{A}_{16} & \tilde{A}_{17} & \tilde{A}_{19} \\ \tilde{A}_{16} & \tilde{A}_{66} & \tilde{A}_{67} & \tilde{A}_{69} \\ \tilde{A}_{17} & \tilde{A}_{67} & \tilde{A}_{77} & \tilde{A}_{79} \\ \tilde{A}_{19} & \tilde{A}_{69} & \tilde{A}_{79} & \tilde{A}_{99} \end{bmatrix}, \quad \mathbf{R}_{\tilde{A}} = \begin{bmatrix} \tilde{A}_{16} & \tilde{A}_{12} & \tilde{A}_{18} & \tilde{A}_{1,10} \\ \tilde{A}_{66} & \tilde{A}_{26} & \tilde{A}_{68} & \tilde{A}_{6,10} \\ \tilde{A}_{67} & \tilde{A}_{27} & \tilde{A}_{78} & \tilde{A}_{7,10} \\ \tilde{A}_{69} & \tilde{A}_{29} & \tilde{A}_{89} & \tilde{A}_{9,10} \end{bmatrix},$$

$$\mathbf{T}_{\tilde{A}} = \begin{bmatrix} \tilde{A}_{66} & \tilde{A}_{26} & \tilde{A}_{68} & \tilde{A}_{6,10} \\ \tilde{A}_{26} & \tilde{A}_{22} & \tilde{A}_{28} & \tilde{A}_{2,10} \\ \tilde{A}_{68} & \tilde{A}_{28} & \tilde{A}_{88} & \tilde{A}_{8,10} \\ \tilde{A}_{6,10} & \tilde{A}_{2,10} & \tilde{A}_{8,10} & \tilde{A}_{10,10} \end{bmatrix}, \quad \mathbf{Q}_{\tilde{B}} = \begin{bmatrix} \tilde{B}_{11} & \frac{1}{2}\tilde{B}_{16} & \frac{1}{2}\tilde{B}_{17} & \frac{1}{2}\tilde{B}_{19} \\ \tilde{B}_{61} & \frac{1}{2}\tilde{B}_{66} & \frac{1}{2}\tilde{B}_{67} & \frac{1}{2}\tilde{B}_{69} \\ \tilde{B}_{71} & \frac{1}{2}\tilde{B}_{76} & \frac{1}{2}\tilde{B}_{77} & \frac{1}{2}\tilde{B}_{79} \\ \tilde{B}_{91} & \frac{1}{2}\tilde{B}_{96} & \frac{1}{2}\tilde{B}_{97} & \frac{1}{2}\tilde{B}_{99} \end{bmatrix},$$

$$\mathbf{R}_{\tilde{B}} = \begin{bmatrix} \frac{1}{2}\tilde{B}_{16} & \tilde{B}_{12} & \frac{1}{2}\tilde{B}_{18} & \frac{1}{2}\tilde{B}_{1,10} \\ \frac{1}{2}\tilde{B}_{66} & \tilde{B}_{62} & \frac{1}{2}\tilde{B}_{68} & \frac{1}{2}\tilde{B}_{6,10} \\ \frac{1}{2}\tilde{B}_{76} & \tilde{B}_{72} & \frac{1}{2}\tilde{B}_{78} & \frac{1}{2}\tilde{B}_{7,10} \\ \frac{1}{2}\tilde{B}_{96} & \tilde{B}_{92} & \frac{1}{2}\tilde{B}_{98} & \frac{1}{2}\tilde{B}_{9,10} \end{bmatrix}, \quad \tilde{\mathbf{R}}_{\tilde{B}} = \begin{bmatrix} \tilde{B}_{61} & \frac{1}{2}\tilde{B}_{66} & \frac{1}{2}\tilde{B}_{67} & \frac{1}{2}\tilde{B}_{69} \\ \tilde{B}_{21} & \frac{1}{2}\tilde{B}_{26} & \frac{1}{2}\tilde{B}_{27} & \frac{1}{2}\tilde{B}_{29} \\ \tilde{B}_{81} & \frac{1}{2}\tilde{B}_{86} & \frac{1}{2}\tilde{B}_{87} & \frac{1}{2}\tilde{B}_{89} \\ \tilde{B}_{10,1} & \frac{1}{2}\tilde{B}_{10,6} & \frac{1}{2}\tilde{B}_{10,7} & \frac{1}{2}\tilde{B}_{10,9} \end{bmatrix},$$

$$\mathbf{T}_{\tilde{B}} = \begin{bmatrix} \frac{1}{2}\tilde{B}_{66} & \tilde{B}_{62} & \frac{1}{2}\tilde{B}_{68} & \frac{1}{2}\tilde{B}_{6,10} \\ \frac{1}{2}\tilde{B}_{26} & \tilde{B}_{22} & \frac{1}{2}\tilde{B}_{28} & \frac{1}{2}\tilde{B}_{2,10} \\ \frac{1}{2}\tilde{B}_{86} & \tilde{B}_{82} & \frac{1}{2}\tilde{B}_{88} & \frac{1}{2}\tilde{B}_{8,10} \\ \frac{1}{2}\tilde{B}_{10,6} & \tilde{B}_{10,2} & \frac{1}{2}\tilde{B}_{10,8} & \frac{1}{2}\tilde{B}_{10,10} \end{bmatrix}, \quad \mathbf{Q}_{\tilde{D}} = \begin{bmatrix} \tilde{D}_{11} & \frac{1}{2}\tilde{D}_{16} & \frac{1}{2}\tilde{D}_{17} & \frac{1}{2}\tilde{D}_{19} \\ \frac{1}{2}\tilde{D}_{16} & \frac{1}{4}\tilde{D}_{66} & \frac{1}{4}\tilde{D}_{67} & \frac{1}{4}\tilde{D}_{69} \\ \frac{1}{2}\tilde{D}_{17} & \frac{1}{4}\tilde{D}_{67} & \frac{1}{4}\tilde{D}_{77} & \frac{1}{4}\tilde{D}_{79} \\ \frac{1}{2}\tilde{D}_{19} & \frac{1}{4}\tilde{D}_{69} & \frac{1}{4}\tilde{D}_{79} & \frac{1}{4}\tilde{D}_{99} \end{bmatrix},$$

$$\mathbf{R}_{\tilde{D}} = \begin{bmatrix} \frac{1}{2}\tilde{D}_{16} & \tilde{D}_{12} & \frac{1}{2}\tilde{D}_{18} & \frac{1}{2}\tilde{D}_{1,10} \\ \frac{1}{4}\tilde{D}_{66} & \frac{1}{2}\tilde{D}_{26} & \frac{1}{4}\tilde{D}_{68} & \frac{1}{4}\tilde{D}_{6,10} \\ \frac{1}{4}\tilde{D}_{67} & \frac{1}{2}\tilde{D}_{27} & \frac{1}{4}\tilde{D}_{78} & \frac{1}{4}\tilde{D}_{7,10} \\ \frac{1}{4}\tilde{D}_{69} & \frac{1}{2}\tilde{D}_{29} & \frac{1}{4}\tilde{D}_{89} & \frac{1}{4}\tilde{D}_{9,10} \end{bmatrix}, \mathbf{T}_{\tilde{D}} = \begin{bmatrix} \frac{1}{4}\tilde{D}_{66} & \frac{1}{2}\tilde{D}_{26} & \frac{1}{4}\tilde{D}_{68} & \frac{1}{4}\tilde{D}_{6,10} \\ \frac{1}{2}\tilde{D}_{26} & \tilde{D}_{22} & \frac{1}{2}\tilde{D}_{28} & \frac{1}{2}\tilde{D}_{2,10} \\ \frac{1}{4}\tilde{D}_{68} & \frac{1}{2}\tilde{D}_{28} & \frac{1}{4}\tilde{D}_{88} & \frac{1}{4}\tilde{D}_{8,10} \\ \frac{1}{4}\tilde{D}_{6,10} & \frac{1}{2}\tilde{D}_{2,10} & \frac{1}{4}\tilde{D}_{8,10} & \frac{1}{4}\tilde{D}_{10,10} \end{bmatrix}, \tag{8.52b}$$

Note that like the electro-elastic laminates, the transformation matrix $\mathbf{I}_t$ of (8.33c) should be changed to that of (8.46a) where

$$\begin{aligned} &\mathbf{I}_1 = \langle 1,1,1,1,0,0,0,0 \rangle, \ \mathbf{I}_2 = \langle 0,0,0,0,1,1,2,2 \rangle, \\ &\mathbf{I}_3 = \langle 0,0,0,0,1,1,0.5,0.5 \rangle. \end{aligned} \tag{8.53}$$

8.2.5 Extended Stroh-Like Formalism for Thermal Stresses in Laminates

Based upon Kirchhoff's assumptions for laminated plates and the constitutive laws (8.22) for anisotropic elastic solids under thermal environment, the kinematic relations, the constitutive laws, and the equilibrium equations for thermal stress analysis of composite laminates can be written as follows (Hwu, 2010):

$$\begin{aligned} &\breve{h}_i = -K_{ij}T^0_{,j} - K^*_{ij}T^*_{,j} - K_{i3}T^*, \quad \varepsilon_{ij} = \frac{1}{2}\left(u_{i,j} + u_{j,i}\right), \quad \kappa_{ij} = \frac{1}{2}\left(\beta_{i,j} + \beta_{j,i}\right), \\ &N_{ij} = A_{ijkl}\varepsilon_{kl} + B_{ijkl}\kappa_{kl} - A^t_{ij}T^0 - B^t_{ij}T^*, \ M_{ij} = B_{ijkl}\varepsilon_{kl} + D_{ijkl}\kappa_{kl} - B^t_{ij}T^0 - D^t_{ij}T^*, \\ &N_{ij,j} = 0, \ M_{ij,ij} + q = 0, \ Q_i = M_{ij,j}, \ \breve{h}_{i,i} = 0, \quad i,j,k,l = 1,2, \end{aligned} \tag{8.54}$$

in which the additional undefined symbols are: T^0 is the midplane temperature; T^* is the rate of temperature changes in thickness direction; $\breve{h}_i$ is the heat flux resultant; $A^t_{ij}, B^t_{ij}, D^t_{ij}$ are the tensors related to thermal moduli β_{ij}; K_{ij} and K^*_{ij} are the coefficients related to heat conduction coefficients k_{ij}. Their definitions are

$$\breve{h}_i = \int_{-h/2}^{h/2} h_i dx_3, \quad K_{ij} = \int_{-h/2}^{h/2} k_{ij} dx_3, \quad K_{ij}^* = \int_{-h/2}^{h/2} k_{ij} x_3 dx_3,$$
$$A_{ij}^t = \int_{-h/2}^{h/2} \beta_{ij} dx_3, \quad B_{ij}^t = \int_{-h/2}^{h/2} \beta_{ij} x_3 dx_3, \quad D_{ij}^t = \int_{-h/2}^{h/2} \beta_{ij} x_3^2 dx_3. \tag{8.55}$$

Since the basic equations stated in (8.54) are quite general, it is not easy to find a solution satisfying all these basic equations. Only two special cases that occur frequently in engineering applications are considered in the literature. One is the case that temperature distribution depends on x_1 and x_2 only, that is, $T^* = 0$ and $T = T^0 = T^0(x_1, x_2)$, and the other is the case that temperature varies linearly in thickness direction, that is, $T = T^0 + x_3 T^*$ in which T^0 and T^* are constants independent of x_1 and x_2.

CASE 1: TEMPERATURE DEPENDS ON X_1 AND X_2 ONLY

If the temperature is assumed to depend on x_1 and x_2 only and the lateral distributed load and heat flux applied on the laminates are neglected, the basic equations (8.54) can be simplified to

$$\breve{h}_{i,i} = -K_{ij} T_{,ij} = 0, \quad N_{ij,j} = A_{ijkl} u_{k,lj} + B_{ijkl} \beta_{k,lj} - A_{ij}^t T_{,j} = 0,$$
$$M_{ij,ij} = B_{ijkl} u_{k,lij} + D_{ijkl} \beta_{k,lij} - B_{ij}^t T_{,ij} = 0, \quad i,j,k,l = 1,2. \tag{8.56}$$

A general solution satisfying (8.56) has been obtained as

$$T = 2\mathrm{Re}\{g'(z_t)\}, \quad \breve{h}_i = -2\mathrm{Re}\{(K_{i1} + \tau K_{i2}) g''(z_t)\},$$
$$\mathbf{u}_d = 2\mathrm{Re}\{\mathbf{A}\mathbf{f}(z) + \mathbf{c} g(z_t)\}, \quad \boldsymbol{\phi}_d = 2\mathrm{Re}\{\mathbf{B}\mathbf{f}(z) + \mathbf{d} g(z_t)\}, \tag{8.57a}$$

where

$$\mathbf{u}_d = \begin{Bmatrix} \mathbf{u} \\ \boldsymbol{\beta} \end{Bmatrix}, \quad \boldsymbol{\phi}_d = \begin{Bmatrix} \boldsymbol{\phi} \\ \boldsymbol{\psi} \end{Bmatrix}, \quad \mathbf{u} = \begin{Bmatrix} u_1 \\ u_2 \end{Bmatrix}, \quad \boldsymbol{\beta} = \begin{Bmatrix} \beta_1 \\ \beta_2 \end{Bmatrix}, \quad \boldsymbol{\phi} = \begin{Bmatrix} \phi_1 \\ \phi_2 \end{Bmatrix}, \quad \psi = \begin{Bmatrix} \psi_1 \\ \psi_2 \end{Bmatrix},$$
$$\mathbf{A} = [\mathbf{a}_1 \ \mathbf{a}_2 \ \mathbf{a}_3 \ \mathbf{a}_4], \quad \mathbf{B} = [\mathbf{b}_1 \ \mathbf{b}_2 \ \mathbf{b}_3 \ \mathbf{b}_4],$$
$$\mathbf{f}(z) = \begin{Bmatrix} f_1(z_1) \\ f_2(z_2) \\ f_3(z_3) \\ f_4(z_4) \end{Bmatrix}, \quad z_\alpha = x_1 + \mu_\alpha x_2, \quad \alpha = 1,2,3,4, \quad z_t = x_1 + \tau x_2. \tag{8.57b}$$

The stress resultants/bending moments are related to the stress functions by (8.32). The eigenrelation for the determination of μ_α and $\mathbf{a}_\alpha$, $\mathbf{b}_\alpha$ is given in (8.33). The relations for the determination of thermal eigenvalues τ and thermal eigenvectors $\mathbf{c}$, $\mathbf{d}$ are

$$K_{22}\tau^2 + 2K_{12}\tau + K_{11} = 0, \quad \mathbf{N}\boldsymbol{\eta} = \tau\boldsymbol{\eta} + \boldsymbol{\gamma}, \tag{8.58a}$$

where

$$\mathbf{N} = \mathbf{I}_t\mathbf{N}_m\mathbf{I}_t, \quad \eta = \begin{Bmatrix} \mathbf{c} \\ \mathbf{d} \end{Bmatrix}, \quad \boldsymbol{\gamma} = -\mathbf{I}_t \begin{bmatrix} \mathbf{0} & (\mathbf{N}_m)_2 \\ \mathbf{I} & (\mathbf{N}_m)_1^T \end{bmatrix} \begin{Bmatrix} \tilde{\boldsymbol{\alpha}}_1 \\ \tilde{\boldsymbol{\alpha}}_2 \end{Bmatrix}, \tag{8.58b}$$

and

$$\tilde{\boldsymbol{\alpha}}_1 = \begin{Bmatrix} \tilde{A}_{11}^t \\ \tilde{A}_{21}^t \\ \tilde{B}_{12}^t \\ \tilde{B}_{22}^t \end{Bmatrix}, \quad \tilde{\boldsymbol{\alpha}}_2 = \begin{Bmatrix} \tilde{A}_{12}^t \\ \tilde{A}_{22}^t \\ -\tilde{B}_{11}^t \\ -\tilde{B}_{21}^t \end{Bmatrix}. \tag{8.58c}$$

$\tilde{A}_{ij}^t$ and $\tilde{B}_{ij}^t$ of (8.58c) are defined by

$$\tilde{A}_{ij}^t = A_{ij}^t - \tilde{B}_{ijkl}B_{kl}^t, \quad \tilde{B}_{ij}^t = \tilde{D}_{ijkl}B_{kl}^t, \tag{8.59}$$

where $\tilde{B}_{ijkl}$ and $\tilde{D}_{ijkl}$ are tensor notations of $\tilde{\boldsymbol{B}}$ and $\tilde{\boldsymbol{D}}$ defined in (8.35), and A_{ij}^t and B_{ij}^t are defined in (8.55).

CASE 2: TEMPERATURE DEPENDS ON X_3 ONLY

If the temperature varies linearly across the thickness direction, the basic equations (8.54) can be simplified as

$$\begin{aligned}
N_{ij} &= A_{ijkl}u_{k,l} + B_{ijkl}\beta_{k,l} - A_{ij}^tT^0 - B_{ij}^tT^*, \\
M_{ij} &= B_{ijkl}u_{k,l} + D_{ijkl}\beta_{k,l} - B_{ij}^tT^0 - D_{ij}^tT^*, \\
N_{ij,j} &= A_{ijkl}u_{k,lj} + B_{ijkl}\beta_{k,lj} = 0, \quad M_{ij,ij} = B_{ijkl}u_{k,lij} + D_{ijkl}\beta_{k,lij} = 0, \\
Q_i &= M_{ij,j}, \quad i,j,k,l = 1,2.
\end{aligned} \tag{8.60}$$

$\breve{h}_i = -K_{i3}T^*$ and hence $\breve{h}_{i,i} = 0$ is satisfied automatically. A general solution satisfying (8.60) has been obtained as

$$\mathbf{u}_d = 2\mathrm{Re}\{\mathbf{A}\mathbf{f}(z)\},\quad \boldsymbol{\phi}_d = 2\mathrm{Re}\{\mathbf{B}\mathbf{f}(z)\} - x_1\boldsymbol{\vartheta}_2 + x_2\boldsymbol{\vartheta}_1, \tag{8.61a}$$

where

$$\boldsymbol{\vartheta}_i = \boldsymbol{\alpha}_i T^0 + \boldsymbol{\alpha}_i^* T^*,\quad i = 1,2, \tag{8.61b}$$

and

$$\boldsymbol{\alpha}_1 = \begin{Bmatrix} A_{11}^t \\ A_{21}^t \\ B_{11}^t \\ B_{21}^t \end{Bmatrix},\quad \boldsymbol{\alpha}_2 = \begin{Bmatrix} A_{12}^t \\ A_{22}^t \\ B_{12}^t \\ B_{22}^t \end{Bmatrix},\quad \boldsymbol{\alpha}_1^* = \begin{Bmatrix} B_{11}^t \\ B_{21}^t \\ D_{11}^t \\ D_{21}^t \end{Bmatrix},\quad \boldsymbol{\alpha}_2^* = \begin{Bmatrix} B_{12}^t \\ B_{22}^t \\ D_{12}^t \\ D_{22}^t \end{Bmatrix}. \tag{8.61c}$$

Note that unlike the general solution shown in (8.57) for case 1, the general solution for case 2 does not include the expressions for temperature and heat flux resultant since they are *known* as linear temperature distributions and constant heat flows in thickness direction.

8.3 THREE-DIMENSIONAL ANALYSIS

As stated in the previous two sections, the Stroh's complex variable formalism is elegant and powerful for problems with two variables, such as two-dimensional and coupled stretching–bending problems. To extend this formalism to problems with three variables such as three-dimensional analysis, a Radon transform (Deans, 1983), which maps a three-dimensional solid to a two-dimensional plane, was suggested to combine with the Stroh formalism (Wu, 1998; Buroni and Denda, 2014; Hsu et al., 2019). In this section, the Radon–Stroh formalism is therefore introduced by following the work stated in Hsu et al. (2019).

8.3.1 Radon–Stroh Formalism for Anisotropic Elastic Materials

The basic equations for three-dimensional linear anisotropic elasticity have been written in (8.1). Substituting $(8.1)_1$ into $(8.1)_2$ with C_{ijkl} fully symmetric, we get

$$\sigma_{ij} = C_{ijkl}u_{k,l}. \tag{8.62}$$

The governing differential equations which the displacements must satisfy can then be obtained by substituting (8.62) into $(8.1)_3$, that is,

$$C_{ijkl}u_{k,lj}=0. \tag{8.63}$$

Consider a Radon transform from the space of rectangular Cartesian coordinate $\mathbf{x}=(x_1, x_2, x_3)$ to the space of cylindrical coordinate (ρ, θ, x_3) by (Figure 8.1)

$$\breve{f}(\mathbf{n},\rho,x_3)=\int_{-\infty}^{\infty} f(\mathbf{x})\delta(\mathbf{n}\cdot\mathbf{x}-\rho)d\mathbf{x}, \tag{8.64}$$

where δ is the Dirac delta function, and $\mathbf{n}=(n_1, n_2, 0)$ with $n_1=\cos\theta$, $n_2=\sin\theta$ is a unit vector normal to the line $\mathbf{n}\cdot\mathbf{x}=\rho$. Note that in this transformation the coordinate variable x_3 keeps unchanged, and the argument written as $(\mathbf{n}, \rho, x_3)$ has the same meaning as (ρ, θ, x_3) since the unit vector $\mathbf{n}$ is a function of θ.

To apply the Radon transform (8.64), the governing equations (8.63) are rewritten by separating the indices into (x_1, x_2) and x_3 as

$$C_{i\alpha k\beta}u_{k,\alpha\beta}+\left(C_{i\alpha k3}+C_{i3k\alpha}\right)u_{k,3\alpha}+C_{i3k3}u_{k,33}=0,\quad i,k=1,2,3,\ \alpha,\beta=1,2. \tag{8.65}$$

Knowing that $R\{f_{,\alpha}\}=n_\alpha\breve{f}_{,\rho}$ and $R\{f_{,3}\}=\breve{f}_{,3}$, application of the Radon transform (8.64) to (8.65) leads to

$$C_{i\alpha k\beta}n_\alpha n_\beta\breve{u}_{k,\rho\rho}+\left(C_{i\alpha k3}+C_{i3k\alpha}\right)n_\alpha\breve{u}_{k,3\rho}+C_{i3k3}\breve{u}_{k,33}=0. \tag{8.66}$$

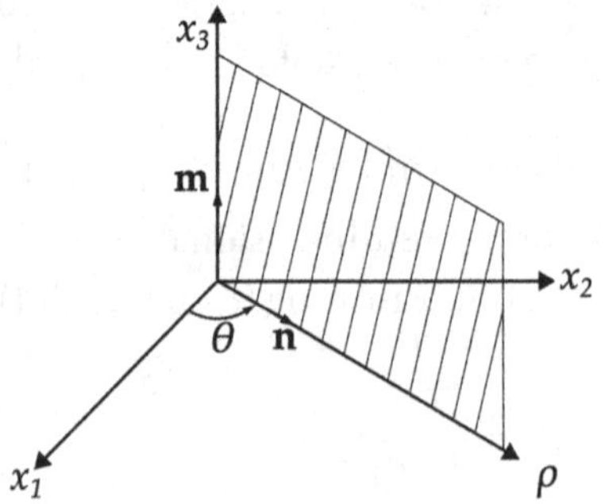

FIGURE 8.1 Example of Radon-plane spanned by $\mathbf{n}=(\cos\theta, \sin\theta, 0)$ and $\mathbf{m}=(0, 0, 1)$.

Since the governing differential equations (8.66) in the domain of Radon transform are a set of homogeneous second-order differential equations which depend only on two variables ρ and x_3, a general solution for $\breve{u}_k$ can be written as

$$\breve{u}_k = a_k \breve{f}(z), \text{ or in matrix form } \mathbf{u} = \mathbf{a}\breve{f}(z) \tag{8.67a}$$

where

$$z = \rho + \mu x_3. \tag{8.67b}$$

In (8.67a), **a** and μ are constants to be determined and $\breve{f}$ is an arbitrary function of z. Substituting (8.67a)$_1$ into (8.66), we get

$$\left\{Q_{ik} + \mu\left(R_{ik} + R_{ki}\right) + \mu^2 T_{ik}\right\} a_k \breve{f}''(z) = 0, \tag{8.68a}$$

or in matrix form

$$\left\{\mathbf{Q} + \mu\left(\mathbf{R} + \mathbf{R}^T\right) + \mu^2\mathbf{T}\right\}\mathbf{a} = \mathbf{0}, \tag{8.68b}$$

where the superscript T stands for the transpose, and

$$\begin{aligned} &Q_{ik} = C_{ijkl} n_j n_l, \quad R_{ik} = C_{ijkl} n_j m_l, \quad T_{ik} = C_{ijkl} m_j m_l, \\ &\mathbf{n} = (\cos\theta, \sin\theta, 0), \mathbf{m} = (0,0,1). \end{aligned} \tag{8.68c}$$

A nontrivial solution of **a** exists if

$$\left\| \mathbf{Q} + \mu\left(\mathbf{R} + \mathbf{R}^T\right) + \mu^2\mathbf{T} \right\| = 0, \tag{8.69}$$

which gives a sextic equation for μ. Note that μ and **a** determined from (8.68), which are generally called material eigenvalue and material eigenvector, now depend not only on the elastic constants C_{ijkl} but also on the variable θ.

After obtaining μ and a_k for the displacements in the transform domain, the stresses in the transform domain can be determined from (8.62) by employing (8.64) and (8.67). The result is

$$\breve{\sigma}_{ij} = \left(C_{ijk\alpha} n_\alpha + \mu C_{ijk3}\right) a_k \breve{f}'(z), \tag{8.70}$$

which then leads the tractions on the surfaces with normal **n**, **m** to

$$\breve{t}_i^n = \breve{\sigma}_{ij} n_j = \left(Q_{ik} + \mu R_{ik}\right) a_k \breve{f}'(z), \; \breve{t}_i^m = \breve{\sigma}_{ij} m_j = \left(R_{ki} + \mu T_{ik}\right) a_k \breve{f}'(z). \tag{8.71}$$

Introduce a new vector **b** as

$$\mathbf{b} = \left(\mathbf{R}^T + \mu\mathbf{T}\right)\mathbf{a} = -\frac{1}{\mu}\left(\mathbf{Q} + \mu\mathbf{R}\right)\mathbf{a}, \tag{8.72}$$

in which the second equality of (8.72) is obtained from (8.68b). With (8.72), equation (8.71) can now be written as

$$\breve{t}_i^n = -\mu b_i \breve{f}'(z), \quad \breve{t}_i^m = b_i \breve{f}'(z). \tag{8.73}$$

Introduce the *stress functions* $\breve{\phi}_i$, $i = 1, 2, 3$, such that

$$\breve{t}_i^n = -\breve{\phi}_{i,3}, \quad \breve{t}_i^m = \breve{\phi}_{i,\rho}, \tag{8.74}$$

we get

$$\breve{\phi}_i = b_i \breve{f}(z), \quad \text{or in matrix form} \quad \breve{\boldsymbol{\phi}} = \mathbf{b}\breve{f}(z). \tag{8.75}$$

In 2D anisotropic elasticity, it has been proved that the six roots of material eigenvalues μ cannot be real and are three pairs of complex conjugates (Hwu, 2010). Thus, by letting $\mathrm{Im}\,\mu_k > 0, \mu_{k+3} = \bar{\mu}_k$, $\mathbf{a}_{k+3} = \bar{\mathbf{a}}_k$, $\mathbf{b}_{k+3} = \bar{\mathbf{b}}_k$, $k = 1, 2, 3$, and assuming that μ_k are distinct, the general solution obtained by superposing six solutions of $(8.67\text{a})_2$ and (8.75) can be written as

$$\breve{\mathbf{u}} = 2\,\mathrm{Re}\left\{\mathbf{A}\breve{\mathbf{f}}(z)\right\}, \quad \breve{\boldsymbol{\phi}} = 2\,\mathrm{Re}\left\{\mathbf{B}\breve{\mathbf{f}}(z)\right\}, \tag{8.76a}$$

where

$$\mathbf{A} = \begin{bmatrix}\mathbf{a}_1 & \mathbf{a}_2 & \mathbf{a}_3\end{bmatrix}, \quad \mathbf{B} = \begin{bmatrix}\mathbf{b}_1 & \mathbf{b}_2 & \mathbf{b}_3\end{bmatrix},$$
$$\breve{\mathbf{f}}(z) = \left[\breve{f}_1(z_1), \breve{f}_2(z_2), \breve{f}_3(z_3)\right]^T, \quad z_k = \rho + \mu_k x_3, \; k = 1,2,3, \tag{8.76b}$$

and Re stands for the real part of a complex number.

In 2D anisotropic elasticity, the traction has a direct relation with the stress function by $\breve{\mathbf{t}} = \partial\breve{\boldsymbol{\phi}}/\partial s$, where s denotes the tangential direction of the surface. Here, in three-dimensional problems, the surface that the traction lies on should be represented by two orthogonal tangential directions instead of a line direction like s only. Thus, it can be expected that the relation between traction and stress function should be expressed in terms of two variables instead of one. Without involving the derivation of the new relation, we now derive the traction on the surface with normal **e** directly by using (8.70) as

$$\breve{t}_i^e = \breve{\sigma}_{ij} e_j = \left(C_{ijk\alpha} n_\alpha e_j + \mu C_{ijk3} e_j\right) a_k \, f'(z). \tag{8.77}$$

Like (8.72), we introduce a new vector $\mathbf{b}^e$ as

$$\mathbf{b}^e = \left(\mathbf{Q}^e + \mu \mathbf{R}^e\right)\mathbf{a}, \tag{8.78}$$

in which the components of the matrices $\mathbf{Q}^e$ and $\mathbf{R}^e$ are defined as

$$Q_{ik}^e = C_{ijkl} e_j n_l, \quad R_{ik}^e = C_{ijkl} e_j m_l. \tag{8.79}$$

Thus,

$$\breve{t}_i^e = b_i^e \, f'(z). \tag{8.80}$$

In the above, all the formulae have been purposely organized into the mathematical form of Stroh formalism for 2D anisotropic elasticity. Thus, some important relations and identities developed in Stroh formalism (Hwu, 2010) can all be employed in the domain of Radon transform for 3D elastostatic analysis. For example, the well-known material eigenrelation $\mathbf{N}\boldsymbol{\xi} = \mu\boldsymbol{\xi}$ shown in (8.4).

After finding the solutions in the transform domain, the 3D solutions for the original problem should be recovered by applying the inverse Radon transform. It's known that (Deans, 1983; Wu, 1998; Buroni and Denda, 2014)

$$f(\mathbf{x}) = \frac{1}{2\pi} \int_0^{\pi} \overset{\frown}{f}(\mathbf{n} \cdot \mathbf{x}, \theta) \, d\theta, \tag{8.81a}$$

where

$$\overset{\frown}{f}(\mathbf{n} \cdot \mathbf{x}, \theta) = \frac{-1}{\pi} \int_{-\infty}^{\infty} \frac{\breve{f}_{,\rho}(\rho, \theta)}{\rho - \mathbf{n} \cdot \mathbf{x}} \, d\rho. \tag{8.81b}$$

The symbol $\widehat{\bullet}$ is a notation for the Hilbert transform, whose integral can be derived analytically by considering the contour integral of $\breve{f}'(z,\theta)/(z-\rho)$ around a very large semi-circle involving a small semi-circular detour around the pole ρ in the upper (or lower) half-plane. If $\breve{f}'(z,\theta)\to 0$ when $|z|\to\infty$ and no pole of $\breve{f}'(z,\theta)$ locates on the upper (or lower) half-plane when $x_3>\hat{x}_3$ (or $x_3<\hat{x}_3$), the Hilbert transform (8.81b) can be obtained as

$$\widehat{\breve{f}}(\mathbf{n}\cdot\mathbf{x},\theta)=-i\,\mathrm{sgn}(\Delta x_3)\breve{f}'(\mathbf{n}\cdot\mathbf{x}), \tag{8.82}$$

where $\mathrm{sgn}(\Delta x_3)=1$ for $\Delta x_3>0$, $\mathrm{sgn}(\Delta x_3)=-1$ for $\Delta x_3<0$, and $\Delta x_3=x_3-\hat{x}_3$. Here, the prime $\bullet'$ denotes the derivative with respect to the argument of the function. Note that $\mathrm{sgn}(\Delta x_3)$ is undefined for $\Delta x_3=0$ and hence all the solutions derived later cannot be applied to the cases where $\Delta x_3=0$. To obtain the solutions on $\Delta x_3=0$ or near $\Delta x_3=0$, alternative Radon-planes are suggested later in (8.86) and a selection criterion for the Radon-plane is stated in (8.90).

Using the relation (8.82), the Hilbert transform of $\breve{\mathbf{u}}$ and $\breve{\boldsymbol{\phi}}$ in (8.76) can be obtained as

$$\widehat{\breve{\mathbf{u}}}=2\,\mathrm{sgn}(\Delta x_3)\mathrm{Im}\{\mathbf{A}\breve{\mathbf{f}}'(z)\},\quad \widehat{\breve{\boldsymbol{\phi}}}=2\,\mathrm{sgn}(\Delta x_3)\mathrm{Im}\{\mathbf{B}\breve{\mathbf{f}}'(z)\}, \tag{8.83}$$

where Im stands for the imaginary part of a complex number, and the prime $\bullet'$ denotes the derivative with respect to z_α. With the results of (8.83), the vectors of displacements and stress functions can be obtained from (8.81a) as

$$\mathbf{u}=\frac{\mathrm{sgn}(\Delta x_3)}{\pi}\mathrm{Im}\left\{\int_0^\pi \mathbf{A}\breve{\mathbf{f}}'(z)\,d\theta\right\},\quad \boldsymbol{\phi}=\frac{\mathrm{sgn}(\Delta x_3)}{\pi}\mathrm{Im}\left\{\int_0^\pi \mathbf{B}\breve{\mathbf{f}}'(z)\,d\theta\right\}. \tag{8.84}$$

Note that due to the sign change at the points when $\Delta x_3=0$, in numerical calculation unstable results may occur at the region around the surface $x_3=\hat{x}_3$. To avoid the calculation near the surface of $x_3=\hat{x}_3$, we can apply the solution obtained by the alternative transformation made by keeping x_1 or x_2 unchanged instead of x_3. In other words, the Radon transformation is made by considering

$$(x_1,x_2,x_3)\to(\rho,x_2,\theta)\quad\text{or}\quad(x_1,x_2,x_3)\to(x_1,\rho,\theta), \tag{8.85}$$

in which the two base vectors **n** and **m**, (8.68c), of the 2D-Radon surfaces should be changed to

$$\mathbf{n}=\left(\cos\theta,0,\sin\theta\right),\ \mathbf{m}=\left(0,1,0\right),$$
$$\text{or }\ \mathbf{n}=\left(0,\cos\theta,\sin\theta\right),\ \mathbf{m}=\left(1,0,0\right). \tag{8.86}$$

Consequently, the complex variable z_k defined in (8.76b) and the solution obtained in (8.84) should also be generalized as

$$z_k=\mathbf{n}\cdot\mathbf{x}+\mu_k\mathbf{m}\cdot\mathbf{x},\ k=1,2,3, \tag{8.87}$$

and

$$\mathbf{u}=\frac{\operatorname{sgn}\left(\mathbf{m}\cdot\Delta\mathbf{x}\right)}{\pi}\operatorname{Im}\left\{\int_0^{\pi}\mathbf{A}\breve{\mathbf{f}}'\left(z\right)d\theta\right\},\quad \boldsymbol{\phi}=\frac{\operatorname{sgn}\left(\mathbf{m}\cdot\Delta\mathbf{x}\right)}{\pi}\operatorname{Im}\left\{\int_0^{\pi}\mathbf{B}\breve{\mathbf{f}}'\left(z\right)d\theta\right\}. \tag{8.88}$$

With (8.88) and (8.80), we have

$$\mathbf{t}^e=\frac{\operatorname{sgn}\left(\mathbf{m}\cdot\Delta\mathbf{x}\right)}{\pi}\operatorname{Im}\left\{\int_0^{\pi}\mathbf{B}^e\breve{\mathbf{f}}''\left(z\right)d\theta\right\}, \tag{8.89a}$$

where

$$\mathbf{B}^e=\left[\mathbf{b}_1^e\quad\mathbf{b}_2^e\quad\mathbf{b}_3^e\right]. \tag{8.89b}$$

Note that based upon the numerical test, a criterion for the selection of proper 2D Radon-plane **n-m** has been suggested as (Hsu et al., 2019)

$$\text{if }\max\left(\left|\Delta x_1\right|,\left|\Delta x_2\right|,\left|\Delta x_3\right|\right)=\left|\Delta x_3\right|,\ \mathbf{n},\mathbf{m}:\left(8.68\text{c}\right)_2,$$
$$\text{if }\max\left(\left|\Delta x_1\right|,\left|\Delta x_2\right|,\left|\Delta x_3\right|\right)=\left|\Delta x_2\right|,\ \mathbf{n},\mathbf{m}:\left(8.86\right)_1,$$
$$\text{if }\max\left(\left|\Delta x_1\right|,\left|\Delta x_2\right|,\left|\Delta x_3\right|\right)=\left|\Delta x_1\right|,\ \mathbf{n},\mathbf{m}:\left(8.86\right)_2. \tag{8.90}$$

Furthermore, like the extension stated in Section 8.1 by suitable transformation and expansion, the Radon–Stroh formalism can also be extended to the three-dimensional analysis with viscoelastic, piezoelectric, or MEE solids. Successful examples on boundary element analysis for piezoelectric and MEE solids can be found in Hsu et al. (2019).

8.3.2 Radon–Stroh Formalism for Anisotropic Piezoelectric and MEE Materials

For electromechanical analysis of piezoelectric materials, the basic equations can be written in an expanded tensor form as (8.15). Similarly, for magneto-electro-mechanical analysis of MEE materials, their basic equations can be written in an expanded tensor form as (8.19b). With this similarity, the Stroh formalism for elastic, piezoelectric, and MEE materials can all be expressed in the same mathematical form as that shown in (8.2), (8.16), and (8.20) for two-dimensional analysis. Since these basic equations are valid for both two- and three-dimensional analysis, the same conclusion can also be made for the extension of Radon–Stroh formalism to the cases with piezoelectric and MEE solids. The only difference between them is the expansion of the associated vectors and matrices from 3 to 4 for piezoelectric solids and from 3 to 5 for MEE solids.

For example, if a MEE material is considered in the three-dimensional boundary element, the vectors of displacements and tractions should be expanded to

$$\mathbf{u}=\left(u_1,u_2,u_3,u_4,u_5\right)^T,\ \mathbf{t}=\left(t_1,t_2,t_3,t_4,t_5\right)^T, \tag{8.91a}$$

where the additional components

$$\begin{aligned}
u_4 &= -\int E_1 dx_1 = -\int E_2 dx_2 = -\int E_3 dx_3,\\
u_5 &= -\int H_1 dx_1 = -\int H_2 dx_2 = -\int H_3 dx_3,\\
t_4 &= \sigma_{4j}e_j = D_j e_j,\ t_5 = \sigma_{5j}e_j = B_j e_j.
\end{aligned} \tag{8.91b}$$

8.4 GREEN'S FUNCTIONS FOR TWO-DIMENSIONAL PROBLEMS

In Sections 3.4.2 and 4.3.2 we have presented the Green's functions for the problems of thick laminated plates and thick laminated beams. In these two cases, no boundaries are set for the plates and beams. Both of them are considered to be infinite. When they are employed for boundary-valued problems, the system of equations is usually constructed through the satisfaction of boundary conditions since all the other basic equations such as constitutive laws, kinematic relations, and equilibrium equations have all been satisfied by the Green's functions. If we now have the problems with additional boundaries such as a straight line, an interface, a hole/crack boundary, extra efforts are needed to derive their associated Green's functions. For two-dimensional anisotropic elastic analysis, the Green's functions for several different problems have been presented in Hwu (2010) such as those for an infinite plane, a half-plane, a bi-material, an elliptical hole, a straight crack, an elliptical rigid inclusion, and an elliptical elastic inclusion. All these Green's functions are expressed by using the solution form (8.2) of Stroh formalism. As stated in Section 8.1, this solution form is applicable for the general anisotropic elastic, piezoelectric, MEE materials as well as in the Laplace domain of viscoelastic materials. For the convenience of our presentation, we now rewrite the general solution (8.2) for the following two different situations:

i. *Problem with only one material*:

$$\mathbf{u}=2\,\mathrm{Re}\{\mathbf{Af}(z)\},\quad \boldsymbol{\phi}=2\,\mathrm{Re}\{\mathbf{Bf}(z)\},\tag{8.92a}$$

ii. *Problem with two dissimilar materials*:

$$\begin{aligned}\mathbf{u}_1&=2\,\mathrm{Re}\{\mathbf{A}_1\mathbf{f}_1(z)\},\quad \boldsymbol{\phi}_1=2\,\mathrm{Re}\{\mathbf{B}_1\mathbf{f}_1(z)\},\\ \mathbf{u}_2&=2\,\mathrm{Re}\{\mathbf{A}_2\mathbf{f}_2(z)\},\quad \boldsymbol{\phi}_2=2\,\mathrm{Re}\{\mathbf{B}_2\mathbf{f}_2(z)\},\end{aligned}\tag{8.92b}$$

The subscripts 1 and 2 in (8.92b) denote, respectively, the value related to the material 1 and 2.

With the solution form of (8.92), in the following only the complex function vector $\mathbf{f}(z)$ (or $\mathbf{f}_1(z)$ and $\mathbf{f}_2(z)$) is presented. The detailed derivation of these Green's functions can be found in Hwu (2010).

8.4.1 An Infinite Anisotropic Elastic Plane

Consider an infinite plane loaded by a point force $\hat{\mathbf{p}}=(\hat{p}_1,\hat{p}_2,\hat{p}_3)$ applied at an internal point $\hat{\mathbf{x}}=(\hat{x}_1,\hat{x}_2)$ (Figure 8.2). The solution of the complex function vector $\mathbf{f}(z)$ to this problem has been found to be

$$\mathbf{f}(z)=\frac{1}{2\pi i}\left\langle \ln(z_\alpha-\hat{z}_\alpha)\right\rangle \mathbf{A}^T\hat{\mathbf{p}},\quad \text{where } \hat{z}_\alpha=\hat{x}_1+\mu_\alpha\hat{x}_2.\tag{8.93}$$

8.4.2 An Anisotropic Elastic Half-Plane

Consider a semi-infinite half-plane subjected to a point force $\hat{\mathbf{p}}=(\hat{p}_1,\hat{p}_2,\hat{p}_3)$ applied at an internal point $\hat{\mathbf{x}}=(\hat{x}_1,\hat{x}_2)$ (Figure 8.3). The half-plane surface

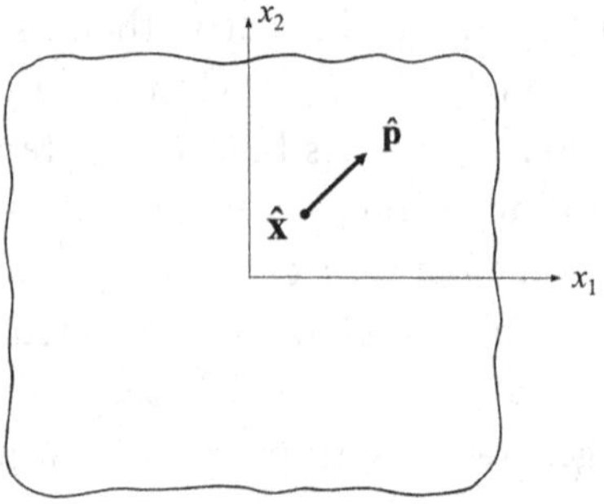

FIGURE 8.2 An infinite plane subjected to a point force.

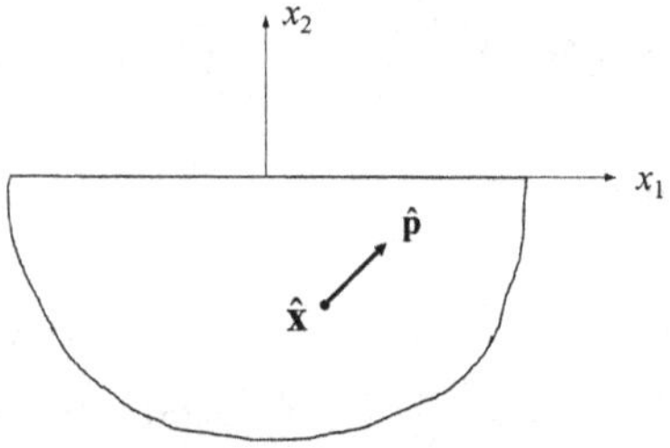

FIGURE 8.3 A half-plane subjected to a point force.

is assumed to be traction-free. The solution of the complex function vector $\mathbf{f}(z)$ to this problem has been found to be

$$\mathbf{f}(z) = \frac{1}{2\pi i}\left\{\left\langle \ln\left(z_\alpha - \hat{z}_\alpha\right)\right\rangle \mathbf{A}^T + \sum_{j=1}^{3}\left\langle \ln\left(z_\alpha - \bar{\hat{z}}_j\right)\right\rangle \mathbf{B}^{-1}\bar{\mathbf{B}}\mathbf{I}_j\bar{\mathbf{A}}^T\right\}\hat{\mathbf{p}}, \quad (8.94)$$

where $\mathbf{I}_j$, $j = 1, 2, 3$, are the diagonal matrices with a unit value at the (jj) component and all the other components are zero.

8.4.3 An Anisotropic Elastic Bi-Material

A bi-material composed of two dissimilar anisotropic elastic half-planes is subjected to a point force $\hat{\mathbf{p}}$ at point $\hat{\mathbf{x}} = (\hat{x}_1, \hat{x}_2)$ of material 1 (Figure 8.4). These two dissimilar materials are assumed to be perfectly bonded along

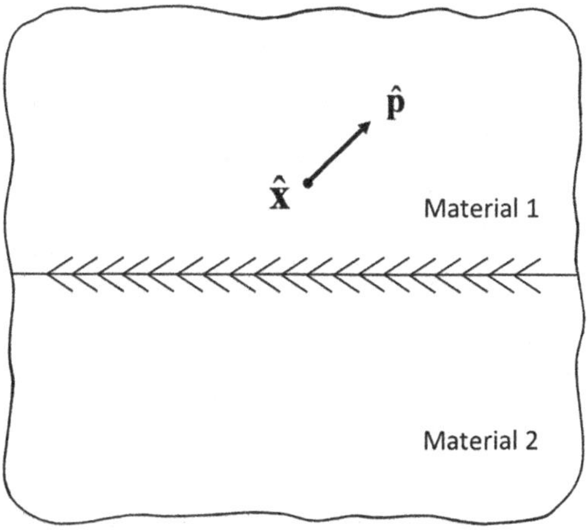

FIGURE 8.4 A bi-material subjected to a point force.

the interface $x_2 = 0$. The solution of the complex function vectors $\mathbf{f}_1(z)$ and $\mathbf{f}_2(z)$ to this problem has been found to be

$$\mathbf{f}_1(z) = \frac{1}{2\pi i}\left\{\left\langle \ln\left(z_\alpha^{(1)} - \hat{z}_\alpha^{(1)}\right)\right\rangle \mathbf{A}_1^T + \sum_{j=1}^{3}\left\langle \ln\left(z_\alpha^{(1)} - \bar{\hat{z}}_j^{(1)}\right)\right\rangle \mathbf{A}_{1f}\mathbf{I}_j\bar{\mathbf{A}}_1^T\right\}\hat{\mathbf{p}},$$

$$\mathbf{f}_2(z) = -\frac{1}{2\pi}\sum_{j=1}^{3}\left\langle \ln\left(z_\alpha^{(2)} - \hat{z}_j^{(1)}\right)\right\rangle \mathbf{A}_{2f}\mathbf{I}_j\mathbf{A}_1^T\hat{\mathbf{p}}, \tag{8.95a}$$

where

$$\mathbf{A}_{1f} = \mathbf{A}_1^{-1}\left(\bar{\mathbf{M}}_2 + \mathbf{M}_1\right)^{-1}\left(\bar{\mathbf{M}}_2 - \bar{\mathbf{M}}_1\right)\bar{\mathbf{A}}_1,\quad \mathbf{A}_{2f} = \mathbf{A}_2^{-1}\left(\mathbf{M}_2 + \bar{\mathbf{M}}_1\right)^{-1}\mathbf{A}_1^{-T},$$
$$\mathbf{M}_1 = -i\mathbf{B}_1\mathbf{A}_1^{-1},\quad \mathbf{M}_2 = -i\mathbf{B}_2\mathbf{A}_2^{-1}. \tag{8.95b}$$

Same as the subscripts 1 and 2, the superscripts (1) and (2) in (8.95a) also denote, respectively, the value related to materials 1 and 2.

8.4.4 A Plane with an Elliptical Hole

Consider an infinite anisotropic elastic plane containing an elliptic hole subjected to a point force $\hat{\mathbf{p}}$ at point $\hat{\mathrm{x}}$ (Figure 8.5). The contour of the hole boundary is represented by

$$x_1 = a\cos\varphi \quad x_2 = b\sin\varphi, \tag{8.96}$$

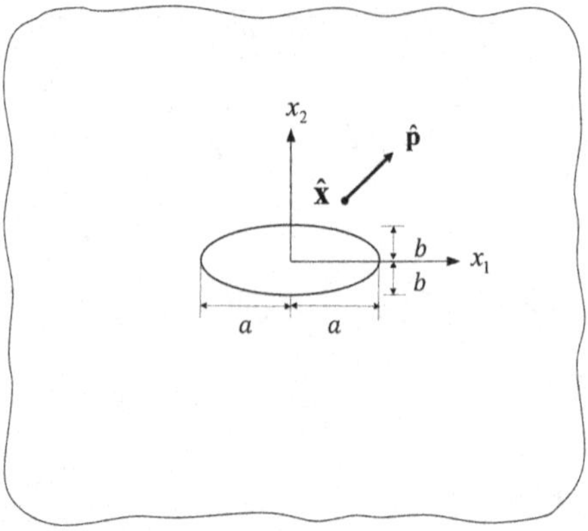

FIGURE 8.5 An anisotropic elastic plane containing an elliptic hole subjected to a point force $\hat{\mathbf{p}}$.

where $2a$, $2b$ are the major and minor axes of the ellipse and φ is a real parameter. The hole boundary is assumed to be traction-free. The solution of the complex function vector $\mathbf{f}(z)$ to this problem has been found to be

$$\mathbf{f}(z) = \frac{1}{2\pi i}\left\{\left\langle \ln\left(\zeta_\alpha - \hat{\zeta}_\alpha\right)\right\rangle \mathbf{A}^T + \sum_{j=1}^{3}\left\langle \ln\left(\zeta_\alpha^{-1} - \bar{\hat{\zeta}}_j\right)\right\rangle \mathbf{B}^{-1}\bar{\mathbf{B}}\mathbf{I}_j\bar{\mathbf{A}}^T\right\}\hat{\mathbf{p}}, \tag{8.97a}$$

where

$$\zeta_\alpha = \frac{z_\alpha + \sqrt{z_\alpha^2 - a^2 - b^2\mu_\alpha^2}}{a - ib\mu_\alpha}, \quad \hat{\zeta}_\alpha = \frac{\hat{z}_\alpha + \sqrt{\hat{z}_\alpha^2 - a^2 - b^2\mu_\alpha^2}}{a - ib\mu_\alpha}, \quad \alpha = 1,2,3. \tag{8.97b}$$

8.4.5 A Plane with a Straight Crack

Consider an infinite anisotropic elastic plane containing a traction-free crack of length $2a$ subjected to a point force $\hat{\mathbf{p}}$ at $\hat{\mathrm{x}}$ (Figure 8.6). The solution of this problem can be obtained directly from (8.97) by letting $b = 0$, that is,

$$\mathbf{f}(z) = \frac{1}{2\pi i}\left\{\left\langle \ln\left(\zeta_\alpha - \hat{\zeta}_\alpha\right)\right\rangle \mathbf{A}^T + \sum_{j=1}^{3}\left\langle \ln\left(\zeta_\alpha^{-1} - \bar{\hat{\zeta}}_j\right)\right\rangle \mathbf{B}^{-1}\bar{\mathbf{B}}\mathbf{I}_j\bar{\mathbf{A}}^T\right\}\hat{\mathbf{p}}, \tag{8.100a}$$

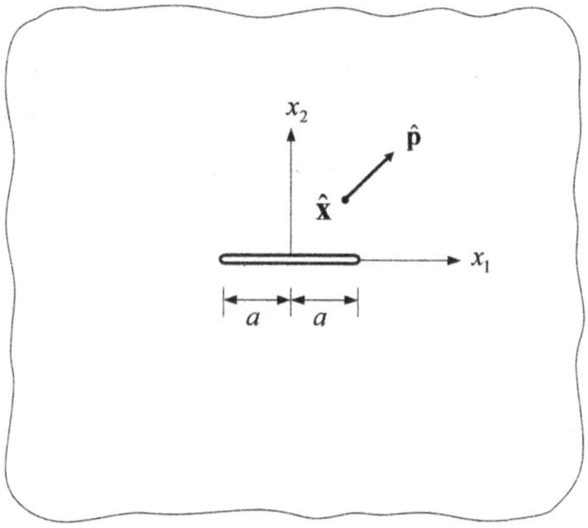

FIGURE 8.6 A crack in an anisotropic elastic plane subjected to a point force $\hat{\mathbf{p}}$ at $\hat{\mathbf{x}}$.

where

$$\zeta_\alpha = \frac{1}{a}\left(z_\alpha + \sqrt{z_\alpha^2 - a^2}\right), \quad \hat{\zeta}_\alpha = \frac{1}{a}\left(\hat{z}_\alpha + \sqrt{\hat{z}_\alpha^2 - a^2}\right), \quad \alpha = 1,2,3. \quad (8.100b)$$

8.4.6 A Plane with an Elliptical Inclusion

Consider an infinite anisotropic elastic plane containing an elliptical inclusion subjected to a point force $\hat{\mathbf{p}} = \left(\hat{p}_1, \hat{p}_2, \hat{p}_3\right)$ at the point $\hat{\mathbf{x}} = \left(\hat{x}_1, \hat{x}_2\right)$ in the matrix (Figure 8.7).

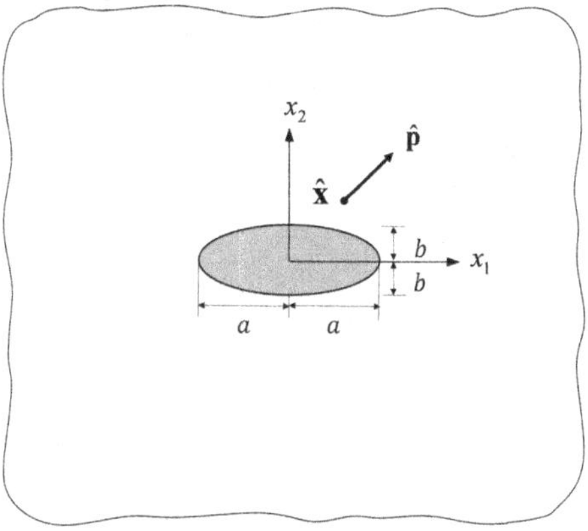

FIGURE 8.7 An anisotropic elastic plane with an elliptical inclusion subjected to a point force at the matrix.

Rigid Inclusion

When the inclusion is absolutely rigid and cannot be deformed, a rigid body rotation ω relative to the matrix may occur. The solution of the complex function vector $\mathbf{f}(z)$ to this problem has been found to be

$$\mathbf{f}(z)=\frac{1}{2\pi i}\left\{\left\langle \ln\left(\zeta_\alpha-\hat{\zeta}_\alpha\right)\right\rangle \mathbf{A}^T+\sum_{j=1}^{3}\left\langle \ln\left(\zeta_\alpha^{-1}-\bar{\hat{\zeta}}_j\right)\right\rangle \mathbf{A}^{-1}\bar{\mathbf{A}}\mathbf{I}_j\bar{\mathbf{A}}^T\right\}\hat{\mathbf{p}}$$
$$+\frac{\omega}{2}\left\langle \zeta_\alpha^{-1}\right\rangle \mathbf{A}^{-1}\bar{\mathbf{k}}, \tag{8.103a}$$

where

$$\omega=\frac{\mathrm{Re}\left\{\bar{\mathbf{k}}^T\mathbf{A}^{-T}\left\langle \hat{\zeta}_\alpha^{-1}\right\rangle \mathbf{A}^T\right\}\hat{\mathbf{p}}}{\pi\mathbf{k}^T\mathbf{M}\bar{\mathbf{k}}},\quad \mathbf{k}=\begin{Bmatrix} ib \\ a \\ 0 \end{Bmatrix},\quad \mathbf{M}=-i\mathbf{B}\mathbf{A}^{-1}. \tag{8.103b}$$

Elastic Inclusion

When the inclusion is elastic and is assumed to be perfectly bonded with the matrix along the interface, the solution of the complex function vector to this problem has been found to be

$$\mathbf{f}_1(z)=\frac{1}{2\pi i}\left\{\left\langle \ln\left(\zeta_\alpha-\hat{\zeta}_\alpha\right)-\ln\left(-\hat{\zeta}_\alpha\right)\right\rangle \mathbf{A}_1^T\right.$$
$$\left.+\sum_{j=1}^{3}\left\langle \ln\left(\zeta_\alpha^{-1}-\bar{\hat{\zeta}}_j\right)-\ln\left(-\bar{\hat{\zeta}}_j\right)\right\rangle \mathbf{D}_j+\sum_{k=1}^{\infty}\left\langle \zeta_\alpha^{-k}\right\rangle \mathbf{E}_k\right\}\hat{\mathbf{p}},$$
$$\mathbf{f}_2(z)=\frac{1}{2\pi i}\sum_{k=1}^{\infty}\left\langle \left(\zeta_\alpha^*\right)^k+\left(\gamma_\alpha^*/\zeta_\alpha^*\right)^k\right\rangle \mathbf{C}_k\hat{\mathbf{p}}, \tag{8.104a}$$

where

$$\mathbf{D}_j=\mathbf{B}_1^{-1}\bar{\mathbf{B}}_1\mathbf{I}_j\bar{\mathbf{A}}_1^T,\quad j=1,2,3;$$
$$\mathbf{C}_k=\left(\mathbf{G}_o-\bar{\mathbf{G}}_k\bar{\mathbf{G}}_0^{-1}\mathbf{G}_k\right)^{-1}\left(\mathbf{T}_k+\bar{\mathbf{G}}_k\bar{\mathbf{G}}_0^{-1}\bar{\mathbf{T}}_k\right),$$
$$\mathbf{E}_k=-\mathbf{B}_1^{-1}\left\{\overline{\mathbf{B}_2\mathbf{C}_k}-\mathbf{B}_2\left\langle \gamma_\alpha^{*k}\right\rangle \mathbf{C}_k\right\},\quad k=1,2,\ldots,\infty, \tag{8.104b}$$

and

$$\mathbf{T}_k = \frac{i}{k}\mathbf{A}_1^{-T}\left\langle \hat{\varsigma}_\alpha^{-k} \right\rangle \mathbf{A}_1^T,$$

$$\mathbf{G}_0 = \left(\bar{\mathbf{M}}_1 + \mathbf{M}_2\right)\mathbf{A}_2, \quad \mathbf{G}_k = \left(\mathbf{M}_1 - \mathbf{M}_2\right)\mathbf{A}_2\left\langle \gamma_\alpha^{*k} \right\rangle, \quad \gamma_\alpha^* = \frac{a + ib\mu_\alpha^*}{a - ib\mu_\alpha^*}. \quad (8.104c)$$

In (8.104), with or without the superscript * denote, respectively, the values related to the inclusion and matrix.

8.5 GREEN'S FUNCTIONS FOR COUPLED STRETCHING–BENDING DEFORMATION

If only linear partial differential equations are involved in the related equations of composite laminated plates such as those presented in (8.30), the superposition principle is applicable to the general loading conditions. Thus, the solutions associated with the concentrated forces and moments applied on an arbitrary point, generally called *Green's functions*, become important in constructing general solutions through superposition. Due to its importance, many analytical closed-form solutions of Green's functions have been obtained and presented in Hwu (2010, 2021) by using the Stroh-like formalism for several different problems such as an infinite laminated plate without or with the presence of an elliptical hole, a straight crack, or an elliptical inclusion, and will be presented in this section.

8.5.1 An Infinite Laminated Plate

Consider an infinite laminated composite plate subjected to a concentrated force $\hat{\mathbf{f}} = \left(\hat{f}_1, \hat{f}_2, \hat{f}_3\right)$ and a concentrated moment $\hat{\mathbf{m}} = \left(\hat{m}_1, \hat{m}_2, \hat{m}_3\right)$ at the point $\hat{\mathbf{x}} = \left(\hat{x}_1, \hat{x}_2\right)$ as shown in Figure 8.8. Cutting a free body enclosing the point $\hat{\mathbf{x}}$ by an anticlockwise closed contour C, the force/moment equilibrium conditions can be written in terms of the stress functions ϕ_i, ψ_i, $i = 1, 2$, as (Hwu, 2004, 2010)

$$\oint_C d\phi_1 = \hat{f}_1, \quad \oint_C d\phi_2 = \hat{f}_2, \quad \oint_C d\eta = \hat{f}_3,$$

$$\oint_C d\psi_1 = \hat{m}_2 + \left(x_1 - \hat{x}_1\right)\hat{f}_3, \quad \oint_C d\psi_2 = -\hat{m}_1 + \left(x_2 - \hat{x}_2\right)\hat{f}_3,$$

$$\oint_C d\left(\left(x_1 - \hat{x}_1\right)\phi_2 - \left(x_2 - \hat{x}_2\right)\phi_1 - \Phi\right) = \hat{m}_3, \quad (8.105a)$$

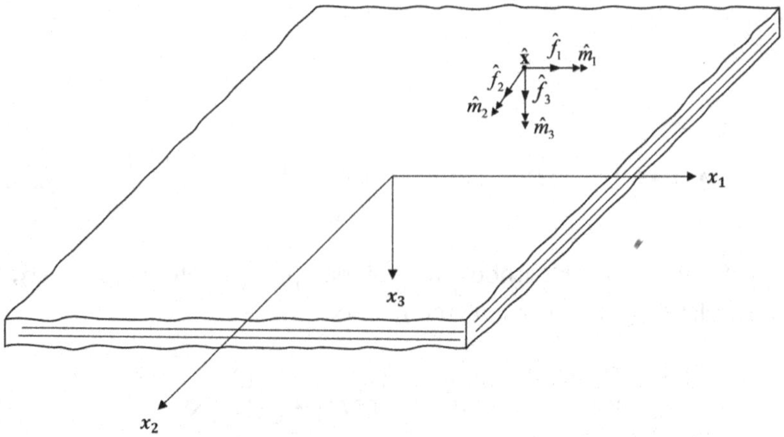

FIGURE 8.8 A laminated composite plate subjected to concentrated forces and moments.

in which η is defined in (8.32) and Φ is related to ϕ_i by

$$\phi_1 = -\Phi_{,2}, \quad \phi_2 = \Phi_{,1}. \tag{8.105b}$$

By using the general solution (8.31) of Stroh-like formalism, a solution satisfying all the basic equations (8.30) of laminated plates under coupled stretching–bending deformation and the point force/moment equilibrium conditions (8.105) has been obtained (Hwu, 2004), in which the complex function vector $\mathbf{f}(z)$ can be expressed as

$$\mathbf{f}(z) = \frac{1}{2\pi i}\left\langle \ln\left(z_\alpha - \hat{z}_\alpha\right)\right\rangle \mathbf{A}^T\hat{\mathbf{p}} + \frac{\hat{f}_3}{2\pi i}\left\langle \left(z_\alpha - \hat{z}_\alpha\right)\left[\ln\left(z_\alpha - \hat{z}_\alpha\right) - 1\right]\right\rangle \mathbf{A}^T\mathbf{i}_3$$
$$-\frac{\hat{m}_3}{2\pi i}\left\langle \frac{1}{z_\alpha - \hat{z}_\alpha}\right\rangle \mathbf{A}^T\mathbf{i}_2, \tag{8.106a}$$

and

$$\hat{\mathbf{p}} = \begin{Bmatrix} \hat{f}_1 \\ \hat{f}_2 \\ \hat{m}_2 \\ -\hat{m}_1 \end{Bmatrix}, \quad \mathbf{i}_2 = \begin{Bmatrix} 0 \\ 1 \\ 0 \\ 0 \end{Bmatrix}, \quad \mathbf{i}_3 = \begin{Bmatrix} 0 \\ 0 \\ 1 \\ 0 \end{Bmatrix}. \tag{8.106b}$$

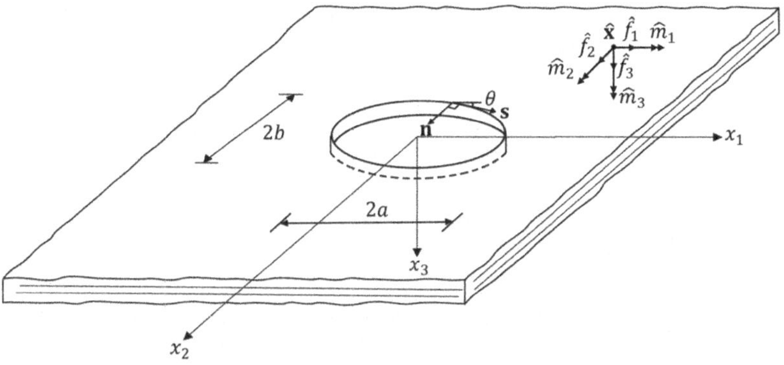

FIGURE 8.9 A laminated composite plate containing an elliptical hole subjected to concentrated forces and moments.

8.5.2 An Infinite Laminated Plate with an Elliptical Hole

Consider an infinite laminated composite plate containing a traction-free elliptical hole subjected to a concentrated force $\hat{\mathbf{f}}$ and a concentrated moment $\hat{\mathbf{m}}$ at the point $\hat{\mathbf{x}}$ (see Figure 8.9). According to the Kirchhoff's hypotheses and classical lamination theory, by $(3.15c)_4$ the traction-free hole boundary conditions can be written as

$$N_{nn} = N_{sn} = M_{nn} = V_n = 0, \quad \text{along the hole boundary.} \qquad (8.107)$$

With the relations of (8.32b), the traction-free conditions of (8.107) can be further expressed in terms of the stress functions as

$$\mathbf{n}^T\boldsymbol{\phi}_{,s} = \mathbf{s}^T\boldsymbol{\phi}_{,s} = \mathbf{n}^T\boldsymbol{\psi}_{,s} = \left(\mathbf{s}^T\boldsymbol{\psi}_{,s}\right)_{,s} = 0, \quad \text{along the hole boundary.} \quad (8.108)$$

It can be proved that (8.108) will be satisfied if

$$\boldsymbol{\phi}_{,s} = \mathbf{0}, \ \boldsymbol{\psi}_{,s} = \eta_0\mathbf{s}, \ \text{ along the hole boundary,} \qquad (8.109a)$$

or

$$\boldsymbol{\phi} = \mathbf{0}, \ \boldsymbol{\psi} = \eta_0\mathbf{x}, \ \text{ along the hole boundary,} \qquad (8.109b)$$

in which η_0 is a constant to be determined through the requirement of single-valued deflection, that is,

$$\oint_C dw = \oint_C \frac{\partial w}{\partial s} ds = -\oint_C \beta_s ds = -\oint_C (\beta_1\cos\theta + \beta_2\sin\theta) ds = 0, \quad (8.110)$$

where C is a closed integral path. The last equality of (8.110) can also be written in a simple matrix form as

$$\oint_C \mathbf{s}^T \boldsymbol{\beta} ds = 0, \quad \text{where } \mathbf{s}^T = (\cos\theta, \sin\theta), \tag{8.111a}$$

or in a generalized matrix form

$$\oint_C \breve{\mathbf{s}}^T \mathbf{u}_d ds = 0, \text{ where } \breve{\mathbf{s}}^T = (0, 0, \mathbf{s}^T). \tag{8.111b}$$

Note that (8.109b) is obtained by integrating (8.109a) with respect to the tangential variable s and ignoring the constant terms which are related to the rigid body motion, where $\mathbf{x}$ is the position vector of field point.

With the traction-free condition described by (8.109b) together with (8.111b), by using the general solution (8.31) and the point force/moment equilibrium conditions (8.105), the solutions of complex function vector $\mathbf{f}(z)$ have been obtained (Hwu, 2005) and corrected as (Hsu and Hwu, 2021a,b)

$$\mathbf{f}(z) = \mathbf{f}_1(\zeta) + \mathbf{f}_2(\zeta) + \mathbf{f}_3(\zeta) + \mathbf{f}_c(\zeta), \tag{8.112a}$$

where

$$\begin{aligned}
\mathbf{f}_1(\zeta) &= \frac{1}{2\pi i}\left\{\left\langle \ln(\zeta_\alpha - \hat{\zeta}_\alpha)\right\rangle \mathbf{A}^T + \sum_{k=1}^{4}\left\langle \ln(\zeta_\alpha^{-1} - \bar{\hat{\zeta}}_k)\right\rangle \mathbf{D}_k \bar{\mathbf{A}}^T\right\}\hat{\mathbf{p}}, \\
\mathbf{f}_2(\zeta) &= \frac{\hat{f}_3}{2\pi i}\left\{\left\langle g(\zeta_\alpha)\right\rangle \mathbf{A}^T + \sum_{k=1}^{4}\left\langle \zeta_\alpha^{-1} - \bar{\hat{\zeta}}_k\right\rangle \mathbf{E}_k(\zeta)\bar{\mathbf{A}}^T\right\}\mathbf{i}_3, \\
\mathbf{f}_3(\zeta) &= -\frac{\hat{m}_3}{2\pi i}\left\{\left\langle \frac{c_{4\alpha}\hat{\zeta}_\alpha}{\zeta_\alpha - \hat{\zeta}_\alpha}\right\rangle \mathbf{A}^T + \sum_{k=1}^{4}\left\langle \frac{1}{\zeta_\alpha^{-1} - \bar{\hat{\zeta}}_k}\right\rangle \mathbf{D}_k \left\langle \overline{c_{4\alpha}\hat{\zeta}_\alpha}\right\rangle \bar{\mathbf{A}}^T\right\}\mathbf{i}_2, \\
\mathbf{f}_c(\zeta) &= <\zeta_\alpha^{-1}>\mathbf{q},
\end{aligned} \tag{8.112b}$$

and

$$\begin{aligned}
g(\zeta_\alpha) &= (z_\alpha - \hat{z}_\alpha)\ln(\zeta_\alpha - \hat{\zeta}_\alpha) + c_{2\alpha}(\zeta_\alpha - \hat{\zeta}_\alpha) - c_{3\alpha}(\zeta_\alpha^{-1} - \hat{\zeta}_\alpha^{-1}), \\
c_{2\alpha} &= c_\alpha(\ln c_\alpha - 1), \quad c_{3\alpha} = c_\alpha \gamma_\alpha \ln(-\hat{\zeta}_\alpha), \quad c_{4\alpha} = \frac{1}{c_\alpha(\hat{\zeta}_\alpha - \gamma_\alpha / \hat{\zeta}_\alpha)},
\end{aligned}$$

$$c_\alpha = \frac{1}{2}\left(a - ib\mu_\alpha\right), \quad \gamma_\alpha = \frac{a + ib\mu_\alpha}{a - ib\mu_\alpha},$$
$$\mathbf{D}_k = \mathbf{B}^{-1}\overline{\mathbf{B}}\mathbf{I}_k,$$
$$\mathbf{E}_k(\zeta) = \mathbf{D}_k < \overline{c}_{2\alpha} > + < \overline{\hat{\zeta}}_k^{-1}\zeta_\alpha > \mathbf{D}_k < \overline{c}_{3\alpha} >$$
$$+ < (1 - \overline{\gamma}_k \overline{\hat{\zeta}}_k^{-1}\zeta_\alpha)\ln(\zeta_\alpha^{-1} - \overline{\hat{\zeta}}_k) > \mathbf{D}_k < \overline{c}_\alpha >,$$
$$\mathbf{q} = \eta_0 \mathbf{B}^{-1}(a\mathbf{i}_3 + ib\mathbf{i}_4). \tag{8.112c}$$

In the above, the overbar stands for the complex conjugate; $\hat{\zeta}_\alpha$ is the mapped variable of $\hat{z}_\alpha \left(= \hat{x}_1 + \mu_\alpha \hat{x}_2\right)$ whose relation is given in (8.97b); the matrix $\mathbf{I}_k$ is a diagonal matrix whose kk^{th} component is one and all the other components are zero, for example, $\mathbf{I}_2$ = diag [0 1 0 0]; the load vector $\hat{\mathbf{p}}$ and the two base vectors $\mathbf{i}_2$ and $\mathbf{i}_3$ are defined in (8.106b), whereas the other base vector $\mathbf{i}_4 = (0, 0, 0, 1)^T$. The constant η_0 appeared in (8.112c) is determined by (8.111b) as

$$\eta_0 = \frac{\mathbf{i}_3^T\mathbf{L}^{-1}\mathbf{p}_I + \lambda\mathbf{i}_4^T\mathbf{L}^{-1}\mathbf{p}_R}{\pi a\left(\mathbf{i}_3^T\boldsymbol{\lambda}_a + \lambda\mathbf{i}_4^T\boldsymbol{\lambda}_b\right)}, \quad \lambda = b/a, \tag{8.113a}$$

where

$$\boldsymbol{\lambda}_a = \mathbf{L}^{-1}\left(\mathbf{i}_3 - \lambda\mathbf{S}^T\mathbf{i}_4\right), \ \boldsymbol{\lambda}_b = \mathbf{L}^{-1}\left(\mathbf{S}^T\mathbf{i}_3 + \lambda\mathbf{i}_4\right). \tag{8.113b}$$

The two real vectors, $\mathbf{p}_R$ and $\mathbf{p}_I$, are related to the loads $\hat{\mathbf{p}}, \hat{f}_3, \hat{m}_3$ by

$$\mathbf{p}_R + i\mathbf{p}_I = \mathbf{P}_1\hat{\mathbf{p}} + \hat{f}_3\mathbf{P}_2\mathbf{i}_3 + \hat{m}_3\mathbf{P}_3\mathbf{i}_2, \tag{8.114a}$$

where

$$\mathbf{P}_1 = -2\mathbf{B}\left\langle \hat{\zeta}_\alpha^{-1} \right\rangle \mathbf{A}^T,$$
$$\mathbf{P}_2 = 2\mathbf{B}\left\langle c_\alpha\left[\ln\left(-c_\alpha\hat{\zeta}_\alpha\right) + \gamma_\alpha\hat{\zeta}_\alpha^{-2}/2\right]\right\rangle \mathbf{A}^T,$$
$$\mathbf{P}_3 = 2\mathbf{B}\left\langle c_{4\alpha}\hat{\zeta}_\alpha^{-1} \right\rangle \mathbf{A}^T. \tag{8.114b}$$

In (8.113), $\mathbf{L}$ and $\mathbf{S}$ are two real matrices, generally called the *Barnett–Lothe Tensors*, which are related to the material eigenvector matrices $\mathbf{A}$ and $\mathbf{B}$ by (Hwu, 2010)

$$\mathbf{L} = -2i\mathbf{B}\mathbf{B}^T, \quad \mathbf{S} = i\left(2\mathbf{A}\mathbf{B}^T - \mathbf{I}\right). \tag{8.115}$$

8.5.3 An Infinite Laminated Plate with a Straight Crack

An elliptic hole can be made into a crack of length $2a$ by letting the minor axis $2b$ be equal to zero. The solutions for crack problems can therefore be obtained from (8.112) with $b = 0$. From the relations given in (8.32a), we know that the stress intensity factors defined by

$$\mathbf{k} = \begin{Bmatrix} K_{II} \\ K_I \\ K_{IIB} \\ K_{IB} \end{Bmatrix} = \lim_{r\to 0}\sqrt{2\pi r}\begin{Bmatrix} N_{12}/h \\ N_{22}/h \\ 6M_{12}/h^2 \\ 6M_{22}/h^2 \end{Bmatrix}, \tag{8.116a}$$

can be expressed as

$$\mathbf{k} = \lim_{r\to 0}\sqrt{2\pi r}\left\langle d_\alpha\right\rangle\left(\boldsymbol{\phi}_{d,1} - m_0\mathbf{i}_3\right), \tag{8.116b}$$

where

$$d_1 = d_2 = 1/h,\ \ d_3 = d_4 = 6/h^2,\ \ m_0 = \left(\mathbf{i}_3^T\boldsymbol{\phi}_{d,1} + \mathbf{i}_4^T\boldsymbol{\phi}_{d,2}\right)/2. \tag{8.116c}$$

In the above, K_I, K_{II}, K_{IB}, and K_{IIB} denote, respectively, the stress intensity factors of opening, shearing, bending, and twisting modes; r is the distance ahead of the crack tip; h is the thickness of the plate. The closed-form solution of stress intensity factors can then be determined if the limits, $\lim_{r\to 0}\sqrt{2\pi r}\boldsymbol{\phi}_{d,1}$ and $\lim_{r\to 0}\sqrt{2\pi r}\boldsymbol{\phi}_{d,2}$, are known. By the substitution of the solutions obtained in (8.112) with $b = 0$ into $(8.31\text{a})_2$, the stress intensity factors $\mathbf{k}$ defined in (8.116b) can be obtained as (Hsu and Hwu, 2021a)

$$\mathbf{k} = \frac{1}{\sqrt{\pi a}}\left\langle d_\alpha\right\rangle\left\{\mathbf{g} - \frac{1}{2}\left(\mathbf{i}_3^T\mathbf{g} + \mathbf{i}_4^T\mathbf{G}_1\mathbf{g}\right)\mathbf{i}_3\right\}, \tag{8.117a}$$

where

$$\mathbf{g} = \sum_{i=1}^{3} 2\mathrm{Im}\left\{\mathbf{B}\left\langle h_i\right\rangle\mathbf{A}^T\right\}\hat{\mathbf{p}}_i - \pi a\eta_0\mathbf{i}_3,$$

$$h_1 = \frac{1}{1-\hat{\zeta}_\alpha},\ h_2 = \frac{a}{2}\left\{\ln\frac{a\left(-\hat{\zeta}_\alpha\right)}{2} - \frac{1}{\hat{\zeta}_\alpha}\right\},\ h_3 = \frac{2\hat{\zeta}_\alpha}{a\left(\hat{\zeta}_\alpha - \hat{\zeta}_\alpha^{-1}\right)\left(1-\hat{\zeta}_\alpha\right)^2},$$

$$\hat{\mathbf{p}}_1 = \hat{\mathbf{p}}, \quad \hat{\mathbf{p}}_2 = \hat{f}_3\mathbf{i}_3, \quad \hat{\mathbf{p}}_3 = \hat{m}_3\mathbf{i}_2,$$
$$\mathbf{G}_1 = \mathbf{N}_1^T - \mathbf{N}_3\mathbf{S}\mathbf{L}^{-1}. \tag{8.117b}$$

In the above, Im stands for the imaginary part of a complex number; $\mathbf{N}_1$ and $\mathbf{N}_3$ are the fundamental elasticity matrices in (8.33b).

8.5.4 An Infinite Laminated Plate with an Elliptical Inclusion

Consider an infinite laminated composite plate containing an elliptical inclusion subjected to a concentrated force $\hat{\mathbf{f}}$ and a concentrated moment $\hat{\mathbf{m}}$ at the point $\hat{\mathbf{x}}$ as shown in Figure 8.10. The inclusion and the matrix are assumed to be perfectly bonded along the interface. The contour of the

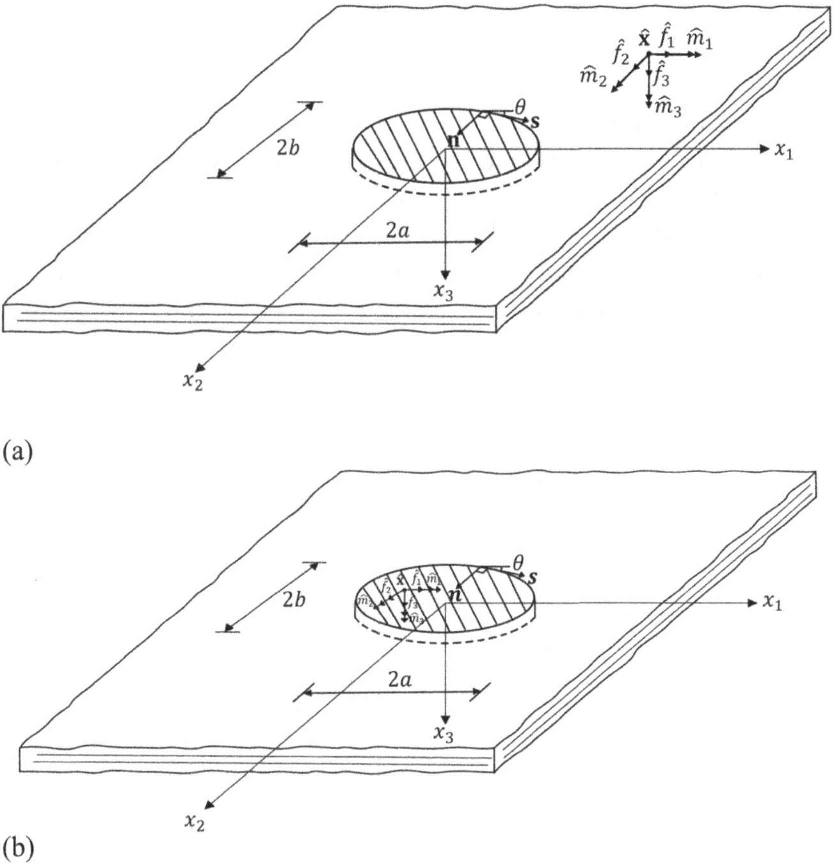

(a)

(b)

FIGURE 8.10 A laminated composite plate containing an elliptical inclusion subjected to concentrated forces and moments (a) outside the inclusion, (b) inside the inclusion.

inclusion boundary is represented by (8.96). Since we now have two different materials, the general solution of (8.31) should be rewritten as

$$\begin{aligned} \mathbf{u}_d^{(1)} &= 2\,\mathrm{Re}\{\mathbf{A}_1\mathbf{f}_1(\zeta)\}, \quad \boldsymbol{\phi}_d^{(1)} = 2\,\mathrm{Re}\{\mathbf{B}_1\mathbf{f}_1(\zeta)\}, \\ \mathbf{u}_d^{(2)} &= 2\,\mathrm{Re}\{\mathbf{A}_2\mathbf{f}_2(\zeta^*)\}, \quad \boldsymbol{\phi}_d^{(2)} = 2\,\mathrm{Re}\{\mathbf{B}_2\mathbf{f}_2(\zeta^*)\}, \end{aligned} \tag{8.118}$$

where the subscripts 1 and 2 or the superscripts (1) and (2) denote, respectively, the values related to the matrix and the inclusion. The variable with the superscript * denotes the value related to the properties of inclusion.

In terms of the vectors of displacements and stress functions, the interface continuity conditions can be expressed as (Hwu and Tan, 2007)

$$\mathbf{u}_d^{(1)} = \mathbf{u}_d^{(2)}, \; \boldsymbol{\phi}_d^{(1)} = \boldsymbol{\phi}_d^{(2)}, \; w^{(1)} = w^{(2)}. \tag{8.119}$$

The solutions of complex function vectors $\mathbf{f}_1(\zeta)$ and $\mathbf{f}_2(\zeta)$ can then be determined through the satisfaction of point force/moment equilibrium conditions (8.105) and continuity conditions (8.119). Their solutions were first derived by Hwu and Tan (2007) using the method of analytical continuation and then corrected by Hsu and Hwu (2020) with the addition of missing constant terms to satisfy the continuity conditions. Depending on the location of concentrated forces/moments, their solutions can be written as follows (Hwu, 2021):

Loads outside the inclusion

$$\begin{aligned} \mathbf{f}_1(\zeta) &= \frac{1}{2\pi i}\sum_{j=1}^{3}\left\langle g_\alpha^{(j)}\right\rangle \mathbf{A}_1^T\hat{\mathbf{p}}_j + \sum_{k=1}^{\infty}\left\langle \zeta_\alpha^{-k}\right\rangle \mathbf{d}_k, \\ \mathbf{f}_2(\zeta^*) &= \sum_{k=1}^{\infty}\left\langle \left(\zeta_\alpha^*\right)^k + \left(\gamma_\alpha^*/\zeta_\alpha^*\right)^k\right\rangle \mathbf{c}_k, \end{aligned} \tag{8.120a}$$

where

$$\begin{aligned} g_\alpha^{(1)} &= \ln\left(\zeta_\alpha - \hat{\zeta}_\alpha\right) - \ln\left(-\hat{\zeta}_\alpha\right), \\ g_\alpha^{(2)} &= g(\zeta_\alpha) - c_\alpha\hat{\zeta}_\alpha\left[1 - \ln\left(-c_\alpha\hat{\zeta}_\alpha\right) - \gamma_\alpha\hat{\zeta}_\alpha^{-2}\right], \\ g_\alpha^{(3)} &= -\frac{c_{4\alpha}\zeta_\alpha}{\zeta_\alpha - \hat{\zeta}_\alpha}. \end{aligned} \tag{8.120b}$$

In the above, $\hat{\mathbf{p}}_1, \hat{\mathbf{p}}_2, \hat{\mathbf{p}}_3$, c_α, γ_α, $g(\zeta_\alpha)$, and $c_{4\alpha}$ are defined in (8.117b) and (8.112c), $\mathbf{c}_k$ and $\mathbf{d}_k$ are defined as

$$\begin{aligned} \mathbf{c}_k &= \left(\mathbf{G}_o - \bar{\mathbf{G}}_k \bar{\mathbf{G}}_0^{-1} \mathbf{G}_k\right)^{-1} \left(\mathbf{t}_k - \bar{\mathbf{G}}_k \bar{\mathbf{G}}_0^{-1} \bar{\mathbf{t}}_k\right), \\ \mathbf{d}_k &= \mathbf{B}_1^{-1} \left\{ \bar{\mathbf{B}}_2 \bar{\mathbf{c}}_k + \mathbf{B}_2 \left\langle \gamma_\alpha^{*k} \right\rangle \mathbf{c}_k - \bar{\mathbf{B}}_1 \bar{\mathbf{e}}_k \right\}, \end{aligned} \tag{8.121a}$$

where

$$\begin{aligned} &\mathbf{G}_0 = \left(\bar{\mathbf{M}}_1 + \mathbf{M}_2\right)\mathbf{A}_2, \quad \mathbf{G}_k = \left(\mathbf{M}_1 - \mathbf{M}_2\right)\mathbf{A}_2 \left\langle \gamma_\alpha^{*k} \right\rangle, \\ &\mathbf{M}_1 = -i\mathbf{B}_1\mathbf{A}_1^{-1}, \quad \mathbf{M}_2 = -i\mathbf{B}_2\mathbf{A}_2^{-1}, \\ &\mathbf{t}_k = -i\mathbf{A}_1^{-T}\mathbf{e}_k, \qquad k = 1, 2, \cdots \infty. \end{aligned} \tag{8.121b}$$

In (8.121b), $\mathbf{e}_k$ is related to the loads by

$$\mathbf{e}_k = \frac{1}{2\pi i}\left\{ \frac{-1}{k}\left\langle \hat{\zeta}_\alpha^{-k} \right\rangle \mathbf{A}_1^T \hat{\mathbf{p}} + \hat{f}_3 \left\langle h_\alpha \right\rangle \mathbf{A}_1^T \mathbf{i}_3 + \hat{m}_3 \left\langle c_{4\alpha} \hat{\zeta}_\alpha^{-k} \right\rangle \mathbf{A}_1^T \mathbf{i}_2 \right\}, \tag{8.122a}$$

where

$$h_\alpha = \begin{cases} \dfrac{c_\alpha}{\hat{\zeta}_\alpha}\left[\hat{\zeta}_\alpha \ln\left(-c_\alpha \hat{\zeta}_\alpha\right) + \dfrac{\gamma_\alpha}{2\hat{\zeta}_\alpha} \right], & \text{when } k = 1, \\ \dfrac{c_\alpha}{k\hat{\zeta}_\alpha^k}\left[\dfrac{-\hat{\zeta}_\alpha}{k-1} + \dfrac{\gamma_\alpha}{(k+1)\hat{\zeta}_\alpha} \right], & \text{when } k > 1. \end{cases} \tag{8.122b}$$

Loads inside the inclusion

$$\begin{aligned} \mathbf{f}_1(\zeta) &= \frac{1}{2\pi i}\left\{ \left\langle \ln\zeta_\alpha \right\rangle \mathbf{A}_1^T \hat{\mathbf{p}} + \hat{f}_3 \left\langle (z_\alpha - \hat{z}_\alpha)\ln\zeta_\alpha \right\rangle \mathbf{A}_1^T \mathbf{i}_3 \right. \\ &\quad \left. + \left\langle \zeta_\alpha \right\rangle \mathbf{k}_1 + \left\langle \zeta_\alpha^{-1} \right\rangle \mathbf{k}_{-1} \right\} + \sum_{k=1}^{\infty} \left\langle \zeta_\alpha^{-k} \right\rangle \mathbf{d}_k, \\ \mathbf{f}_2(\zeta^*) &= \frac{1}{2\pi i}\sum_{j=1}^{3} \left\langle g_\alpha^{*(j)} \right\rangle \mathbf{A}_2^T \hat{\mathbf{p}}_j + \sum_{k=1}^{\infty} \left\langle \left(\zeta_\alpha^*\right)^k + \left(\gamma_\alpha^* / \zeta_\alpha^*\right)^k \right\rangle \mathbf{c}_k, \end{aligned} \tag{8.123a}$$

where

$$
\begin{aligned}
g_\alpha^{*(1)} &= \ln\left(z_\alpha^* - \hat{z}_\alpha^*\right) - \ln c_\alpha^*,\\
g_\alpha^{*(2)} &= \left(z_\alpha^* - \hat{z}_\alpha^*\right)\left[\ln\left(z_\alpha^* - \hat{z}_\alpha^*\right) - 1\right] + \hat{z}_\alpha^* \ln c_\alpha^*,\\
g_\alpha^{*(3)} &= -\left(z_\alpha^* - \hat{z}_\alpha^*\right)^{-1}.
\end{aligned}
\tag{8.123b}
$$

In the above, the superscript $*$ denotes the value related to the inclusion; $\mathbf{c}_k$ is defined in $(8.121\text{a})_1$ with $\mathbf{t}_k$ of $(8.121\text{b})_5$ changed to

$$
\mathbf{t}_k = -\left(\bar{\mathbf{M}}_1 - \bar{\mathbf{M}}_2\right)\bar{\mathbf{A}}_2 \bar{\mathbf{e}}_k^+.
\tag{8.124}
$$

The coefficient vector $\mathbf{d}_k$ of (8.123a) is defined as

$$
\begin{aligned}
\mathbf{d}_k &= \mathbf{B}_1^{-1}\left\{\bar{\mathbf{B}}_2\bar{\mathbf{c}}_k + \mathbf{B}_2\left\langle \gamma_\alpha^{*k}\right\rangle \mathbf{c}_k + \mathbf{B}_2\mathbf{e}_k^+\right\}\\
&= \mathbf{A}_1^{-1}\left\{\bar{\mathbf{A}}_2\bar{\mathbf{c}}_k + \mathbf{A}_2\left\langle \gamma_\alpha^{*k}\right\rangle \mathbf{c}_k + \mathbf{A}_2\mathbf{e}_k^+\right\}.
\end{aligned}
\tag{8.125}
$$

The other vectors $\mathbf{k}_{-1}$, $\mathbf{k}_1$, and $\mathbf{e}_k^+$ are defined by

$$
\begin{aligned}
&\mathbf{k}_1 = \hat{f}_3\left(\mathbf{B}_1^T\mathbf{Q}_2 + \mathbf{A}_1^T\mathbf{Q}_1^T\right)\mathbf{i}_3,\quad \mathbf{k}_{-1} = -\hat{f}_3\left(\mathbf{B}_1^T\bar{\mathbf{Q}}_2 + \mathbf{A}_1^T\bar{\mathbf{Q}}_1^T\right)\mathbf{i}_3,\\
&\mathbf{e}_k^+ = \frac{1}{2\pi i}\sum_{j=1}^{3}\left\langle q_j\left(\zeta_\alpha^*\right)\right\rangle \mathbf{A}_2^T\hat{\mathbf{p}}_j,
\end{aligned}
\tag{8.126a}
$$

where

$$
\begin{aligned}
q_1(\zeta_\alpha^*) &= \frac{-1}{k}\left\{\hat{\zeta}_\alpha^{*k} + \left(\gamma_\alpha^*/\hat{\zeta}_\alpha^*\right)^k\right\},\\
q_2(\zeta_\alpha^*) &= \begin{cases}
c_\alpha^*\left\{\dfrac{1}{2}\left[\hat{\zeta}_\alpha^{*2} + (\gamma_\alpha^*/\hat{\zeta}_\alpha^*)^2\right] + \gamma_\alpha^*\left(\ln c_\alpha^* + 1\right)\right\}, & \text{when } k=1,\\
\dfrac{c_\alpha^*}{k}\left\{\begin{aligned}&\dfrac{1}{k+1}\left[\hat{\zeta}_\alpha^{*k+1} + (\gamma_\alpha^*/\hat{\zeta}_\alpha^*)^{k+1}\right]\\ &-\dfrac{\gamma_\alpha^*}{k-1}\left[\hat{\zeta}_\alpha^{*k-1} + (\gamma_\alpha^*/\hat{\zeta}_\alpha^*)^{k-1}\right]\end{aligned}\right\}, & \text{when } k>1,
\end{cases}\\
q_3(\zeta_\alpha^*) &= -\frac{\hat{\zeta}_\alpha^{*k} - (\gamma_\alpha^*/\hat{\zeta}_\alpha^*)^k}{c_\alpha^*[\hat{\zeta}_\alpha^* - (\gamma_\alpha^*/\hat{\zeta}_\alpha^*)]}.
\end{aligned}
\tag{8.126b}
$$

In (8.126a), $\mathbf{Q}_1$ and $\mathbf{Q}_2$ are defined as

$$\mathbf{Q}_1 = \mathbf{A}_2 \left\langle c_\alpha^* \left(\ln c_\alpha^* - 1 \right) \right\rangle \mathbf{B}_2^T, \ \mathbf{Q}_2 = \mathbf{A}_2 \left\langle c_\alpha^* \left(\ln c_\alpha^* - 1 \right) \right\rangle \mathbf{A}_2^T. \quad (8.127)$$

8.6 GREEN'S FUNCTIONS FOR THREE-DIMENSIONAL PROBLEMS

Consider a point force $\hat{\mathbf{p}} = (\hat{p}_1, \hat{p}_2, \hat{p}_3)$ applied at $\hat{\mathbf{x}} = (\hat{x}_1, \hat{x}_2, \hat{x}_3)$ of the infinite 3D linear anisotropic elastic solids. The solution satisfying all the basic equations stated in (8.1) and the point force equilibrium condition is usually called *Green's function.* In order to get this solution from Radon–Stroh formalism stated in Section 8.3.1, we employ its corresponding 2D solutions (8.93) to $\breve{\mathbf{f}}(z)$ (Hsu et al., 2019),

$$\breve{\mathbf{f}}(z) = \frac{1}{2\pi i} \left\langle \ln(z_\alpha - \hat{z}_\alpha) \right\rangle \mathbf{A}^T \hat{\mathbf{p}}, \quad (8.128a)$$

where

$$\begin{aligned} &z_\alpha - \hat{z}_\alpha = \mathbf{n} \cdot \Delta\mathbf{x} + \mu_\alpha \mathbf{m} \cdot \Delta\mathbf{x}, \ \alpha = 1,2,3, \\ &\Delta\mathbf{x} = (\Delta x_1, \Delta x_2, \Delta x_3) = (x_1 - \hat{x}_1, x_2 - \hat{x}_2, x_3 - \hat{x}_3). \end{aligned} \quad (8.128b)$$

Substituting (8.128) into (8.88) and (8.89), we get

$$\begin{aligned} \mathbf{u} &= \frac{-\operatorname{sgn}(\mathbf{m}\cdot\Delta\mathbf{x})}{2\pi^2} \int_0^\pi \operatorname{Re}\left\{ \mathbf{A} \left\langle \frac{1}{z_\alpha - \hat{z}_\alpha} \right\rangle \mathbf{A}^T \right\} d\theta \hat{\mathbf{p}}, \\ \boldsymbol{\phi} &= \frac{-\operatorname{sgn}(\mathbf{m}\cdot\Delta\mathbf{x})}{2\pi^2} \int_0^\pi \operatorname{Re}\left\{ \mathbf{B} \left\langle \frac{1}{z_\alpha - \hat{z}_\alpha} \right\rangle \mathbf{A}^T \right\} d\theta \hat{\mathbf{p}}, \\ \mathbf{t}^e &= \frac{\operatorname{sgn}(\mathbf{m}\cdot\Delta\mathbf{x})}{2\pi^2} \int_0^\pi \operatorname{Re}\left\{ \mathbf{B}^e \left\langle \frac{1}{(z_\alpha - \hat{z}_\alpha)^2} \right\rangle \mathbf{A}^T \right\} d\theta \hat{\mathbf{p}}. \end{aligned} \quad (8.129)$$

With the solution (8.129) not only the numerical integration is necessary but also the calculation of Stroh's eigenvalues and eigenvector matrices cannot be avoided. To have a possible improvement on the numerical calculation, we employ the following identities obtained for 2D anisotropic elasticity (Hwu, 2010):

$$2r\mathbf{A}\left\langle \left(z_\alpha - \hat{z}_\alpha\right)^{-1} \right\rangle \mathbf{A}^T = -\sin\psi \mathbf{N}_2(\psi) + i(\ldots),$$
$$2r\mathbf{B}\left\langle \left(z_\alpha - \hat{z}_\alpha\right)^{-1} \right\rangle \mathbf{A}^T = \cos\psi \mathbf{I} - \sin\psi \mathbf{N}_1^T(\psi) + i(\ldots), \quad (8.130)$$

where r and ψ are related to 2D rectangular coordinates by

$$\mathbf{n}\cdot\Delta\mathbf{x} = r\cos\psi, \quad \mathbf{m}\cdot\Delta\mathbf{x} = r\sin\psi, \quad (8.131a)$$

that is,

$$r = \sqrt{\left(\mathbf{n}\cdot\Delta\mathbf{x}\right)^2 + \left(\mathbf{m}\cdot\Delta\mathbf{x}\right)^2}, \; \psi = \tan^{-1}\frac{\mathbf{m}\cdot\Delta\mathbf{x}}{\mathbf{n}\cdot\Delta\mathbf{x}}. \quad (8.131b)$$

In other words, r and ψ are determined by the positions $\mathbf{x}$, $\hat{\mathbf{x}}$ and the location of Radon-plane **n-m**.

Substituting (8.130) into $(8.129)_{1,2}$, we obtain

$$\mathbf{u} = \frac{\operatorname{sgn}\left(\mathbf{m}\cdot\Delta\mathbf{x}\right)}{4\pi^2}\int_0^\pi \frac{1}{r}\sin\psi \mathbf{N}_2(\psi)\, d\theta\hat{\mathbf{p}},$$
$$\boldsymbol{\phi} = \frac{-\operatorname{sgn}\left(\mathbf{m}\cdot\Delta\mathbf{x}\right)}{4\pi^2}\int_0^\pi \frac{1}{r}\left[\cos\psi \mathbf{I} - \sin\psi \mathbf{N}_1^T(\psi)\right] d\theta\hat{\mathbf{p}}. \quad (8.132)$$

Since $\mathbf{B}^e$ of $(8.129)_3$ is a matrix related to the surface normal $\mathbf{e}$ and does not have the identities like (8.130), to get the real form expression of $\mathbf{t}^e$ we use the relation of $(8.1)_{1,2}$ which lead to

$$t_i^e = \sigma_{ij}e_j = C_{ijkl}u_{k,l}e_j. \quad (8.133)$$

Differentiating r and ψ of (8.131b) with respect to x_l, we have

$$r_{,l} = \frac{c_l}{r}, \; \psi_{,l} = \frac{d_l}{r^2}, \quad (8.134a)$$

where

$$c_l = \left(\mathbf{n}\cdot\Delta\mathbf{x}\right)n_l + \left(\mathbf{m}\cdot\Delta\mathbf{x}\right)m_l, \; d_l = \left(\mathbf{n}\cdot\Delta\mathbf{x}\right)m_l - \left(\mathbf{m}\cdot\Delta\mathbf{x}\right)n_l. \quad (8.134b)$$

Substituting $(8.132)_1$ and (8.134) into (8.133), the real form expression of the traction can be obtained and written in matrix form as

$$\mathbf{t}^e = \frac{\operatorname{sgn}(\mathbf{m}\cdot\Delta\mathbf{x})}{4\pi^2}\int_0^{\pi}\frac{1}{r^3}\left\{\mathbf{Q}_c^e\mathbf{N}_2(\psi)+\mathbf{Q}_d^e\mathbf{N}_2'(\psi)\right\}d\theta\hat{\mathbf{p}}, \tag{8.135a}$$

where

$$\mathbf{Q}_c^e = -x^*\mathbf{Q}^e + y^*\mathbf{R}^e,\ \ \mathbf{Q}_d^e = y_s^*\mathbf{Q}^e + x_s^*\mathbf{R}^e, \tag{8.135b}$$

and

$$\begin{aligned} & x^* = x_c^* + x_s^*,\ y^* = y_c^* + y_s^*, \\ & x_s^* = (\mathbf{n}\cdot\Delta\mathbf{x})\sin\psi,\quad y_s^* = -(\mathbf{m}\cdot\Delta\mathbf{x})\sin\psi. \\ & x_c^* = (\mathbf{m}\cdot\Delta\mathbf{x})\cos\psi,\quad y_c^* = (\mathbf{n}\cdot\Delta\mathbf{x})\cos\psi. \end{aligned} \tag{8.135c}$$

8.6.1 Derivatives of the Green's Functions

In boundary element formulation, to find the internal strains and stresses from the associated boundary integral equations we need to know the derivatives of displacements and tractions with respect to the source point (the position where the point force locates, i.e., $\hat{\mathbf{x}}$). From the solutions shown in (8.129) we know that $\partial/\partial\hat{x}_i = -\partial/\partial x_i$. Thus, in order to be comparable with the solutions presented in the literature, in the following the derivatives are presented by the form of $\mathbf{u}_{,i}$ and $\mathbf{t}_{,i}$, that is, $\partial\mathbf{u}/\partial x_i$ and $\partial\mathbf{t}/\partial x_i$.

With the complex variable z_α defined in (8.128b), differentiating $\mathbf{u}$ and $\mathbf{t}^e$ of (8.129) with respect to x_i we obtain

$$\begin{aligned} \mathbf{u}_{,i} &= \frac{\operatorname{sgn}(\mathbf{m}\cdot\Delta\mathbf{x})}{2\pi^2}\int_0^{\pi}\operatorname{Re}\left\{\mathbf{A}\left\langle\frac{n_i+\mu_\alpha m_i}{(z_\alpha-\hat{z}_\alpha)^2}\right\rangle\mathbf{A}^T\right\}d\theta\hat{\mathbf{p}}, \\ \mathbf{t}_{,i}^e &= \frac{\operatorname{sgn}(\mathbf{m}\cdot\Delta\mathbf{x})}{2\pi^2}\int_0^{\pi}\operatorname{Re}\left\{\mathbf{B}^e\left\langle\frac{-2(n_i+\mu_\alpha m_i)}{(z_\alpha-\hat{z}_\alpha)^3}\right\rangle\mathbf{A}^T\right\}d\theta\hat{\mathbf{p}}. \end{aligned} \tag{8.136}$$

In order to get the real form expressions of $\mathbf{u}_{,i}$ and $\mathbf{t}^e_{,i}$, we differentiate $\mathbf{u}$ of (8.132) and $\mathbf{t}^e$ of (8.135a) with respect to x_i and use the relation of (8.134). The final simplified results can be expressed as

$$\mathbf{u}_{,i} = \frac{\operatorname{sgn}(\mathbf{m}\cdot\Delta\mathbf{x})}{4\pi^2}\int_0^{\pi}\frac{1}{r^3}\left\{c_i^*\mathbf{N}_2(\psi)+d_i^*\mathbf{N}_2'(\psi)\right\}d\theta\hat{\mathbf{p}},$$

$$\mathbf{t}^e_{,i} = -\frac{\operatorname{sgn}(\mathbf{m}\cdot\Delta\mathbf{x})}{4\pi^2}\int_0^{\pi}\frac{1}{r^5}\left\{\mathbf{Q}^e_{1,i}\mathbf{N}_2(\psi)+\mathbf{Q}^e_{2,i}\mathbf{N}_2'(\psi)+\mathbf{Q}^e_{3,i}\mathbf{N}_2''(\psi)\right\}d\theta\hat{\mathbf{p}},$$

(8.137a)

where

$$c_i^* = -c_i\sin\psi + d_i\cos\psi,\quad d_i^* = d_i\sin\psi,$$
$$\mathbf{Q}^e_{1,i} = \left(-3c_ix^* + d_iy^* + m_i^*r^2\right)\mathbf{Q}^e + \left(3c_iy^* + d_ix^* - n_i^*r^2\right)\mathbf{R}^e,$$
$$\mathbf{Q}^e_{2,i} = \left[3c_iy_s^* + d_i\left(x_c^* + x^*\right) + m_ir^2\sin\psi\right]\mathbf{Q}^e$$
$$+[3c_ix_s^* - d_i\left(y_c^* + y^*\right) - n_ir^2\sin\psi]\mathbf{R}^e,$$
$$\mathbf{Q}^e_{3,i} = -d_i\left(y_s^*\mathbf{Q}^e + x_s^*\mathbf{R}^e\right), \qquad (8.137b)$$

and

$$m_i^* = m_i\cos\psi + n_i\sin\psi,\quad n_i^* = -m_i\sin\psi + n_i\cos\psi. \qquad (8.137c)$$

8.6.2 Computation of Green's Functions and Their Derivatives

Two different kinds of expressions of the Green's functions are presented in this section. One is complex form shown in (8.129) and (8.136), and the other is real form shown in (8.132), (8.135), and (8.137). In order to know which expression is more efficient than the other, in the following we show the calculation procedure for these two expressions.

i. *Complex form solution, (8.129) and (8.136)*

   1. Given the elastic properties C_{ijkl}, the positions of $\mathbf{x}$ and $\hat{\mathbf{x}}$, the point force $\hat{\mathbf{p}}$, and the surface normal e_j.

2. Select the proper 2D Radon-plane **n-m** by the following criterion:

$$\begin{aligned}
&\text{if } \max\left(\left|\Delta x_1\right|,\left|\Delta x_2\right|,\left|\Delta x_3\right|\right)=\left|\Delta x_3\right|,\ \mathbf{n},\mathbf{m}\text{: }\left(8.68c\right)_2,\\
&\text{if } \max\left(\left|\Delta x_1\right|,\left|\Delta x_2\right|,\left|\Delta x_3\right|\right)=\left|\Delta x_2\right|,\ \mathbf{n},\mathbf{m}\text{: }\left(8.86\right)_1,\\
&\text{if } \max\left(\left|\Delta x_1\right|,\left|\Delta x_2\right|,\left|\Delta x_3\right|\right)=\left|\Delta x_1\right|,\ \mathbf{n},\mathbf{m}\text{: }\left(8.86\right)_2.
\end{aligned} \quad (8.138)$$

3. Transform each integrand of (8.129) and (8.136) to the standard form required by the Gaussian quadrature rule and find the associated angle θ_g of the Gaussian point t_g, which ranges from −1 to 1, by

$$\theta_g = \frac{\pi\left(t_g+1\right)}{2}. \quad (8.139)$$

4. Calculate **Q**, **R**, **T**, $\mathbf{Q}^e$, and $\mathbf{R}^e$ by (8.68c) and (8.79) with **n** and **m** determined by (8.138).

5. Determine $\mu_k, \mathbf{a}_k, \mathbf{b}_k, \mathbf{b}_k^e, \mathbf{A}, \mathbf{B}, \mathbf{B}^e$ by (8.68b), (8.72), (8.78), $(8.76b)_1$, and (8.89b).

6. Repeat steps 3–5 for each Gaussian point and store the values of μ_k, **A**, **B**, $\mathbf{B}^e$, which are independent of the position of **x**, for the following calculation.

7. Integrate each integral of (8.129) and (8.136) by Gaussian quadrature rule.

8. Determine **u**, $\mathbf{t}^e$, $\mathbf{u}_{,i}$, and $\mathbf{t}_i^e$ by (8.129) and (8.136).

ii. *Real form solution, (8.132), (8.135) and (8.137)*

1-4. Same as steps 1–4 of (i).

5. Calculate r and ψ by (8.131b).

6. Calculate $\mathbf{Q}(\psi)$, $\mathbf{R}(\psi)$, $\mathbf{T}(\psi)$, and $\mathbf{N}_1(\psi)$, $\mathbf{N}_2(\psi)$, $\mathbf{N}_3(\psi)$ by the generalized form of (8.5) and (8.4), that is, (Hwu, 2010; Hsu et al., 2019)

$$\begin{aligned}
\mathbf{Q}(\psi)&=\mathbf{Q}\cos^2\psi+\left(\mathbf{R}+\mathbf{R}^T\right)\sin\psi\cos\psi+\mathbf{T}\sin^2\psi,\\
\mathbf{R}(\psi)&=\mathbf{R}\cos^2\psi+\left(\mathbf{T}-\mathbf{Q}\right)\sin\psi\cos\psi-\mathbf{R}^T\sin^2\psi,\\
\mathbf{T}(\psi)&=\mathbf{T}\cos^2\psi-\left(\mathbf{R}+\mathbf{R}^T\right)\sin\psi\cos\psi+\mathbf{Q}\sin^2\psi.
\end{aligned} \quad (8.140a)$$

$$\mathbf{N}_1(\psi) = -\mathbf{T}^{-1}(\psi)\mathbf{R}^T(\psi), \quad \mathbf{N}_2(\psi) = \mathbf{T}^{-1}(\psi) = \mathbf{N}_2^T(\psi),$$
$$\mathbf{N}_3(\psi) = \mathbf{R}(\psi)\mathbf{T}^{-1}(\psi)\mathbf{R}^T(\psi) - \mathbf{Q}(\psi). \tag{8.140b}$$

7. Calculate $\mathbf{N}_2'(\psi)$ and $\mathbf{N}_2''(\psi)$ by the following relations (Hwu, 2010):

$$\mathbf{N}_2'(\psi) = -\{\mathbf{N}_1(\psi)\mathbf{N}_2(\psi) + \mathbf{N}_2(\psi)\mathbf{N}_1^T(\psi)\},$$
$$\mathbf{N}_2''(\psi) = -\{\mathbf{N}_1(\psi)\mathbf{N}_2(\psi) + \mathbf{N}_2(\psi)\mathbf{N}_1^T(\psi)\}',$$
$$\text{in which } \mathbf{N}_1'(\psi) = -\{\mathbf{I} + \mathbf{N}_1^2(\psi) + \mathbf{N}_2(\psi)\mathbf{N}_3(\psi)\}, \tag{8.141}$$

8. Calculate c_i, d_i, c_i^*, d_i^*, $\mathbf{Q}_c^e, \mathbf{Q}_d^e$, $\mathbf{Q}_{1,i}^e, \mathbf{Q}_{2,i}^e, \mathbf{Q}_{3,i}^e$ by (8.134b), (8.135b), and (8.137b).
9. Repeat steps 3–8 for each Gaussian point, in which the values obtained in step 4 are independent of the position of **x** and can be stored in advance for the convenience of the following calculation. However, all the other values calculated from steps 5–8 depend on the position of **x** and should be re-calculated for each Gaussian point every time when the position of **x** changes.
10. Integrate each integral of (8.132), (8.135), and (8.137) by Gaussian quadrature rule.
11. Determine $\mathbf{u}$, $\mathbf{t}^e$, $\mathbf{u}_i$ and $\mathbf{t}_i^e$ by (8.132), (8.135), and (8.137).

From the calculation procedure stated for two different expressions, we see that the advantage of the complex form solution is that the material eigenvalues μ_k and eigenvector matrices **A**, **B**, $\mathbf{B}^e$ depend only on the material properties and the angle θ of Radon plane, and are independent of the position **x**. And hence, all these values can be evaluated in advance and stored for the calculation of the Green's functions for each different point **x**. On the other hand, the advantage of the real form solution is that the calculation of material eigenvalues and eigenvector matrices can be avoided. However, all the other values needed for the calculation of the Green's functions are functions of r and ψ which depend on the position of **x**. In this viewpoint, the real form solution may not be more efficient than the complex form solution no matter which kind of programming languages, Matlab or Fortran or others, are used.

Since the common steps for both solution forms are the calculation of **Q**, **R**, **T**, $\mathbf{Q}^e$, and $\mathbf{R}^e$, it may be useful to have their matrix expressions instead of the tensor definition given in (8.68c) and (8.79). If **n** and **m** are selected to be those of $(8.68c)_2$, we have

$$\begin{aligned}
&\mathbf{Q} = \cos^2\theta\mathbf{Q}_0 + \cos\theta\sin\theta\left(\mathbf{R}_0 + \mathbf{R}_0^T\right) + \sin^2\theta\mathbf{T}_0,\\
&\mathbf{R} = \cos\theta\mathbf{R}_1 + \sin\theta\mathbf{R}_2, \quad \mathbf{T} = \mathbf{R}_3,\\
&\mathbf{Q}^e = e_1\left(\cos\theta\mathbf{Q}_0 + \sin\theta\mathbf{R}_0\right) + e_2\left(\cos\theta\mathbf{R}_0^T + \sin\theta\mathbf{T}_0\right)\\
&\qquad + e_3\left(\cos\theta\mathbf{R}_1^T + \sin\theta\mathbf{R}_2^T\right),\\
&\mathbf{R}^e = e_1\mathbf{R}_1 + e_2\mathbf{R}_2 + e_3\mathbf{R}_3.
\end{aligned} \tag{8.142a}$$

and

$$\begin{aligned}
&\mathbf{Q}_0 = \begin{bmatrix} C_{11} & C_{16} & C_{15} \\ C_{16} & C_{66} & C_{56} \\ C_{15} & C_{56} & C_{55} \end{bmatrix}, \quad \mathbf{R}_0 = \begin{bmatrix} C_{16} & C_{12} & C_{14} \\ C_{66} & C_{26} & C_{46} \\ C_{56} & C_{25} & C_{45} \end{bmatrix},\\
&\mathbf{T}_0 = \begin{bmatrix} C_{66} & C_{26} & C_{46} \\ C_{26} & C_{22} & C_{24} \\ C_{46} & C_{24} & C_{44} \end{bmatrix}, \quad \mathbf{R}_1 = \begin{bmatrix} C_{15} & C_{14} & C_{13} \\ C_{65} & C_{64} & C_{63} \\ C_{55} & C_{54} & C_{53} \end{bmatrix},\\
&\mathbf{R}_2 = \begin{bmatrix} C_{65} & C_{64} & C_{63} \\ C_{25} & C_{24} & C_{23} \\ C_{45} & C_{44} & C_{43} \end{bmatrix}, \quad \mathbf{R}_3 = \begin{bmatrix} C_{55} & C_{54} & C_{53} \\ C_{45} & C_{44} & C_{43} \\ C_{35} & C_{34} & C_{33} \end{bmatrix}.
\end{aligned} \tag{8.142b}$$

In the above, C_{ij}, $i, j = 1, 2, \ldots, 6$ are the contracted notation of elastic tensor C_{ijkl}.

REFERENCES

Buroni, F.C. and Denda, M., 2014, "Radon-Stroh formalism for 3D theory of anisotropic elasticity," In: Mallardo V, Aliabadi MH, editors. *Conference Proceedings of International Conference on Boundary Element and Meshless Techniques XV (Beteq 2014)*, Advances in boundary element and meshless techniques XV, Florence: EC Ltd; pp. 295–300.

Deans, S.R., 1983, *The Radon Transform and Some of Its Applications*, John Wiley & Sons, New York.

Hsieh, M.C. and Hwu, C., 2003, "Extended Stroh-like formalism for magneto-electro-elastic composite laminates," In: *International Conference on Computational Mesomechanics Associated with Development and Fabrication of Use-Specific Materials*, Tokyo, Japan, pp. 325–332.

Hsu, C.L., Hwu, C. and Shiah, Y.C., 2019, "Three-Dimensional Boundary Element Analysis for Anisotropic Elastic Solids and its Extension to Piezoelectric and

Magnetoelectroelastic Solids," *Engineering Analysis with Boundary Elements*, Vol. 98, pp. 265–280.

Hsu, C.W. and Hwu C., 2020, "Green's functions for Unsymmetric Composite Laminates with Inclusions," *Proceedings of the Royal Society A: Mathematical, Physical and Engineering Sciences*, Vol. 476, 20190437.

Hsu, C.W. and Hwu C., 2021a, "Correction of the Existing Solutions for Hole/Crack Problems of Composite Laminates Under Coupled Stretching-Bending Deformation," *Composite Structures*, Vol. 260, 113154.

Hsu, C.W. and Hwu C., 2021b, "A special Boundary Element for Holes/Cracks in Composite Laminates Under Coupled Stretching-Bending Deformation," *Engineering Analysis with Boundary Elements*, Vol. 133, pp. 30–48.

Hwu, C., 2003, "Stroh-Like Formalism for the Coupled Stretching-Bending Analysis of Composite Laminates," *International Journal of Solids and Structures*, Vol. 40, No. 13–14, pp. 3681–3705.

Hwu, C., 2004, "Green's Function for the Composite Laminates with Bending Extension Coupling," *Composite Structures*, Vol. 63, pp. 283–292.

Hwu C., 2005, "Green's Functions for Holes/Cracks in Laminates with Stretching-Bending Coupling," ASME *Journal of Applied Mechanics*, Vol. 72, pp. 282–289.

Hwu, C. and Tan, C.J., 2007, "In-Plane/Out-of-Plane Concentrated Forces and Moments on Composite Laminates with Elliptical Elastic Inclusions," *International Journal of Solids and Structures*, Vol. 44, pp. 6584–6606.

Hwu, C., 2010, *Anisotropic Elastic Plates*, Springer, New York.

Hwu, C., 2021, *Anisotropic Elasticity with Matlab*, Springer, Cham.

Hwu, C. and Becker, W., 2022, "Stroh Formalism for Various Types of Materials and Deformations," *Journal of Mechanics*, Vol. 38, pp. 433–444.

Lekhnitskii, S.G., 1963, *Theory of Elasticity of an Anisotropic Body*, MIR, Moscow.

Lekhnitskii, S.G., 1968, *Anisotropic Plates*, Gordon and Breach Science Publishers, New York.

Rogacheva, N.N., 1994, *The Theory of Piezoelectric Shells and Plates*, CRC, London.

Schapery R.A., 1962, "Approximate Methods of Transform Inversion for Viscoelastic Stress Analysis," *Proceeding of the 4th US National Congress on Applied Mechanics*, ASME, pp. 1075–1084.

Soh, A.K. and Liu, J.X., 2005, "On the Constitutive Equations of Magneto-electroelastic Solids," *Journal of Intelligent Material Systems and Structures*, Vol. 16, pp. 597–602.

Stehfest, H., 1970, "Algorithm 368: Numerical Inversion of Laplace Transforms," *Communications of the ACM*, Vol. 13(1), pp. 47–49.

Ting, T.C.T., 1996, *Anisotropic Elasticity: Theory and Applications*, Oxford Science Publications, New York.

Wu, K.C., 1998, "Generalization of the Stroh Formalism to 3-Dimensional Anisotropic Elasticity," *Journal of Elasticity*, Vol. 51, pp. 213–225.

Xie, L., Zhang, C., Hwu, C., Pan, E., 2016, "On Novel Explicit Expressions of Green's Function and Its Derivatives for Magnetoelectroelastic Materials," *European Journal of Mechanics - A/Solids*, Vol. 60, pp. 134–144.

CHAPTER 9

Numerical Methods

Currently, most of the commercial codes for mechanical analysis are developed based upon the finite element method (FEM). Boundary element method (BEM) and meshless method (or called meshfree method, MM) are two other important and popular techniques for practical engineering problems. The main advantage of BEM over FEM is the reduction of the problem dimension by one, whereas that of MM is the free set of nodes. To know how these methods be applied to solve the problems of laminated composite structures, we briefly introduce FEM, BEM, and MM in Sections 9.1 through 9.3, respectively. Only brief introduction will be provided in Section 9.1 for FEM since its detailed introduction can be found in many other textbooks. For BEM, brief introduction is provided sequentially for beam analysis, two-dimensional analysis, coupled stretching–bending analysis, three-dimensional analysis, and boundary-based finite element method. Most of the details of BEM can be found in my recently published book (Hwu, 2021). As to MM, only two representative methods, element-free Galerkin (EFG) method and meshless local Petrov–Galerkin (MLPG), are introduced in Section 9.3. The last section will then present some numerical examples for the analysis of laminated composite beams and plates.

9.1 FINITE ELEMENT METHOD

In FEM the continuum is divided into a series of elements which are connected at a finite number of points known as nodal points. If the finite element displacement method is considered, the variation of displacement within each element is assumed to be described by means of proper shape

DOI: 10.1201/9781003470465-9

functions together with unknown displacement values at the nodal points such as

$$\mathbf{u} = \boldsymbol{N}\mathbf{u}_e, \tag{9.1}$$

where **u** is the vector of displacements at any point of the element; ***N*** is the set of shape functions, and $\mathbf{u}_e$ is the vector of nodal displacements of the element.

For two-dimensional deformation, the use of eight-node isoparametric element with the local coordinate (ξ, η) allows us to make elements with curvilinear shapes. The geometry of each element can be expressed by

$$x(\xi,\eta) = \sum_{i=1}^{8} \varpi_i(\xi,\eta)x_i, \quad y(\xi,\eta) = \sum_{i=1}^{8} \varpi_i(\xi,\eta)y_i, \tag{9.2a}$$

where (x_i, y_i) are the coordinates of node i, and $\varpi_i(\xi, \eta)$ are the two-dimensional quadratic shape functions given by

$$\begin{aligned}
\varpi_1 &= -\frac{1}{4}(1-\xi)(1-\eta)(1+\xi+\eta), & \varpi_2 &= \frac{1}{2}(1-\xi^2)(1-\eta), \\
\varpi_3 &= -\frac{1}{4}(1+\xi)(1-\eta)(1-\xi+\eta), & \varpi_4 &= \frac{1}{2}(1+\xi)(1-\eta^2), \\
\varpi_5 &= -\frac{1}{4}(1+\xi)(1+\eta)(1-\xi-\eta), & \varpi_6 &= \frac{1}{2}(1-\xi^2)(1+\eta), \\
\varpi_7 &= -\frac{1}{4}(1-\xi)(1+\eta)(1+\xi-\eta), & \varpi_8 &= \frac{1}{2}(1-\xi)(1-\eta^2).
\end{aligned} \tag{9.2b}$$

Same shape functions are used for the interpolation of deformation. In other words, the detailed expressions of (9.1) can be written as

$$\mathbf{u} = \begin{Bmatrix} u \\ v \\ w \end{Bmatrix}, \quad \mathbf{N} = \begin{bmatrix} \mathbf{N}_1\, \mathbf{N}_2 \ldots \mathbf{N}_8 \end{bmatrix}, \quad \mathbf{u}_e = \begin{Bmatrix} \mathbf{u}_1 \\ \mathbf{u}_2 \\ \vdots \\ \mathbf{u}_8 \end{Bmatrix}, \tag{9.3a}$$

where

$$\mathbf{N}_i = \varpi_i \mathbf{I}, \quad \mathbf{u}_i = \begin{Bmatrix} u_i \\ v_i \\ w_i \end{Bmatrix}, \quad i = 1,2,\ldots,8. \tag{9.3b}$$

In (9.3b), $\mathbf{I}$ is a 3×3 unit matrix, and u, v, and w are the displacements in the directions of x, y, and z axes.

By the strain–displacement relation, the strains within the element can also be expressed in terms of the element nodal displacements as

$$\boldsymbol{\varepsilon} = \hat{\mathbf{N}}\mathbf{u}_e, \tag{9.4a}$$

where

$$\boldsymbol{\varepsilon} = \begin{Bmatrix} \varepsilon_{11} \\ \varepsilon_{22} \\ 2\varepsilon_{23} \\ 2\varepsilon_{31} \\ 2\varepsilon_{12} \end{Bmatrix} = \begin{Bmatrix} \partial u / \partial x \\ \partial v / \partial y \\ \partial w / \partial y \\ \partial w / \partial x \\ \partial u / \partial y + \partial v / \partial x \end{Bmatrix}, \quad \hat{\mathbf{N}} = \left[\hat{\mathbf{N}}_1 \, \hat{\mathbf{N}}_2 \ldots \hat{\mathbf{N}}_8 \right], \tag{9.4b}$$

and

$$\hat{\mathbf{N}}_i = \begin{bmatrix} \partial \varpi_i / \partial x & 0 & 0 \\ 0 & \partial \varpi_i / \partial y & 0 \\ 0 & 0 & \partial \varpi_i / \partial y \\ 0 & 0 & \partial \varpi_i / \partial x \\ \partial \varpi_i / \partial y & \partial \varpi_i / \partial x & 0 \end{bmatrix}. \tag{9.4c}$$

Note that in (9.4b) ε_3 is neglected because the displacement field considered is a function of x and y only and $\varepsilon_3 = \partial w/\partial z = 0$. With this constraint, the reduced stress–strain law can be expressed as

$$\boldsymbol{\sigma} = \mathbf{C}\boldsymbol{\varepsilon} = \mathbf{C}\hat{\mathbf{N}}\mathbf{u}_e, \tag{9.5a}$$

where

$$\boldsymbol{\sigma} = \begin{Bmatrix} \sigma_{11} \\ \sigma_{22} \\ \sigma_{23} \\ \sigma_{31} \\ \sigma_{12} \end{Bmatrix}, \quad \mathbf{C} = \begin{bmatrix} C_{11} & C_{12} & C_{14} & C_{15} & C_{16} \\ C_{12} & C_{22} & C_{24} & C_{25} & C_{26} \\ C_{14} & C_{24} & C_{44} & C_{45} & C_{46} \\ C_{15} & C_{25} & C_{45} & C_{55} & C_{56} \\ C_{16} & C_{26} & C_{46} & C_{56} & C_{66} \end{bmatrix}. \tag{9.5b}$$

In (9.5b), C_{ij} are the contracted notation of the elastic tensor C_{ijkl}.

The total potential energy Π of the elastic body will be the sum of the energy contributions of the individual elements. Thus, $\Pi = \sum \Pi_e$, where Π_e represents the potential energy of element e, which can be written as

$$\Pi_e = \frac{1}{2}\int_{\Omega_e} \boldsymbol{\sigma}^T \boldsymbol{\varepsilon} d\Omega - \int_{\Omega_e} \mathbf{u}^T \mathbf{b} d\Omega - \int_{\Gamma_e} \mathbf{u}^T \mathbf{t} d\Gamma. \tag{9.6}$$

Here, **b** and **t** denote the body forces and the surface tractions, respectively. $\boldsymbol{\Omega}_e$ is the element volume and Γ_e is the loaded element surface area. If the body forces are neglected, substituting (9.1), (9.4), and (9.5) into (9.6) leads to

$$\Pi_e = \frac{1}{2}\int_{\Omega_e} \mathbf{u}_e^T \hat{N}^T \mathbf{C} \hat{N} \mathbf{u}_e d\Omega - \int_{\Gamma_e} \mathbf{u}_e^T N^T \mathbf{t} d\Gamma. \tag{9.7}$$

Performance of the minimization for element e with respect to the nodal displacement $\mathbf{u}_e$ results in

$$\mathbf{K}_e \mathbf{u}_e = \mathbf{f}_e, \tag{9.8a}$$

where

$$\mathbf{K}_e = \int_{\Omega_e} \hat{N}^T \mathbf{C} \hat{N} d\Omega, \quad \mathbf{f}_e = \int_{\Gamma_e} N^T \mathbf{t} d\Gamma. \tag{9.8b}$$

$\mathbf{K}_e$ is termed the element stiffness matrix and $\mathbf{f}_e$ is the equivalent nodal force. The summation of the terms in (9.8) over all the elements results in a system of equilibrium equations for the complete continuum. These equations are then solved by any standard technique to yield the nodal displacements. The stresses and strains within each element can then be calculated from the nodal displacements using (9.4) and (9.5).

Based upon the above description for the standard finite element method, several extended works have been developed in our following studies such as coupling of boundary element and finite element (Yen and Hwu, 1993), a global-local finite element model (Hwu, 1998), Stroh finite element (Hwu et al., 2001), and a comprehensive finite element model for tapered composite wing structures (Hwu and Yu, 2010), etc.

9.2 BOUNDARY ELEMENT METHOD

The main advantages of BEM over FEM are the reduction of the problem dimension by one and the exact satisfaction of certain boundary conditions for particular problems if their associated fundamental solutions are embedded in boundary element formulation. Based upon the works presented in our recent studies such as Hwu (2010a, 2010b), Hwu (2012), Hsu et al. (2019), Hwu (2021), Hsu and Hwu (2021, 2022a, 2022b, 2023a, 2023b) and (Huang et al., 2024a, 2024b) in this section we will present the BEM for beam analysis, two-dimensional analysis, coupled stretching–bending analysis, and three-dimensional analysis.

9.2.1 Beam Analysis

The boundary integral equations (BIE) are the basis of BEM. To derive the boundary integral equations for the thick laminated composite beams (refer to Section 4.3.1 for the general formulation of thick laminated composite beams), we employ the reciprocal theorem of Betti and Rayleigh (Sokolnikoff, 1956). Consider two different loading systems applied on the same laminated composite beam. One is a system of the actual load, and the other is the one associated with the concentrated forces/moments, whose solutions can be obtained from the Green's functions presented in Section 4.3.2. The reciprocal theorem of Betti and Rayleigh gives us

$$\int_0^\ell \left(N_x\varepsilon_{x0}^* + M_x\kappa_x^* + Q_x\gamma_{xz}^*\right)dx = \int_0^\ell \left(N_x^*\varepsilon_{x0} + M_x^*\kappa_x + Q_x^*\gamma_{xz}\right)dx, \quad (9.9)$$

where the superscript $*$ denotes the value corresponding to the Green's functions and ℓ is the beam length. By taking integration by parts term by term of (9.9), for example,

$$\begin{aligned}
&\int_0^\ell N_x\varepsilon_{x0}^* dx = \left[N_x u_0^*\right]_0^\ell - \int_0^\ell \frac{\partial N_x}{\partial x} u_0^* dx,\\
&\int_0^\ell Q_x\gamma_{xz}^* dx = \int_0^\ell Q_x\left(\beta_x^* + \frac{\partial w^*}{\partial x}\right)dx = \int_0^\ell Q_x\beta_x^* dx + \left[Q_x w^*\right]_0^\ell - \int_0^\ell \frac{\partial Q_x}{\partial x} w^* dx,\\
&\int_0^\ell M_x\kappa_{x0}^* dx = \left[M_x\beta_x^*\right]_0^\ell - \int_0^\ell \frac{\partial M_x}{\partial x}\beta_x^* dx,
\end{aligned} \quad (9.10)$$

and using the equilibrium equations of the beams $(4.47)_{7-9}$, Equation (9.9) leads to

$$\left[N_x u_0^* + M_x \beta_x^* + Q_x w^*\right]_0^\ell + \int_0^\ell \left(q_x u_0^* + m_x \beta_x^* + q_z w^*\right) dx$$

$$= \left[N_x^* u_0 + M_x^* \beta_x + Q_x^* w\right]_0^\ell + \int_0^\ell \left(q_x^* u_0 + m_x^* \beta_x + q_z^* w\right) dx. \tag{9.11}$$

Consider the following three independent unit point load or moment applied at the point ξ of the beam:

$$\begin{aligned} &(1): q_x^* = \delta(x-\xi), m_x^* = q_z^* = 0, \\ &(2): q_z^* = \delta(x-\xi), q_x^* = m_x^* = 0, \\ &(3): m_x^* = \delta(x-\xi), q_x^* = q_z^* = 0, \end{aligned} \tag{9.12}$$

where $\delta(\xi - x)$ represents the Dirac delta function. Substituting each point load of (9.12) independently into (9.11), we get three equations which can be written in tensor notation as

$$u_i(\xi) + \left[t_{ij}^*(\xi,x) u_j(x)\right]_0^\ell = \left[u_{ij}^*(\xi,x) t_j(x)\right]_0^\ell$$

$$+ \int_0^\ell u_{ij}^*(\xi,x) q_j(x) dx, \; i,j = 1,2,3, \tag{9.13}$$

where u_i and t_i are the generalized displacements and tractions defined as

$$u_1 = u_0, u_2 = w, u_3 = \beta_x, t_1 = N_x, t_2 = Q_x, t_3 = M_x. \tag{9.14}$$

$u_{ij}^*(\xi,x)$ and $t_{ij}^*(\xi,x)$ denote, respectively, the displacements u_j and tractions t_j at the field point x corresponding to a unit point force q_i($q_1 = q_x$, $q_2 = q_z$, $q_3 = m_x$) applied at the source point ξ, and are generally called *fundamental solutions of displacements and tractions.*

For the convenience of computer programming, Equations (9.13) are usually expressed in matrix form as

$$\mathbf{u}(\xi) + \left[\mathbf{T}^*(\xi,x)\mathbf{u}(x)\right]_0^\ell = \left[\mathbf{U}^*(\xi,x)\mathbf{t}(x)\right]_0^\ell + \int_0^\ell \mathbf{U}^*(\xi,x)\mathbf{q}(x) dx, \tag{9.15a}$$

where

$$\mathbf{u} = \begin{Bmatrix} u_0 \\ w \\ \beta_x \end{Bmatrix}, \quad \mathbf{t} = \begin{Bmatrix} N_x \\ Q_x \\ M_x \end{Bmatrix}, \quad \mathbf{q} = \begin{Bmatrix} q_x \\ q_z \\ m_x \end{Bmatrix}. \tag{9.15b}$$

With the Green's functions given in (4.73) and (4.74), the explicit solutions of $\mathbf{T}^*$ and $\mathbf{U}^*$ can be written explicitly as

$$\mathbf{T}^*(\xi,x) = \frac{-1}{2}\begin{bmatrix} r' & 0 & 0 \\ 0 & r' & r \\ 0 & 0 & r' \end{bmatrix},$$

$$\mathbf{U}^*(\xi,x) = \frac{1}{4}\begin{bmatrix} -2D_{11}^* r & -B_{11}^* r^2 r' & 2B_{11}^* r \\ B_{11}^* r^2 r' & \dfrac{-2}{A_{55}} r + \dfrac{A_{11}^*}{3} r^3 & -A_{11}^* r^2 r' \\ 2B_{11}^* r & A_{11}^* r^2 r' & -2A_{11}^* r \end{bmatrix}. \tag{9.16}$$

Take the source point ξ to be the point near one of the end points of the beam, that is, $\xi = \varepsilon$ or $\xi = \ell - \varepsilon$ where ε approaches to zero, six equations can be constructed from (9.15a) and be expressed as

$$\begin{bmatrix} \mathbf{I} - \mathbf{T}_{11}^* & \mathbf{T}_{12}^* \\ -\mathbf{T}_{21}^* & \mathbf{I} + \mathbf{T}_{22}^* \end{bmatrix} \begin{Bmatrix} \mathbf{u}_1 \\ \mathbf{u}_2 \end{Bmatrix} = \begin{bmatrix} -\mathbf{U}_{11}^* & \mathbf{U}_{12}^* \\ -\mathbf{U}_{21}^* & \mathbf{U}_{22}^* \end{bmatrix} \begin{Bmatrix} \mathbf{t}_1 \\ \mathbf{t}_2 \end{Bmatrix} + \begin{Bmatrix} \mathbf{p}_1 \\ \mathbf{p}_2 \end{Bmatrix}, \tag{9.17a}$$

where

$$\begin{aligned}
&\mathbf{u}_1 = \mathbf{u}(0), \mathbf{u}_2 = \mathbf{u}(\ell), \mathbf{t}_1 = \mathbf{t}(0), \mathbf{t}_2 = \mathbf{t}(\ell), \\
&\mathbf{T}_{11}^* = \mathbf{T}^*(\varepsilon,0),\ \mathbf{T}_{12}^* = \mathbf{T}^*(\varepsilon,\ell),\ \mathbf{T}_{21}^* = \mathbf{T}^*(\ell-\varepsilon,0),\ \mathbf{T}_{22}^* = \mathbf{T}^*(\ell-\varepsilon,\ell), \\
&\mathbf{U}_{11}^* = \mathbf{U}^*(\varepsilon,0),\ \mathbf{U}_{12}^* = \mathbf{U}^*(\varepsilon,\ell),\ \mathbf{U}_{21}^* = \mathbf{U}^*(\ell-\varepsilon,0),\ \mathbf{U}_{22}^* = \mathbf{U}^*(\ell-\varepsilon,\ell), \\
&\mathbf{p}_1 = \int_0^\ell \mathbf{U}^*(\varepsilon,x)\mathbf{q}(x)dx, \\
&\mathbf{p}_2 = \int_0^\ell \mathbf{U}^*(\ell-\varepsilon,x)\mathbf{q}(x)dx, \text{where } \varepsilon > 0 \text{ and } \varepsilon \to 0.
\end{aligned} \tag{9.17b}$$

With (9.16), $\mathbf{T}_{ij}^*$ and $\mathbf{U}_{ij}^*$ defined in (9.17b) can be obtained as

$$\mathbf{T}_{11}^* = -\mathbf{T}_{22}^* = \frac{1}{2}\begin{bmatrix} 1 & 0 & 0 \\ 0 & 1 & 0 \\ 0 & 0 & 1 \end{bmatrix}, \mathbf{T}_{12}^* = \frac{-1}{2}\begin{bmatrix} 1 & 0 & 0 \\ 0 & 1 & \ell \\ 0 & 0 & 1 \end{bmatrix}, \mathbf{T}_{21}^* = \frac{1}{2}\begin{bmatrix} 1 & 0 & 0 \\ 0 & 1 & -\ell \\ 0 & 0 & 1 \end{bmatrix}, \tag{9.18a}$$

$$\mathbf{U}_{11}^* = \mathbf{U}_{22}^* = \mathbf{0},\ \mathbf{U}_{12}^* = \frac{\ell}{4}\begin{bmatrix} -2D_{11}^* & -B_{11}^*\ell & 2B_{11}^* \\ B_{11}^*\ell & -\dfrac{2}{A_{55}} + \dfrac{A_{11}^*}{3}\ell^2 & -A_{11}^*\ell \\ 2B_{11}^* & A_{11}^*\ell & -2A_{11}^* \end{bmatrix},$$

$$\mathbf{U}_{21}^* = \frac{\ell}{4}\begin{bmatrix} -2D_{11}^* & B_{11}^*\ell & 2B_{11}^* \\ -B_{11}^*\ell & -\dfrac{2}{A_{55}} + \dfrac{A_{11}^*}{3}\ell^2 & A_{11}^*\ell \\ 2B_{11}^* & -A_{11}^*\ell & -2A_{11}^* \end{bmatrix}. \tag{9.18b}$$

To evaluate $\mathbf{p}_1$ and $\mathbf{p}_2$ by the fourth set of (9.17b), we need to know the loads $\mathbf{q}$ applied on the beams. In general, two different kinds of loads are considered in the practical engineering problems. One is a concentrated load, and the other is a distributed load. The former can be represented by the Dirac delta function, and the latter can be expressed in terms of polynomial functions. Combine them together, we may let

$$\mathbf{q}(x) = \sum_{k=1}^{n} \delta(x - \xi_k)\hat{\mathbf{q}}_k + \sum_{j=0}^{m} \mathbf{q}_j x^j, \tag{9.19}$$

in which $\hat{\mathbf{q}}_k$ is the kth concentrated forces/moment vector applied at ξ_k, $k = 1,2,\ldots, n$, and $\mathbf{q}_j$, $j = 0, 1, 2, \ldots, m$, are the coefficient vectors of the distributed loads applied on the beam. Both $\hat{\mathbf{q}}_k$ and $\mathbf{q}_j$ are supplied by the users and are treated as known vectors. Substituting (9.19) into the fourth set equations of (9.17b), we get

$$\mathbf{p}_1 = \sum_{k=1}^{n} \mathbf{V}_k^{(1)}\hat{\mathbf{q}}_k + \sum_{j=0}^{m} \mathbf{W}_k^{(1)}\mathbf{q}_j,\quad \mathbf{p}_2 = \sum_{k=1}^{n} \mathbf{V}_k^{(2)}\hat{\mathbf{q}}_k + \sum_{j=0}^{m} \mathbf{W}_k^{(2)}\mathbf{q}_j, \tag{9.20a}$$

where

$$\mathbf{V}_k^{(1)} = \frac{1}{4}\begin{bmatrix} -2D_{11}^*\xi_k & -B_{11}^*\xi_k^2 & 2B_{11}^*\xi_k \\ B_{11}^*\xi_k^2 & \dfrac{-2}{A_{55}}\xi_k + \dfrac{A_{11}^*}{3}\xi_k^3 & -A_{11}^*\xi_k^2 \\ 2B_{11}^*\xi_k & A_{11}^*\xi_k^2 & -2A_{11}^*\xi_k \end{bmatrix},$$

$$\mathbf{V}_k^{(2)} = \frac{1}{4}\begin{bmatrix} -2D_{11}^*(\ell-\xi_k) & B_{11}^*(\ell-\xi_k)^2 & 2B_{11}^*(\ell-\xi_k) \\ -B_{11}^*(\ell-\xi_k)^2 & \dfrac{-2}{A_{55}}(\ell-\xi_k) + \dfrac{A_{11}^*}{3}(\ell-\xi_k)^3 & A_{11}^*(\ell-\xi_k)^2 \\ 2B_{11}^*(\ell-\xi_k) & -A_{11}^*(\ell-\xi_k)^2 & -2A_{11}^*(\ell-\xi_k) \end{bmatrix}, \tag{9.20b}$$

$$\mathbf{W}_j^{(1)} = \frac{\ell^{j+2}}{4}\begin{bmatrix} \dfrac{-2D_{11}^*}{j+2} & \dfrac{-B_{11}^*\ell}{j+3} & \dfrac{2B_{11}^*}{j+2} \\ \dfrac{B_{11}^*\ell}{j+3} & \dfrac{-2}{A_{55}(j+2)} + \dfrac{A_{11}^*\ell^2}{3(j+4)} & \dfrac{-A_{11}^*\ell}{j+3} \\ \dfrac{2B_{11}^*}{j+2} & \dfrac{A_{11}^*\ell}{j+3} & \dfrac{-2A_{11}^*}{j+2} \end{bmatrix},$$

$$\mathbf{W}_j^{(2)} = \frac{\ell^{j+2}}{4(j+1)(j+2)}\begin{bmatrix} -2D_{11}^* & \dfrac{2B_{11}^*\ell}{j+3} & 2B_{11}^* \\ \dfrac{-2B_{11}^*\ell}{j+3} & \dfrac{-2}{A_{55}} + \dfrac{2A_{11}^*\ell^2}{(j+3)(j+4)} & \dfrac{2A_{11}^*\ell}{j+3} \\ 2B_{11}^* & \dfrac{-2A_{11}^*\ell}{j+3} & -2A_{11}^* \end{bmatrix}. \tag{9.20c}$$

For the convenience of the latter discussion, we now rewrite (9.17a) in the standard form of BEM, that is,

$$\mathbf{Y}_e\mathbf{u}_e = \mathbf{G}_e\mathbf{t}_e + \mathbf{p}_e, \tag{9.21a}$$

where

$$\mathbf{Y}_e = \begin{bmatrix} \mathbf{I}-\mathbf{T}_{11}^* & \mathbf{T}_{12}^* \\ -\mathbf{T}_{21}^* & \mathbf{I}+\mathbf{T}_{22}^* \end{bmatrix},\ \mathbf{G}_e = \begin{bmatrix} -\mathbf{U}_{11}^* & \mathbf{U}_{12}^* \\ -\mathbf{U}_{21}^* & \mathbf{U}_{22}^* \end{bmatrix},$$

$$\mathbf{u}_e = \begin{Bmatrix} \mathbf{u}_1 \\ \mathbf{u}_2 \end{Bmatrix},\ \mathbf{t}_e = \begin{Bmatrix} \mathbf{t}_1 \\ \mathbf{t}_2 \end{Bmatrix},\ \mathbf{p}_e = \begin{Bmatrix} \mathbf{p}_1 \\ \mathbf{p}_2 \end{Bmatrix}, \tag{9.21b}$$

and the subscript e denotes the value of one element of the beam.

If a straight beam without any internal supports is considered, by setting the boundary conditions on the two ends of the beam, each end will have only three unknowns (since either u_0 or N_x, w or Q_x, β_x or M_x are prescribed). Totally, there will be six unknowns to be determined by the six equations of (9.21). Upon having all the boundary physical quantities $\mathbf{u}_1$, $\mathbf{u}_2$, $\mathbf{t}_1$, $\mathbf{t}_2$, the generalized displacements $\mathbf{u}$ at the internal point ξ can be obtained from (9.15a). By differentiating both sides of (9.15a) with respect to ξ and using the constitutive laws (the second set of equations of (4.47)), the axial strains, curvatures, transverse shear strains, axial forces, bending moments, and transverse shear forces of internal points can then be calculated.

Since in our derivation of BIE all the stiffnesses like A_{11}, B_{11}, D_{11}, A_{55} are treated as constants, the final system of equations can only be applied to the cases of a straight beam with uniform thickness. In order to extend its application to more general cases such as a curved non-uniform thickness beam, like the traditional FEM we discretize the entire beam into several segments and each segment may have different orientation and different thickness or lamination (see Figure 9.1). To combine all the beam segments together in the global coordinate system, we introduce a local coordinate for each segment. The transformation relation between global and local coordinates is

$$\begin{Bmatrix} x^* \\ z^* \end{Bmatrix} = \begin{bmatrix} \cos\theta & \sin\theta \\ -\sin\theta & \cos\theta \end{bmatrix} \begin{Bmatrix} x-x_0 \\ z-z_0 \end{Bmatrix}, \tag{9.22}$$

where (x_0, z_0) and θ denote, respectively, the origin and orientation of the local coordinate (see Figure 9.1). With the relation of (9.22), the vectors of displacements, tractions, and loads between global and local coordinates are transformed by

$$\mathbf{u}_e^* = \mathbf{\Omega}\mathbf{u}_e,\ \mathbf{t}_e^* = \mathbf{\Omega}\mathbf{t}_e,\ \mathbf{p}_e^* = \mathbf{\Omega}\mathbf{p}_e, \tag{9.23a}$$

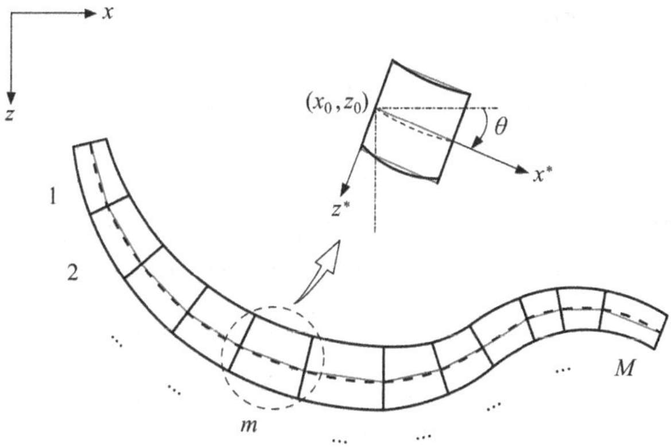

FIGURE 9.1 Discretization of curvilinear beam, and the transformation between global (x–z) and local (x*–z*) coordinates.

where

$$\boldsymbol{\Omega} = \begin{bmatrix} \boldsymbol{\Omega}_0 & \mathbf{0} \\ \mathbf{0} & \boldsymbol{\Omega}_0 \end{bmatrix}, \quad \boldsymbol{\Omega}_0 = \begin{bmatrix} \cos\theta & \sin\theta & 0 \\ -\sin\theta & \cos\theta & 0 \\ 0 & 0 & 1 \end{bmatrix}. \tag{9.23b}$$

For each segment of the curved beam, the system of equations is constructed based upon the local coordinate. Therefore, (9.21) should be written as

$$\mathbf{Y}_e^*\mathbf{u}_e^* = \mathbf{G}_e^*\mathbf{t}_e^* + \mathbf{p}_e^*, \tag{9.24}$$

where the influence matrices $\mathbf{Y}_e^*$ and $\mathbf{G}_e^*$ are evaluated by $(9.21\text{b})_{1,2}$ with $\mathbf{T}_{ij}^*$ and $\mathbf{U}_{ij}^*$ determined by (9.18) using the properties of the *beam segment in the local coordinate*. Note that in (9.24), $\mathbf{t}_e^*$ is a traction vector in terms of local coordinate, whose sign convention follows the definition of stresses and its associated surface normal. Knowing that the normal of first end-point of the beam is opposite to that of the second end-point, its associated nodal forces $\mathbf{f}_e^*$ is related to $\mathbf{t}_e^*$ by

$$\mathbf{f}_e^* = \begin{Bmatrix} \mathbf{f}_1^* \\ \mathbf{f}_2^* \end{Bmatrix} = \begin{Bmatrix} -\mathbf{t}_1^* \\ \mathbf{t}_2^* \end{Bmatrix} = \mathbf{I}_f\mathbf{t}_e^*, \quad \text{where } \mathbf{I}_f = \begin{bmatrix} -\mathbf{I} & \mathbf{0} \\ \mathbf{0} & \mathbf{I} \end{bmatrix}, \tag{9.25}$$

and $\mathbf{I}$ is a 3 × 3 identity matrix. Or inversely,

$$\mathbf{t}_e^* = \mathbf{I}_f \mathbf{f}_e^*, \text{ where } \mathbf{f}_e^* = \boldsymbol{\Omega} \mathbf{f}_e. \tag{9.26}$$

For the convenience of the assembly of all beam segments, like the boundary-based finite element method (BFEM) described later in Section 9.2.6, by using (9.23a) and (9.26) we can now transform (9.24) into the standard form of FEM in the global coordinate as

$$\mathbf{K}_e \mathbf{u}_e = \mathbf{f}_e + \tilde{\mathbf{p}}_e, \tag{9.27a}$$

where

$$\mathbf{K}_e = \boldsymbol{\Omega}^{-1} \left(\mathbf{G}_e^* \mathbf{I}_f \right)^{-1} \mathbf{Y}_e^* \boldsymbol{\Omega}, \quad \tilde{\mathbf{p}}_e = \boldsymbol{\Omega}^{-1} \left(\mathbf{G}_e^* \mathbf{I}_f \right)^{-1} \boldsymbol{\Omega} \mathbf{p}_e. \tag{9.27b}$$

To describe the element assembly through nodal connections, we may rewrite (9.27a) in terms of nodal displacements and nodal forces as

$$\begin{bmatrix} \mathbf{K}_{11}^{(m)} & \mathbf{K}_{12}^{(m)} \\ \mathbf{K}_{21}^{(m)} & \mathbf{K}_{22}^{(m)} \end{bmatrix} \begin{Bmatrix} \mathbf{u}_1^{(m)} \\ \mathbf{u}_2^{(m)} \end{Bmatrix} = \begin{Bmatrix} \mathbf{f}_1^{(m)} + \tilde{\mathbf{p}}_1^{(m)} \\ \mathbf{f}_2^{(m)} + \tilde{\mathbf{p}}_2^{(m)} \end{Bmatrix}, \quad m = 1,2,3,\ldots, M, \tag{9.28}$$

where $\mathbf{K}_{ij}^{(m)}, \mathbf{u}_i^{(m)}, \mathbf{f}_i^{(m)} + \tilde{\mathbf{p}}_i^{(m)}$; $i, j = 1,2$, are the sub-matrices/vectors of $\mathbf{K}_e$, $\mathbf{u}_e$ and $\mathbf{f}_e + \tilde{\mathbf{p}}_e$ of the mth element, and M is the number of elements. The requirements of displacement continuity and force equilibrium give us

$$\mathbf{u}_2^{(m-1)} = \mathbf{u}_1^{(m)}, \quad \mathbf{f}_2^{(m-1)} + \mathbf{f}_1^{(m)} = \begin{cases} \mathbf{0}, & \text{without support,} \\ \tilde{\mathbf{f}}_1^{(m)}, & \text{with support,} \end{cases} \quad m = 2,3,\ldots, M, \tag{9.29}$$

in which $\tilde{\mathbf{f}}_1^{(m)}$ is the unknown reaction force vector to be determined. By assembly (9.28) of all beam segments and using (9.29), the system of equations for the entire curvilinear non-uniform thickness beam can be expressed as

$$\begin{aligned} &\mathbf{K}_{11}^{(1)} \mathbf{u}_1^{(1)} + \mathbf{K}_{12}^{(1)} \mathbf{u}_1^{(2)} = \mathbf{f}_1^{(1)} + \tilde{\mathbf{p}}_1^{(1)}, \\ &\mathbf{K}_{21}^{(m-1)} \mathbf{u}_1^{(m-1)} + \left[\mathbf{K}_{22}^{(m-1)} + \mathbf{K}_{11}^{(m)} \right] \mathbf{u}_1^{(m)} + \mathbf{K}_{12}^{(m)} \mathbf{u}_1^{(m+1)} = \tilde{\mathbf{f}}_1^{(m)} + \tilde{\mathbf{p}}^{(m)}, m = 2,3,\ldots, M, \\ &\mathbf{K}_{21}^{(M)} \mathbf{u}_1^{(M)} + \mathbf{K}_{22}^{(M)} \mathbf{u}_2^{(M)} = \mathbf{f}_2^{(M)} + \tilde{\mathbf{p}}_2^{(M)}, \end{aligned} \tag{9.30a}$$

where

$$\tilde{\mathbf{p}}^{(m)} = \tilde{\mathbf{p}}_2^{(m-1)} + \tilde{\mathbf{p}}_1^{(m)}, \quad \mathbf{u}_1^{(M+1)} = \mathbf{u}_2^{(M)}. \tag{9.30b}$$

In (9.30a), we have $3(M+1)$ equations and $6(M+1)$ variables, that is, $\mathbf{u}_1^{(1)}, \mathbf{f}_1^{(1)}, \mathbf{u}_2^{(M)}, \mathbf{f}_2^{(M)}, \mathbf{u}_1^{(m)}, \tilde{\mathbf{f}}_1^{(m)}, m = 2, 3, \ldots, M$. No matter which kind of boundary valued problems (displacement-prescribed, traction-prescribed, or mixed; with or without internal supports) are considered, they will all provide $3(M+1)$ known values. Thus, all the other unknown variables can be determined by solving the system of $3(M+1) \times 3(M+1)$ algebraic equations. In other words, all the nodal displacements and nodal forces can be solved by (9.30). With the relation (9.26), the tractions of each segment can also be determined. Upon having all the boundary physical quantities of each segment, the internal solutions can be evaluated by (9.15a).

9.2.2 Two-Dimensional Analysis

Similar to the derivation of BIE for beam analysis, the BIE for two-dimensional linear anisotropic elasticity can be written as (Brebbia et al., 1984)

$$c_{ij}(\boldsymbol{\xi})u_j(\boldsymbol{\xi}) + \int_{\Gamma}^{*} t_{ij}^{*}(\boldsymbol{\xi},\mathbf{x})u_j(\mathbf{x})d\Gamma(\mathbf{x}) = \int_{\Gamma} u_{ij}^{*}(\boldsymbol{\xi},\mathbf{x})t_j(\mathbf{x})d\Gamma(\mathbf{x}), i, j = 1,2,3. \tag{9.31}$$

In (9.31), $\boldsymbol{\xi} = (\hat{x}_1, \hat{x}_2)$ and $\mathbf{x} = (x_1, x_2)$ represent, respectively, the source point and field point of the boundary integral equations. Γ and Ω are the boundary and body of the elastic solid. The symbol $\int_{\Gamma}^{*}$ denotes an integral taken in the sense of *Cauchy principal value.* $c_{ij}(\boldsymbol{\xi})$ are the free term coefficients dependent on the location of the source point $\boldsymbol{\xi}$, which equals to $\delta_{ij}/2$ for a smooth boundary and $c_{ij} = \delta_{ij}$ for an internal point. The symbol δ_{ij} is the Kronecker delta, that is, $\delta_{ij} = 1$ when $i = j$ and $\delta_{ij} = 0$ when $i \neq j$. $u_j(\mathbf{x})$ and $t_j(\mathbf{x})$ are the displacements and surface tractions at the field point $\mathbf{x}$. $u_{ij}^{*}(\boldsymbol{\xi},\mathbf{x})$ and $t_{ij}^{*}(\boldsymbol{\xi},\mathbf{x})$ are, respectively, the displacements and tractions in the x_j direction at point $\mathbf{x}$ corresponding to a unit point force acting in the x_i direction applied at point $\boldsymbol{\xi}$. In practical applications, $c_{ij}(\boldsymbol{\xi})$ can be computed by considering rigid body motion. For example, if a unit rigid body movement in the direction of x_j is considered, we have $u_j = 1$ which will not induce any stresses and hence $t_j = 0$. Substituting this condition into (9.31), we get

$$c_{ij}\left(\xi\right) = -\int_{\Gamma} t_{ij}^{*}\left(\xi, \mathbf{x}\right) \, d\Gamma\left(\mathbf{x}\right). \tag{9.32}$$

The boundary integral equations given in (9.31) have now three unknown functions, that is, u_j or t_j, $j = 1, 2, 3$, and three equations, which constitute the basis of the boundary element formulation. Written in matrix form, we have

$$\mathbf{C}\left(\xi\right)\mathbf{u}\left(\xi\right) + \int_{\Gamma} \mathbf{T}^{*}\left(\xi, \mathbf{x}\right)\mathbf{u}\left(\mathbf{x}\right) d\Gamma\left(\mathbf{x}\right) = \int_{\Gamma} \mathbf{U}^{*}\left(\xi, \mathbf{x}\right)\mathbf{t}\left(\mathbf{x}\right) d\Gamma\left(\mathbf{x}\right), \tag{9.33a}$$

where

$$\mathbf{C}\left(\xi\right) = \begin{bmatrix} c_{11}\left(\xi\right) & c_{12}\left(\xi\right) & c_{13}\left(\xi\right) \\ c_{21}\left(\xi\right) & c_{22}\left(\xi\right) & c_{23}\left(\xi\right) \\ c_{31}\left(\xi\right) & c_{32}\left(\xi\right) & c_{33}\left(\xi\right) \end{bmatrix}, \quad \mathbf{u}\left(\mathbf{x}\right) = \begin{Bmatrix} u_1\left(\mathbf{x}\right) \\ u_2\left(\mathbf{x}\right) \\ u_3\left(\mathbf{x}\right) \end{Bmatrix}, \quad \mathbf{t}\left(\mathbf{x}\right) = \begin{Bmatrix} t_1\left(\mathbf{x}\right) \\ t_2\left(\mathbf{x}\right) \\ t_3\left(\mathbf{x}\right) \end{Bmatrix},$$

$$\mathbf{T}^{*}\left(\xi, \mathbf{x}\right) = \begin{bmatrix} t_{11}^{*}\left(\xi, \mathbf{x}\right) & t_{12}^{*}\left(\xi, \mathbf{x}\right) & t_{13}^{*}\left(\xi, \mathbf{x}\right) \\ t_{21}^{*}\left(\xi, \mathbf{x}\right) & t_{22}^{*}\left(\xi, \mathbf{x}\right) & t_{23}^{*}\left(\xi, \mathbf{x}\right) \\ t_{31}^{*}\left(\xi, \mathbf{x}\right) & t_{32}^{*}\left(\xi, \mathbf{x}\right) & t_{33}^{*}\left(\xi, \mathbf{x}\right) \end{bmatrix},$$

$$\mathbf{U}^{*}\left(\xi, \mathbf{x}\right) = \begin{bmatrix} u_{11}^{*}\left(\xi, \mathbf{x}\right) & u_{12}^{*}\left(\xi, \mathbf{x}\right) & u_{13}^{*}\left(\xi, \mathbf{x}\right) \\ u_{21}^{*}\left(\xi, \mathbf{x}\right) & u_{22}^{*}\left(\xi, \mathbf{x}\right) & u_{23}^{*}\left(\xi, \mathbf{x}\right) \\ u_{31}^{*}\left(\xi, \mathbf{x}\right) & u_{32}^{*}\left(\xi, \mathbf{x}\right) & u_{33}^{*}\left(\xi, \mathbf{x}\right) \end{bmatrix}. \tag{9.33b}$$

In boundary element formulation, the boundary Γ is approximated by a series of elements, and the points $\mathbf{x}$, displacements $\mathbf{u}$, and tractions $\mathbf{t}$ on the boundary are approximated by the nodal points $\mathbf{x}_n$, nodal displacement $\mathbf{u}_n$, and nodal traction $\mathbf{t}_n$ through different interpolation functions. If a quadratic element is considered, we let

$$\begin{aligned} \mathbf{x} &= \varpi_1 \mathbf{x}_m^{(1)} + \varpi_2 \mathbf{x}_m^{(2)} + \varpi_3 \mathbf{x}_m^{(3)} \\ \mathbf{u} &= \varpi_1 \mathbf{u}_m^{(1)} + \varpi_2 \mathbf{u}_m^{(2)} + \varpi_3 \mathbf{u}_m^{(3)}, \\ \mathbf{t} &= \varpi_1 \mathbf{t}_m^{(1)} + \varpi_2 \mathbf{t}_m^{(2)} + \varpi_3 \mathbf{t}_m^{(3)}, \quad \text{on the } m\text{th element}, \end{aligned} \tag{9.34a}$$

where

$$\varpi_1 = \frac{1}{2}\varsigma(\varsigma - 1), \varpi_2 = (1+\varsigma)(1-\varsigma), \varpi_3 = \frac{1}{2}\varsigma(1+\varsigma). \tag{9.34b}$$

In the above, a symbol with subscript m denotes the value related to the mth element, and the subscripts 1, 2, and 3 (or the superscripts (1), (2), and (3)) denote the values of nodes 1, 2, and 3 of the mth element. The variable ς is the dimensionless local coordinate ranging from −1 to 1. The evaluation of the integrals along the boundary of each element requires the use of a Jacobian since all functions are now expressed in terms of ς, and

$$d\Gamma_m = |J_m| d\varsigma, \text{ where } |J_m| = \sqrt{\left(\frac{\partial x_1}{\partial \varsigma}\right)^2 + \left(\frac{\partial x_2}{\partial \varsigma}\right)^2}. \tag{9.35}$$

Substituting the interpolation functions given in (9.34) into (9.35), we get

$$|J_m| = \sqrt{(a_m + \varsigma b_m)^2 + (c_m + \varsigma d_m)^2}, \tag{9.36a}$$

where

$$\begin{aligned} a_m &= \frac{1}{2}\left(x_m^{(3)} - x_m^{(1)}\right), \quad b_m = x_m^{(1)} - 2x_m^{(2)} + x_m^{(3)}, \\ c_m &= \frac{1}{2}\left(y_m^{(3)} - y_m^{(1)}\right), \quad d_m = y_m^{(1)} - 2y_m^{(2)} + y_m^{(3)}. \end{aligned} \tag{9.36b}$$

In (9.36b), for the convenience of writing we use (x, y) to stand for the coordinate variable (x_1, x_2).

If the boundary Γ is discretized into M linear elements with N nodes, substitution of (9.34) and (9.36) into (9.33a) yields

$$\begin{aligned} &\mathbf{C}(\boldsymbol{\xi})\mathbf{u}(\boldsymbol{\xi}) + \sum_{m=1}^{M}\left\{\hat{\mathbf{Y}}_m^{(1)}(\boldsymbol{\xi})\mathbf{u}_m^{(1)} + \hat{\mathbf{Y}}_m^{(2)}(\boldsymbol{\xi})\mathbf{u}_m^{(2)} + \hat{\mathbf{Y}}_m^{(3)}(\boldsymbol{\xi})\mathbf{u}_m^{(3)}\right\} \\ &= \sum_{m=1}^{M}\left\{\mathbf{G}_m^{(1)}(\boldsymbol{\xi})\mathbf{t}_m^{(1)} + \mathbf{G}_m^{(2)}(\boldsymbol{\xi})\mathbf{t}_m^{(2)} + \mathbf{G}_m^{(3)}(\boldsymbol{\xi})\mathbf{t}_m^{(3)}\right\}, \end{aligned} \tag{9.37}$$

in which $\hat{\mathbf{Y}}_m^{(i)}(\xi)$ and $\mathbf{G}_m^{(i)}(\xi)$, $i = 1, 2, 3$, are the *matrices of influence coefficients* defining the interaction between the point ξ and the particular node (1 or 2 or 3) on element m, and are defined as

$$\hat{\mathbf{Y}}_m^{(i)}(\xi) = \int_{-1}^{1} \mathbf{T}^*(\xi, \mathbf{x})\varpi_i(\varsigma)|J_m|d\varsigma,$$

$$\mathbf{G}_m^{(i)}(\xi) = \int_{-1}^{1} \mathbf{U}^*(\xi, \mathbf{x})\varpi_i(\varsigma)|J_m|d\varsigma, \quad i = 1,2,3. \tag{9.38}$$

Since the integrands involved in $\hat{\mathbf{Y}}_m^{(i)}(\xi)$ and $\mathbf{G}_m^{(i)}(\xi)$ may be a regular function or a weakly singular function with $f(x)\ln x$ or a strongly singular function with $f(x)/x$, one should be careful about their numerical integration. If the connecting elements, for example the $(m - 1)$th and the mth element, are continuous at the connecting nodal points, the third node of the $(m - 1)$th element will be the first node of the mth element and can be named as the nth node of the whole boundary element. To write (9.37) corresponding to point ξ in discrete form, we need to add the contribution from two adjoining elements, m and m-1, into one term. We let

$$\hat{\mathbf{Y}}_{m-1}^{(3)} + \hat{\mathbf{Y}}_m^{(1)} = \hat{\mathbf{Y}}_n, \quad \mathbf{G}_{m-1}^{(3)} + \mathbf{G}_m^{(1)} = \mathbf{G}_n, \ldots, \text{etc.} \tag{9.39}$$

Equation (9.37) can then be rewritten as

$$\mathbf{C}(\xi)\mathbf{u}(\xi) + \sum_{n=1}^{N} \hat{\mathbf{Y}}_n(\xi)\mathbf{u}_n = \sum_{n=1}^{N} \mathbf{G}_n(\xi)\mathbf{t}_n. \tag{9.40}$$

Consider ξ to be the location of node i and use $\mathbf{C}_i, \mathbf{u}_i, \hat{\mathbf{Y}}_{in}, \mathbf{G}_{in}$ to denote the values of $\mathbf{C}, \mathbf{u}, \hat{\mathbf{Y}}_n, \mathbf{G}_n$ at node i. Equation (9.40) can now be expressed as

$$\sum_{n=1}^{N} \mathbf{Y}_{in}\mathbf{u}_n = \sum_{n=1}^{N} \mathbf{G}_{in}\mathbf{t}_n, \quad i = 1,2,\ldots,N, \tag{9.41a}$$

in which

$$\begin{aligned} \mathbf{Y}_{in} &= \hat{\mathbf{Y}}_{in}, \quad \text{for } i \neq n, \\ \mathbf{Y}_{in} &= \hat{\mathbf{Y}}_{in} + \mathbf{C}_i, \quad \text{for } i = n. \end{aligned} \tag{9.41b}$$

When all the nodes are taken into consideration, Equation (9.41) produces a $3N \times 3N$ system of equations. By applying the boundary condition such that either u_i or t_i at each node is prescribed, the system of equations (9.41) can be reordered in such a way that the final system of equations can be expressed as $\mathbf{Kv} = \mathbf{p}$ where $\mathbf{K}$ is a fully populated matrix, $\mathbf{v}$ is a vector containing all the boundary unknowns and $\mathbf{p}$ is a vector containing all the prescribed values given on the boundary. Note that through the system of equations shown in (9.41), a rigid body movement represented by $\mathbf{u}_1 = \mathbf{u}_2 = \ldots = \mathbf{u}_N = \mathbf{i}_k$, $\mathbf{t}_1 = \mathbf{t}_2 = \ldots = \mathbf{t}_N = \mathbf{0}$, $k = 1, 2, 3$, where $\mathbf{i}_1 = (1, 0, 0)^T$, $\mathbf{i}_2 = (0, 10)^T$ and $\mathbf{i}_3 = (0, 0, 1)^T$, can give us the following relation:

$$\mathbf{Y}_{ii} = -\sum_{j \neq i} \mathbf{Y}_{ij}, \tag{9.42}$$

which can be used to calculate $\mathbf{Y}_{ii}$ directly without knowing the coefficient matrix $\mathbf{C}_i$ and its associated influence coefficients $\hat{\mathbf{Y}}_{ii}$. And hence, the trouble of singular integral occurred in $\hat{\mathbf{Y}}_{ii}$ and $\mathbf{C}_i$, whose source point locates at one node of the element, can be circumvented.

9.2.3 Coupled Stretching–Bending Analysis – Thin Plate

As described in Section 8.2, if an unsymmetric laminated plate is considered, it will be stretched as well as bent even under pure in-plane forces or pure bending moments. To cover both in-plane and plate bending deformation, the associated boundary integral equations written in (9.31) for two-dimensional analysis should be modified as (Hwu, 2012)

$$\begin{aligned} &c_{ip}(\xi)u_p(\xi) + \overset{*}{\int_\Gamma} t_{ij}^*(\xi,\mathbf{x})\hat{u}_j(\mathbf{x})d\Gamma(\mathbf{x}) + \sum_{k=1}^{N_c^*} t_{ic}^*(\xi,\mathbf{x}_k)\hat{u}_3(\mathbf{x}_k) \\ &= \int_\Gamma u_{ij}^*(\xi,\mathbf{x})t_j(\mathbf{x})d\Gamma(\mathbf{x}) + \int_\Omega u_{ij}^*(\xi,\mathbf{x})q_j(\mathbf{x})d\Omega(\mathbf{x}) + \sum_{k=1}^{N_c^*} u_{i3}^*(\xi,\mathbf{x}_k)t_c(\mathbf{x}_k), \\ &\qquad i,j = 1,2,3,4,\ p = 1,2,3,4,5. \end{aligned} \tag{9.43}$$

The difference between equations (9.43) and (9.31) are: (1) the appearance of the symbol $\hat{\bullet}$, (2) the physical meaning of u_j and t_j, and the range of the sub-indices, (3) the surface integral related to the loads q_i on the plate surface, and (4) the summation terms related to the corner force t_c and deflection u_3. These are explained as follows:

$$\hat{u}_i(\mathbf{x}) = u_i(\mathbf{x}), \quad i = 1,2,4, \quad \hat{u}_3(\mathbf{x}) = u_3(\mathbf{x}) - u_3(\boldsymbol{\xi}), \tag{9.44a}$$

and

$$\begin{aligned} &u_1 = u_0, \quad u_2 = v_0, \quad u_3 = w, \quad u_4 = \beta_n, \quad u_5 = \beta_s, \\ &t_1 = T_x, \quad t_2 = T_y, \quad t_3 = V_n, \quad t_4 = M_n, \quad t_c = M_{ns}^+ - M_{ns}^-, \\ &q_1 = q_x, \quad q_2 = q_y, \quad q_3 = q, \quad q_4 = m_n, \end{aligned} \tag{9.44b}$$

where β_n and β_s are the negative slopes of deflection in normal and tangential directions; T_x and T_y are the x and y components of surface traction; V_n and M_n are the effective transverse shear force and bending moment on the surface with normal direction $\mathbf{n}$; q_x, q_y, q and m_n represent the distributed loads in x, y, z directions and the moment in n direction; t_c is the corner force related to the twisting moments M_{ns} ahead (+) and behind (−) of the corner. $u_{ij}^*(\boldsymbol{\xi}, \mathbf{x}), t_{ij}^*(\boldsymbol{\xi}, \mathbf{x})$, and $t_{ic}^*(\boldsymbol{\xi}, \mathbf{x})$, $i = 1, 2, 3, j = 1, 2, 3, 4$, are the fundamental solutions which represent, respectively, u_j, t_j, and t_c at point $\mathbf{x}$ corresponding to a unit point force acting in the x_i direction applied at point $\boldsymbol{\xi}$, whereas $u_{4j}^*(\boldsymbol{\xi}, \mathbf{x}), t_{4j}^*(\boldsymbol{\xi}, \mathbf{x})$, and $t_{4c}^*(\boldsymbol{\xi}, \mathbf{x})$, $j = 1,2,3,4$ are the fundamental solutions which represent u_j, t_j, and t_c at point $\mathbf{x}$ corresponding to a unit point moment acting on the surface with normal $\mathbf{n}$ applied at point $\boldsymbol{\xi}$. N_c^* of the summation terms is related to the number of corners N_c by

$$N_c^* = \begin{cases} N_c, & \text{if the source point } \boldsymbol{\xi} \text{ is not a corner point;} \\ N_c - 1, & \text{if the source point } \boldsymbol{\xi} \text{ is a corner point.} \end{cases} \tag{9.45}$$

In other words, when the source point $\boldsymbol{\xi}$ is a corner, the location $\mathbf{x}_k$ of the corner of the summation terms in (9.43) does not include the source point itself. Thus, no singularity occurs in the terms of summation.

Similar to the procedure stated in Section 9.2.2 for two-dimensional analysis, the use of the interpolation functions for position $\mathbf{x}$, traction $\mathbf{t}$, and displacement $\mathbf{u}$ in each element, and assembly of all elements will lead (9.43) to (Hwu, 2012)

$$\sum_{j=1}^{N} \left\{ \mathbf{Y}_{ij}\mathbf{u}_j - \mathbf{G}_{ij}\mathbf{t}_j \right\} + \sum_{k=1}^{N_c^*} \left\{ \mathbf{t}_{ic}^*(\mathbf{x}_k) w^{(k)} - \mathbf{u}_{i3}^*(\mathbf{x}_k) t_c^{(k)} \right\} = \mathbf{q}_i^*, \; i = 1,2,\ldots,N, \tag{9.46a}$$

in which

$$\mathbf{u}_j = \begin{Bmatrix} u_0 \\ v_0 \\ w \\ \beta_n \end{Bmatrix}_j, \mathbf{t}_j = \begin{Bmatrix} T_x \\ T_y \\ V_n \\ M_n \end{Bmatrix}_j. \tag{9.46b}$$

$\mathbf{t}_{ic}^*(\mathbf{x}_k)$ and $\mathbf{u}_{i3}^*(\mathbf{x}_k)$ are the vectors of fundamental solutions for corner forces and deflection at the corner $\mathbf{x}_k$. The first three components correspond to a unit point force acting in the direction of x_1, x_2, x_3 applied at the boundary node i, and the fourth component corresponds to a unit moment also applied at the boundary node i but acting on the surface with normal $\mathbf{n}$ of the boundary node i.

In (9.46a), the additional term $\mathbf{q}_i^*$ is associated with the transverse load q_j applied on the plate surface and related to the fundamental solution of displacement $u_{ij}^*(\xi, \mathbf{x})$ by

$$q_i^*(\xi) = \int_A u_{ij}^*(\xi, \mathbf{x}) q_j(\mathbf{x}) dA. \tag{9.47}$$

As to the definition given in (9.38) for the two-dimensional analysis, the matrices of influence coefficients $\mathbf{Y}_{ij}$ and $\mathbf{G}_{ij}$ in (9.46a) are also related to the fundamental solutions and interpolation functions. Moreover, since the derivative of the deflection β_s in the tangential direction is not an independent variable for thin plate analysis, during the derivation for the system of equations (9.46) conversion from β_s to w or β_n of the neighboring nodes should be made if the original formulation treats β_s as an independent variable. One may refer to Hwu (2012) for a detailed description of $\mathbf{Y}_{ij}$ and $\mathbf{G}_{ij}$.

Since Equation (9.46) produces only $4N$ equations through N boundary nodes, to solve the equations with $4N + N_c$ unknowns we need another N_c equations. The additional N_c equations can be set by considering the third equation of (9.46) with the load being applied at the location of corner node. It has been shown that if the three nodes of a corner are selected to be located at the same position, their associated equations may be dependent on each other, and equation can be simply replaced by (Hwu and Chang, 2015)

$$w^{(k)-} - w^{(k)} = 0. \tag{9.48}$$

If a displacement-prescribed boundary condition is considered, Equation (9.48) will vanish and the auxiliary equations introduced in Hwu and Chang (2015) become necessary.

9.2.4 Coupled Stretching–Bending Analysis – Thick Plate

In the previous section a thin plate is considered and the transverse shear deformation is neglected. If a thick plate is considered, the transverse shear strains cannot be ignored. By using the first-order shear deformation theory presented in Section 3.4, the associated boundary integral equations for the coupled stretching–bending analysis of thick laminated composite plates have been derived and written as (Hsu and Hwu, 2023a, 2023b), that is,

$$c_{ij}(\hat{\mathbf{x}})u_j(\hat{\mathbf{x}})+\overset{*}{\int_\Gamma} t_{ij}^*(\hat{\mathbf{x}},\mathbf{x})u_j(\mathbf{x})d\Gamma(\mathbf{x})$$
$$=\int_\Gamma u_{ij}^*(\hat{\mathbf{x}},\mathbf{x})t_j(\mathbf{x})d\Gamma(\mathbf{x})+\int_\Omega u_{ij}^*(\hat{\mathbf{x}},\mathbf{x})q_j(\mathbf{x})d\Omega(\mathbf{x}),\ i,j=1,2,\cdots,5. \quad (9.49)$$

The main difference between (9.49) and (9.43) for thin plate is the deletion of the summation terms related to the corner forces, which make the analysis of thick plates simpler than thin plates. Moreover, the mathematical form of (9.49) is almost the same as the conventional one for two-dimensional or three-dimensional analysis. Their differences are (1) the physical meaning of u_j, t_j, (2) the index ranges, and (3) the surface integral related to the distributed loads q_j. The physical meaning of each variable is explained as follows:

$$\begin{aligned}&u_1=u_0,\ u_2=v_0,\ u_3=\beta_x,\ u_4=\beta_y,\ u_5=w,\\&t_1=T_x,\ t_2=T_y,\ t_3=H_x,\ t_4=H_y,\ t_5=Q_n,\\&q_1=q_x,\ q_2=q_y,\ q_3=m_x,\ q_4=m_y,\ q_5=q_z,\end{aligned} \quad (9.50a)$$

where

$$\begin{aligned}&T_x=N_xn_1+N_{xy}n_2,\ T_y=N_{xy}n_1+N_yn_2,\\&H_x=M_xn_1+M_{xy}n_2,\ H_y=M_{xy}n_1+M_yn_2,\\&Q_n=Q_xn_1+Q_yn_2,\end{aligned} \quad (9.50b)$$

and n_1, n_2 are the components of unit outward normal vector. From (9.50), we see the fundamental solutions $u_{ij}^*(\hat{\mathbf{x}},\mathbf{x})$ and $t_{ij}^*(\hat{\mathbf{x}},\mathbf{x})$ required in (9.49) can be obtained by the following way whose details can be found in (Hsu and Hwu, 2023b):

$$\begin{aligned}&\mathbf{v}\text{ of }(3.117)\rightarrow\mathbf{u},\boldsymbol{\beta},w\text{ of }(3.101\text{b})\rightarrow u_i\text{ of }(9.50\text{a}),\\&\mathbf{N},\mathbf{M},\mathbf{Q}\text{ of }(3.126)\rightarrow N_{ij},M_{ij},Q_i\text{ of }(9.50\text{b})\rightarrow t_i\text{ of }(9.50\text{a}).\end{aligned}\quad(9.51)$$

9.2.5 Three-Dimensional Analysis

Like the two-dimensional analysis presented in Section 9.2.2, the boundary integral equations for three-dimensional analysis can also be expressed in matrix form as

$$\mathbf{C}(\boldsymbol{\xi})\mathbf{u}(\boldsymbol{\xi})+\int_{\Gamma}\mathbf{T}^*(\boldsymbol{\xi},\mathbf{x})\mathbf{u}(\mathbf{x})d\Gamma(\mathbf{x})=\int_{\Gamma}\mathbf{U}^*(\boldsymbol{\xi},\mathbf{x})\mathbf{t}(\mathbf{x})d\Gamma(\mathbf{x}),\quad(9.52)$$

in which the source point and field point are $\boldsymbol{\xi}=(\hat{x}_1,\hat{x}_2,\hat{x}_3)$ and $\mathbf{x}=(x_1, x_2, x_3)$, respectively. If a quadratic quadrilateral element is considered, we let

$$\mathbf{x}=\sum_{k=1}^{8}\varpi_k\mathbf{x}_m^{(k)},\ \mathbf{u}=\sum_{k=1}^{8}\varpi_k\mathbf{u}_m^{(k)},\ \mathbf{t}=\sum_{k=1}^{8}\varpi_k\mathbf{t}_m^{(k)},\ \text{on the }m\text{th element}\quad(9.53)$$

where the interpolation function $\varpi_k=\varpi_k(\xi,\eta)$ can be written as (9.2b). The surface normal $\mathbf{s}_e$ for the point on the mth element can then be determined by the tangential directions $\mathbf{s}_\xi$ and $\mathbf{s}_\eta$ as

$$\mathbf{s}_e=\mathbf{s}_\xi\times\mathbf{s}_\eta=\frac{\partial\mathbf{x}}{\partial\xi}\times\frac{\partial\mathbf{x}}{\partial\eta}=(j_1,j_2,j_3),\quad(9.54\text{a})$$

where

$$j_1=\frac{\partial x_2}{\partial\xi}\frac{\partial x_3}{\partial\eta}-\frac{\partial x_3}{\partial\xi}\frac{\partial x_2}{\partial\eta},\ j_2=\frac{\partial x_3}{\partial\xi}\frac{\partial x_1}{\partial\eta}-\frac{\partial x_1}{\partial\xi}\frac{\partial x_3}{\partial\eta},\ j_3=\frac{\partial x_1}{\partial\xi}\frac{\partial x_2}{\partial\eta}-\frac{\partial x_2}{\partial\xi}\frac{\partial x_1}{\partial\eta}.\quad(9.54\text{b})$$

With the Green's function obtained by the Radon–Stroh formalism introduced in Section 8.6, $\mathbf{U}^*$ and $\mathbf{T}^*$ can be expressed as (Hsu, et al., 2019)

$$\mathbf{U}^* = \frac{-\text{sgn}(\mathbf{m}\cdot\Delta\mathbf{x})}{2\pi^2}\int_0^\pi \text{Re}\left\{\mathbf{A}\left\langle\frac{1}{z_\alpha-\hat{z}_\alpha}\right\rangle\mathbf{A}^T\right\}d\theta,$$

$$\mathbf{T}^* = \frac{\text{sgn}(\mathbf{m}\cdot\Delta\mathbf{x})}{2\pi^2}\int_0^\pi \text{Re}\left\{\mathbf{A}\left\langle\frac{1}{(z_\alpha-\hat{z}_\alpha)^2}\right\rangle(\mathbf{B}^e)^T\right\}d\theta. \tag{9.55}$$

Substituting (9.53) and (9.55) into (9.52), and following the standard approach of boundary element method, we can obtain the same expression as (9.41) for the three-dimensional analysis. The Jacobian required in the integration of influence matrices (9.38) now becomes

$$\left|J_m(\xi,\eta)\right| = \left|\mathbf{s}_\xi\times\mathbf{s}_\eta\right| = \sqrt{j_1^2+j_2^2+j_3^2}. \tag{9.56}$$

9.2.6 Boundary-Based Finite Element Method

A special boundary element can be established by using a special fundamental solution such as that for the problem with a single hole, crack, or inclusion. If there are multiple holes, cracks, and inclusions inside the body, subregion technique can be applied to make each subregion contain only one hole, crack, or inclusion (Hwu and Liao, 1994). The final system of equations for the whole region is then obtained by adding the set of equations for each subregion together with compatibility and equilibrium conditions between their interfaces. Therefore, if several holes, cracks, and inclusions appear in the anisotropic elastic solid, the system of equations will become complicated due to the requirement of compatibility and equilibrium along all the interfaces of subregions. To avoid the trouble caused by the subregion, a boundary-based finite element method has been introduced by Hwu et al. (2017) through the transformation of the special boundary element into a super finite element.

If we treat the subregion enclosed by the boundary elements as a super finite element, the system of simultaneous linear algebraic equations can be written in the form of

$$\mathbf{Y}_e\mathbf{u}_e = \mathbf{G}_e\mathbf{t}_e + \mathbf{q}_e, \tag{9.57}$$

where $\mathbf{Y}_e$ and $\mathbf{G}_e$ are the influence matrices of the entire body of the subregion, $\mathbf{u}_e$ and $\mathbf{t}_e$ are, respectively, vectors for the nodal displacements and tractions of the entire subregion; $\mathbf{q}_e$ is the extra term contributed by thermal effects, inertia effects, or surface loads.

As to the finite element method, if the variation of displacement within each element is assumed to be (9.1), that is,

$$\mathbf{u} = \boldsymbol{N}\mathbf{u}_e, \tag{9.58}$$

the minimization of the total potential energy of element with respect to the nodal displacements results in a standard finite element system (9.8), that is,

$$\mathbf{K}_e\mathbf{u}_e = \mathbf{f}_e, \tag{9.59}$$

in which the nodal force $\mathbf{f}_e$ is related to the surface traction by

$$\mathbf{f}_e = \int_{\Gamma_e} \boldsymbol{N}^T \mathbf{t} d\Gamma, \tag{9.60}$$

where Γ_e is the boundary of the element. By comparing the system of equations (9.57) with (9.59), we see that the key approach to transform the results of boundary element into an equivalent finite element is deriving the relation between element nodal force $\mathbf{f}_e$ and traction $\mathbf{t}_e$ through (9.60). If the displacements and tractions along the mth element of the boundary are approximated as

$$\mathbf{u} = \boldsymbol{N}\mathbf{u}_m, \ \mathbf{t} = \boldsymbol{N}\mathbf{t}_m. \tag{9.61}$$

Substituting $(9.61)_2$ into (9.60), we obtain

$$\mathbf{f}_e = \sum_{m=1}^{M} \int_{\Gamma_m} \boldsymbol{N}^T (\boldsymbol{N}\mathbf{t}_m) d\Gamma_m = \sum_{m=1}^{M} \mathbf{W}_m \mathbf{t}_m = \mathbf{W}_e \mathbf{t}_e, \tag{9.62}$$

where M is the number of boundary element in the subregion, and

$$\mathbf{W}_m = \int_{\Gamma_m} \boldsymbol{N}^T \boldsymbol{N} d\Gamma_m. \tag{9.63}$$

Substituting (9.62) into (9.57) with $\mathbf{q}_e = \mathbf{0}$, we get

$$\mathbf{K}_e\mathbf{u}_e = \mathbf{f}_e, \tag{9.64a}$$

where

$$\mathbf{K}_e = \mathbf{W}_e \mathbf{G}_e^{-1} \mathbf{Y}_e. \tag{9.64b}$$

In general, unlike the finite element formulation, the equivalent stiffness matrix $\mathbf{K}_e$ obtained from (9.64b) is asymmetric. In the literature, there are some discussions about symmetrization (Brebbia *et al.*, 1984) such as the replacement by

$$\mathbf{K}_e \rightarrow \frac{1}{2}\left(\mathbf{K}_e + \mathbf{K}_e^T\right). \tag{9.65}$$

In this formulation, we only transform the boundary element to finite element and will not consider the combination of finite element method and boundary element method. And hence, *the equivalent stiffness matrix* $\mathbf{K}_e$ *is constructed by (9.64b) and no further symmetrization like (9.65) is made.*

9.3 MESHLESS METHOD

Meshes are required in both FEM and BEM to provide a certain relationship between nodes, which become the building blocks of the formulation procedure of these two conventional numerical methods. The meshless method instead establishes a system of algebraic equations for the entire problem domain by a set of nodes scattered within the problem domain without the confinement of predefined meshes. A number of meshless methods have been proposed in the literature, such as element-free Galerkin (EFG) method, meshless local Petrov–Galerkin (MLPG) method, point interpolation method (PIM), etc. Each of them has its own characters and detailed presentation can be found in the related published papers or books, for example, Liu (2009). Here, we just briefly introduce EFG and MLPG to see their difference with FEM and BEM.

9.3.1 Element-Free Galerkin (EFG) Method

In the EFG method the entire domain is discretized by a set of nodes scattered in the problem domain and on the boundaries of the domain. To remove the confinement made by the predefined meshes of FEM, the displacement field is approximated by moving least square (MLS) shape functions, that is,

$$\mathbf{u}^h(\mathbf{x}) = \sum_{i \in S_n} \varphi_i(\mathbf{x}) \mathbf{u}_i, \tag{9.66}$$

where $\mathbf{u}^h(\mathbf{x})$ represents the approximation of the displacement function vector $\mathbf{u}(\mathbf{x})$; $\mathbf{u}_i$ is the displacement vector of node i; S_n is the set of nodes in the support domain of the point $\mathbf{x}$ for constructing MLS shape function; $\varphi_i(\mathbf{x})$ is the shape function defined by

$$\varphi_i(\mathbf{x}) = w(\mathbf{x}-\mathbf{x}_i)\mathbf{p}^T(\mathbf{x})\left[\sum_{i=1}^{n} w(\mathbf{x}-\mathbf{x}_i)\mathbf{p}(\mathbf{x}_i)\mathbf{p}^T(\mathbf{x}_i)\right]^{-1}\mathbf{p}(\mathbf{x}_i), \quad (9.67)$$

in which $w(\mathbf{x} - \mathbf{x}_i)$ is a weight function and $\mathbf{p}(\mathbf{x})$ is a $(2m + 1) \times 1$ vector containing the terms for mth power of polynomial if $\mathbf{x} = (x, y)$ is a two-dimensional variable. The following are the commonly used examples of these two functions:

$$w(\mathbf{x}-\mathbf{x}_i) = \begin{cases} c_0 \exp\left(-\alpha\overline{\|\mathbf{x}-\mathbf{x}_i\|}^2\right) + c_1, & \text{when } \overline{\|\mathbf{x}-\mathbf{x}_i\|} \le 1, \\ 0, & \text{when } \overline{\|\mathbf{x}-\mathbf{x}_i\|} > 1, \end{cases}$$
$$\mathbf{p}^T(\mathbf{x}_i) = \left[1, \bar{x}_i, \bar{y}_i, \ldots, \bar{x}_i^k, \bar{y}_i^k, \ldots, \bar{x}_i^m, \bar{y}_i^m\right], \quad (9.68a)$$

where

$$c_0 = 1/\left(1-e^{-\alpha}\right), \quad c_1 = -e^{-\alpha}c_0, \quad \alpha = 1, \quad m = 1,$$
$$\overline{\|\mathbf{x}-\mathbf{x}_i\|} = \|\mathbf{x}-\mathbf{x}_i\|/r_c, \quad \bar{x}_i = x_i/r_c, \quad \bar{y}_i = y_i/r_c, \quad r_c = r/4, \quad (9.68b)$$

and r is the radius of circular integration path.

Note that the MLS shape functions $\varphi_i(\mathbf{x})$ do not possess the Kronecker delta function property, that is,

$$\varphi_i(\mathbf{x}_j) \neq \delta_{ij}, \quad (9.69)$$

which results in

$$\mathbf{u}^h(\mathbf{x}_i) \neq \mathbf{u}_i. \quad (9.70)$$

This implies that the essential boundary conditions cannot be exactly satisfied via enforcing

$$\mathbf{u}_j = \hat{\mathbf{u}}_j, \text{ for node } j \text{ on } \Gamma_u. \quad (9.71)$$

To enforce the satisfaction of $\mathbf{u} = \hat{\mathbf{u}}$ on the essential boundary Γ_u, the principle of minimum potential energy is considered by adding the constraint $\mathbf{u} - \hat{\mathbf{u}} = \mathbf{0}$ and constructing a Lagrange function L as

$$L = \Pi + \int_{\Gamma_u} \boldsymbol{\lambda}^T \left(\mathbf{u} - \hat{\mathbf{u}}\right) d\Gamma, \tag{9.72}$$

where $\boldsymbol{\lambda}$ is a vector of Lagrange multipliers and Π is the total potential energy of the elastic body which can be written as

$$\Pi = \frac{1}{2}\int_{\Omega} \boldsymbol{\sigma}^T \boldsymbol{\varepsilon} d\Omega - \int_{\Omega} \mathbf{u}^T \mathbf{b} d\Omega - \int_{\Gamma} \mathbf{u}^T \mathbf{t} d\Gamma. \tag{9.73}$$

Unlike FEM in which $\Pi = \sum \Pi_e$ is summed up from all the individual element Π_e as shown in (9.6), here the integral of (9.73) is for the entire body, and its integration is related to the support domain set for MLS shape function.

By the method of Lagrange function, the Lagrange multiplier $\boldsymbol{\lambda}$ is an additional unknown function which can be treated as a field variable and interpolated by using the nodes on the essential boundary, that is,

$$\boldsymbol{\lambda}(\mathbf{x}) = \sum_{i \in S_\lambda} N_i(s) \boldsymbol{\lambda}_i, \quad \mathbf{x} \in \Gamma_u, \tag{9.74}$$

where $\boldsymbol{\lambda}_i$ is the vector of Lagrange multipliers at node i on the essential boundary; S_λ is the set of nodes used for this interpolation; $N_i(s)$ can be a Lagrange polynomial used in the conventional FEM in which s is the curvilinear coordinate along the essential boundary. For example, consider a Lagrange polynomial of order k, which is constructed by k+1 data points,

$$N_i(s) = \frac{\left(s - s_0\right)}{\left(s_i - s_0\right)} \cdots \frac{\left(s - s_{i-1}\right)}{\left(s_i - s_{i-1}\right)} \frac{\left(s - s_{i+1}\right)}{\left(s_i - s_{i+1}\right)} \cdots \frac{\left(s - s_k\right)}{\left(s_i - s_k\right)}. \tag{9.75}$$

Taking the variation on the Lagrange function L, we get

$$\begin{aligned} \delta L = & \int_{\Omega} \delta\left(\mathbf{D}\mathbf{u}\right)^T \mathbf{C}\mathbf{D}\mathbf{u} d\Omega - \int_{\Omega} \left(\delta \mathbf{u}\right)^T \mathbf{b} d\Omega - \int_{\Gamma_t} \left(\delta \mathbf{u}\right)^T \hat{\mathbf{t}} d\Gamma \\ & + \int_{\Gamma_u} \left(\delta \boldsymbol{\lambda}\right)^T \left(\mathbf{u} - \hat{\mathbf{u}}\right) d\Gamma + \int_{\Gamma_u} \left(\delta \mathbf{u}\right)^T \boldsymbol{\lambda} d\Gamma = 0, \end{aligned} \tag{9.76}$$

in which $\mathbf{C}$ is a 3×3 stiffness matrix containing the components of C_{ij}, $i, j = 1,2,6$, of (9.5b) for 2D problems, and $\mathbf{D}$ is a matrix of differential operator such that $\boldsymbol{\varepsilon} = \mathbf{Du}$. For 2D problems,

$$\boldsymbol{\varepsilon} = \begin{Bmatrix} \varepsilon_{11} \\ \varepsilon_{22} \\ 2\varepsilon_{12} \end{Bmatrix} = \begin{bmatrix} \dfrac{\partial}{\partial x} & 0 \\ 0 & \dfrac{\partial}{\partial y} \\ \dfrac{\partial}{\partial y} & \dfrac{\partial}{\partial x} \end{bmatrix} \begin{Bmatrix} u_1 \\ u_2 \end{Bmatrix} = \mathbf{Du}. \tag{9.77}$$

Substituting (9.66) and (9.74) into (9.76), we obtain

$$\begin{aligned} \delta L = & \sum_{i \in S_n} \sum_{j \in S_n} (\delta \mathbf{u}_i)^T \mathbf{K}_{ij} \mathbf{u}_j - \sum_{i \in S_n} (\delta \mathbf{u}_i)^T \mathbf{f}_i + \sum_{i \in S_\lambda} \sum_{j \in S_n} (\delta \boldsymbol{\lambda}_i)^T \mathbf{G}_{ji}^T \mathbf{u}_j \\ & - \sum_{i \in S_\lambda} (\delta \boldsymbol{\lambda}_i)^T \mathbf{q}_i + \sum_{i \in S_n} \sum_{j \in S_\lambda} (\delta \mathbf{u}_i)^T \mathbf{G}_{ij} \boldsymbol{\lambda}_j = 0, \end{aligned} \tag{9.78a}$$

where

$$\begin{aligned} \mathbf{K}_{ij} &= \int_\Omega \mathbf{B}_i^T \mathbf{C} \mathbf{B}_j d\Omega, \quad \mathbf{G}_{ij} = \int_{\Gamma_u} \boldsymbol{\Phi}_i^T \mathbf{N}_j d\Gamma, \\ \mathbf{f}_i &= \int_\Omega \boldsymbol{\Phi}_i^T \mathbf{b} d\Omega + \int_{\Gamma_t} \boldsymbol{\Phi}_i^T \hat{\mathbf{t}} d\Gamma, \quad \mathbf{q}_i = \int_{\Gamma_u} \mathbf{N}_i^T \hat{\mathbf{u}} d\Gamma. \end{aligned} \tag{9.78b}$$

and

$$\mathbf{B}_i = \begin{bmatrix} \dfrac{\partial \varphi_i}{\partial x} & 0 \\ 0 & \dfrac{\partial \varphi_i}{\partial y} \\ \dfrac{\partial \varphi_i}{\partial y} & \dfrac{\partial \varphi_i}{\partial x} \end{bmatrix}, \quad \boldsymbol{\Phi}_i = \begin{bmatrix} \varphi_i(\mathbf{x}) & 0 \\ 0 & \varphi_i(\mathbf{x}) \end{bmatrix}, \quad \mathbf{N}_i = \begin{bmatrix} N_i(s) & 0 \\ 0 & N_i(s) \end{bmatrix}. \tag{9.78c}$$

Let n and m be the total number of nodes in the entire domain and on the essential boundary, respectively. Assembly of all the related matrices and vectors of (9.78a) leads to

$$\delta L = \left(\delta \mathbf{u}_T\right)^T \left[\mathbf{K}_T \mathbf{u}_T - \mathbf{f}_T + \mathbf{G}_T \boldsymbol{\lambda}_T\right] + \left(\delta \boldsymbol{\lambda}_T\right)^T \left[\mathbf{G}_T^T \mathbf{u}_T - \mathbf{q}_T\right] = 0, \quad (9.79a)$$

where

$$\mathbf{K}_T = \begin{bmatrix} \mathbf{K}_{11} & \mathbf{K}_{12} & \cdots & \mathbf{K}_{1n} \\ \mathbf{K}_{21} & \mathbf{K}_{22} & \cdots & \mathbf{K}_{2n} \\ \vdots & \vdots & \ddots & \vdots \\ \mathbf{K}_{n1} & \mathbf{K}_{n2} & \cdots & \mathbf{K}_{nn} \end{bmatrix}, \quad \mathbf{G}_T = \begin{bmatrix} \mathbf{G}_{11} & \mathbf{G}_{12} & \cdots & \mathbf{G}_{1m} \\ \mathbf{G}_{21} & \mathbf{G}_{22} & \cdots & \mathbf{G}_{2m} \\ \vdots & \vdots & \ddots & \vdots \\ \mathbf{G}_{n1} & \mathbf{G}_{n2} & \cdots & \mathbf{G}_{nm} \end{bmatrix},$$

$$\mathbf{f}_T = \begin{Bmatrix} \mathbf{f}_1 \\ \mathbf{f}_2 \\ \vdots \\ \mathbf{f}_n \end{Bmatrix}, \quad \mathbf{q}_T = \begin{Bmatrix} \mathbf{q}_1 \\ \mathbf{q}_2 \\ \vdots \\ \mathbf{q}_m \end{Bmatrix}, \quad \mathbf{u}_T = \begin{Bmatrix} \mathbf{u}_1 \\ \mathbf{u}_2 \\ \vdots \\ \mathbf{u}_n \end{Bmatrix}, \quad \boldsymbol{\lambda}_T = \begin{Bmatrix} \boldsymbol{\lambda}_1 \\ \boldsymbol{\lambda}_2 \\ \vdots \\ \boldsymbol{\lambda}_m \end{Bmatrix}. \quad (9.79b)$$

Since $\delta \mathbf{u}_T$ and $\delta \boldsymbol{\lambda}_T$ are arbitrary, Equation (9.79a) can be satisfied only when

$$\mathbf{K}_T \mathbf{u}_T - \mathbf{f}_T + \mathbf{G}_T \boldsymbol{\lambda}_T = 0, \quad \mathbf{G}_T^T \mathbf{u}_T - \mathbf{q}_T = 0, \quad (9.80)$$

which can be written in matrix form as

$$\begin{bmatrix} \mathbf{K}_T & \mathbf{G}_T \\ \mathbf{G}_T^T & \mathbf{0} \end{bmatrix} \begin{Bmatrix} \mathbf{u}_T \\ \boldsymbol{\lambda}_T \end{Bmatrix} = \begin{Bmatrix} \mathbf{f}_T \\ \mathbf{q}_T \end{Bmatrix}. \quad (9.81)$$

The displacements of the entire domain can then be solved through (9.81) and the MLS approximation (9.66), whose accuracy and efficiency rely on the proper choice of the computation method for the nodal stiffness matrices $\mathbf{K}_{ij}$, $\mathbf{G}_{ij}$ and nodal force vectors $\mathbf{f}_i$ and $\mathbf{q}_i$. One may refer to Liu (2009) for detailed discussions and numerical examples.

9.3.2 Meshless Local Petrov–Galerkin (MLPG) Method

In the EFG method, a background mesh of cells is required in performing the numerical integration for (9.78b). The main reason for this is the use of minimum total potential energy principle, which leads to weak form formulation expressed in integral form implying the physical phenomena are satisfied only in an integral (averaged) sense. To avoid using the background mesh, some researchers suggested the use of the weighted residual method to satisfy the strong form governing equations point by point using information in a local domain of the point. Thus, the integration can be implemented locally by carrying out numerical integration over the local domain. With this consideration, in this section we introduce the other meshless method called MLPG method originated by Atluri and Zhu (1998).

The strong form of the governing equations of elasticity can be expressed by the equilibrium equations together with the associated boundary conditions as

$$\begin{aligned} &\sigma_{ij,j} + b_i = 0, \text{ for all points in the entire domain } \Omega, \\ &u_i = \hat{u}_i, \qquad \text{on } \Gamma_u: \text{ essential boundary conditions}, \\ &t_i = \sigma_{ij} n_j = \hat{t}_i, \text{ on } \Gamma_t: \text{ natural boundary conditions}, \end{aligned} \tag{9.82}$$

in which the stresses σ_{ij} are related to the strains ε_{ij} and displacements u_i by

$$\sigma_{ij} = C_{ijkl}\varepsilon_{kl}, \quad \varepsilon_{kl} = \frac{1}{2}\left(u_{k,l} + u_{l,k}\right) \Rightarrow \sigma_{ij} = C_{ijkl}u_{k,l}. \tag{9.83}$$

A generalized local weak form of (9.82) over a local sub-domain Ω_s of node I can be stated by using the weighted residual method as

$$\int_{\Omega_s} \left(\sigma_{ij,j} + b_i\right) v_I d\Omega - \alpha \int_{\Gamma_{su}} \left(u_i - \hat{u}_i\right) v_I d\Gamma = 0, \tag{9.84}$$

in which v_I is the *test function* or called *weight function* for node I; α is a *penalty parameter*, which is used to impose the essential boundary conditions; Γ_{su} is the part of the essential boundary Γ_u that intersects with the sub-domain Ω_s. In general, the boundary of the local sub-domain Ω_s denoted by $\partial\Omega_s$ can be divided into two parts, that is, $\partial\Omega_s = \Gamma_s + L_s$ where Γ_s denotes the part of global boundary which may vanish when Ω_s is totally inside the body, and L_s denotes the other part of local boundary inside the body. Along the global boundary, Γ_s may be further divided into two parts, that is, $\Gamma_s = \Gamma_{su} + \Gamma_{st}$, and displacements and tractions are prescribed, respectively, along Γ_{su} and Γ_{st}.

Using the divergence theorem

$$\int_{\Omega_s} (\ldots)_{,j}\, d\Omega = \int_{\partial\Omega_s} (\ldots) n_j d\Gamma = \int_{L_s+\Gamma_{su}+\Gamma_{st}} (\ldots) n_j d\Gamma, \tag{9.85}$$

imposing the natural boundary conditions $t_i = \sigma_{ij} n_j = \hat{t}_i$ on Γ_{st}, and purposely choosing a weight function v_I that vanishes on L_s, we can simplify the expression (9.84) to

$$\int_{\Omega_s} \sigma_{ij} v_{I,j} d\Omega + \alpha \int_{\Gamma_{su}} u_i v_I d\Gamma - \int_{\Gamma_{su}} t_i v_I d\Gamma = \int_{\Gamma_{st}} \hat{t}_i v_I d\Gamma + \alpha \int_{\Gamma_{su}} \hat{u}_i v_I d\Gamma + \int_{\Omega_s} b_i v_I d\Gamma, \tag{9.86}$$

which is the local Petrov–Galerkin weak form. Rewrite (9.86) in matrix form, for 2D problems we have

$$\int_{\Omega_s} \mathbf{V}_I^* \boldsymbol{\sigma} d\Omega + \alpha \int_{\Gamma_{su}} \mathbf{V}_I \mathbf{u} d\Gamma - \int_{\Gamma_{su}} \mathbf{V}_I \mathbf{t} d\Gamma = \int_{\Gamma_{st}} \mathbf{V}_I \hat{\mathbf{t}} d\Gamma + \alpha \int_{\Gamma_{su}} \mathbf{V}_I \hat{\mathbf{u}} d\Gamma + \int_{\Omega_s} \mathbf{V}_I \mathbf{b} d\Gamma, \tag{9.87a}$$

where

$$\mathbf{V}_I^* = \begin{bmatrix} v_{I,1} & 0 & v_{I,2} \\ 0 & v_{I,2} & v_{I,1} \end{bmatrix}, \boldsymbol{\sigma} = \begin{Bmatrix} \sigma_{11} \\ \sigma_{22} \\ \sigma_{12} \end{Bmatrix}, \mathbf{V}_I = \begin{bmatrix} v_I & 0 \\ 0 & v_I \end{bmatrix},$$

$$\mathbf{u} = \begin{Bmatrix} u_1 \\ u_2 \end{Bmatrix}, \mathbf{t} = \begin{Bmatrix} t_1 \\ t_2 \end{Bmatrix}, \mathbf{b} = \begin{Bmatrix} b_1 \\ b_2 \end{Bmatrix}. \tag{9.87b}$$

By choosing the MLS trial function as that shown in (9.66) for the displacement vector **u**, using the relation (9.83) for the stresses, and using the relation $(9.82)_3$ for the tractions, we can further simplify (9.87a) to

$$\int_{\Omega_s} \mathbf{V}_I^* \mathbf{C} \sum_{j\in S_n} \mathbf{B}_j \mathbf{u}_j d\Omega + \alpha \int_{\Gamma_{su}} \mathbf{V}_I \sum_{j\in S_n} \boldsymbol{\Phi}_j \mathbf{u}_j d\Gamma - \int_{\Gamma_{su}} \mathbf{V}_I \mathbf{L}_n \mathbf{C} \sum_{j\in S_n} \mathbf{B}_j \mathbf{u}_j d\Gamma$$

$$= \int_{\Gamma_{st}} \mathbf{V}_I \hat{\mathbf{t}} d\Gamma + \alpha \int_{\Gamma_{su}} \mathbf{V}_I \hat{\mathbf{u}} d\Gamma + \int_{\Omega_s} \mathbf{V}_I \mathbf{b} d\Gamma, \tag{9.88}$$

in which $\mathbf{C}, \mathbf{B}_j, \mathbf{\Phi}_j$ are defined in (9.5b) and (9.78c), $\mathbf{V}_I^*, \mathbf{V}_I, \mathbf{u}, \mathbf{t}, \mathbf{b}$ are defined in (9.87b), and $\mathbf{L}_n$ is a matrix related to the unit outward normal vector (n_1, n_2) of the boundary and is defined as

$$\mathbf{L}_n = \begin{bmatrix} n_1 & 0 & n_2 \\ 0 & n_2 & n_1 \end{bmatrix}. \tag{9.89}$$

Equation (9.88) can be rewritten as

$$\sum_{j \in S_n} \mathbf{K}_{Ij} \mathbf{u}_j = \mathbf{f}_I, \tag{9.90a}$$

where the nodal stiffness $\mathbf{K}_{Ij}$ and nodal force vector $\mathbf{f}_I$ are

$$\begin{aligned} \mathbf{K}_{Ij} &= \int_{\Omega_s} \mathbf{V}_I^* \mathbf{C} \mathbf{B}_j d\Omega + \alpha \int_{\Gamma_{su}} \mathbf{V}_I \mathbf{\Phi}_j d\Gamma - \int_{\Gamma_{su}} \mathbf{V}_I \mathbf{L}_n \mathbf{C} \mathbf{B}_j d\Gamma, \\ \mathbf{f}_I &= \int_{\Gamma_{st}} \mathbf{V}_I \hat{\mathbf{t}} d\Gamma + \alpha \int_{\Gamma_{su}} \mathbf{V}_I \hat{\mathbf{u}} d\Gamma + \int_{\Omega_s} \mathbf{V}_I \mathbf{b} d\Gamma. \end{aligned} \tag{9.90b}$$

Equation (9.90) provides two linear equations for node I. Assembling all the equations generated for each node, we can then construct a $2n*2n$ system of equations where n is the total number of nodes in the entire problem domain.

From (9.90), we see that the accuracy and efficiency of the MLPG method depend on the choice of the weight function, penalty parameter, MLS trial function, and the number and size of the sub-domain. One may refer to the discussions presented in the literature such as Liu (2009).

9.4 NUMERICAL EXAMPLES

Starting from the fundamental concept of FEM, BEM, and MM presented in the previous three sections, several computer programs were designed and coded in different communities. The power of simulation was then created through its associated software development. Currently, most of the popular software is developed by using FEM due to its simplicity and generality. On the other hand, several research works are still undergoing in the academic community to improve BEM, MM, and their combined use with FEM. To show their applicability to practical problems, in this section, we will present some numerical examples by using the commercial

FEM software ANSYS and our currently developed BEM software AEPH (Hwu, 2021). Two examples for laminated composite beams are adapted from our recently published papers (Huang, et al. 2024a, 2024b), whereas the other three examples for laminated composite plates are adapted from Hsu and Hwu (2021, 2022b, 2023b).

9.4.1 Laminated Composite Beams

To verify the correctness and demonstrate the wide applicability of the newly developed BFEM for curvilinear non-uniform thickness laminated composite beams presented in Sections 9.2.1 and 9.2.6, two representative numerical examples are considered: (1) a simply supported laminated beam with two internal supports, (2) a cantilever curved composite sandwich beam with uniform, stepped, or smoothly varying thickness.

EXAMPLE 1: A SIMPLY SUPPORTED LAMINATED BEAM WITH TWO INTERNAL SUPPORTS

Consider a four-layered cross-ply laminated composite beam with both ends pinned and two internal roller supports subjected to a linearly distributed transverse force $q_0 = 1$ kN/m (see Figure 9.2 without P_0). The beam length is $3\ell = 1.8$ m, and the beam width is $b = 0.03$ m. Two stacking sequences, symmetric $[0_3/90_3]_s$ and unsymmetric $[0_6/90_6]$, are both investigated. The laminates are made by the graphite/epoxy fiber-reinforced composite whose material properties are given in (3.91). The thickness of each lamina is 5 mm.

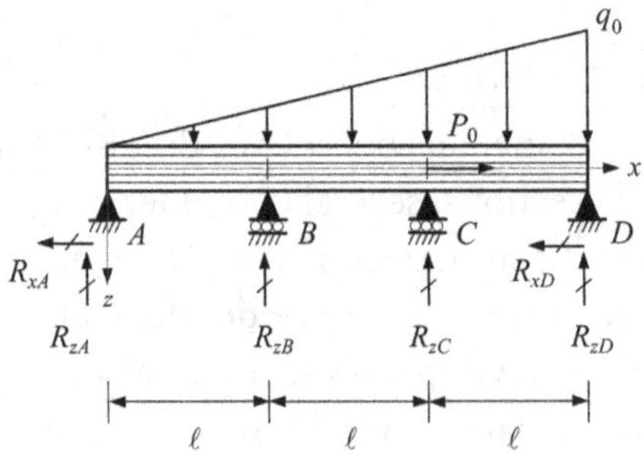

FIGURE 9.2 A laminated composite beam with pinned ends and two internal supports.

Since a straight beam with internal supports is considered, only three elements separated by two internal supports are required for BFEM to get the exact solutions. On the other hand, the convergence test is necessary for getting the approximated solutions by ANSYS. Here, 180 beam elements and 181 nodes were used in ANSYS with BEAM188 beam element to get the convergent solutions. The axial displacement and deflection of the entire beam obtained by BFEM and ANSYS are shown in Figure 9.3. From these results, we see great agreement between these two approaches. Note that the unsymmetric lamination indeed induces coupling responses, leading to nonzero axial displacement and axial reaction forces for the pure transverse load. On the other hand, no axial displacement and reaction are evoked in the symmetric laminated beam, where the stretching and bending are decoupled.

After the simple check on the coupling effect, we now implement the combined loads by further adding an extra concentrated axial force $P_0 = 1$ kN applied at $x = 2\ell$ of the unsymmetric laminated beam (see Figure 9.2). The force and moment diagrams are presented in Figure 9.4 which shows that the axial force jumps at the location of concentrated load, and the

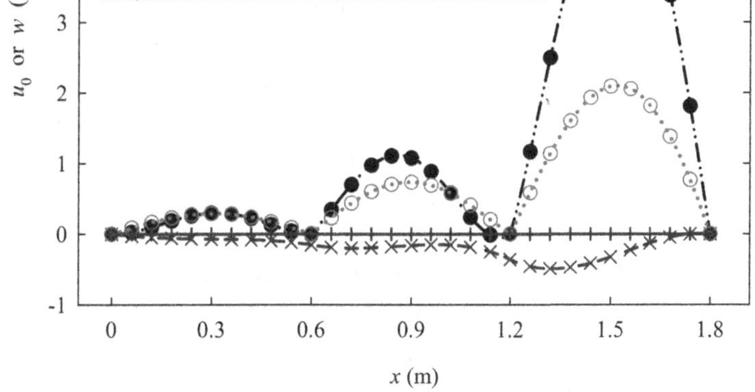

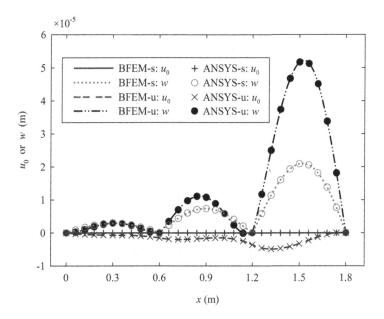

FIGURE 9.3 Axial displacement and deflection of a simply supported beam with internal supports subjected to linearly distributed load. (-s and -u denote, respectively, the symmetric $[0_3/90_3]_s$ and unsymmetric $[0_6/90_6]$ laminated beams.)

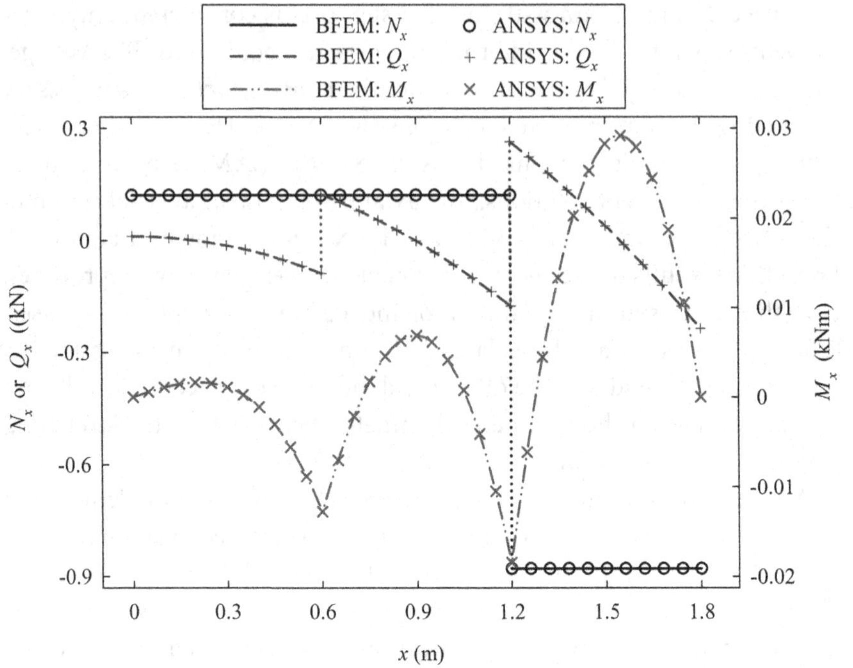

FIGURE 9.4 Axial force, transverse shear force, and bending moment of a simply supported $[0_6/90_6]$ beam with internal supports subjected to combined loads.

transverse shear force is stepwise with discontinuities at internal roller supports. Since the supports provide no rotating reaction, the bending moment is continuously but piecewise distributed. In addition, under the linearly distributed transverse load, the transverse shear force and bending moment are, respectively, quadratically and cubically. Hence, their variations become greater as the linear loading function grows. While the lack of distributed axial load makes the axial force constant in each beam segment. Similar to the case of pure transverse load, we can again observe the well agreement between BFEM and ANSYS.

EXAMPLE 2: A CANTILEVER CURVED COMPOSITE SANDWICH BEAM WITH UNIFORM, STEPPED, OR SMOOTHLY VARYING THICKNESS

Consider a quarter circular [-45/0/core/45/-45] composite sandwich beam with width $b = 0.02$ m. The faces of the sandwich are made by stacking the graphite/epoxy fiber-reinforced laminae whose material properties are given in (3.91). The thickness of each lamina is 0.01 m. The sandwich core is constructed by aluminum and its effective shear modulus is G = 146 MPa. As shown in Figure 9.5(a–c), three different kinds of thickness variation (uniform, stepped, or smoothly varying) are considered, and all of them under the same loading and supported conditions. This beam is fixed at end A and subjected to an axial force F_0 = 5 kN, a transverse force P_0 = 2 kN and a bending moment m_0 = 1 kNm at the other end B, and a vertical force W_0 = 1 kN at the middle point ($\theta = \pi/4$). The radius of the quarter circle is set to be r = 1 m. The associated values of thickness are: (1) $h_u = (h_A + h_B)/2$ for the uniform beam, (2) h_A, h_u, h_B for the stepped beam, and (3) $h(\theta) = h_A + 2\theta(h_B - h_A)/\pi$ for the smoothly varying beam, where h_A = 0.15m and h_B = 0.06m. Under this thickness arrangement, these three different beams possess the same weight with the same volume $V = \pi bRh_u/2 = 3.299 \times 10^{-3}\text{m}^3$.

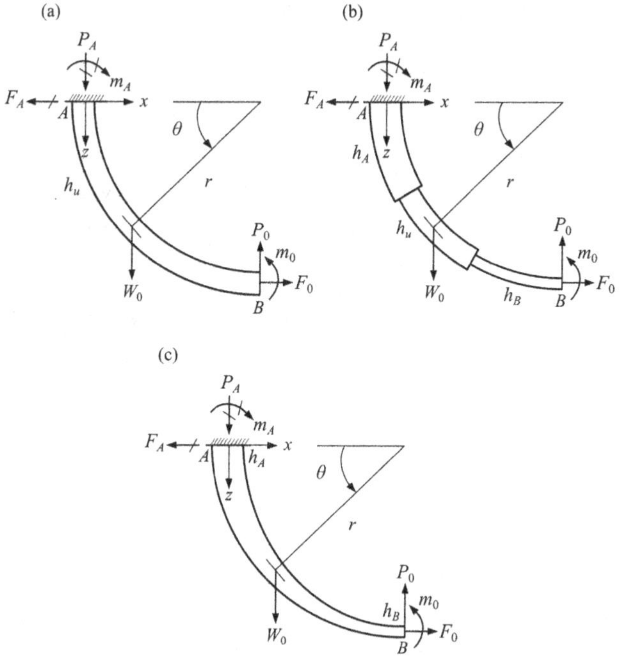

FIGURE 9.5 A cantilever curved composite sandwich beam subjected to combined loads: (a) uniform thickness, (b) stepped thickness, and (c) smoothly varying thickness.

In the BFEM we use 12 discretized segments to simulate the quarter circular beam. To have a fair comparison, we also present the associated finite element results obtained by ANSYS using BEAM188 with 90 elements and 91 nodes. Figure 9.6 shows the beam displacements u_x and u_z along the curved beam, where θ is related to the point location by $x = 1 - r \cos \theta$ and $z = r \sin \theta$. Here, u_x and u_z denote, respectively, the mid-plane displacement in x and z directions of Figure 9.5. From Figure 9.6 we can observe that the results of BFEM are almost the same as ANSYS. Note that for a cantilever beam, which is a kind of static determinate structures, the reaction forces and moment at the fixed end (point A) should be equilibrated with the combined loads applied at the other points (center and free end B), *and are nothing to do with thickness variation.* By using the boundary nodal forces and moment calculated from BFEM, we obtain the reactions at point A as $F_A = 5$ kN, $P_A = 1$ kN, $m_A = 7.707$ kNm for all three different thickness arrangements, which are identical to the exact values provided by equilibrium of combined loads.

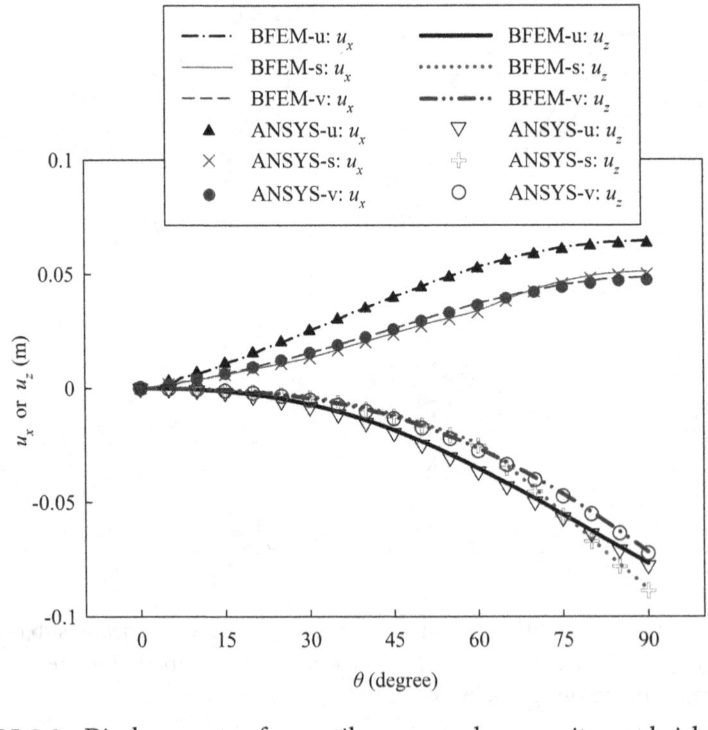

FIGURE 9.6 Displacements of a cantilever curved composite sandwich beam. (-u, -s, and -v denote, respectively, uniform, stepped, and smoothly varying thickness beams.)

Interesting results can also be observed from the results of Figure 9.6 that both of the displacements in x and z directions of uniform beam are larger than those of the smoothly varying beam. While for the stepped beam it induces the largest deflection and its axial displacement at the free end is larger than the smoothly varying beam. Thus, the performance of smoothly varying beam is better than the others if the requirement of the same weight is set for the curved beam, which agrees with our engineering intuition.

9.4.2 Laminated Composite Plates

To verify the correctness and illustrate the generality of BEM presented in Sections 9.2.3 and 9.2.4 for the coupled stretching–bending analysis, in this section we consider three different types of laminated composite plates: thin plates, MEE thin plates, and thick plates. Their respective examples are: (1) a cantilever laminated thin plate with an inclined hole; (2) a cantilever laminated MEE thin plate with an inclined hole and crack; (3) a cantilever quarter circular laminated thick plate with a semicircular hole.

EXAMPLE 3: A CANTILEVER LAMINATED THIN PLATE WITH AN INCLINED HOLE

Consider a cantilever square unsymmetric [45/0/45/-45] laminated plate containing an inclined elliptical hole subjected to a uniform bending moment M_0 = 1 kN on the right edge (see Figure 9.7). The elliptical hole is oriented 120° clockwise from the x_1-axis. The laminates are made by the graphite/epoxy fiber-reinforced composite whose material properties are given in (3.91). The thickness of each lamina is 1 mm. The major and minor axis lengths of the elliptical hole are $2a$ = 0.4 m and $2b$ = 0.2 m, and the plate size is $W = 10a$.

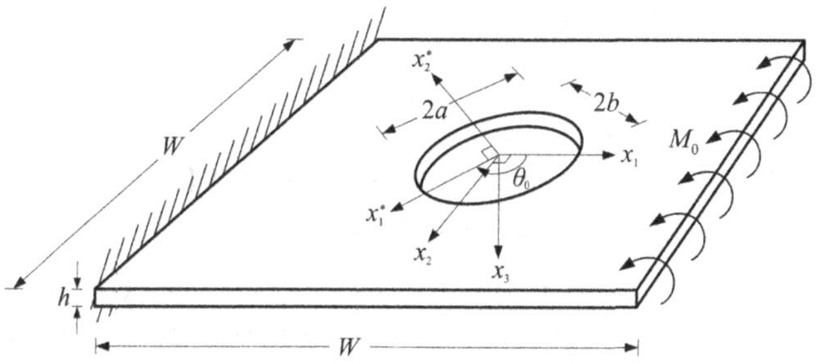

FIGURE 9.7 A clamped square laminate with an inclined elliptical hole subjected to bending moment.

TABLE 9.1 Stress Resultants and Bending Moments along the Hole of [45/0/45/-45] Unsymmetric Laminate Subjected to Uniform Bending Moment (All Columns Except for φ Are in Unit of kN)

φ	$N_{ss}h$	$N_{nn}h$	$N_{sn}h$	M_{ss}	M_{nn}	M_{sn}
Numerical Results from SBEM						
0°	3.972	0	0	3.403	0	0.515
45°	−0.298	0	0	0.272	0	−0.191
90°	2.931	0	0	2.019	0	0.912
135°	−0.75	0	0	1.448	0	0.308
180°	4.877	0	0	3.869	0	0.747
225°	−0.593	0	0	0.193	0	−0.13
270°	2.447	0	0	1.793	0	0.872
315°	0.465	0	0	1.709	0	0.18
Numerical Results from ANSYS						
0°	4.366	0.022	0.79	3.168	−0.019	0.399
45°	−0.288	0.004	−0.003	0.296	0.047	−0.217
90°	2.74	0.008	0.069	1.938	−0.002	0.874
135°	−0.907	−0.006	−0.385	1.551	0.012	0.319
180°	4.916	0.027	0.519	3.689	−0.044	0.66
225°	−0.546	0.002	0.013	0.222	0.044	−0.147
270°	2.332	0.007	0.078	1.743	−0.001	0.841
315°	0.195	−0.003	−0.329	1.772	0.006	0.193

After the convergence test on the proposed SBEM (with special fundamental solution presented in Section 8.5.2), the outer square boundary is meshed with 40 equidistant elements and 48 nodes (including 4 corner nodes), and no mesh is made on the hole boundary. The numerical integration is performed by Gaussian quadrature rule with 32 Gaussian points. Table 9.1 shows the stress resultants and bending moments along the hole boundary, from which we see that the satisfaction of traction-free conditions is perfect in SBEM with simple mesh but imprecise in ANSYS with complicated mesh (40599 nodes and 40179 elements).

EXAMPLE 4: A CANTILEVER LAMINATED MEE THIN PLATE WITH AN INCLINED HOLE AND CRACK

Consider a cantilever [M/45/0/P] unsymmetric laminate with an inclined elliptical hole and an inclined straight crack subjected to a uniform tensile load $N_0 = 1$ kN/m along the lower edge, as shown in Figure 9.8. The plate size is $W = 10a$, and the centers of hole and crack are located at $(-d, 0)$ and $(d, 0)$. The lengths of major and minor axes of elliptical hole

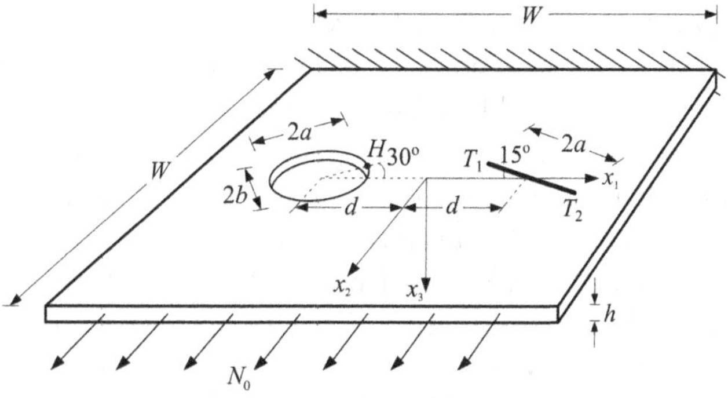

FIGURE 9.8 A cantilever MEE laminate with an inclined elliptical hole and an inclined straight crack subjected to tensile load.

are $2a$ = 0.2 m and b = 0.5a, whereas the crack length is $2a$ = 0.2 m. The material properties of each lamina (piezoelectric, fiber-reinforced, and MEE) are given in (3.91)–(3.93), and the thickness of each lamina is 1 mm.

TABLE 9.2 Magnetic Stress Resultants and Bending Moments at Point H and Stress Intensity Factors at Crack Tips of the Cantilever MEE Laminate Subjected to Uniform Tensile Load

Point	***H***				T_1	T_2
d/a	$\hat{N}_{5s}$	$\hat{N}_{5n}$	$\hat{M}_{5s}$	$\hat{M}_{5n}$	$\hat{K}_I$	$\hat{K}_I$
1.2	3.655	0	−1.37	0	1.293	1.093
1.3	3.248	0	−1.218	0	1.212	1.071
1.4	2.994	0	−1.122	0	1.16	1.056
1.5	2.829	0	−1.06	0	1.124	1.043
1.6	2.654	0	−0.994	0	1.105	1.041
1.8	2.511	0	−0.94	0	1.068	1.024
2	2.409	0	−0.902	0	1.047	1.015

Table 9.2 shows the magnetic stress resultants and bending moments at point H and the SIFs at tips T_1 and T_2. In this table, the nondimensionalization is made by

$$\hat{N}_{5j} = \frac{N_{5j}e_{22}}{N_0 m_{22}} \times 10^3, \ \hat{M}_{5j} = \frac{M_{5j}e_{22}}{N_0 h m_{22}}, \quad j = s, n,$$
$$\hat{K}_I = \frac{K_I}{K_0}, \ K_0 = \frac{N_0}{h}\sqrt{\pi a}, \tag{9.91}$$

where the values of e_{22} and m_{22} are given in (3.93). From the results of Table 9.2, we see that the closer between hole and crack, the larger the stress intensity factor K_I at both crack tips T_1 and T_2, and the larger magnetic stress resultants and bending moments at the inner point H of the hole. Moreover, since the crack tip T_1 is closer to the hole, the K_I at tip T_1 are larger than those at the other tip T_2. And the difference of K_I at T_1 and T_2 decreases as the distance $2d$ increases. The above-mentioned behavior agrees with our engineering intuition for interaction between hole and crack. Besides, as shown in Table 9.2 the traction-free conditions ($N_{5n} = M_{5n} = 0$) are always satisfied for all distance variation. Here, the stress intensity factor K_I related to the tensile response is defined by Hsu and Hwu (2022a)

$$K_I = \frac{1}{h} \lim_{\substack{r\to 0 \\ \theta=0}} \sqrt{2\pi r} N_{22}, \tag{9.92}$$

where (r, θ) is the polar coordinate based upon the crack tip and crack orientation, and h is the thickness of the plate.

EXAMPLE 5: A CANTILEVER QUARTER CIRCULAR LAMINATED THICK PLATE WITH A SEMICIRCULAR HOLE

Consider a cantilever quarter circular unsymmetric [45/0/45/-45] laminated plate of radius $R = 1$ m with a semicircular hole located at (x_c, y_c) = (0.4m, 0.4m) subjected to a uniformly distributed transverse load $Q_0 = 1$ kN/m on the outer arc edge, as shown in Figure 9.9(a). The material properties of each lamina are given in (3.91).

The thickness of each lamina is 1 cm. To investigate the influence of hole size, we consider two cases with different radius r of semicircular hole. One is a small hole represented by $r = 0.05$ m, and the other is a big hole with $r = 0.25$ m. The mesh of BEM contains 105 elements (35 for the outer boundary and 70 for the inner hole) with 215 nodes for both cases, as shown in Figure 9.9(b). In the ANSYS models we use the finite element meshes with more than 23800 elements and 72400 nodes. Figure 9.10 shows the stress resultant N_y and transverse shear force Q_y along the straight line AB ($x = 0.6$, $0.1 \le y \le 0.7$, see Figure 9.9(a)). For both cases of big and small hole, well agreement of internal solutions obtained by our BEM and ANSYS can be observed for N_y, Q_y in Figure 9.10. Figure 9.10 also indicates that the variation of stress resultant N_y for big hole is more drastic than the one for small hole. Both the distributions of N_y and Q_y are strongly influenced by the hole size.

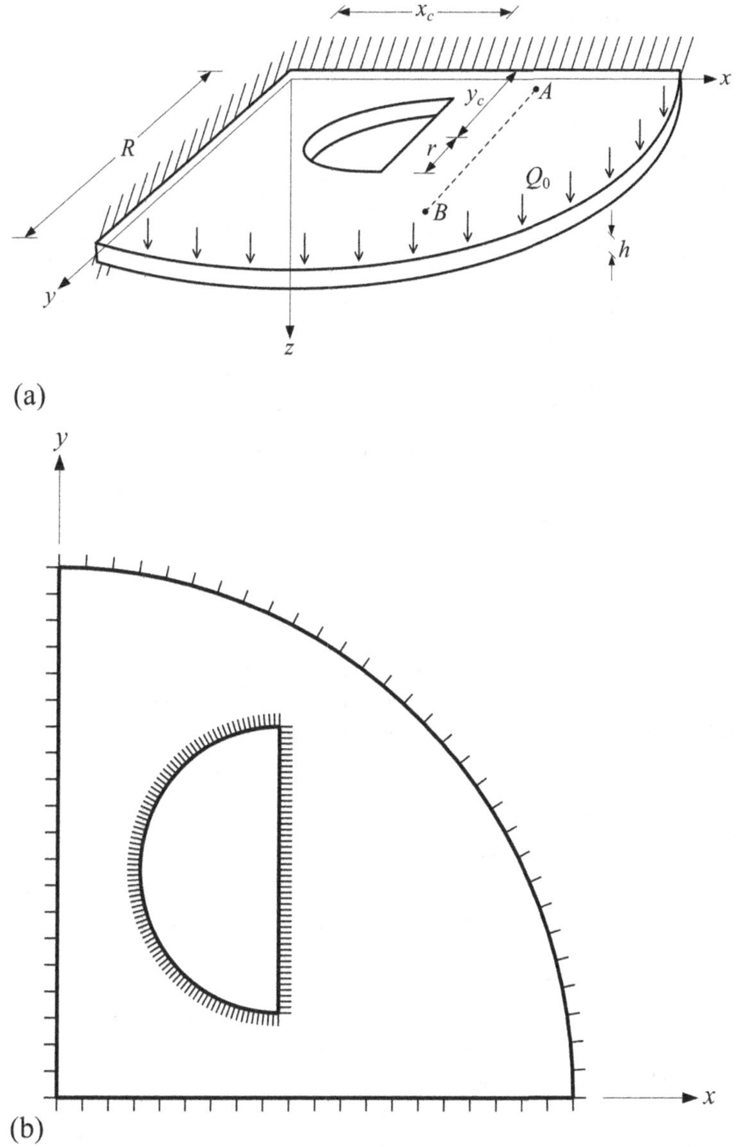

FIGURE 9.9 (a) A cantilever quarter circular laminated plate with a semicircular hole subjected to transverse load; (b) diagram of boundary element mesh.

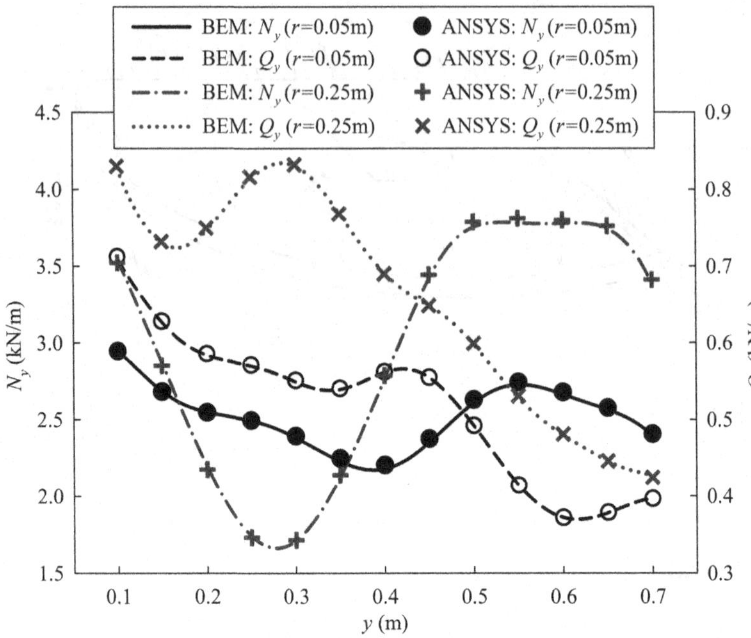

FIGURE 9.10 Stress resultant and transverse shear force along line AB of Figure 9.9(a) for a cantilever quarter circular laminated plate with a semicircular hole subjected to transverse load.

REFERENCES

Atluri, S. and Zhu, T., 1998, "A new Meshless Local Petrov-Galerkin (MLPG) approach in computational mechanics," *Computational Mechanics*, Vol. 22, pp. 117–127.

Brebbia, C.A., Telles, J.C.F. and Wrobel, L.C., 1984, *Boundary Element Techniques*, Springer-Verlag, New York.

Hsu, C.L., Hwu, C. and Shiah, Y.C., 2019, "Three-Dimensional Boundary Element Analysis for Anisotropic Elastic Solids and its Extension to Piezoelectric and Magnetoelectroelastic Solids," *Engineering Analysis with Boundary Elements*, Vol. 98, pp. 265–280.

Hsu, C.W. and Hwu, C., 2021, "A Special Boundary Element for Holes/Cracks in Composite Laminates Under Coupled Stretching-Bending Deformation," *Engineering Analysis with Boundary Elements*, Vol. 133, pp. 30–48.

Hsu, C.W. and Hwu, C., 2022a, "Holes/Cracks/Inclusions in Magneto- Electro-Elastic Composite Laminates under Coupled Stretching-Bending Deformation," *Composite Structures*, Vol. 297, 115960.

Hsu, C.W. and Hwu, C., 2022b, "Coupled Stretching-Bending Boundary Element Analysis for Unsymmetric Magneto-Electro-Elastic Laminates with Multiple Holes, Cracks and Inclusions," *Engineering Analysis with Boundary Elements*, Vol. 139, pp. 137–151.

Hsu, C.W. and Hwu, C., 2023a, "Green's Functions for Thick Laminated Composite Plates with Coupled Stretching-Bending and Transverse Shear Deformation," *Composite Structures*, Vol. 320, 117179.

Hsu, C.W. and Hwu, C., 2023b, "An Isoparametric Quadratic Boundary Element for Coupled Stretching-Bending Analysis of Thick Laminated Composite Plates with Transverse Shear Deformation," *Engineering Analysis with Boundary Elements*, Vol. 156, pp. 175–188.

Huang, W.Y., Hwu, C. and Hsu, C.W., 2024a, "Bending-extension coupling analysis of shear deformable laminated composite curved beams with non-uniform thickness,"*Engineering Structures*, Vol. 305, 117696.

Huang, W.Y., Hwu, C. and Hsu, C.W., 2024b, "Explicit Analytical Solutions for Arbitrarily Laminated Composite Beams with Coupled Stretching-Bending and Transverse Shear Deformation," *European Journal of Mechanics - A/Solids*, Vol. 103, 105147.

Hwu, C. and Liao, C.Y., 1994, "A Special Boundary Element for the Problems of Multi-Holes, Cracks and Inclusions," *Computers and Structures*, Vol. 51, No. 1, pp. 23–31.

Hwu, C., 1998, "A Global-Local Finite Element Model for Multi-Layer Composite Laminates," *Transactions of the Aeronautical and Astronautical Society of the Republic of China*, Vol. 31, No. 1, pp. 27–37.

Hwu, C., Wu, J.Y., Fan, C.W. and Hsieh, M.C., 2001, "Stroh Finite Element for Two-Dimensional Linear Anisotropic Elastic Solids," *ASME Journal of Applied Mechanics*, Vol. 68, No. 3, pp. 468–475.

Hwu, C. and Yu, M.C., 2010, "A Comprehensive Finite Element Model for Tapered Composite Wing Structures," *Computer Modeling in Engineering & Sciences*, Vol. 67, No. 2, pp. 151–173.

Hwu, C., 2010a, *Anisotropic Elastic Plates*, Springer, New York.

Hwu, C., 2010b, "Boundary Integral Equations for General Laminated Plates with Coupled Stretching-Bending Deformation," *Proceedings of the Royal Society, Series A*, Vol. 466, pp. 1027–1054.

Hwu, C., 2012, "Boundary Element Formulation for the Coupled Stretching-Bending Analysis of Thin Laminated Plates," *Engineering Analysis with Boundary Elements*, Vol. 36, pp. 1027–1039.

Hwu, C. and Chang, H.W., 2015, "Coupled Stretching-Bending Analysis of Laminated Plates with Corners via Boundary Elements," *Composite Structures*, Vol. 120, pp. 300–314.

Hwu, C., Huang S.T., and Li, C. C., 2017, "Boundary-Based Finite Element Method for Two-Dimensional Anisotropic Elastic Solids with Multiple Holes and Cracks," *Engineering Analysis with Boundary Elements*, Vol. 79, pp. 13–22.

Hwu, C., 2021, *Anisotropic Elasticity with Matlab*, Springer, Cham.

Liu, G.R., 2009, *Meshfree Methods: Moving Beyond the Finite Element Method*, 2nd Ed., CRC Press, Boca Raton.

Sokolnikoff, I.S., 1956, *Mathematical Theory of Elasticity*, McGraw-Hill, New York.

Yen, W. J. and Hwu, C., 1993, "Interlaminar Stresses Around Hole Boundaries of Composite Laminates Subjected to In-Plane Loading," *Composite Structures*, Vol. 24, pp. 299–310.

Author Index

A

Agarwal, B.D., 2, 22–24, 28, 36
Allen, H.G., 181
Altenbach, H., 2
Altenbach, J., 2
Ambartsumyan, S.A., 143, 145, 150, 166
Antes, H., 129
Asheley, H., 272
Ashton, J.E., 52
Atluri, S., 377

B

Babcock, C.D., 238
Baker, A.A., 2, 4
Becker, W., 295
Bert, C.W., 181
Bi, T.Y., 248
Bisplinghoff, R.L., 272
Brebbia, C.A., 361, 372
Bronson, R., 98–100
Broutman, L.J., 2, 22–24, 28, 36
Budiansky, B., 272
Buroni, F.C., 317, 321
Burton, W.S., 181

C

Cahill, W.F., 103
Carlsson, L.A., 25
Carrer, J.A.M., 138
Chai, H., 238
Chang, H.W., 367
Chang, W.C., 225
Chou, T.W., 1
Chow, T.S., 121
Christensen, R.M., 1, 9, 50
Clyne, T.W., 2
Cody, W.J., 103
Cowper, G. R., 87, 121, 186

D

Deans, S.R., 317, 321
Denda, M., 317, 321
Didriksson, T., 182
Diederich, F.W., 272
dos Reis A., 104

F

Fan, C.W., 352
Fliischfresser, S.A., 138
Franklin, J.N., 77, 96

G

Gai, H.S., 193, 253, 269–271
Gautschi, W., 103
Gel'fand, I.M., 94, 103
Gere, J.M., 118, 138
Gibson, R.F., 2–3
Goodno, B.J., 138

H

Halfman, R.L., 272
Halpin, J.C., 20
Heath, W.G., 181
Herakovich, C. T., 2
Hill, R., 28
Hoff, N. J., 181

Horn, R.A., 98
Hoskin, B.C., 2, 4
Hsieh, C.H., 237
Hsieh, M.C., 311
Hsu, C.L., 317, 323, 341, 345, 353, 369
Hsu, C.W., 60, 67, 77, 79–80, 85, 98, 103–104, 106, 334, 336, 338, 353, 368–369, 380, 388
Hsu, H.W., 225
Hu, J.S., 185, 233, 237–238, 248
Huang, M., 104, 106
Huang, S.T., 129, 380
Huang, W.Y., 120–121, 135, 137, 139–140, 353, 380
Hull, D., 2
Hunsaker, J. C., 269
Hwu, C., 2, 15, 20, 60, 67, 72, 74, 77, 79–80, 85, 98, 103–104, 106, 129, 181, 185, 188, 192–193, 196, 203, 220, 225, 232–233, 237–238, 248–249, 253, 269–271, 276, 295, 300, 302, 304, 309, 311, 314, 320–321, 325, 331–338, 341, 345–349, 352–353, 365–370, 380, 388

J

Johnson, C.R, 98
Jones, R.M., 1, 18, 20, 24, 28–29, 57, 254

K

Karpouzian, G., 258, 272
Kassapoglou, C., 2
Khdeir, A. A., 258, 263, 272
Kissing, W., 2
Knauss, W.G., 238
Kollar, L. P., 2
Kraus, H., 143, 146, 150, 154, 156, 166, 169, 173, 177–178
Krone, Jr. N. J., 272

L

Lahtinen, H., 121
Laitinen, M., 121, 141
Lekhnitskii, S.G., 2, 294–295
Lerner, E., 272
Li, C. C., 129, 370
Liao, C.Y., 370
Librescu, L., 184, 258, 263, 272
Lima Albuquerque É., 104
Lin, Y.H., 198
Liu, G.R., 372, 376, 379
Liu, J.X., 300
Lo, K.H., 50
Lottati, I., 272
Love, A.E.H., 143, 145

M

Mansur, W.J., 138
Markowitz, J., 272
Megson, T. H. G., 253
Meirovitch, L., 214, 224–225, 267
Mendelson, A., 28
Mindlin, R.D., 184
Moh, J.S., 181, 188, 192–193
Morozov, E. V., 2
Mujumdar, P.M., 242, 244

N

Noor, A.K., 181
Nosier, A., 79

O

Oyibo, G. A., 272

P

Palermo Júnior, L., 104
Pan, E., 300
Penzien, J., 182
Pipes, R.B., 25, 41
Plantema, F.J., 181, 183
Powell, P.C., 2

R

Reddy, J.N., 2, 79, 87, 90, 276
Reissner, E., 122, 143, 184
Rogacheva, N.N., 298
Rosen, B.W., 23, 41
Roy, A.K., 186

S

Sallam, S.N., 250
Schapery, R.A., 298
Scuciato, R.F., 138
Shiah, Y.C., 345, 369
Shilov, G.E., 94
Sierakowski, R.L., 2, 52, 57, 168
Simitses, G.L., 238
Simovich, J., 258, 272
Sjölind, S.G., 121
Soh, A.K., 300
Sokolnikoff, I.S., 145, 353
Springer, G. S., 2
Stehfest, H., 298
Suryanarayan, S., 242, 244
Swanson, S.R., 114
Szilard, R., 57

T

Tan, C.J., 338
Telles, J.C.F., 104
Thacher, H.C., 103
Thangjitham, S., 258, 272
Timoshenko, S.P., 118
Ting, T.C.T., 2, 302
Tsai, S.W., 1, 20, 28
Tsai, Z.S., 253, 255, 276

V

Vasiliev, V.V., 2, 9
Verchery, G., 186
Vinson, J.R., 1–2, 9, 52, 57, 168
Vlachoutsis, S., 121

W

Wang, J., 104, 106
Weisshaar, T.A., 272
Whitney, J.M., 52, 121
Wrobel, L.C., 372
Wu, E.M., 28, 50
Wu, J.Y., 352
Wu, K.C., 317, 321

X

Xie, L., 300

Y

Yen, W. J., 352
Yin, W.L., 238
Yu, M.C., 276, 352

Z

Zhang, C., 300
Zhu, T., 377

Subject Index

Pages in *italics* refer to figures and pages in **bold** refer to tables.

A

aeroelastic divergence, 272, 275, 285
anisotropic
  elasticity, 1–2, 7, 294–295, 298, 303, 317, 320–321, 341, 361
  plate, 303

B

beam
  narrow, 112–115, 118, 205
  wide, 109, 114–115, 118, 204–205
bending boundary layer, 168–169
boundary
  essential, 373–374, 376–377
  natural, 377–378
boundary-based finite element, 129, 349, 360, 370
boundary element method, 20, 43, 80, 127, 129, 349, 353, 370, 372
boundary integral equation, 80, 343, 353, 361–362, 365, 368–369
buckling
  analysis, 180–181, 188, 203, 206, 233, 239, 242
  load, 58, 119, 190–191, 209, 211, 233–234, 236–240
  postbuckling analysis, 238–239

C

Cauchy principal value, 103, 361
classical lamination theory, 29–30, 35, 43, 46, 50, 62, 142, 184, 294, 333
column, 7, 109
composite
  chopped fiber, 4
  fiber-reinforced, 1, 3–4, 10–11, *16*, 18–19, 28, 60, 80, 269, 294, 300, 308, 380, 385
  fibrous, 3–5, 10, 21–22
  hybrid, 4
  laminated, 3, 7, 10–12, 33, 38, 42–43, 46, 59, 88, 93, 106, 109–110, 116–122, 127–142, 150, 156, 180–181, 188, 204, 214, 226, 254–255, 264, 331–333, 337, 349, 353, 368, 380, 385
  particulate, 3, 20
  sandwich, 4, 50, 87, 122, 180–181, 185–188, 190–191, 193, 195–200, 202–203, 206, 208–213, 220–226, 229–239, 242–244, 248–252, 255, 258, 262, 264, 271, 380–383, *384*
comprehensive wing model, 276, 280
constitutive law, 44, 46, 54, 62, 64, 78, 86–88, 97, 110–114, 120, 142–143, 145, 153, 156, 158, 160, 162–163, 165, 167, 170, 175, 177, 185, 200, 203–204, 206–207, 226–227, 229, 252, 285, 304, 308, 311, 314, 324, 358
correspondence principle, 298
coupled stretching-bending analysis, 295, 303–304, 349, 353, 365, 368, 385
curvilinear coordinate, 142–143, 145–146, 374

D

damage, 21, 23
delamination, 3, 6, 186, 232–233, 237–249
divergence speed, 275–276

E

effective transverse shear force, 48, 51, 74, 305, 309, 366
electro-elastic laminate, 72, 304, 308, 310, 312, 314
element-free Galerkin method, 349, 372
equilibrium equation, 44–49, 57, 63–64, 87–88, 110–112, 116, 120, 122, 142, 145–146, 154, 156, 158, 160, 162–163, 165, 167, 170, 174–176, 183, 186–188, 191, 193, 203–207, 211, 252, 256–258, 284–286, 294–295, 304, 308, 311, 314, 324, 352, 354, 377

F

failure criterion, 28, 40–41
feedback control, 222–223, 225, 271
fiber volume fraction, 22–23, *23*
finite element method, 20, 80, 129, 252, 284, 289, 349, 352, 360, 371–372
fundamental elasticity matrix, 296–297, 303, 308
fundamental solution, 127, 353–354, 366–367, 369–370, 386

G

Gauss-Codazzi relation, 147, 157
generalized coordinate, 178–179, 218–219, 221, 270–271
generalized force, 219, 222, 270
Green's function, 43, 85, 93, 97–98, 103–104, 106, 110, 120, 127, 129, 140, 295, 324–325, 331, 341, 343–344, 346, 353, 355, 369

H

Halpin-Tsai equation, 20–21
Hamilton's principle, 90, 200, 276–277, 280, 291
high-order plate theory, 50
Hilbert transform, 322
Hill's criterion, 28
hygrothermal effect, 33

I

influence coefficient, 364–365, 367
interpolation function, 281–282, 362–363, 366–367, 369

K

Kalman filter gain matrix, 224
kinematic relation, 7, 64, 191, 200, 226, 229, 311, 314, 324
kinetic energy, 91, 200, 277–278
Kirchhoff's assumption, 314

L

Lagrange multiplier, 374
Lagrangian function, 90, 200, 202, 276, 282, 284–285
Lame coefficient, 143–144, 147, 154
lamina failure criterion, 40–41
laminate
  angle-ply, 37–38
  anti-symmetric, 37, 40
  balanced, 38, 47
  cross-ply, 37–38, 52–54, 57, 59, 113, 115, 188, 196, 198, 200, 229–230, 232, 380
  quasi-isotropic, 38
  specially orthotropic, 37–38
  symmetric, 36–37, 42, 47, 50, 60, 109, 115–116, 118, 140, 185, 188, 304, 307, 381
Laplace transform, 274, 297–298
Lekhnitskii formalism, 294
Levy solution, 52, 55

M

macromechanics, 6
magneto-electro-elastic, 59, 81, 300
material eigenvalue, 296, 298, 305, 308–309, 313, 319–320, 346
material eigenvector, 296, 302, 319, 335

matrix
metal, 6
thermoplastic resin, 6
thermosetting resin, 5
maximum stress theory, 25, 27–28
maximum strain theory, 27
maximum work theory, 28
meshless local Petrov-Galerkin, 349, 372, 376
meshless method, 349, 372, 377
micromechanics, 6, 15–16
modal analysis, 178, 214, 218, 267, 270
moving least square, 372

N

natural frequency, 120, 177–178, 195, 199, 212–214, 232–233, 242, 245–246, 249, 265, 267, 270, 284–285
natural mode, 195, 212, 215–216, 218, 265, 267, 270
Navier solution, 52–*53*, 55, 58

O

orthogonality condition, 179, 202, 214, 218–219, 267, 269, 271
orthotropic
generally, 12, 44
specially, 12, 37–38, 65, 80, 115, 153

P

penalty parameter, 377, 379
piezoelectric
actuator, 221–222
material, 60, 80–81, 220, 298–301, 303, 308, 324
sensor, 220–221, 271
plate bending analysis, 304, 307
potential energy, 19, 90–91, 191–193, 200, 240–242, 276–278, 284, 289, 352, 371, 374, 377
principal material axes, 12, 25, 30

R

Radon-Stroh formalism, 295, 317, 323–324, 341, 369
Radon transform, 317–319, 321–322
Rayleigh-Ritz method, 52, 191
Riccati equation, 224–225
relaxation function, 297

S

sandwich structure, 7, 180
shear deformation theory, 43, 50, 85, 87, 120, 187, 286, 368
shell
conical, 142, 148, *149*, 159, 161
cylindrical, 142, 149, 159, 161, 166, 168–169, 178, 180, 225, 229–232
ellipsoidal, 149
membrane, 142, 169–175
spherical, 142, 149, 159, 162–164, *164*, 178
singular function, 364
stiffness
bending, 31, 35, 38, 62, 65, 71, 81, 86, 114, 121, 180, 185, 227, 275, 294, 304, 307–308, 311
coupling, 36, 47, 50, 52, 54, 113, 121, 307
extensional, 38, 71, 114, 121
strain energy density, 91, 192, 277
strength
lamina, 21, 25
laminate, 38–39, *39*, 41
longitudinal compressive, 23–25
longitudinal tensile, 22, *23*, 23, 25
off-axis shear, 27
off-axis tensile, 26, 28
shear, 25–28
transverse, 24
ultimate, 22
stress
couple, 151–153, 156, 177
function, 166, 294, 296, 298, 300–301, 305, 307, 309, 312, 316, 320, 322, 331, 333, 338
resultant, 2, 40, 57, 62, 65, 71, 80, **82**, 83, 86, 97, 151–153, 156, 169, 172–173, 175, 177, 227, *228*, 255, 258, 304, 307, 309, 311–312, 316, 386–388, **386–387**, *390*
Stroh formalism, 294–295, 297–302, 308, 317, 321, 323–325, 341, 369

Stroh-like formalism, 295, 304, 307–308, 311, 314, 331–332
Stroh's eigenvalue, 296, 341
Stroh's eigenvector, 296
Strong form, 289, 377

T

test function, 377
thermal
eigenvalue, 302, 316
eigenvector, 302, 316
expansion coefficient, 302
moduli, 302, 314
stress, 304, 314
transformation law, 25–26, 48
transverse isotropic, 11
transverse shear deformation, 49–50, 61, 85, 109, 120, 135, 138–139, 182, 184, 213, 251, 253, 262, 264, 270, 276, 368
transverse shear stiffness, 50, 87, 121, 186, 227, 233–234, 237, 249
trial function, 378–379
Tsai-Hill theory, 28
Tsai-Wu tensor theory, 28

V

vibration
forced, 178–180, 203, 218, 270–271
free, 42, 57–59, 79, 109, 118–119, 177–181, 193, 195–196, 200, 203, 211–212, 225, 230, 242, 244, 248, 265, 284, 286
suppression, 180, 203, 220, 271
viscoelastic material, 297, 325
von Mises' isotropic yield criterion, 28

W

warping restraint effect, 258, 262–263, 270, 272
weak form, 289, 377–378
weight function, 373, 377–379
wing structure, 251–256, 261, 265, 267, 269, 271–272, 275–276, 284–285, 287, 352

Y

Young's modulus, 18–19, 81, 122, 182, 254

Printed in the United States
by Baker & Taylor Publisher Services